THERMODYNAMICS

Edited by **Tadashi Mizutani**

Thermodynamics
Edited by Tadashi Mizutani

Published by InTech
Janeza Trdine 9, 51000 Rijeka, Croatia

Publishing Process Manager Ana Nikolic

Technical Editor Teodora Smiljanic

Cover Designer InTech Design Team

Additional hard copies can be obtained from orders@intechopen.com

Thermodynamics, Edited by Tadashi Mizutani
p. cm.
ISBN 978-953-307-544-0

For EU product safety concerns :
IN TECH d.o.o., Prolaz Marije Krucifikse Kozulić 3, 51000 Rijeka, Croatia,
info@intechopen.com or visit our website at intechopen.com.

Meet the editor

Tadashi Mizutani is a professor of Biofunctional Chemistry, Faculty of Science and Engineering at Doshisha University, Japan. He was born in Osaka, Japan in 1957. He received a Doctor of Engineering degree from Kyoto University. He joined Toyota Central Research Laboratories, Inc., in 1986. After research and education in Tottori University and Kyoto University, he moved to Doshisha University in 2003. He has conducted experimental research on synthetic receptors, synthesis and supramolecular chemistry of tetrapyrrolic dyes, synthesis of clay minerals, nanoscience of organic-inorganic interfacial structures, and organic-inorganic hybrid materials. He has published over 100 papers in these fields including reviews and book contributions.

Contents

Preface

Progress of thermodynamics has been stimulated by the findings of a variety of fields of science and technology. In the nineteenth century, studies on engineering problems, efficiency of thermal machines, lead to the discovery of the second law of thermodynamics. Following development of statistical mechanics and quantum mechanics allowed us to understand thermodynamics on the basis of the properties of constituent molecules. Thermodynamics and statistical mechanics provide a bridge between microscopic systems composed of molecules and quantum particles and their macroscopic properties. Therefore, in the era of the mesoscopic science, it is time that various aspects of state-of-the-art thermodynamics are reviewed in this book.

In modern science a number of researchers are interested in nanotechnology, surface science, molecular biology, and environmental science. In order to gain insight into the principles of various phenomena studied in such fields, thermodynamics should offer solid theoretical frameworks and valuable tools to analyse new experimental observations. Classical thermodynamics can only treat equilibrium systems. However, thermodynamics should be extended to non-equilibrium systems, because understanding of transport phenomena and the behaviour of non-equilibrium systems is essential in biological and materials research. Extension of thermodynamics to a system at the mesoscopic scale is also important due to recent progress in nanotechnology. The principles of thermodynamics are so general that the application is widespread to such fields as solid state physics, chemistry, biology, astronomical science, materials science, information science, and chemical engineering. These are also major topics in the book.

The first section of the book covers the fundamentals of thermodynamics, that is, theoretical framework of thermodynamics, foundations of statistical mechanics and quantum statistical mechanics, limits of standard thermodynamics, macroscopic fluctuations, extension of equilibrium thermodynamics to non-equilibrium systems, astronomical problems, quantum fluids, and information theory. The second section covers application of thermodynamics to solid state physics, materials science/engineering, surface science, environmental science, and information science. Readers can expect coverages from theoretical aspects of thermodynamics to applications to science and engineering. The content should be of help to many scientists and engineers of such field as physics, chemistry, biology, nanoscience, materials science, computer science, and chemical engineering.

Tadashi Mizutani
Doshisha University, Kyoto
Japan

Part 1

Fundamentals of Thermodynamics

1

New Microscopic Connections of Thermodynamics

A. Plastino[1] and M. Casas[2]
[1]*Facultad de C. Exactas, Universidad Nacional de La Plata*
IFLP-CONICET, C.C. 727, 1900 La Plata
[2]*Physics Departament and IFISC-CSIC, University of Balearic Islands*
07122 Palma de Mallorca
[1]*Argentina*
[2]*Spain*

1. Introduction

This is a work that discusses the foundations of statistical mechanics (SM) by revisiting its postulates in the case of the two main extant versions of the theory. A third one will here we added, motivated by the desire for an axiomatics that possesses some thermodynamic "flavor", which does not happen with neither of the two main SM current formulations, namely, those of Gibbs' (1; 2), based on the ensemble notion, and of Jaynes', centered on MaxEnt (3; 4; 5).
One has to mention at the outset that we "rationally understand" some physical problem when we are able to place it within the scope and context of a specific "*Theory*". In turn, we have a theory when we can both derive all the known interesting results *and* successfully predict new ones starting from a small set of axioms. Paradigmatic examples are von Neumann's axioms for Quantum Mechanics, Maxwell's equations for electromagnetism, Euclid's axioms for classical geometry, etc. (1; 3).
Boltzmann's main goal in inventing statistical mechanics during the second half of the XIX century was to explain thermodynamics. However, he did not reach the axiomatic stage described above. The first successful SM theory was that of Gibbs (1902) (2), formulated on the basis of four ensemble-related postulates (1). The other great SM theory is that of Jaynes' (4), based upon the MaxEnt axiom (derived from Information Theory): *ignorance is to be extremized (with suitable constraints).*
Thermodynamics (TMD) itself has also been axiomatized, of course, using four macroscopic postulates (6). Now, the axioms of SM and of thermodynamics belong to different worlds altogether. The former speak of either "ensembles" (Gibbs), which are *mental* constructs, or of "observers' ignorance" (Jaynes), concepts germane to thermodynamics' language, that refers to laboratory-parlance. In point of fact, TMD enjoys a very particular status in the whole of science, as the one and only theory whose axioms are empirical statements (1).
Of course, there is nothing to object to the two standard SM-axiomatics referred to above. However, a natural question emerges: would it be possible to have a statistical mechanics derived from axioms that speak, as far as possible, the same language as that of thermodynamics? To what an extent is this feasible? It is our intention here that of attempting a serious discussion of such an issue and try to provide answers to the query, following ideas developed in (7; 8; 9; 10; 11; 12; 13).

2. Thermodynamics' axioms

Thermodynamics can be thought of as a formal logical structure whose *axioms* are empirical facts, which gives it a unique status among the scientific disciplines (1). The four postulates we state below are entirely equivalent to the celebrated three laws of thermodynamics (6):

1. For every system there exists a quantity E, called the internal energy, such that a unique $E-$value is associated to each of its states. The difference between such values for two different states in a closed system is equal to the work required to bring the system, while adiabatically enclosed, from one state to the other.
2. There exist particular states of a system, called the equilibrium ones, that are uniquely determined by E and a set of extensive (macroscopic) parameters A_ν, $\nu = 1, \ldots, M$. The number and characteristics of the A_ν depends on the nature of the system (14).
3. For every system there exists a state function $S(E, \forall A_\nu)$ that (i) always grows if internal constraints are removed and (ii) is a monotonously (growing) function of E. S remains constant in quasi-static adiabatic changes.
4. S and the temperature $T = [\frac{\partial E}{\partial S}]_{A_1,\ldots,A_M}$ vanish for the state of minimum energy and are ≥ 0 for all other states.

From the second and 3rd. Postulates we will extract and highlight the following two assertions, that are essential for our purposes

- Statement 3a) for every system there exists a state function S, a function of E and the A_ν

$$S = S(E, A_1, \ldots, A_M). \tag{1}$$

- Statement 3b) S is a monotonous (growing) function of E, so that one can interchange the roles of E and S in (1) and write

$$E = E(S, A_1, \ldots, A_M), \tag{2}$$

Eq. (2) clearly indicates that

$$dE = \frac{\partial E}{\partial S} dS + \sum_\nu \frac{\partial E}{\partial A_\nu} dA_\nu \Rightarrow dE = TdS + \sum_\nu P_\nu dA_\nu, \tag{3}$$

with P_ν generalized pressures and the temperature T defined as (6)

$$T = \left(\frac{\partial E}{\partial S}\right)_{[\forall A_\nu]}. \tag{4}$$

Eq. (3) will play a central role in our considerations, as discussed below.
If we know $S(E, A_1, \ldots, A_n)$ (or, equivalently because of monotonicity, $E(S, A_1, \ldots, A_n)$) we have a *complete* thermodynamic description of a system. It is often experimentally more convenient to work with *intensive* variables.
Let define $S \equiv A_0$. The intensive variable associated to the extensive A_i, to be called P_i is:

$$P_0 \equiv T = [\frac{\partial E}{\partial S}]_{A_1,\ldots,A_n}, \; 1/T = \beta$$

$$P_j \equiv \lambda_j / T = [\frac{\partial E}{\partial A_j}]_{S, A_1, \ldots, A_{j-1}, A_{j+1}, \ldots, A_n}$$

Any one of the Legendre transforms that replaces any s extensive variables by their associated intensive ones (β, λ's will be Lagrange multipliers in SM)

$$L_{r_1,\ldots,r_s} = E - \sum_j P_j A_j,\ (j = r_1,\ldots,r_s)$$

contains the same information as either S or E. The transform $L_{r_1,\ldots,r_s}$ is a function of $n-s$ extensive and s intensive variables. This is called the *Legendre invariant structure of thermodynamics*.

3. Gibbs' approach to statistical mechanics

In 1903 Gibbs formulated the first axiomatic theory for statistical mechanics (1), that revolves around the basic physical concept of phase space. Gibbs calls the "phase of the system" to its phase space (PS) precise location, given by generalized coordinates and momenta. His postulates refer to the notion of ensemble (a mental picture), an extremely great collection of N independent systems, all identical in nature with the one under scrutiny, but differing in phase. One imagines the original system to be repeated many times, each of them with a different arrangement of generalized coordinates and momenta. Liouville's celebrated theorem of volume conservation in phase space for Hamiltonian motion applies. The ensemble amounts to a distribution of N PS-points, representative of the "true" system. N is so large that one can speak of a density D at the PS-point $\phi = q_1,\ldots,q_N;p_1,\ldots,p_N$, with $D = D(q_1,\ldots,q_N;p_1,\ldots,p_N,t) \equiv D(\phi)$, with t the time, and, if we agree to call $d\phi$ the pertinent volume element,

$$N = \int d\phi\, D;\ \ \forall t. \tag{5}$$

If a system were to be extracted randomly from the ensemble, the probability of selecting one whose phase lies in a neighborhood of ϕ would be simply

$$P(\phi) = D(\phi)/N. \tag{6}$$

Consequently,

$$\int P\,d\phi = 1. \tag{7}$$

Liouville's theorem follows from the fact that, since phase-space points can not be "destroyed", if

$$N_{12} = \int_{\phi_1}^{\phi_2} D\,d\phi, \tag{8}$$

then

$$\frac{dN_{12}}{dt} = 0. \tag{9}$$

An appropriate analytical manipulation involving Hamilton's canonical equations of motion then yields the theorem in the form (1)

$$\dot{D} + \sum_i^N \frac{\partial D}{\partial p_i}\dot{p}_i + \sum_i^N \frac{\partial D}{\partial q_i}\dot{q}_i = 0, \tag{10}$$

entailing what Gibbs calls the conservation of density-in-phase.

Equilibrium is simply the statement $\dot{D} = 0$, i. e.,

$$\sum_i^N \frac{\partial D}{\partial p_i} \dot{p}_i + \sum_i^N \frac{\partial D}{\partial q_i} \dot{q}_i = 0. \tag{11}$$

3.1 Gibbs' postulates for statistical mechanics

The following statements wholly and thoroughly explain in microscopic fashion the corpus of equilibrium thermodynamics (1).

- The probability that at time t the system will be found in the dynamical state characterized by ϕ equals the probability $P(\phi)$ that a system randomly selected from the ensemble shall possess the phase ϕ will be given by (6).
- All phase-space neighborhoods (cells) have the same a priori probability.
- D depends only upon the system's Hamiltonian.
- The time-average of a dynamical quantity F equals its average over the ensemble, evaluated using D.

4. Information theory (IT)

The IT-father, Claude Shannon, in his celebrated foundational paper (15), associates a degree of knowledge (or ignorance) to any normalized probability distribution $p(i)$, $(i = 1, \ldots, N)$, determined by a functional of the $\{p_i\}$ called the information measure $I[\{p_i\}]$, giving thus birth to a new branch of mathematics, that was later axiomatized by Kinchin (16), on the basis of four axioms, namely,

- I is a function ONLY of the $p(i)$,
- I is an absolute maximum for the uniform probability distribution,
- I is not modified if an $N+1$ event of probability zero is added,
- Composition law.

4.1 Composition

Consider two sub-systems $[\Sigma^1, \{p^1(i)\}]$ and $[\Sigma^2, \{p^2(j)\}]$ of a composite system $[\Sigma, \{p(i,j)\}]$ with $p(i,j) = p^1(i)\, p^2(j)$. Assume further that the conditional probability distribution (PD) $Q(j|i)$ of realizing the event j in system 2 for a fixed $i-$event in system 1. To this PD one associates the information measure $I[Q]$. Clearly,

$$p(i,j) = p^1(i)\, Q(j|i). \tag{12}$$

Then Kinchin's fourth axiom states that

$$I(p) = I(p^1) + \sum_i p^1(i)\, I(Q(j|i)). \tag{13}$$

An important consequence is that, out of the four Kinchin axioms one finds that Shannons's measure

$$S = -\sum_{i=1}^N p(i) \ln[p(i)], \tag{14}$$

is the one and only measure complying with them.

5. Information theory and statistical mechanics

Information theory (IT) entered physics via Jaynes' Maximum Entropy Principle (MaxEnt) in 1957 with two papers in which statistical mechanics was re-derived à la IT (5; 17; 18), without appeal to Gibbs' ensemble ideas. Since IT's central concept is that of information measure (IM) (5; 15; 17; 19), a proper understanding of its role must at the outset be put into its proper perspective.
In the study of Nature, scientific truth is established through the agreement between two *independent* instances that can neither bribe nor suborn each other: analysis (pure thought) and experiment (20). The analytic part employs mathematical tools and concepts. The following scheme thus ensues:

WORLD OF MATHEMATICAL ENTITIES ⇔ LABORATORY

The mathematical realm was called by Plato Topos Uranus (TP). Science in general, and physics in particular, is thus *primarily* (although not exclusively, of course) to be regarded as a TP⇔"Experiment" two-way bridge, in which TP concepts are related to each other in the form of "laws" that are able to adequately describe the relationships obtaining among suitable chosen variables that describe the phenomenon one is interested in. In many instances, although not in all of them, these laws are integrated into a comprehensive theory (e.g., classical electromagnetism, based upon Maxwell's equations) (1; 21; 22; 23; 24).
If recourse is made to MaxEnt ideas in order to describe thermodynamics, the above scheme becomes now:

IT as a part of TP⇔ Thermal Experiment,

or in a more general scenario:

IT ⇔ Phenomenon to be described.

It should then be clear that the relation between an information measure and entropy is:

IM ⇔ Entropy S.

One can then state that an IM is *not* necessarily an entropy! How could it be? The first belongs to the Topos Uranus, because it is a mathematical concept. The second to the laboratory, because it is a measurable physical quantity. All one can say is that, at most, in some special cases, an association $IM \Leftrightarrow$ entropy S can be made. As shown by Jaynes (5), this association is both useful and proper in very many situations.

6. MaxEnt rationale

The central *IM idea* is that of giving quantitative form to the everyday concept of *ignorance* (17). If, in a given scenario, N distinct outcomes ($i = 1,\ldots,N$) are possible, then *three* situations may ensue (17):

1. Zero ignorance: predict with certainty the actual outcome.
2. Maximum ignorance: Nothing can be said in advance. The N outcomes are equally likely.
3. Partial ignorance: we are given the probability distribution $\{P_i\}$; $i = 1,\ldots,N$.

The underlying philosophy of the application of IT ideas to physics via the celebrated Maximum Entropy Principle (MaxEnt) of Jaynes' (4) is that originated by Bernoulli and

Laplace (the fathers of Probability Theory) (5), namely: *the concept of probability refers to an state of knowledge.* An information measure quantifies the information (or ignorance) content of a probability distribution (5). If our state of knowledge is appropriately represented by a set of, say, M expectation values, then the "best", least unbiased probability distribution is the one that

- reflects just what we know, without "inventing" unavailable pieces of knowledge (5; 17) and, additionally,
- maximizes ignorance: the truth, all the truth, *nothing but* the truth.

Such is the MaxEnt rationale (17). It should be then patently clear that, in using MaxEnt, one is NOT maximizing a physical entropy. One is maximizing ignorance in order to obtain the least biased distribution compatible with the a priori knowledge.

6.1 Jaynes mathematical formulation

As stated above, Statistical Mechanics and thereby Thermodynamics can be formulated on the basis of Information Theory if the statistical operator $\hat{\rho}$ is obtained by recourse to the MAXIMUM ENTROPY PRINCIPLE (MaxEnt). Consequently, we have the MaxEnt principle: *MaxEnt: Assume your prior knowledge about the system is given by the values of M expectation values* $< A_1 >, \ldots, < A_M >$*. Then* $\hat{\rho}$ *is uniquely fixed by extremizing* $I(\hat{\rho})$ *subject to the constraints given by the M conditions*

$$< A_j >= Tr[\hat{\rho}\, \hat{A}_j]$$

(entailing the introduction of M associated Lagrange multipliers λ_i*) plus normalization of* $\hat{\rho}$ *(entailing a normalization Lagrange multiplier* ξ*.)* In the process one discovers that $I \equiv S$, the equilibrium Boltzmann's entropy, if our prior knowledge $< A_1 >, \ldots, < A_M >$ refers to extensive quantities. Such $I-$value, once determined, *yields complete thermodynamical information with respect to the system of interest.*

7. Possible new axioms for SM

Both Gibbs' and MaxEnt are beautiful, elegant theories that satisfactorily account for equilibrium thermodynamics. Whys should we be looking for still another axiomatics? Precisely because, following Jaynes IT-spirit, one should be endeavoring to use all information actually available to us in building up our theoretic foundations, and this is not done in MaxEnt, as we are about to explicitate.

Our main argument revolves around the possibility of giving Eq. (3), an empirical statement, the status of an axiom, actually employing thus a piece of information available to us without any doubt. This constitutes the first step in our present discourse. More explicitly, in order to concoct a new SM-axiomatics, we start by establishing as a theoretic postulate the following macroscopic assertion:

Axiom (1)

$$dE = TdS + \sum_{\nu} P_{\nu} dA_{\nu}. \tag{15}$$

Since this is a macroscopic postulate in a microscopic axiomatics' corpus, it is pertinent now to ask ourselves which is the minimum amount of microscopic information that we would have to add to such an axiomatics in order to get all the microscopic results of equilibrium statistical mechanics. Since we know about Kinchin's postulates, we borrow from him his first one. Consequently, we conjecture at this point, and will prove below, that the following statements meets the bill:

Axiom (2) If there are $\mathcal{N}$ microscopic accessible states labelled by i, of microscopic probability p_i, then

$$S = S(p_1, p_2, \ldots, p_{\mathcal{N}}). \tag{16}$$

In what follows, the number of microstates will also be denoted by W.
Now, we will take as a postulate something that we actually know form both quantum and classical mechanics.

Axiom (3) The internal energy E and the external parameters A_ν are to be regarded as expectation values of suitable operators, respectively the hamiltonian H and $\mathcal{R}_\nu$ (i.e., $A_\nu \equiv <\mathcal{R}_\nu>$).

Thus the A_ν (and also E) will depend on the eigenvalues of these operators *and* on the probability set. (The energy eigenvalues depend of course upon the $\mathcal{R}_\nu$.) The reader will immediately realize that **Axiom (2)** is just a form of Boltzmann's "atomic" conjecture, pure and simple. In other words, macroscopic quantities are statistical averages evaluated using a microscopic probability distribution (25). It is important to realize that our three new axioms are statements of fact in the sense that they are borrowed either from experiment or from pre-existent theories. In fact, the 3 axioms do not incorporate any knew knowledge at all!
In order to prove that our above three postulates do allow one to build up the mighty SM-edifice we will show below that they are equivalent to Jaynes' SM-axiomatics (4).
Of course, the main SM-goal is that of ascertaining which is the PD (or the density operator) that best describes the system of interest. Jaynes appeals in this respect to his MaxEnt postulate, the only one needed in this SM-formulation. We restate it below for the sake of fixing notation.
MaxEnt axiom: assume your prior knowledge about the system is given by the values of M expectation values

$$A_1 \equiv <\mathcal{R}_1>, \ldots, A_R \equiv <\mathcal{R}_M>. \tag{17}$$

Then, ρ is uniquely fixed by extremizing the information measure $I(\rho)$ subject to $\rho-$normalization plus the constraints given by the M conditions constituting our assumed foreknowledge

$$A_\nu = <\mathcal{R}_\nu> = Tr[\rho\,\mathcal{R}_\nu]. \tag{18}$$

This leads, after a Lagrange-constrained extremizing process, to the introduction of M Lagrange multipliers λ_ν, that one assimilates to the generalized pressures P_ν. The truth, the whole truth, nothing but the truth (17). If the entropic measure that reflects our ignorance were not maximized, we would be inventing information that we do not actually possess.
In performing the variational process Jaynes discovers that, provided one multiplies the right-hand-side of the information measure expression by Boltzmann's constant k_B, the IM equals the entropic one. Thus, $I \equiv S$, the equilibrium thermodynamic entropy, with the caveat that our prior knowledge $A_1 = <\mathcal{R}_1>, \ldots, A_M = <\mathcal{R}_M>$ must refer just to extensive quantities. Once ρ is at hand, $I(\rho)$ *yields complete microscopic information with respect to the system of interest*. Our goal should be clear now. We need to prove that our new axiomatics, encapsulated by (15) and (16), is equivalent to MaxEnt.

8. Equivalence between MaxEnt and our new axiomatics

We will here deal with the classical instance only. The quantal extension is of a straightforward nature. Consider a generic change $p_i \to p_i + dp_i$ constrained by Eq. (15), that is, the change dp_i must be of such nature that (15) is verified. Obviously, S, A_j, and E will change with dp_i and, let us insist, these changes are constrained by (15). We will *not* specify the information measure, as several possibilities exist (26). For a detailed discussion of this issue see (27). In this endeavor our ingredients are

- an arbitrary, smooth function $f(p)$ that allows us to express the information measure in the fashion
$$I \equiv S(\{p_i\}) = \sum_i p_i f(p_i), \tag{19}$$
such that $S(\{p_i\})$ is a concave function,
- M quantities A_ν that represent mean values of extensive physical quantities $\langle \mathcal{R}_\nu \rangle$, that take, for the micro-state i, the value a_i^ν with probability p_i,
- another arbitrary smooth, monotonic function $g(p_i)$ ($g(0) = 0$; $g(1) = 1$). It is in order to use generalized, non-Shannonian entropies that we have slightly generalized mean-value definitions using the function g.

We deal then with (we take $A_1 \equiv E$), using the function g to evaluate (generalized) expectation values,
$$A_\nu \equiv \langle \mathcal{R}_\nu \rangle = \sum_i^W a_i^\nu g(p_i); \; \nu = 2, \ldots, M, \tag{20}$$
$$E = \sum_i^W \epsilon_i g(p_i), \tag{21}$$
where ϵ_i is the energy associated to the microstate i. The probability variations dp_i will now generate corresponding changes dS, dA_ν, and dE in, respectively, S, the A_ν, and E.

8.1 Proof, part I

The essential point of our present methodology is to enforce *obedience to*
$$dE - TdS + \sum_{\nu=1}^W dA_\nu \lambda_\nu = 0, \tag{22}$$
with T the temperature and λ_ν generalized pressures. We use now the expressions (19), (20), and (21) so as to cast (22) in terms of the probabilities, according to an infinitesimal probabilities' change
$$p_i \to p_i + dp_i. \tag{23}$$
If we expand the resulting equation up to first order in the dp_i, it is immediately found, after a little algebra, that the following set of equations ensues (7; 8; 9; 10; 11; 12; 13) (remember that the Lagrange multipliers λ_ν are identical to the generalized pressures P_ν of Eq. (3))
$$C_i^{(1)} = [\textstyle\sum_{\nu=1}^M \lambda_\nu a_i^\nu + \epsilon_i]$$
$$C_i^{(2)} = -T \frac{\partial S}{\partial p_i}$$

$$\Sigma_i [C_i^{(1)} + C_i^{(2)}] dp_i \equiv \Sigma_i K_i dp_i = 0. \tag{24}$$

We can rearrange matters in the fashion

$$T_i^{(1)} = f(p_i) + p_i f'(p_i)$$
$$T_i^{(2)} = -\beta[(\Sigma_{\nu=1}^{M} \lambda_\nu a_i^\nu + \epsilon_i) g'(p_i) - K],$$
$$(\beta \equiv 1/kT), \tag{25}$$

so that we can recast (24) as

$$T_i^{(1)} + T_i^{(2)} = 0; \ (for\ any\ i), \tag{26}$$

a relation whose importance will become manifest in Appendix I.

We wish that Eqs. (24) or (26) should yield one and just one p_i-expression, which it indeed does (7; 8; 9; 10; 11; 12; 13). We do not need here, however, for our demonstration, an explicit expression for this probability distribution, as will be immediately realized below.

8.2 Proof, part II: follow Jaynes' procedure

Alternatively, proceed à la MaxEnt. This requires extremizing the entropy S subject to the usual constraints in E, A_ν, and normalization. The ensuing, easy to carry out Jaynes' variational treatment, can be consulted in (7; 8; 9; 10; 11; 12; 13), that is (we set $\lambda_1 \equiv \beta = 1/T$)

$$\delta_{p_i}[S - \beta\langle H\rangle - \sum_{\nu=2}^{M} \lambda_\nu \langle \mathcal{R}_\nu \rangle - \xi \sum_i p_i] = 0, \tag{27}$$

(we need also a normalization Lagrange multiplier ξ) is easily seen to yield as a solution the very set of Eqs. (24) as well! (see Appendix I for the proof). These equations arise then out of two clearly separate treatments: (I) our methodology, based on Eqs. (15) and (16), and (II), following the MaxEnt prescriptions. This entails that MaxEnt and our axiomatics co-imply each other, becoming thus equivalent ways of building up statistical mechanics. An important point is to be here emphasized with respect to the functional $S-$form.

The specific form of $S[p_i]$ is not needed neither in Eqs. (24) nor in (27)!

9. What does all of this mean?

We have already formally proved that our axiomatics is equivalent to MaxEnt, and serves thus as a foundation for equilibrium statistical mechanics. We wish now to dwell in deeper fashion into the meaning of our new SM-formulation. First of al it is to be emphasized that, in contrast to both Gibbs' and Jaynes' postlates, ours have zero new informational content, since they are borrowed either from experiment or from pre-existing theories, namely, information theory and quantum mechanics. In particular, we wish to dwell to a larger extent on both the informational and physical contents of our all-important Eqs. (24) or (26).

The first and second laws of thermodynamics are two of physics' most important statements. They constitute strong pillars of our present understanding of Nature. Of course, statistical mechanics (SM) adds an underlying microscopic substratum that is able to explain not only these two laws but the whole of thermodynamics itself (6; 17; 28; 29; 30; 31). One of SM's basic ingredients is a microscopic probability distribution (PD) that controls the population of microstates of the system under consideration (28). Since we were here restricting our considerations to equilibrium situations, what we have been really doing here was to mainly concern ourselves with obtaining a detailed picture, from a new perspective (7; 8; 9; 10; 11; 12; 13), of how changes in the independent external parameters - thermodynamic parameters - affect this micro-state population and, consequently, the entropy and the internal energy, i.e.,

reversible changes in external parameters $\Delta_{param} \rightarrow$ changes in the microscopic probability distribution $\rightarrow$ entropic (dS) and internal energy (dU) changes.

We regarded as independent external parameters both extensive and intensive quantities defining the macroscopic thermodynamic state of the system. It is well-known that the extensive parameters, always known with some (experimental) uncertainty, help to define the Hilbert space (HS) in which the system can be represented. The intensive parameters are associated with some physical quantities of which only the average value is known. They are related to the mean values of operators acting on the HS previously defined. The eigenvalues of these operators are, therefore, functions of the extensive parameters defining the HS. The microscopic equilibrium probability distribution (PD) is an *explicit* function of the intensive parameters and an *implicit* function - via the eigenvalues of the above referred to operators (known in average) - of the extensive parameters defining the HS.
What is the hard core of the new view-point of (7; 8; 9; 10; 11; 12; 13)? It consists, as will be detailed below, in

- enforcing the relation $dU = TdS + \sum_\nu P_\nu dA_\nu$ in an infinitesimal *microscopic* change $p_i \rightarrow p_i + dp_i$ of the probability distribution (PD) that describes the equilibrium properties of an arbitrary system and ascertaining that
- this univocally determines the PD, and furthermore,
- that the ensuing $\{p_i\}$ coincides with that obtained following the maximum entropy principle (MaxEnt) tenet of extremizing the entropy S subject to an assumedly known mean value U of the system's energy.

Consider now only infinitesimal *macroscopic* parameter-changes (as opposite to the microscopic PD-ones dealt with in (7)), according to the scheme below.

Reversible changes in parameters $\Delta_{param} \rightarrow$ PD-changes $\rightarrow$ entropic (dS) and internal energy (dU) changes + some work effected (δW). Forcing now that Δ_{param} be of such nature that $dU = TdS + \delta W$ one gets an univocal expression for the PD.

That is, we study variations in both the (i) intensive and (ii) extensive parameters of the system and wish to ascertain just how these variations materialize themselves into concrete thermal relations.

9.1 Homogeneous, isotropic, one-component systems

For simplicity, consider just *simple*, one-component systems (6) composed by a single chemical species, macroscopically homogeneous, and isotropic (6). The macroscopic equilibrium thermal state of such a simple, one-component system is described, in self-explanatory notation, by T, V, N (6). Focus attention upon a quite general information measure S that, according to Kinchin's axioms for information theory depends exclusively on of the probability distribution $\{p_i\}$. We use again the *specific* but rather general form given above for S, viz.,

$$S = k \sum_{i=1}^{W} p_i f(p_i), \tag{28}$$

with W the number of microscopic states, $k =$ Boltzmann's constant, and the sum running over a set of quantum numbers, collectively denoted by i (characterizing levels of energy ϵ_i), that specify an appropriate basis in Hilbert's space (f is an arbitrary smooth function of the p_i such that $pf(p)$ is concave).
Remember that the quantity U represents the mean value of the Hamiltonian, and, as befits an homogeneous, isotropic, one-component system in the Helmholtz free energy representation (6) we have

1. as *external parameter* the volume (V) and the number of particles (N) ("exactly" known and used to define the Hilbert space),
2. as *intensive variable* the temperature T, associated with the mean value U of the internal energy E, i.e., $U = \langle E \rangle$.

The energy eigenvalues of the Hamiltonian ϵ_i are, obviously, functions of the volume and of the number of particles, namely, $\{\epsilon_i\} = \{\epsilon_i(V,N)\}$. From now on, for simplicity, we take N as fixed, and drop thereby the dependence of the energy eigenvalues on N, i.e., $\{\epsilon_i\} = \{\epsilon_i(V)\}$. The probability distribution (PD) depends, then, on the external parameters in the fashion

$$p_i = p_i(T, \epsilon_i(V)). \tag{29}$$

Remind that the mean energy $U = \langle E \rangle$ is given by

$$U = \langle E \rangle = \sum_{i=1}^{W} g(p_i)\epsilon_i. \tag{30}$$

The critical difference between what we attempt to do now and what was related above [Cf. Eq. (23)] is to be found in the following assumption, *on which we entirely base our considerations in this Section*:

> the temperature T and the volume V reversibly change in the fashion
>
> $$T \rightarrow T + dT \text{ and } V \rightarrow V + dV. \tag{31}$$

As a consequence of (31), corresponding changes dp_i, dS, $d\epsilon_i$, and dU are generated in, respectively, p_i, S, ϵ_i, and U. Variations in, respectively, p_i, S, and U write

$$dp_i = \frac{\partial p_i}{\partial T} dT + \sum_{j=1}^{W} \frac{\partial p_i}{\partial \epsilon_j} \frac{\partial \epsilon_j}{\partial V} dV, \tag{32}$$

$$dS = \sum_{i=1}^{W} \frac{\partial S}{\partial p_i} \frac{\partial p_i}{\partial T} dT + \sum_{i,j=1}^{W} \frac{\partial S}{\partial p_i} \frac{\partial p_i}{\partial \epsilon_j} \frac{\partial \epsilon_j}{\partial V} dV, \tag{33}$$

and, last but not least,

$$dU = \sum_{i=1}^{W} \frac{\partial g}{\partial p_i} \frac{\partial p_i}{\partial T} \epsilon_i dT + \sum_{i,j=1}^{W} \frac{\partial g}{\partial p_i} \frac{\partial p_i}{\partial \epsilon_j} \frac{\partial \epsilon_j}{\partial V} \epsilon_i dV + \sum_{i=1}^{W} g(p_i) \frac{\partial \epsilon_i}{\partial V} dV, \tag{34}$$

where, for simplicity, we have considered non-degenerate levels. Clearly, on account of normalization, the changes in p_i must satisfy the relation

$$\sum_i dp_i = 0. \tag{35}$$

Note that if we deal with three thermodynamic parameters and one equation of state we can completely describe our system with *any* two of them (32). Here, we are choosing, as the two independent thermodynamic parameters, T and V. It is important to remark that independent thermodynamic parameters *do not mean natural parameters*. For example, if T and V are now the independent thermodynamic parameters, the internal energy can be written as function of these parameters, i.e., $U(T,V)$. Clearly, T and V are not the natural parameters of the internal energy. These are S and V. However, our developments require only independent parameters, that are not necessarily the natural ones (32).

9.2 Macroscopic considerations

Thermodynamics states that, in the present scenario, for a reversible process one has

$$dU = \delta Q + \delta W = TdS + \delta W, \tag{36}$$

where we have used the Clausius relation $\delta Q = TdS$. Multiplying Eq. (33) by T we can recast Eq. (36) in the *microscopic* fashion (involving the microstates' PD)

$$\boxed{dU = T\left(\sum_{i=1}^{W} \frac{\partial S}{\partial p_i}\frac{\partial p_i}{\partial T}dT + \sum_{i,j=1}^{W} \frac{\partial S}{\partial p_i}\frac{\partial p_i}{\partial \epsilon_j}\frac{\partial \epsilon_j}{\partial V}dV\right) + \delta W,} \tag{37}$$

which is to be compared with (34).

9.3 Changes in the temperature

Eqs. (34) and (37) must be equal for arbitrary changes in T and V. We take this equality as the basis of our future considerations. As T and V can be changed in an independent way, let us first consider just changes in T. Enforcing equality in the coefficients of dT appearing in Eqs. (34) and (37) we obtain (we are assuming, as it is obvious, that the mechanical δW does not depend on the temperature)

$$\sum_{i=1}^{W} \frac{\partial g}{\partial p_i}\frac{\partial p_i}{\partial T}\epsilon_i dT = T\sum_{i=1}^{W} \frac{\partial S}{\partial p_i}\frac{\partial p_i}{\partial T}dT, \tag{38}$$

that must be satisfied together with [Cf. (32)]

$$\sum_i dp_i = \sum_i \frac{\partial p_i}{\partial T}dT = 0. \tag{39}$$

We recast now (38) in the fashion

$$\sum_{i=1}^{W}\left(\frac{\partial g}{\partial p_i}\epsilon_i - T\frac{\partial S}{\partial p_i}\right)\frac{\partial p_i}{\partial T}dT \equiv \sum_i K_i \frac{\partial p_i}{\partial T}dT = 0. \tag{40}$$

Since the W p_i's are not independent ($\sum_{i=1}^{W} p_i = 1$), we can separate the sum in (40) into two parts, i.e.,

$$\sum_{i=1}^{W-1}\left(\frac{\partial g}{\partial p_i}\epsilon_i - T\frac{\partial S}{\partial p_i}\right)\frac{\partial p_i}{\partial T}dT + \left(\frac{\partial g}{\partial p_W}\epsilon_W - T\frac{\partial S}{\partial p_W}\right)\frac{\partial p_W}{\partial T}dT = 0. \tag{41}$$

Picking out level W for special attention is arbitrary. Any other $i-$level could have been chosen as well, as the example given below will illustrate. Taking into account now that, from Eq. (39),

$$\frac{\partial p_W}{\partial T} = -\sum_{i=1}^{W-1}\frac{\partial p_i}{\partial T}, \tag{42}$$

we see that Eq. (41) can be rewritten as

$$\sum_{i=1}^{W-1}\left[\left(\frac{\partial g}{\partial p_i}\epsilon_i - T\frac{\partial S}{\partial p_i}\right) - \left(\frac{\partial g}{\partial p_W}\epsilon_W - T\frac{\partial S}{\partial p_W}\right)\right]\frac{\partial p_i}{\partial T}dT = 0. \tag{43}$$

As the $W-1$ p_i's are now independent, the term into brackets should vanish, which entails

$$\frac{\partial g}{\partial p_i}\epsilon_i - T\frac{\partial S}{\partial p_i} - \left(\frac{\partial g}{\partial p_W}\epsilon_W - T\frac{\partial S}{\partial p_W}\right) = 0, \tag{44}$$

for all $i = 1,\cdots,W-1$. Let us call the term into parentheses as

$$K_W = \frac{\partial g}{\partial p_W}\epsilon_W - T\frac{\partial S}{\partial p_W} \equiv K = \text{constant}. \tag{45}$$

Finally, we cast Eqs. (44) and (45) as

$$\frac{\partial g}{\partial p_i}\epsilon_i - T\frac{\partial S}{\partial p_i} - K = 0;\ (i = 1,\cdots,W), \tag{46}$$

an equation that we have encountered before [Cf. Eq. (24) with $g(x) \equiv x$] and that should yield a definite expression for any of the W p_i's. We did not care above about such an expression, but we do now.

Example 1 Consider the Shannon orthodox instance

$$\begin{aligned} S &= -k\sum_i p_i \ln p_i \\ g(p_i) &= p_i \\ \partial S/\partial p_i &= -k[\ln p_i + 1] = k[\beta\epsilon_i + \ln Z - 1]]. \end{aligned} \tag{47}$$

Here equation (46) yields the well known MaxEnt (and also Gibbs?) result

$$\begin{aligned} \ln p_i &= -[\beta\epsilon_i + \ln Z];\ \text{i.e.,} \\ p_i &= Z^{-1}e^{-\epsilon_i/kT} \\ \ln Z &= 1 - K/kT,\ \text{and, finally,} \end{aligned} \tag{48}$$

$$\partial S/\partial p_i = k\beta(\epsilon_i - K),$$

$$\frac{\partial \ln Z}{\partial \epsilon_i} = -\beta p_i;\ \frac{\partial p_i}{\partial \epsilon_j} = -\beta p_i\,(\delta_{ij} - p_j);\ T\frac{\partial S}{\partial p_i}\frac{\partial p_i}{\partial \epsilon_i} = -\beta(\epsilon_i - K)p_i, \tag{49}$$

showing, as anticipated, that we could have selected any $i-$level among the $W-$ones without affecting the final result.

Thus, changes $\delta\beta$ in the inverse temperature β completely specify the microscopic probability density $\{p_{MaxEnt}\}$ if they are constrained to obey the relation $dU = TdS + \delta W$, for any reasonable choice of the information measure S. This equivalence, however, can not be established in similar fashion if the *extensive* variable V also changes. This is our next topic.

9.4 Changes in the extensive parameter

Let us now deal with the effect of changes in the extensive parameters that define the Hilbert space in which our system "lives" and notice that Eq. (37) can be written in the fashion $dU = \delta Q + \delta W = TdS + \delta W \Rightarrow$

$$dU = T\left(dT\sum_{i=1}^{W}\frac{\partial S}{\partial p_i}\frac{\partial p_i}{\partial T} + dV\sum_{i,j=1}^{W}\frac{\partial S}{\partial p_i}\frac{\partial p_i}{\partial \epsilon_j}\frac{\partial \epsilon_j}{\partial V}\right) + \delta W. \tag{50}$$

That is, there are two ingredients entering TdS, namely,

$$TdS = Q_T dT + Q_V dV;\ \text{with}\ Q_T = T\sum_{i=1}^{W}\frac{\partial S}{\partial p_i}\frac{\partial p_i}{\partial T}. \tag{51}$$

Our interest now lies in the second term. What is Q_V? Clearly we have

$$Q_V = T \sum_{i,j=1}^{W} \frac{\partial S}{\partial p_i} \frac{\partial p_i}{\partial \epsilon_j} \frac{\partial \epsilon_j}{\partial V}. \tag{52}$$

Next, substitute the expression for $(\partial g/\partial p_i)\epsilon_i$ given by Eqs. (45) and (46),

$$\frac{\partial g}{\partial p_i}\epsilon_i = T\frac{\partial S}{\partial p_i} + K; \ (i = 1, \ldots, W), \tag{53}$$

into the second term of the R.H.S. of Eq. (34),

$$\begin{aligned}
\sum_{i,j=1}^{W} \frac{\partial g}{\partial p_i} \frac{\partial p_i}{\partial \epsilon_j} \frac{\partial \epsilon_j}{\partial V} \epsilon_i \, dV &= \sum_{i,j=1}^{W} [T\frac{\partial S}{\partial p_i} + K] \frac{\partial p_i}{\partial \epsilon_j} \frac{\partial \epsilon_j}{\partial V} dV \\
&= T \sum_{i,j=1}^{W} \frac{\partial S}{\partial p_i} \frac{\partial p_i}{\partial \epsilon_j} \frac{\partial \epsilon_j}{\partial V} dV + K \sum_{i,j=1}^{W} \frac{\partial p_i}{\partial \epsilon_j} \frac{\partial \epsilon_j}{\partial V} dV \\
&= \left(T \sum_{i,j=1}^{W} \frac{\partial S}{\partial p_i} \frac{\partial p_i}{\partial \epsilon_j} \frac{\partial \epsilon_j}{\partial V} \right) dV = Q_V \, dV,
\end{aligned} \tag{54}$$

on account of the fact that

$$K \sum_{i,j=1}^{W} \frac{\partial p_i}{\partial \epsilon_j} \frac{\partial \epsilon_j}{\partial V} dV = 0; \ \text{since} \ (\partial/\partial V) \sum_i p_i = 0. \tag{55}$$

We recognize in the term $Q_V \, dV$ of the last line of (54) the *m*icroscopic interpretation of a rather unfamiliar "volume contribution" to Clausius' relation $\delta Q = TdS$ (dQ-equations (32)). Notice that we are *not* explicitly speaking here of phase-changes. We deal with reversible processes. If the change in volume were produced by a phase-change one would reasonably be tempted to call the term $Q_V \, dV$ a "latent" heat.

Thus, associated with a change of state in which the volume is modified, we find in the term $Q_V \, dV$ the microscopic expression of a "heat" contribution for that transformation, i.e., the heat given up or absorbed during it. It we wish to call it "latent", the reason would be that it is not associated with a change in temperature. Thus, we saw just how changes in the equilibrium PD caused by modifications in the extensive parameter defining the Hilbert space of the system give also a contribution to the "heat part" of the $dU = TdS + \delta W$ relation.

Example 2: In the Shannon instance discussed in Example 1 one has [Cf. (48) and (49)]

$$\frac{\partial p_i}{\partial \epsilon_i} = -\beta p_i (1 - Z^{-1}), \tag{56}$$

$$T \frac{\partial S}{\partial p_i} \frac{\partial p_i}{\partial \epsilon_i} = -\beta(\epsilon_i - K) p_i, \tag{57}$$

$$Q_V = -\sum_i \beta(\epsilon_i - K) p_i \frac{\partial \epsilon_i}{\partial V} (1 - Z^{-1}). \tag{58}$$

Since the origin of the energy scale is arbitrary, in summing over i we can omit the $K-$term by changing the energy-origin and one may write

$$Q_V = -\sum_i \beta \epsilon_i p_i \frac{\partial \epsilon_i}{\partial V} (1 - Z^{-1}). \tag{59}$$

Foe a particle of mass m in an ideal gas (N particles) the energy ϵ_i is given by (29)

$$\epsilon_i = \tau V^{-2/3} \vec{n}_i^{\,2};\ \tau = \frac{(\pi\hbar)^2}{2m};\ \vec{n}_i^{\,2} \equiv (n_x^2, n_y^2, n_z^2)$$
n_x, n_y, n_z a set of three integers

$$\frac{\partial \epsilon_i}{\partial V} = -(2/3)\epsilon_i / V. \tag{60}$$

Thus,the microscopic expression for Q_V turns out to be

$$Q_V = (2\beta/3V)\langle E^2 \rangle (1 - Z^{-1}), \tag{61}$$

which indeed has dimension of (energy/volume).

Finally, for Eq. (34) to become equal to Eq. (50) we have to demand, in view of the above developments,

$$\delta W = dV \left[\sum_i g(p_i) \frac{\partial \epsilon_i}{\partial V} \right], \tag{62}$$

the quantity within the brackets being the mean value,

$$\left\langle \frac{\partial E}{\partial V} \right\rangle = \sum_i g(p_i) \frac{\partial \epsilon_i}{\partial V}, \tag{63}$$

usually associated in the textbooks with the work done by the system.
Summing up, our analysis of simple systems in the present Section has shown that

- by considering changes dT and dV and how they influence the microscopic probability distribution if these variations are forced to comply with the relation (36) $dU = TdS + \delta W$ we ascertain that
- changes in the intensive parameter give contributions only related to heat and lead to the attaining the equilibrium PD (an alternative way to the MaxEnt principle) and
- changes in the extensive-Hilbert-space-determining parameter lead to two contributions
  1. one related to heat and
  2. the other related to work.

10. Other entropic forms

We illustrate now our procedure with reference to information measures not of the Shannon logarithmic form. We use mostly the relationship (46), namely,

$$\begin{gathered} K = \epsilon_i g'(p_i) - kT\,[f(p_i) + p_i f'(p_i)] \Rightarrow \\ [f(p_i) + p_i f'(p_i)] - \beta[\epsilon_i g'(p_i) - K] = 0, \\ \beta \equiv 1/kT. \end{gathered} \tag{64}$$

10.1 Tsallis measure with linear constraints

We have, for any real number q the information measure (28) built up with (26; 33; 34)

$$f(p_i) = \frac{(1 - p_i^{q-1})}{q-1}, \tag{65}$$

and, in the energy-constraint of Eq. (30)

$$g(p_i) = p_i, \tag{66}$$

so that $f'(p_i) = -p_i^{q-2}$ and Eq. (64) becomes, with $\beta = (1/kT)$,

$$q\,p_i^{q-1} = 1 + (q-1)\beta K - (q-1)\,\beta\epsilon_i, \tag{67}$$

which after normalization yields a distribution often referred to as the Tsallis' one (33)

$$\begin{aligned} p_i &= Z_q^{-1}\,[1 - (q-1)\,\beta'\,\epsilon_i]^{1/(q-1)} \\ Z_q &= \sum_i [1 - (q-1)\,\beta'\,\epsilon_i]^{1/(q-1)}, \end{aligned} \tag{68}$$

where $\beta' \equiv \beta/(1 + (q-1)\beta K)$.

10.2 Tsallis measure with non-linear constraints

The information measure is still the one built up with the function $f(p_i)$ of (65), but we use now the so-called Curado-Tsallis constraints (35) that arise if one uses

$$U = \langle E \rangle = \sum_{i=1}^{W} g(p_i)\epsilon_i, \tag{69}$$

with

$$g(p_i) = p_i^q \Rightarrow g'(p_i) = q\,p_i^{q-1}. \tag{70}$$

Eq. (64) leads to

$$p_i = \left(\frac{1}{q}\right)^{1/(q-1)} [1 - (1-q)\,\beta\epsilon_i]^{1/(1-q)}, \tag{71}$$

and, after normalization, one is led to the Curado-Tsallis distribution (35)

$$\begin{aligned} p_i &= (Z_q)^{-1}\,[1 - (1-q)\,\beta\epsilon_i]^{1/(1-q)} \\ Z_q &= \sum_i [1 - (1-q)\,\beta\epsilon_i]^{1/(1-q)}. \end{aligned} \tag{72}$$

10.3 Exponential entropic form

This measure is given in (36; 37) and also used in (38). One has

$$f(p_i) = \frac{1 - \exp(-bp_i)}{p_i} - S_0, \tag{73}$$

where b is a positive constant and $S_0 = 1 - \exp(-b)$, together with

$$g(p_i) = \frac{1 - e^{-bp_i}}{S_0} \Rightarrow g'(p_i) = \frac{be^{-bp_i}}{S_0}, \tag{74}$$

which, inserted into (64), after a little algebra, leads to

$$p_i = \frac{1}{b}\left[\ln \frac{b}{S_0 - \beta K} + \ln(1 - \frac{\beta \epsilon_i}{S_0})\right]. \tag{75}$$

which, after normalization, gives the correct answer (37).

11. Conclusions

We have seen that the set of equations

$$\sum_i [C_i^{(1)} + C_i^{(2)}] dp_i = 0,$$

$$C_i^{(1)} = [\sum_{\nu=1}^{M} P_\nu a_i^\nu + \epsilon_i] g'(p_i)$$

$$C_i^{(2)} = -T \frac{\partial S}{\partial p_i}$$

yields a probability distribution that coincides with the PD provided by either

- the MaxEnt's, SM axiomatics of Jaynes'
- our two postulates (15) and (16).

We remind the reader that in our instance the postulates start with

1. the macroscopic thermodynamic relation $dE = TdS + \sum_\nu P_\nu dA_\nu$,, adding to it
2. Boltzmann's conjecture of an underlying microscopic scenario ruled by microstate probability distributions.

The two postulates combine then (i) a well-tested macroscopic result with (ii) a by now un uncontestable microscopic state of affairs (which was not the case in Boltzmann's times). Thus we may dare to assert that the two axioms we are here advancing are intuitively intelligible from a physical laboratory standpoint. This cannot be said neither for Gibbs' ensemble nor for Jaynes' extremizing of the Observer's ignorance, their extraordinary success notwithstanding, since they introduce concepts like ensemble or ignorance that are not easily assimilated to laboratory equipment. We must insist: there is nothing wrong with making use of these concepts, of course. We just tried to see whether they could be eliminated from the axioms of the theory.

Summing up, we have revisited the foundations of statistical mechanics and shown that it is possible to reformulate it on the basis of just a basic thermodynamics' relation plus Boltzmann's "atomic" hypothesis. The latter entails (1) the (obvious today, but not in 1866) existence of a microscopic realm ruled by probability distributions.

12. Appendix I

Here we prove that Eqs. (24) are obtained via the MaxEnt variational problem (27). Assume now that you wish to extremize S subject to the constraints of fixed valued for i) U and ii) the M values A_ν. This is achieved via Lagrange multipliers (1) β and (2) M γ_ν. We need also a normalization Lagrange multiplier ξ. Recall that

$$A_\nu = \langle \mathcal{R}_\nu \rangle = \sum_i p_i a_i^\nu, \tag{76}$$

with $a_i^\nu = \langle i|\mathcal{R}_\nu|i\rangle$ the matrix elements in the chosen basis $\langle i \rangle$ of $\mathcal{R}_\nu$. The MaxEnt variational problem becomes now ($U = \sum_i p_i \epsilon_i$)

$$\delta_{\{p_i\}} \left[S - \beta U - \sum_{\nu=1}^{M} \gamma_\nu A_\nu - \xi \sum_i p_i \right] = 0, \tag{77}$$

leading, with $\gamma_\nu = \beta \lambda_\nu$, to the vanishing of

$$\delta_{p_m} \sum_i \left(p_i f(p_i) - [\beta p_i (\sum_{\nu=1}^{M} \lambda_\nu a_i^\nu + \epsilon_i) + \xi p_i] \right), \tag{78}$$

so that the 2 quantities below vanish

$$\begin{aligned} & f(p_i) + p_i f'(p_i) - [\beta(\textstyle\sum_{\nu=1}^{M} \lambda_\nu a_i^\nu + \epsilon_i) + \xi] \\ & \Rightarrow \text{ if } \xi \equiv \beta K, \\ & f(p_i) + p_i f'(p_i) - \beta p_i (\textstyle\sum_{\nu=1}^{M} \lambda_\nu a_i^\nu + \epsilon_i) + K] \\ & \Rightarrow 0 = T_i^{(1)} + T_i^{(2)}. \end{aligned} \tag{79}$$

Clearly, (26) and the last equality of (79) are one and the same equation! Our equivalence is thus proven.

13. Acknowledgments

This work is founded by the Spain Ministry of Science and Innovation (Project FIS2008-00781) and by FEDER funds (EU).

14. References

[1] R. B. Lindsay and H. Margenau, *Foundations of physics*, NY, Dover, 1957.
[2] J. Willard Gibbs, *Elementary Principles in Statistical Mechanics*, New Haven, Yale University Press, 1902.
[3] E.T. Jaynes, Probability Theory: The Logic of Science, Cambridge University Press, Cambridge, 2005.
[4] W.T. Grandy Jr. and P. W. Milonni (Editors), *Physics and Probability. Essays in Honor of Edwin T. Jaynes*, NY, Cambridge University Press, 1993.
[5] E. T. Jaynes *Papers on probability, statistics and statistical physics*, edited by R. D. Rosenkrantz, Dordrecht, Reidel, 1987.
[6] E. A. Desloge, *Thermal physics* NY, Holt, Rhinehart and Winston, 1968.
[7] E. Curado, A. Plastino, Phys. Rev. E 72 (2005) 047103.
[8] A. Plastino, E. Curado, Physica A 365 (2006) 24

[9] A. Plastino, E. Curado, International Journal of Modern Physics B 21 (2007) 2557
[10] A. Plastino, E. Curado, Physica A 386 (2007) 155
[11] A. Plastino, E. Curado, M. Casas, Entropy A 10 (2008) 124
[12] International Journal of Modern Physics B 22, (2008) 4589
[13] E. Curado, F. Nobre, A. Plastino, Physica A 389 (2010) 970.
[14] The MaxEnt treatment assumes that these macrocopic parameters are the expectation values of appropiate operators.
[15] C. E. Shannon, Bell System Technol. J. 27 (1948) 379-390.
[16] A. Plastino and A. R. Plastino in *Condensed Matter Theories*, Volume 11, E. Ludeña (Ed.), Nova Science Publishers, p. 341 (1996).
[17] A. Katz, *Principles of Statistical Mechanics, The information Theory Approach*, San Francisco, Freeman and Co., 1967.
[18] D. J. Scalapino in *Physics and probability. Essays in honor of Edwin T. Jaynes* edited by W. T. Grandy, Jr. and P. W. Milonni (Cambridge University Press, NY, 1993), and references therein.
[19] T. M. Cover and J. A. Thomas, *Elements of information theory*, NY, J. Wiley, 1991.
[20] B. Russell, *A history of western philosophy* (Simon & Schuster, NY, 1945).
[21] P. W. Bridgman *The nature of physical theory* (Dover, NY, 1936).
[22] P. Duhem *The aim and structure of physical theory* (Princeton University Press, Princeton, New Jersey, 1954).
[23] R. B. Lindsay *Concepts and methods of theoretical physics* (Van Nostrand, NY, 1951).
[24] H. Weyl *Philosophy of mathematics and natural science* (Princeton University Press, Princeton, New Jersey, 1949).
[25] D. Lindley, *Boltzmann's atom*, NY, The free press, 2001.
[26] M. Gell-Mann and C. Tsallis, Eds. *Nonextensive Entropy: Interdisciplinary applications*, Oxford, Oxford University Press, 2004.
[27] G. L. Ferri, S. Martinez, A. Plastino, Journal of Statistical Mechanics, P04009 (2005).
[28] R.K. Pathria, *Statistical Mechanics* (Pergamon Press, Exeter, 1993).
[29] F. Reif, *Statistical and thermal physics* (McGraw-Hill, NY, 1965).
[30] J. J.Sakurai, *Modern quantum mechanics* (Benjamin, Menlo Park, Ca., 1985).
[31] B. H. Lavenda, *Statistical Physics* (J. Wiley, New York, 1991); B. H. Lavenda, *Thermodynamics of Extremes* (Albion, West Sussex, 1995).
[32] K. Huang, *Statistical Mechanics*, 2nd Edition. (J. Wiley, New York, 1987). Pages 7-8.
[33] C. Tsallis, Braz. J. of Phys. 29, 1 (1999); A. Plastino and A. R. Plastino, Braz. J. of Phys. 29, 50 (1999).
[34] A. R. Plastino and A. Plastino, Phys. Lett. A 177, 177 (1993).
[35] E. M. F. Curado and C. Tsallis, J. Phys. A, 24, L69 (1991).
[36] E. M. F. Curado, Braz. J. Phys. 29, 36 (1999).
[37] E. M. F. Curado and F. D. Nobre, Physica A 335, 94 (2004).
[38] N. Canosa and R. Rossignoli, Phys. Rev. Lett. 88, 170401 (2002).

2

Rigorous and General Definition of Thermodynamic Entropy

Gian Paolo Beretta[1] and Enzo Zanchini[2]
[1]*Università di Brescia, Via Branze 38, Brescia*
[2]*Università di Bologna, Viale Risorgimento 2, Bologna*
Italy

1. Introduction

Thermodynamics and Quantum Theory are among the few sciences involving fundamental concepts and universal content that are controversial and have been so since their birth, and yet continue to unveil new possible applications and to inspire new theoretical unification. The basic issues in Thermodynamics have been, and to a certain extent still are: the range of validity and the very formulation of the Second Law of Thermodynamics, the meaning and the definition of entropy, the origin of irreversibility, and the unification with Quantum Theory (Hatsopoulos & Beretta, 2008). The basic issues with Quantum Theory have been, and to a certain extent still are: the meaning of complementarity and in particular the wave-particle duality, understanding the many faces of the many wonderful experimental and theoretical results on entanglement, and the unification with Thermodynamics (Horodecki et al., 2001).
Entropy has a central role in this situation. It is astonishing that after over 140 years since the term entropy has been first coined by Clausius (Clausius, 1865), there is still so much discussion and controversy about it, not to say confusion. Two recent conferences, both held in October 2007, provide a state-of-the-art scenario revealing an unsettled and hard to settle field: one, entitled *Meeting the entropy challenge* (Beretta et al., 2008), focused on the many physical aspects (statistical mechanics, quantum theory, cosmology, biology, energy engineering), the other, entitled *Facets of entropy* (Harremöes, 2007), on the many different mathematical concepts that in different fields (information theory, communication theory, statistics, economics, social sciences, optimization theory, statistical mechanics) have all been termed *entropy* on the basis of some analogy of behavior with the *thermodynamic entropy*.
Following the well-known Statistical Mechanics and Information Theory interpretations of thermodynamic entropy, the term *entropy* is used in many different contexts wherever the relevant *state description* is in terms of a probability distribution over some set of possible events which characterize the *system description*. Depending on the context, such events may be *microstates*, or *eigenstates*, or *configurations*, or *trajectories*, or *transitions*, or *mutations*, and so on. Given such a probabilistic description, the term entropy is used for some functional of the probabilities chosen as a quantifier of their *spread* according to some reasonable set of defining axioms (Lieb & Yngvason, 1999). In this sense, the use of a common name for all the possible different state functionals that share such broad defining features, may have some unifying advantage from a broad conceptual point of view, for example it may suggest analogies and inter-breeding developments between very different fields of research sharing similar probabilistic descriptions.

However, from the physics point of view, entropy — the *thermodynamic entropy* — is a single definite property of every well-defined material system that can be measured in every laboratory by means of standard measurement procedures. Entropy is a property of paramount practical importance, because it turns out (Gyftopoulos & Beretta, 2005) to be monotonically related to the difference $E - \Psi$ between the energy E of the system, above the lowest-energy state, and the adiabatic availability Ψ of the system, *i.e.*, the maximum work the system can do in a process which produces no other external effects. It is therefore very important that whenever we talk or make inferences about physical (*i.e.*, thermodynamic) entropy, we first agree on a precise definition.

In our opinion, one of the most rigorous and general axiomatic definitions of thermodynamic entropy available in the literature is that given in (Gyftopoulos & Beretta, 2005), which extends to the nonequilibrium domain one of the best traditional treatments available in the literature, namely that presented by Fermi (Fermi, 1937).

In this paper, the treatment presented in (Gyftopoulos & Beretta, 2005) is assumed as a starting point and the following improvements are introduced. The basic definitions of system, state, isolated system, environment, process, separable system, and parameters of a system are deepened, by developing a logical scheme outlined in (Zanchini, 1988; 1992). Operative and general definitions of these concepts are presented, which are valid also in the presence of internal semipermeable walls and reaction mechanisms. The treatment of (Gyftopoulos & Beretta, 2005) is simplified, by identifying the minimal set of definitions, assumptions and theorems which yield the definition of entropy and the principle of entropy non-decrease. In view of the important role of entanglement in the ongoing and growing interplay between Quantum Theory and Thermodynamics, the effects of correlations on the additivity of energy and entropy are discussed and clarified. Moreover, the definition of a reversible process is given with reference to a given *scenario*; the latter is the largest isolated system whose subsystems are available for interaction, for the class of processes under exam. Without introducing the quantum formalism, the approach is nevertheless compatible with it (and indeed, it was meant to be so, see, *e.g.*, Hatsopoulos & Gyftopoulos (1976); Beretta et al. (1984; 1985); Beretta (1984; 1987; 2006; 2009)); it is therefore suitable to provide a basic logical framework for the recent scientific revival of thermodynamics in Quantum Theory [quantum heat engines (Scully, 2001; 2002), quantum Maxwell demons (Lloyd, 1989; 1997; Giovannetti et al., 2003), quantum erasers (Scully et al., 1982; Kim et al., 2000), etc.] as well as for the recent quest for quantum mechanical explanations of irreversibility [see, *e.g.*, Lloyd (2008); Bennett (2008); Hatsopoulos & Beretta (2008); Maccone (2009)].

The paper is organized as follows. In Section 2 we discuss the drawbacks of the traditional definitions of entropy. In Section 3 we introduce and discuss a full set of basic definitions, such as those of system, state, process, etc. that form the necessary unambiguous background on which to build our treatment. In Section 4 we introduce the statement of the First Law and the definition of energy. In Section 5 we introduce and discuss the statement of the Second Law and, through the proof of three important theorems, we build up the definition of entropy. In Section 6 we briefly complete the discussion by proving in our context the existence of the fundamental relation for the stable equilibrium states and by defining temperature, pressure, and other generalized forces. In Section 7 we extend our definitions of energy and entropy to the model of an open system. In Section 8 we prove the existence of the fundamental relation for the stable equilibrium states of an open system. In Section 9 we draw our conclusions and, in particular, we note that nowhere in our construction we use or need to define the concept of *heat*.

2. Drawbacks of the traditional definitions of entropy

In traditional expositions of thermodynamics, entropy is defined in terms of the concept of heat, which in turn is introduced at the outset of the logical development in terms of heuristic illustrations based on mechanics. For example, in his lectures on physics, Feynman (Feynman, 1963) describes heat as one of several different forms of energy related to the jiggling motion of particles stuck together and tagging along with each other (pp. 1-3 and 4-2), a form of energy which really is just kinetic energy — internal motion (p. 4-6), and is measured by the random motions of the atoms (p. 10-8). Tisza (Tisza, 1966) argues that such slogans as "heat is motion", in spite of their fuzzy meaning, convey intuitive images of pedagogical and heuristic value.
There are at least three problems with these illustrations. First, work and heat are not stored in a system. Each is a mode of transfer of energy from one system to another. Second, concepts of mechanics are used to justify and make plausible a notion — that of heat — which is beyond the realm of mechanics; although at a first exposure one might find the idea of *heat as motion* harmless, and even natural, the situation changes drastically when the notion of heat is used to define entropy, and the logical loop is completed when entropy is shown to imply a host of results about energy availability that contrast with mechanics. Third, and perhaps more important, heat is a mode of energy (and entropy) transfer between systems that are very close to thermodynamic equilibrium and, therefore, any definition of entropy based on heat is bound to be valid only at thermodynamic equilibrium.
The first problem is addressed in some expositions. Landau and Lifshitz (Landau & Lifshitz, 1980) define heat as the part of an energy change of a body that is not due to work done on it. Guggenheim (Guggenheim, 1967) defines heat as an exchange of energy that differs from work and is determined by a temperature difference. Keenan (Keenan, 1941) defines heat as the energy transferred from one system to a second system at lower temperature, by virtue of the temperature difference, when the two are brought into communication. Similar definitions are adopted in most other notable textbooks that are too many to list.
None of these definitions, however, addresses the basic problem. The existence of exchanges of energy that differ from work is not granted by mechanics. Rather, it is one of the striking results of thermodynamics, namely, of the existence of entropy as a property of matter. As pointed out by Hatsopoulos and Keenan (Hatsopoulos & Keenan, 1965), without the Second Law heat and work would be indistinguishable; moreover, the most general kind of interaction between two systems which are very far from equilibrium is neither a heat nor a work interaction. Following Guggenheim it would be possible to state a rigorous definition of heat, with reference to a very special kind of interaction between two systems, and to employ the concept of heat in the definition of entropy (Guggenheim, 1967). However, Gyftopoulos and Beretta (Gyftopoulos & Beretta, 2005) have shown that the concept of heat is unnecessarily restrictive for the definition of entropy, as it would confine it to the equilibrium domain. Therefore, in agreement with their approach, we will present and discuss a definition of entropy where the concept of heat is not employed.
Other problems are present in most treatments of the definition of entropy available in the literature:

1. many basic concepts, such as those of system, state, property, isolated system, environment of a system, adiabatic process are not defined rigorously;

2. on account of unnecessary assumptions (such as, the use of the concept of quasistatic process), the definition holds only for stable equilibrium states (Callen, 1985), or for systems which are in local thermodynamic equilibrium (Fermi, 1937);

3. in the traditional logical scheme (Tisza, 1966; Landau & Lifshitz, 1980; Guggenheim, 1967; Keenan, 1941; Hatsopoulos & Keenan, 1965; Callen, 1985; Fermi, 1937), some proofs are incomplete.

To illustrate the third point, which is not well known, let us refer to the definition in (Fermi, 1937), which we consider one of the best traditional treatments available in the literature. In order to define the thermodynamic temperature, Fermi considers a reversible cyclic engine which absorbs a quantity of heat Q_2 from a source at (empirical) temperature T_2 and supplies a quantity of heat Q_1 to a source at (empirical) temperature T_1. He states that if the engine performs n cycles, the quantity of heat subtracted from the first source is $n\,Q_2$ and the quantity of heat supplied to the second source is $n\,Q_1$. Thus, Fermi assumes implicitly that the quantity of heat exchanged in a cycle between a source and a reversible cyclic engine is independent of the initial state of the source. In our treatment, instead, a similar statement is made explicit, and proved.

3. Basic definitions

Level of description, constituents, amounts of constituents, deeper level of description. We will call *level of description* a class of physical models whereby all that can be said about the matter contained in a given region of space R, at a time instant t, can be described by assuming that the matter consists of a set of elementary building blocks, that we call *constituents*, immersed in the electromagnetic field. Examples of constituents are: atoms, molecules, ions, protons, neutrons, electrons. Constituents may combine and/or transform into other constituents according to a set of model-specific *reaction mechanisms*.
For instance, at the *chemical level of description* the constituents are the different chemical species, i.e., atoms, molecules, and ions; at the *atomic level of description* the constituents are the atomic nuclei and the electrons; at the *nuclear level of description* they are the protons, the neutrons, and the electrons.
The particle-like nature of the constituents implies that a counting measurement procedure is always defined and, when performed in a region of space delimited by impermeable walls, it is *quantized* in the sense that the measurement outcome is always an integer number, that we call the *number of particles*. If the counting is selective for the i-th type of constituent only, we call the resulting number of particles the *amount of constituent* i and denote it by n_i. When a number-of-particle counting measurement procedure is performed in a region of space delimited by at least one ideal-surface patch, some particles may be found across the surface. Therefore, an outcome of the procedure must also be the sum, for all the particles in this boundary situation, of a suitably defined fraction of their spatial extension which is within the given region of space. As a result, the *number of particles* and the *amount of constituent* i will not be quantized but will have continuous spectra.
A level of description L_2 is called *deeper* than a level of description L_1 if the amount of every constituent in L_2 is conserved for all the physical phenomena considered, whereas the same is not true for the constituents in L_1. For instance, the atomic level of description is deeper than the chemical one (because chemical reaction mechanisms do not conserve the number of molecules of each type, whereas they conserve the number of nuclei of each type as well as the number of electrons).
Levels of description typically have a hierarchical structure whereby the constituents of a given level are aggregates of the constituents of a deeper level.
Region of space which contains particles of the i-th constituent. We will call region of space which contains particles of the i-th constituent a connected region R_i of physical space (the

three-dimensional Euclidean space) in which particles of the i-th constituent are contained. The boundary surface of R_i may be a patchwork of *walls*, i.e., surfaces impermeable to particles of the i-th constituent, and ideal surfaces (permeable to particles of the i-th constituent). The geometry of the boundary surface of R_i and its permeability topology nature (walls, ideal surfaces) can vary in time, as well as the number of particles contained in R_i.

Collection of matter, composition. We will call *collection of matter*, denoted by C^A, a set of particles of one or more constituents which is described by specifying the allowed reaction mechanisms between different constituents and, at any time instant t, the set of r connected regions of space, $\boldsymbol{R}^A = R_1^A, \ldots, R_i^A, \ldots, R_r^A$, each of which contains n_i^A particles of a single kind of constituent. The regions of space $\boldsymbol{R}^A$ can vary in time and overlap. Two regions of space may contain the same kind of constituent provided that they do not overlap. Thus, the i-th constituent could be identical with the j-th constituent, provided that R_i^A and R_j^A are disjoint. If, due to time changes, two regions of space which contain the same kind of constituent begin to overlap, from that instant a new collection of matter must be considered.

Comment. This method of description allows to consider the presence of internal walls and/or internal *semipermeable* membranes, *i.e.*, surfaces which can be crossed only by some kinds of constituents and not others. In the simplest case of a collection of matter without internal partitions, the regions of space $\boldsymbol{R}^A$ coincide at every time instant.

The amount n_i of the constituent in the i-th region of space can vary in time for two reasons:

- *matter exchange*: during a time interval in which the boundary surface of R_i is not entirely a wall, particles may be transferred into or out of R_i; we denote by $\dot{\boldsymbol{n}}^{A\leftarrow}$ the set of rates at which particles are transferred in or out of each region, assumed positive if inward, negative if outward;
- *reaction mechanisms*: in a portion of space where two or more regions overlap, the allowed reaction mechanisms may transform, according to well specified proportions (*e.g.*, stoichiometry), particles of one or more regions into particles of one or more other regions.

Compatible compositions, set of compatible compositions. We say that two compositions, $\boldsymbol{n}^{1A}$ and $\boldsymbol{n}^{2A}$ of a given collection of matter C^A are *compatible* if the change between $\boldsymbol{n}^{1A}$ and $\boldsymbol{n}^{2A}$ or viceversa can take place as a consequence of the allowed reaction mechanisms without matter exchange. We will call *set of compatible compositions* for a system A the set of all the compositions of A which are compatible with a given one. We will denote a set of compatible compositions for A by the symbol $(\boldsymbol{n}^{0A}, \boldsymbol{\nu}^A)$. By this we mean that the set of τ allowed reaction mechanisms is defined like for chemical reactions by a matrix of stoichiometric coefficients $\boldsymbol{\nu}^A = [\nu_k^{(\ell)}]$, with $\nu_k^{(\ell)}$ representing the stoichiometric coefficient of the k-th constituent in the ℓ-th reaction. The set of compatible compositions is a τ-parameter set defined by the reaction coordinates $\boldsymbol{\varepsilon}^A = \varepsilon_1^A, \ldots, \varepsilon_\ell^A, \ldots, \varepsilon_\tau^A$ through the proportionality relations

$$\boldsymbol{n}^A = \boldsymbol{n}^{0A} + \boldsymbol{\nu}^A \cdot \boldsymbol{\varepsilon}^A , \tag{1}$$

where $\boldsymbol{n}^{0A}$ denotes the composition corresponding to the value zero of all the reaction coordinates $\boldsymbol{\varepsilon}^A$. To fix ideas and for convenience, we will select $\boldsymbol{\varepsilon}^A = 0$ at time $t = 0$ so that $\boldsymbol{n}^{0A}$ is the composition at time $t = 0$ and we may call it the *initial composition*.

In general, the rate of change of the amounts of constituents is subject to the *amounts balance equations*

$$\dot{\boldsymbol{n}}^A = \dot{\boldsymbol{n}}^{A\leftarrow} + \boldsymbol{\nu}^A \cdot \dot{\boldsymbol{\varepsilon}}^A . \tag{2}$$

External force field. Let us denote by $\mathbf{F}$ a force field given by the superposition of a gravitational field $\mathbf{G}$, an electric field $\mathbf{E}$, and a magnetic induction field $\mathbf{B}$. Let us denote by

Σ_t^A the union of all the regions of space $\boldsymbol{R}_t^A$ in which the constituents of C^A are contained, at a time instant t, which we also call region of space occupied by C^A at time t. Let us denote by Σ^A the union of the regions of space Σ_t^A, *i.e.*, the union of all the regions of space occupied by C^A during its time evolution.
We call *external force field for* C^A *at time* t, denoted by $\mathbf{F}_{e,t}^A$, the spatial distribution of $\mathbf{F}$ which is measured at time t in Σ_t^A if all the constituents and the walls of C^A are removed and placed far away from Σ_t^A. We call *external force field for* C^A, denoted by $\mathbf{F}_e^A$, the spatial and time distribution of $\mathbf{F}$ which is measured in Σ^A if all the constituents and the walls of C^A are removed and placed far away from Σ^A.
System, properties of a system. We will call *system* A a collection of matter C^A defined by the initial composition $\boldsymbol{n}^{0A}$, the stoichiometric coefficients $\boldsymbol{\nu}^A$ of the allowed reaction mechanisms, and the possibly time-dependent specification, *over the entire time interval of interest*, of:

- the geometrical variables and the nature of the boundary surfaces that define the regions of space $\boldsymbol{R}_t^A$,
- the rates $\dot{\boldsymbol{n}}_t^{A\leftarrow}$ at which particles are transferred in or out of the regions of space, and
- the external force field distribution $\mathbf{F}_{e,t}^A$ for C^A,

provided that the following conditions apply:

1. an ensemble of identically prepared replicas of C^A can be obtained at any instant of time t, according to a specified set of instructions or preparation scheme;
2. a set of measurement procedures, $P_1^A, \ldots, P_n^A$, exists, such that when each P_i^A is applied on replicas of C^A at any given instant of time t: each replica responds with a numerical outcome which may vary from replica to replica; but either the time interval Δt employed to perform the measurement can be made arbitrarily short so that the measurement outcomes considered for P_i^A are those which correspond to the limit as $\Delta t \to 0$, or the measurement outcomes are independent of the time interval Δt employed to perform the measurement;
3. the arithmetic mean $\langle P_i^A \rangle_t$ of the numerical outcomes of repeated applications of any of these procedures, P_i^A, at an instant t, on an ensemble of identically prepared replicas, is a value which is the same for every subensemble of replicas of C^A (the latter condition guarantees the so-called statistical *homogeneity* of the ensemble); $\langle P_i^A \rangle_t$ is called the *value of* P_i^A for C^A at time t;
4. the set of measurement procedures, $P_1^A, \ldots, P_n^A$, is *complete* in the sense that the set of values $\{\langle P_1^A \rangle_t, \ldots, \langle P_n^A \rangle_t\}$ allows to predict the value of any other measurement procedure satisfying conditions 2 and 3.

Then, each measurement procedure satisfying conditions 2 and 3 is called a *property* of system A, and the set $P_1^A, \ldots, P_n^A$ a *complete set of properties* of system A.
Comment. Although in general the amounts of constituents, $\boldsymbol{n}_t^A$, and the reaction rates, $\dot{\boldsymbol{\varepsilon}}_t$, are properties according to the above definition, we will list them separately and explicitly whenever it is convenient for clarity. In particular, in typical chemical kinetic models, $\dot{\boldsymbol{\varepsilon}}_t$ is assumed to be a function of $\boldsymbol{n}_t^A$ and other properties.
State of a system. Given a system A as just defined, we call *state of system* A *at time* t, denoted by A_t, the set of the values *at time* t of

- all the properties of the system or, equivalently, of a complete set of properties, $\{\langle P_1\rangle_t,\ldots,\langle P_n\rangle_t\}$,
- the amounts of constituents, $\boldsymbol{n}_t^A$,
- the geometrical variables and the nature of the boundary surfaces of the regions of space $\boldsymbol{R}_t^A$,
- the rates $\dot{\boldsymbol{n}}_t^{A\leftarrow}$ of particle transfer in or out of the regions of space, and
- the external force field distribution in the region of space Σ_t^A occupied by A at time t, $\mathbf{F}_{e,t}^A$.

With respect to the chosen complete set of properties, we can write

$$A_t \equiv \left\{\langle P_1\rangle_t,\ldots,\langle P_n\rangle_t;\boldsymbol{n}_t^A;\boldsymbol{R}_t^A;\dot{\boldsymbol{n}}_t^{A\leftarrow};\mathbf{F}_{e,t}^A\right\} \ . \tag{3}$$

For shorthand, states A_{t_1}, A_{t_2},..., are denoted by A_1, A_2,.... Also, when the context allows it, the value $\langle P^A\rangle_{t_1}$ of property P^A of system A at time t_1 is denoted depending on convenience by the symbol P_1^A, or simply P_1.

Closed system, open system. A system A is called a *closed system* if, at every time instant t, the boundary surface of every region of space R_{it}^A is a wall. Otherwise, A is called an *open system*.

Comment. For a closed system, in each region of space R_i^A, the number of particles of the i-th constituent can change only as a consequence of allowed reaction mechanisms.

Composite system, subsystems. Given a system C in the external force field $\mathbf{F}_e^C$, we will say that C is the *composite* of systems A and B, denoted AB, if: (a) there exists a pair of systems A and B such that the external force field which obtains when both A and B are removed and placed far away coincides with $\mathbf{F}_e^C$; (b) no region of space R_i^A overlaps with any region of space R_j^B; and (c) the $r_C = r_A + r_B$ regions of space of C are $\boldsymbol{R}^C = R_1^A,\ldots,R_i^A,\ldots,R_{r_A}^A,R_1^B,\ldots,R_j^B,\ldots,R_{r_B}^B$. Then we say that A and B are *subsystems* of the *composite system* C, and we write $C = AB$ and denote its state at time t by $C_t = (AB)_t$.

Isolated system. We say that a closed system I is an isolated system in the stationary external force field $\mathbf{F}_e^I$, or simply an *isolated system*, if, during the whole time evolution of I: (a) only the particles of I are present in Σ^I; (b) the external force field for I, $\mathbf{F}_e^I$, is stationary, *i.e.*, time independent, and conservative.

Comment. In simpler words, a system I is isolated if, at every time instant: no other material particle is present in the whole region of space Σ^I which will be crossed by system I during its time evolution; if system I is removed, only a stationary (vanishing or non-vanishing) conservative force field is present in Σ^I.

Separable closed systems. Consider a composite system AB, with A and B closed subsystems. We say that systems A and B are *separable* at time t if:

- the force field external to A coincides (where defined) with the force field external to AB, i.e., $\mathbf{F}_{e,t}^A = \mathbf{F}_{e,t}^{AB}$;
- the force field external to B coincides (where defined) with the force field external to AB, i.e., $\mathbf{F}_{e,t}^B = \mathbf{F}_{e,t}^{AB}$.

Comment. In simpler words, system A is separable from B at time t, if at that instant the force field produced by B is vanishing in the region of space occupied by A and viceversa. During the subsequent time evolution of AB, A and B need not remain separable at all times.

Subsystems in uncorrelated states. Consider a composite system AB such that at time t the states A_t and B_t of the two subsystems fully determine the state $(AB)_t$, *i.e.*, the values of all the properties of AB can be determined by *local* measurements of properties of systems A and B. Then, at time t, we say that the states of subsystems A and B are *uncorrelated from each other*, and we write the state of AB as $(AB)_t = A_t B_t$. We also say, for brevity, that A and B are *systems uncorrelated from each other* at time t.

Correlated states, correlation. If at time t the states A_t and B_t do not fully determine the state $(AB)_t$ of the composite system AB, we say that A_t and B_t are *states correlated with each other*. We also say, for brevity, that A and B are *systems correlated with each other* at time t.

Comment. Two systems A and B which are uncorrelated from each other at time t_1 can undergo an interaction such that they are correlated with each other at time $t_2 > t_1$.

Comment. Correlations between isolated systems. Let us consider an isolated system $I = AB$ such that, at time t, system A is separable and uncorrelated from B. This circumstance does not exclude that, at time t, A and/or B (or both) may be correlated with a system C, even if the latter is isolated, *e.g.* it is far away from the region of space occupied by AB. Indeed our definitions of separability and correlation are general enough to be fully compatible with the notion of quantum correlations, *i.e.*, *entanglement*, which plays an important role in modern physics. In other words, assume that an isolated system U is made of three subsystems A, B, and C, *i.e.*, $U = ABC$, with C isolated and AB isolated. The fact that A is uncorrelated from B, so that according to our notation we may write $(AB)_t = A_t B_t$, does not exclude that A and C may be entangled, in such a way that the states A_t and C_t do not determine the state of AC, *i.e.*, $(AC)_t \neq A_t C_t$, *nor* we can write $U_t = (A)_t (BC)_t$.

Environment of a system, scenario. If for the time span of interest a system A is a subsystem of an isolated system $I = AB$, we can choose AB as the isolated system to be studied. Then, we will call B the *environment* of A, and we call AB the *scenario* under which A is studied.

Comment. The chosen scenario AB contains as subsystems all and only the systems that are allowed to interact with A; thus all the remaining systems in the universe, even if correlated with AB, are considered as not available for interaction.

Comment. A system uncorrelated from its environment in one scenario, may be correlated with its environment in a broader scenario. Consider a system A which, in the scenario AB, is uncorrelated from its environment B at time t. If at time t system A is entangled with an isolated system C, in the scenario ABC, A is correlated with its environment BC.

Process, cycle. We call *process* for a system A from state A_1 to state A_2 in the scenario AB, denoted by $(AB)_1 \to (AB)_2$, the change of state from $(AB)_1$ to $(AB)_2$ of the isolated system AB which defines the scenario. We call *cycle* for a system A a process whereby the final state A_2 coincides with the initial state A_1.

Comment. In every process of any system A, the force field $\mathbf{F}_e^{AB}$ external to AB, where B is the environment of A, cannot change. In fact, AB is an isolated system and, as a consequence, the force field external to AB is stationary. Thus, in particular, for all the states in which a system A is separable:

- the force field $\mathbf{F}_e^{AB}$ external to AB, where B is the environment of A, is the same;
- the force field $\mathbf{F}_e^{A}$ external to A coincides, where defined, with the force field $\mathbf{F}_e^{AB}$ external to AB, *i.e.*, the force field produced by B (if any) has no effect on A.

Process between uncorrelated states, external effects. A process in the scenario AB in which the end states of system A are both uncorrelated from its environment B is called *process between uncorrelated states* and denoted by $\Pi_{12}^{A,B} \equiv (A_1 \to A_2)_{B_1 \to B_2}$. In such a process, the change of state of the environment B from B_1 to B_2 is called *effect external to A*. Traditional expositions of thermodynamics consider only this kind of process.

Composite process. A time-ordered sequence of processes between uncorrelated states of a system A with environment B, $\Pi_{1k}^{A,B} = (\Pi_{12}^{A,B}, \Pi_{23}^{A,B},\ldots, \Pi_{(i-1)i}^{A,B},\ldots, \Pi_{(k-1)k}^{A,B})$ is called a *composite process* if the final state of AB for process $\Pi_{(i-1)i}^{A,B}$ is the initial state of AB for process $\Pi_{i(i+1)}^{A,B}$, for $i = 1,2,\ldots,k-1$. When the context allows the simplified notation Π_i for $i = 1,2,\ldots,k-1$ for the processes in the sequence, the *composite process* may also be denoted by $(\Pi_1, \Pi_2,\ldots, \Pi_i,\ldots, \Pi_{k-1})$.

Reversible process, reverse of a reversible process. A process for A in the scenario AB, $(AB)_1 \to (AB)_2$, is called a *reversible process* if there exists a process $(AB)_2 \to (AB)_1$ which restores the initial state of the isolated system AB. The process $(AB)_2 \to (AB)_1$ is called *reverse* of process $(AB)_1 \to (AB)_2$. With different words, a process of an isolated system $I = AB$ is reversible if it can be reproduced as a part of a cycle of the isolated system I. For a reversible process between uncorrelated states, $\Pi_{12}^{A,B} \equiv (A_1 \to A_2)_{B_1 \to B_2}$, the *reverse* will be denoted by $-\Pi_{12}^{A,B} \equiv (A_2 \to A_1)_{B_2 \to B_1}$.

Comment. The reverse process may be achieved in more than one way (in particular, not necessarily by retracing the sequence of states $(AB)_t$, with $t_1 \le t \le t_2$, followed by the isolated system AB during the forward process).

Comment. The reversibility in one scenario does not grant the reversibility in another. If the smallest isolated system which contains A is AB and another isolated system C exists in a different region of space, one can choose as environment of A either B or BC. Thus, the time evolution of A can be described by the process $(AB)_1 \to (AB)_2$ in the scenario AB or by the process $(ABC)_1 \to (ABC)_2$ in the scenario ABC. For instance, the process $(AB)_1 \to (AB)_2$ could be irreversible, however by broadening the scenario so that interactions between AB and C become available, a reverse process $(ABC)_2 \to (ABC)_1$ may be possible. On the other hand, a process $(ABC)_1 \to (ABC)_2$ could be irreversible on account of an irreversible evolution $C_1 \to C_2$ of C, even if the process $(AB)_1 \to (AB)_2$ is reversible.

Comment. A reversible process need not be slow. In the general framework we are setting up, it is noteworthy that nowhere we state nor we need the concept that a process to be reversible needs to be *slow* in some sense. Actually, as well represented in (Gyftopoulos & Beretta, 2005) and clearly understood within dynamical systems models based on linear or nonlinear master equations, the time evolution of the state of a system is the result of a competition between (hamiltonian) mechanisms which are reversible and (dissipative) mechanisms which are not. So, to design a reversible process in the nonequilibrium domain, we most likely need a *fast* process, whereby the state is changed quickly by a fast hamiltonian dynamics, leaving negligible time for the dissipative mechanisms to produce irreversible effects.

Weight. We call *weight* a system M always separable and uncorrelated from its environment, such that:

- M is closed, it has a single constituent contained in a single region of space whose shape and volume are fixed,
- it has a constant mass m;
- in any process, the difference between the initial and the final state of M is determined uniquely by the change in the position z of the center of mass of M, which can move only along a straight line whose direction is identified by the unit vector $\mathbf{k} = \nabla z$;
- along the straight line there is a uniform stationary external gravitational force field $\mathbf{G}_e = -g\mathbf{k}$, where g is a constant gravitational acceleration.

As a consequence, the difference in potential energy between any initial and final states of M is given by $mg(z_2 - z_1)$.
Weight process, work in a weight process. A process between states of a closed system A in which A is separable and uncorrelated from its environment is called a *weight process*, denoted by $(A_1 \to A_2)_W$, if the only effect external to A is the displacement of the center of mass of a weight M between two positions z_1 and z_2. We call *work performed by A (or, done by A) in the weight process*, denoted by the symbol $W_{12}^{A\rightarrow}$, the quantity

$$W_{12}^{A\rightarrow} = mg(z_2 - z_1) \ . \tag{4}$$

Clearly, the *work done by A* is positive if $z_2 > z_1$ and negative if $z_2 < z_1$. Two equivalent symbols for the opposite of this work, called *work received by A*, are $-W_{12}^{A\rightarrow} = W_{12}^{A\leftarrow}$.
Equilibrium state of a closed system. A state A_t of a closed system A, with environment B, is called an *equilibrium state* if:

- A is a separable system at time t;
- state A_t does not change with time;
- state A_t can be reproduced while A is an isolated system in the external force field $\mathbf{F}_e^A$, which coincides, where defined, with $\mathbf{F}_e^{AB}$.

Stable equilibrium state of a closed system. An equilibrium state of a closed system A in which A is uncorrelated from its environment B, is called a *stable equilibrium state* if it cannot be modified by any process between states in which A is separable and uncorrelated from its environment such that neither the geometrical configuration of the walls which bound the regions of space $\boldsymbol{R}^A$ where the constituents of A are contained, nor the state of the environment B of A have net changes.
Comment. The stability of equilibrium in one scenario does not grant the stability of equilibrium in another. Consider a system A which, in the scenario AB, is uncorrelated from its environment B at time t and is in a stable equilibrium state. If at time t system A is entangled with an isolated system C, then in the scenario ABC, A is correlated with its environment BC, therefore, our definition of stable equilibrium state is not satisfied.

4. Definition of energy for a closed system

First Law. Every pair of states (A_1, A_2) of a closed system A in which A is separable and uncorrelated from its environment can be interconnected by means of a weight process for A. The works performed by the system in any two weight processes between the same initial and final states are identical.
Definition of energy for a closed system. Proof that it is a property. Let (A_1, A_2) be any pair of states of a closed system A in which A is separable and uncorrelated from its environment. We call *energy difference* between states A_2 and A_1 either the work $W_{12}^{A\leftarrow}$ received by A in any weight process from A_1 to A_2 or the work $W_{21}^{A\rightarrow}$ done by A in any weight process from A_2 to A_1; in symbols:

$$E_2^A - E_1^A = W_{12}^{A\leftarrow} \quad \text{or} \quad E_2^A - E_1^A = W_{21}^{A\rightarrow} . \tag{5}$$

The first law guarantees that at least one of the weight processes considered in Eq. (5) exists. Moreover, it yields the following consequences:
(*a*) if both weight processes $(A_1 \to A_2)_W$ and $(A_2 \to A_1)_W$ exist, the two forms of Eq. (5) yield the same result ($W_{12}^{A\leftarrow} = W_{21}^{A\rightarrow}$);
(*b*) the energy difference between two states A_2 and A_1 in which A is separable and

uncorrelated from its environment depends only on the states A_1 and A_2;
(c) (*additivity of energy differences for separable systems uncorrelated from each other*) consider a pair of closed systems A and B; if A_1B_1 and A_2B_2 are states of the composite system AB such that AB is separable and uncorrelated from its environment and, in addition, A and B are separable and uncorrelated from each other, then

$$E_2^{AB} - E_1^{AB} = E_2^A - E_1^A + E_2^B - E_1^B \;; \tag{6}$$

(d) (*energy is a property for every separable system uncorrelated from its environment*) let A_0 be a reference state of a closed system A in which A is separable and uncorrelated from its environment, to which we assign an arbitrarily chosen value of energy E_0^A; the value of the energy of A in any other state A_1 in which A is separable and uncorrelated from its environment is determined uniquely by the equation

$$E_1^A = E_0^A + W_{01}^{A\leftarrow} \quad \text{or} \quad E_1^A = E_0^A + W_{10}^{A\rightarrow} \tag{7}$$

where $W_{01}^{A\leftarrow}$ or $W_{10}^{A\rightarrow}$ is the work in any weight process for A either from A_0 to A_1 or from A_1 to A_0; therefore, energy is a property of A.
Rigorous proofs of these consequences can be found in (Gyftopoulos & Beretta, 2005; Zanchini, 1986), and will not be repeated here. In the proof of Eq. (6), the restrictive condition of the absence of correlations between AB and its environment as well as between A and B, implicit in (Gyftopoulos & Beretta, 2005) and (Zanchini, 1986), can be released by means of an assumption (Assumption 3) which is presented and discussed in the next section. As a result, Eq. (6) holds also if $(AB)_1$ e $(AB)_2$ are arbitrarily chosen states of the composite system AB, provided that AB, A and B are separable systems.

5. Definition of thermodynamic entropy for a closed system

Assumption 1: restriction to normal system. We will call *normal system* any system A that, starting from every state in which it is separable and uncorrelated from its environment, can be changed to a non-equilibrium state with higher energy by means of a weight process for A in which the regions of space $\boldsymbol{R}^A$ occupied by the constituents of A have no net change (and A is again separable and uncorrelated from its environment).
From here on, we consider only normal systems; even when we say only *system* we mean a *normal system*.
Comment. For a normal system, the energy is unbounded from above; the system can accommodate an indefinite amount of energy, such as when its constituents have translational, rotational or vibrational degrees of freedom. In traditional treatments of thermodynamics, Assumption 1 is *not stated explicitly, but it is used*, for example when one states that any amount of work can be transferred to a thermal reservoir by a stirrer. Notable exceptions to this assumption are important quantum theoretical model systems, such as spins, qubits, qudits, etc. whose energy is bounded from above. The extension of our treatment to such so-called *special systems* is straightforward, but we omit it here for simplicity.
Theorem 1. Impossibility of a PMM2. If a normal system A is in a stable equilibrium state, it is impossible to lower its energy by means of a weight process for A in which the regions of space $\boldsymbol{R}^A$ occupied by the constituents of A have no net change.
Proof. Suppose that, starting from a stable equilibrium state A_{se} of A, by means of a weight process Π_1 with positive work $W^{A\rightarrow} = W > 0$, the energy of A is lowered and the regions of space $\boldsymbol{R}^A$ occupied by the constituents of A have no net change. On account of Assumption 1,

it would be possible to perform a weight process Π_2 for A in which the regions of space $\boldsymbol{R}^A$ occupied by the constituents of A have no net change, the weight M is restored to its initial state so that the positive amount of energy $W^{A\leftarrow} = W > 0$ is supplied back to A, and the final state of A is a nonequilibrium state, namely, a state clearly different from A_{se}. Thus, the zero-work composite process (Π_1, Π_2) would violate the definition of stable equilibrium state.

Comment. Kelvin-Planck statement of the Second Law. As noted in (Hatsopoulos & Keenan, 1965) and (Gyftopoulos & Beretta, 2005, p.64), the impossibility of a perpetual motion machine of the second kind (PMM2), which is also known as the *Kelvin-Planck statement of the Second Law*, is a corollary of the definition of stable equilibrium state, provided that we adopt the (usually implicitly) restriction to normal systems (Assumption 1).

Second Law. Among all the states in which a closed system A is separable and uncorrelated from its environment and the constituents of A are contained in a given set of regions of space $\boldsymbol{R}^A$, there is a stable equilibrium state for every value of the energy E^A.

Lemma 1. Uniqueness of the stable equilibrium state. There can be no pair of different stable equilibrium states of a closed system A with identical regions of space $\boldsymbol{R}^A$ and the same value of the energy E^A.

Proof. Since A is closed and in any stable equilibrium state it is separable and uncorrelated from its environment, if two such states existed, by the first law and the definition of energy they could be interconnected by means of a zero-work weight process. So, at least one of them could be changed to a different state with no external effect, and hence would not satisfy the definition of stable equilibrium state.

Comment. Recall that for a closed system, the composition $\boldsymbol{n}^A$ belongs to the set of compatible compositions $(\boldsymbol{n}^{0A}, \boldsymbol{\nu}^A)$ fixed once and for all by the definition of the system.

Comment. Statements of the Second Law. The combination of our statement of the Second Law and Lemma 1 establishes, for a closed system whose matter is constrained into given regions of space, the existence and uniqueness of a stable equilibrium state for every value of the energy; this proposition is known as the *Hatsopoulos-Keenan statement of the Second Law* (Hatsopoulos & Keenan, 1965). Well-known historical statements of the Second Law, in addition to the Kelvin-Planck statement discussed above, are due to Clausius and to Carathéodory. In (Gyftopoulos & Beretta, 2005, p.64, p.121, p.133) it is shown that each of these historical statements is a logical consequence of the Hatsopoulos-Keenan statement combined with a further assumption, essentially equivalent to our Assumption 2 below.

Lemma 2. Any stable equilibrium state A_s of a closed system A is accessible via an irreversible zero-work weight process from any other state A_1 in which A is separable and uncorrelated with its environment and has the same regions of space $\boldsymbol{R}^A$ and the same value of the energy E^A.

Proof. By the first law and the definition of energy, A_s and A_1 can be interconnected by a zero-work weight process for A. However, a zero-work weight process from A_s to A_1 would violate the definition of stable equilibrium state. Therefore, the process must be in the direction from A_1 to A_s. The absence of a zero-work weight process in the opposite direction, implies that any zero-work weight process from A_1 to A_s is irreversible.

Corollary 1. Any state in which a closed system A is separable and uncorrelated from its environment can be changed to a unique stable equilibrium state by means of a zero-work weight process for A in which the regions of space $\boldsymbol{R}^A$ have no net change.

Proof. The thesis follows immediately from the Second Law, Lemma 1 and Lemma 2.

Mutual stable equilibrium states. We say that two stable equilibrium states A_{se} and B_{se} are *mutual stable equilibrium states* if, when A is in state A_{se} and B in state B_{se}, the composite system

AB is in a stable equilibrium state. The definition holds also for a pair of states of the same system: in this case, system AB is composed of A and of a duplicate of A.

Identical copy of a system. We say that a system A^d, always separable from A and uncorrelated with A, is an *identical copy* of system A (or, a *duplicate* of A) if, at every time instant:

- the difference between the set of regions of space $\boldsymbol{R}^{A^d}$ occupied by the matter of A^d and that $\boldsymbol{R}^A$ occupied by the matter of A is only a rigid translation $\Delta\mathbf{r}$ with respect to the reference frame considered, and the composition of A^d is compatible with that of A;
- the external force field for A^d at any position $\mathbf{r} + \Delta\mathbf{r}$ coincides with the external force field for A at the position $\mathbf{r}$.

Thermal reservoir. We call *thermal reservoir* a system R with a single constituent, contained in a fixed region of space, with a vanishing external force field, with energy values restricted to a finite range such that in any of its stable equilibrium states, R is in mutual stable equilibrium with an identical copy of R, R^d, in any of its stable equilibrium states.

Comment. Every single-constituent system without internal boundaries and applied external fields, and with a number of particles of the order of one mole (so that the *simple system* approximation as defined in (Gyftopoulos & Beretta, 2005, p.263) applies), when restricted to a fixed region of space of appropriate volume and to the range of energy values corresponding to the so-called *triple-point* stable equilibrium states, is an excellent approximation of a thermal reservoir.

Reference thermal reservoir. A thermal reservoir chosen once and for all, will be called a *reference thermal reservoir*. To fix ideas, we will choose as our reference thermal reservoir one having water as constituent, with a volume, an amount, and a range of energy values which correspond to the so-called *solid-liquid-vapor triple-point* stable equilibrium states.

Standard weight process. Given a pair of states (A_1, A_2) of a closed system A, in which A is separable and uncorrelated from its environment, and a thermal reservoir R, we call *standard weight process* for AR from A_1 to A_2 a weight process for the composite system AR in which the end states of R are stable equilibrium states. We denote by $(A_1R_1 \rightarrow A_2R_2)^{\text{sw}}$ a standard weight process for AR from A_1 to A_2 and by $(\Delta E^R)^{\text{sw}}_{A_1A_2}$ the corresponding energy change of the thermal reservoir R.

Assumption 2. Every pair of states (A_1, A_2) in which a closed system A is separable and uncorrelated from its environment can be interconnected by a reversible standard weight process for AR, where R is an arbitrarily chosen thermal reservoir.

Theorem 2. For a given closed system A and a given reservoir R, among all the standard weight processes for AR between a given pair of states (A_1, A_2) in which system A is separable and uncorrelated from its environment, the energy change $(\Delta E^R)^{\text{sw}}_{A_1A_2}$ of the thermal reservoir R has a lower bound which is reached if and only if the process is reversible.

Proof. Let Π_{AR} denote a standard weight process for AR from A_1 to A_2, and $\Pi_{AR\text{rev}}$ a reversible one; the energy changes of R in processes Π_{AR} and $\Pi_{AR\text{rev}}$ are, respectively, $(\Delta E^R)^{\text{sw}}_{A_1A_2}$ and $(\Delta E^R)^{\text{swrev}}_{A_1A_2}$. With the help of Figure 1, we will prove that, regardless of the initial state of R:

a) $(\Delta E^R)^{\text{swrev}}_{A_1A_2} \leq (\Delta E^R)^{\text{sw}}_{A_1A_2}$;

b) if also Π_{AR} is reversible, then $(\Delta E^R)^{\text{swrev}}_{A_1A_2} = (\Delta E^R)^{\text{sw}}_{A_1A_2}$;

c) if $(\Delta E^R)^{\text{swrev}}_{A_1A_2} = (\Delta E^R)^{\text{sw}}_{A_1A_2}$, then also Π_{AR} is reversible.

Proof of a). Let us denote by R_1 and by R_2 the initial and the final states of R in process $\Pi_{AR\text{rev}}$. Let us denote by R^d the duplicate of R which is employed in process Π_{AR}, by R^d_3

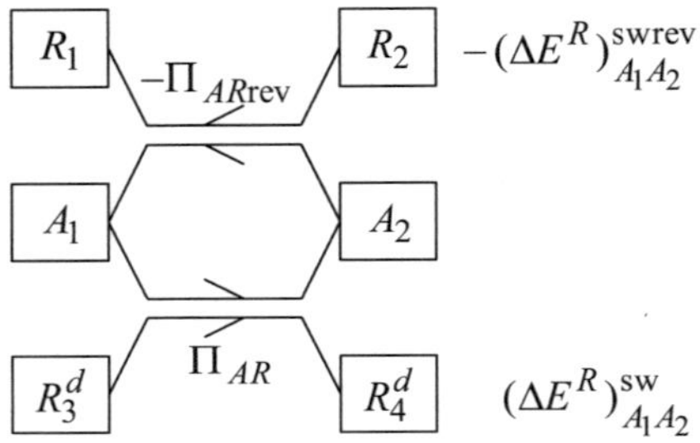

Fig. 1. Illustration of the proof of Theorem 2: standard weight processes $\Pi_{AR\text{rev}}$ (reversible) and Π_{AR}; R^d is a duplicate of R; see text.

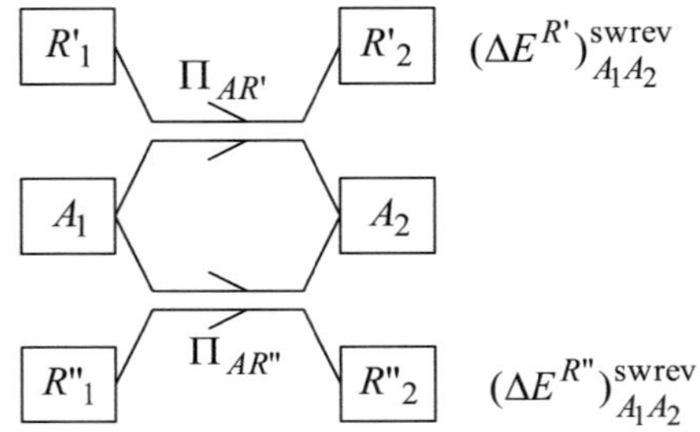

Fig. 2. Illustration of the proof of Theorem 3, part a): reversible standard weight processes $\Pi_{AR'}$ and $\Pi_{AR''}$, see text.

and by R_4^d the initial and the final states of R^d in this process. Let us suppose, *ab absurdo*, that $(\Delta E^R)^{\text{swrev}}_{A_1A_2} > (\Delta E^R)^{\text{sw}}_{A_1A_2}$. Then, the composite process $(-\Pi_{AR\text{rev}}, \Pi_{AR})$ would be a weight process for RR^d in which, starting from the stable equilibrium state $R_2R_3^d$, the energy of RR^d is lowered and the regions of space occupied by the constituents of RR^d have no net change, in contrast with Theorem 1. Therefore, $(\Delta E^R)^{\text{swrev}}_{A_1A_2} \le (\Delta E^R)^{\text{sw}}_{A_1A_2}$.

Proof of b). If Π_{AR} is reversible too, then, in addition to $(\Delta E^R)^{\text{swrev}}_{A_1A_2} \le (\Delta E^R)^{\text{sw}}_{A_1A_2}$, the relation $(\Delta E^R)^{\text{sw}}_{A_1A_2} \le (\Delta E^R)^{\text{swrev}}_{A_1A_2}$ must hold too. Otherwise, the composite process $(\Pi_{AR\text{rev}}, -\Pi_{AR})$ would be a weight process for RR^d in which, starting from the stable equilibrium state $R_1R_4^d$, the energy of RR^d is lowered and the regions of space occupied by the constituents of RR^d have no net change, in contrast with Theorem 1. Therefore, $(\Delta E^R)^{\text{swrev}}_{A_1A_2} = (\Delta E^R)^{\text{sw}}_{A_1A_2}$.

Proof of c). Let Π_{AR} be a standard weight process for AR, from A_1 to A_2, such that $(\Delta E^R)^{\text{sw}}_{A_1A_2} = (\Delta E^R)^{\text{swrev}}_{A_1A_2}$, and let R_1 be the initial state of R in this process. Let $\Pi_{AR\text{rev}}$ be a reversible standard weight process for AR, from A_1 to A_2, with the same initial state R_1 of R. Thus, R_3^d coincides with R_1 and R_4^d coincides with R_2. The composite process $(\Pi_{AR}, -\Pi_{AR\text{rev}})$ is a cycle for the isolated system ARB, where B is the environment of AR. As a consequence, Π_{AR} is reversible, because it is a part of a cycle of the isolated system ARB.

Theorem 3. Let R' and R'' be any two thermal reservoirs and consider the energy changes, $(\Delta E^{R'})^{\text{swrev}}_{A_1A_2}$ and $(\Delta E^{R''})^{\text{swrev}}_{A_1A_2}$ respectively, in the reversible standard weight processes $\Pi_{AR'} = (A_1R'_1 \to A_2R'_2)^{\text{swrev}}$ and $\Pi_{AR''} = (A_1R''_1 \to A_2R''_2)^{\text{swrev}}$, where (A_1, A_2) is an arbitrarily chosen pair of states of any closed system A in which A is separable and uncorrelated from its environment. Then the ratio $(\Delta E^{R'})^{\text{swrev}}_{A_1A_2} / (\Delta E^{R''})^{\text{swrev}}_{A_1A_2}$:

a) is positive;

b) depends only on R' and R'', *i.e.*, it is independent of *(i)* the initial stable equilibrium states of R' and R'', *(ii)* the choice of system A, and *(iii)* the choice of states A_1 and A_2.

Proof of a). With the help of Figure 2, let us suppose that $(\Delta E^{R'})^{\text{swrev}}_{A_1A_2} < 0$. Then, $(\Delta E^{R''})^{\text{swrev}}_{A_1A_2}$ cannot be zero. In fact, in that case the composite process $(\Pi_{AR'}, -\Pi_{AR''})$, which is a cycle for A, would be a weight process for R' in which, starting from the stable equilibrium state R'_1, the energy of R' is lowered and the regions of space occupied by the constituents of R' have no net change, in contrast with Theorem 1. Moreover, $(\Delta E^{R''})^{\text{swrev}}_{A_1A_2}$ cannot be positive. In fact, if it were positive, the work performed by $R'R''$ as a result of the overall weight process

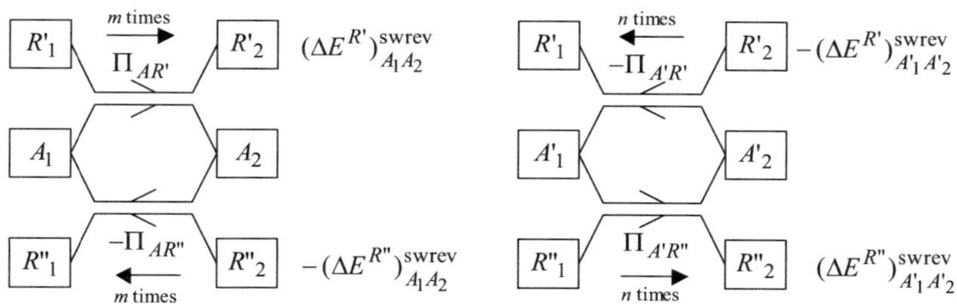

Fig. 3. Illustration of the proof of Theorem 3, part b): composite processes Π_A and $\Pi_{A'}$), see text.

$(\Pi_{AR'}, -\Pi_{AR''})$ for $R'R''$ would be

$$W^{R'R''\rightarrow} = -(\Delta E^{R'})^{\text{swrev}}_{A_1A_2} + (\Delta E^{R''})^{\text{swrev}}_{A_1A_2} \; , \tag{8}$$

where both terms are positive. On account of Assumption 1 and Corollary 1, after the process $(\Pi_{AR'}, -\Pi_{AR''})$, one could perform a weight process $\Pi_{R''}$ for R'' in which a positive amount of energy equal to $(\Delta E^{R''})^{\text{swrev}}_{A_1A_2}$ is given back to R'' and the latter is restored to its initial stable equilibrium state. As a result, the composite process $(\Pi_{AR'}, -\Pi_{AR''}, \Pi_{R''})$ would be a weight process for R' in which, starting from the stable equilibrium state R'_1, the energy of R' is lowered and the region of space occupied by occupied by R' has no net change, in contrast with Theorem 1. Therefore, the assumption $(\Delta E^{R'})^{\text{swrev}}_{A_1A_2} < 0$ implies $(\Delta E^{R''})^{\text{swrev}}_{A_1A_2} < 0$.
Let us suppose that $(\Delta E^{R'})^{\text{swrev}}_{A_1A_2} > 0$. Then, for process $-\Pi_{AR'}$ one has $(\Delta E^{R'})^{\text{swrev}}_{A_2A_1} < 0$. By repeating the previous argument, one proves that for process $-\Pi_{AR''}$ one has $(\Delta E^{R''})^{\text{swrev}}_{A_2A_1} < 0$. Therefore, for process $\Pi_{AR''}$ one has $(\Delta E^{R''})^{\text{swrev}}_{A_1A_2} > 0$.
Proof of b). Given a pair of states (A_1, A_2) of a closed system A, consider the reversible standard weight process $\Pi_{AR'} = (A_1R'_1 \rightarrow A_2R'_2)^{\text{swrev}}$ for AR', with R' initially in state R'_1, and the reversible standard weight process $\Pi_{AR''} = (A_1R''_1 \rightarrow A_2R''_2)^{\text{swrev}}$ for AR'', with R'' initially in state R''_1. Moreover, given a pair of states (A'_1, A'_2) of another closed system A', consider the reversible standard weight process $\Pi_{A'R'} = (A'_1R'_1 \rightarrow A'_2R'_2)^{\text{swrev}}$ for $A'R'$, with R' initially in state R'_1, and the reversible standard weight process $\Pi_{A'R''} = (A'_1R''_1 \rightarrow A'_2R''_2)^{\text{swrev}}$ for $A'R''$, with R'' initially in state R''_1.
With the help of Figure 3, we will prove that the changes in energy of the reservoirs in these processes obey the relation

$$\frac{(\Delta E^{R'})^{\text{swrev}}_{A_1A_2}}{(\Delta E^{R''})^{\text{swrev}}_{A_1A_2}} = \frac{(\Delta E^{R'})^{\text{swrev}}_{A'_1A'_2}}{(\Delta E^{R''})^{\text{swrev}}_{A'_1A'_2}} \; . \tag{9}$$

Let us assume: $(\Delta E^{R'})^{\text{swrev}}_{A_1A_2} > 0$ and $(\Delta E^{R'})^{\text{swrev}}_{A'_1A'_2} > 0$, which implies, $(\Delta E^{R''})^{\text{swrev}}_{A_1A_2} > 0$ and $(\Delta E^{R''})^{\text{swrev}}_{A'_1A'_2} > 0$ on account of part a) of the proof. This is not a restriction, because it is possible to reverse the processes under exam. Now, as is well known, any real number can be approximated with an arbitrarily high accuracy by a rational number. Therefore, we will assume that the energy changes $(\Delta E^{R'})^{\text{swrev}}_{A_1A_2}$ and $(\Delta E^{R'})^{\text{swrev}}_{A'_1A'_2}$ are rational numbers, so that whatever is the value of their ratio, there exist two positive integers m and n such that $(\Delta E^{R'})^{\text{swrev}}_{A_1A_2} / (\Delta E^{R'})^{\text{swrev}}_{A'_1A'_2} = n/m$, *i.e.*,

$$m \, (\Delta E^{R'})^{\text{swrev}}_{A_1A_2} = n \, (\Delta E^{R'})^{\text{swrev}}_{A'_1A'_2} \; . \tag{10}$$

Therefore, as sketched in Figure 3, let us consider the composite processes Π_A and Π'_A defined as follows. Π_A is the following composite weight process for system $AR'R''$: starting from the initial state R'_1 of R' and R''_2 of R'', system A is brought from A_1 to A_2 by a reversible standard weight process for AR', then from A_2 to A_1 by a reversible standard weight process for AR''; whatever the new states of R' and R'' are, again system A is brought from A_1 to A_2 by a reversible standard weight process for AR' and back to A_1 by a reversible standard weight process for AR'', until the cycle for A is repeated m times. Similarly, $\Pi_{A'}$ is a composite weight processes for system $A'R'R''$ whereby starting from the end states of R' and R'' reached by Π_A, system A' is brought from A'_1 to A'_2 by a reversible standard weight process for $A'R''$, then from A'_2 to A'_1 by a reversible standard weight process for $A'R'$; and so on until the cycle for A' is repeated n times.

Clearly, the whole composite process $(\Pi_A, \Pi_A{}')$ is a cycle for AA'. Moreover, it is a cycle also for R'. In fact, on account of Theorem 2, the energy change of R' in each process $\Pi_{AR'}$ is equal to $(\Delta E^{R'})^{\text{swrev}}_{A_1A_2}$ regardless of its initial state, and in each process $-\Pi_{A'R'}$ the energy change of R' is equal to $-(\Delta E^{R'})^{\text{swrev}}_{A'_1A'_2}$. Therefore, the energy change of R' in the composite process (Π_A, Π'_A) is $m\,(\Delta E^{R'})^{\text{swrev}}_{A_1A_2} - n\,(\Delta E^{R'})^{\text{swrev}}_{A'_1A'_2}$ and equals zero on account of Eq. (10). As a result, after (Π_A, Π'_A), reservoir R' has been restored to its initial state, so that (Π_A, Π'_A) is a reversible weight process for R''.

Again on account of Theorem 2, the overall energy change of R'' in (Π_A, Π'_A) is $-m\,(\Delta E^{R''})^{\text{swrev}}_{A_1A_2} + n\,(\Delta E^{R''})^{\text{swrev}}_{A'_1A'_2}$. If this quantity were negative, Theorem 1 would be violated. If this quantity were positive, Theorem 1 would also be violated by the reverse of the process, $(-\Pi'_A, -\Pi_A)$. Therefore, the only possibility is that $-m\,(\Delta E^{R''})^{\text{swrev}}_{A_1A_2} + n\,(\Delta E^{R''})^{\text{swrev}}_{A'_1A'_2} = 0$, *i.e.*,

$$m\,(\Delta E^{R''})^{\text{swrev}}_{A_1A_2} = n\,(\Delta E^{R''})^{\text{swrev}}_{A'_1A'_2}\ . \tag{11}$$

Finally, taking the ratio of Eqs. (10) and (11), we obtain Eq. (9) which is our conclusion.

Temperature of a thermal reservoir. Let R be a given thermal reservoir and R^o a reference thermal reservoir. Select an arbitrary pair of states (A_1, A_2) in which an arbitrary closed system A is separable and uncorrelated from its environment, and consider the energy changes $(\Delta E^R)^{\text{swrev}}_{A_1A_2}$ and $(\Delta E^{R^o})^{\text{swrev}}_{A_1A_2}$ in two reversible standard weight processes from A_1 to A_2, one for AR and the other for AR^o, respectively. We call *temperature* of R the positive quantity

$$T_R = T_{R^o}\,\frac{(\Delta E^R)^{\text{swrev}}_{A_1A_2}}{(\Delta E^{R^o})^{\text{swrev}}_{A_1A_2}}\ , \tag{12}$$

where T_{R^o} is a positive constant associated arbitrarily with the reference thermal reservoir R^o. If for R^o we select a thermal reservoir having water as constituent, with energy restricted to the solid-liquid-vapor triple-point range, and we set $T_{R^o} = 273.16$ K, we obtain the unit kelvin (K) for the thermodynamic temperature, which is adopted in the International System of Units (SI). Clearly, the temperature T_R of R is defined only up to an arbitrary multiplicative constant.

Corollary 2. The ratio of the temperatures of two thermal reservoirs, R' and R'', is independent of the choice of the reference thermal reservoir and can be measured directly as

$$\frac{T_{R'}}{T_{R''}} = \frac{(\Delta E^{R'})^{\text{swrev}}_{A_1A_2}}{(\Delta E^{R''})^{\text{swrev}}_{A_1A_2}}\ , \tag{13}$$

where $(\Delta E^{R'})^{\rm swrev}_{A_1A_2}$ and $(\Delta E^{R''})^{\rm swrev}_{A_1A_2}$ are the energy changes of R' and R'' in two reversible standard weight processes, one for AR' and the other for AR'', which interconnect the same but otherwise arbitrary pair of states (A_1, A_2) in which a closed system A is separable and uncorrelated from its environment.

Proof. Let $(\Delta E^{R^o})^{\rm swrev}_{A_1A_2}$ be the energy change of the reference thermal reservoir R^o in any reversible standard weight process for AR^o which interconnects the same states (A_1, A_2) of A. From Eq. (12) we have

$$T_{R'} = T_{R^o} \frac{(\Delta E^{R'})^{\rm swrev}_{A_1A_2}}{(\Delta E^{R^o})^{\rm swrev}_{A_1A_2}} \,, \tag{14}$$

$$T_{R''} = T_{R^o} \frac{(\Delta E^{R''})^{\rm swrev}_{A_1A_2}}{(\Delta E^{R^o})^{\rm swrev}_{A_1A_2}} \,, \tag{15}$$

therefore the ratio of Eqs. (14) and (15) yields Eq. (13).

Corollary 3. Let (A_1, A_2) be any pair of states in which a closed system A is separable and uncorrelated from its environment, and let $(\Delta E^{R})^{\rm swrev}_{A_1A_2}$ be the energy change of a thermal reservoir R with temperature T_R, in any reversible standard weight process for AR from A_1 to A_2. Then, for the given system A, the ratio $(\Delta E^{R})^{\rm swrev}_{A_1A_2}/T_R$ depends only on the pair of states (A_1, A_2), *i.e.*, it is independent of the choice of reservoir R and of its initial stable equilibrium state R_1.

Proof. Let us consider two reversible standard weight processes from A_1 to A_2, one for AR' and the other for AR'', where R' is a thermal reservoir with temperature $T_{R'}$ and R'' is a thermal reservoir with temperature $T_{R''}$. Then, equation (13) yields

$$\frac{(\Delta E^{R'})^{\rm swrev}_{A_1A_2}}{T_{R'}} = \frac{(\Delta E^{R''})^{\rm swrev}_{A_1A_2}}{T_{R''}} \,. \tag{16}$$

Definition of (thermodynamic) entropy for a closed system. Proof that it is a property. Let $(A_1\,, A_2)$ be any pair of states in which a closed system A is separable and uncorrelated from its environment B, and let R be an arbitrarily chosen thermal reservoir placed in B. We call *entropy difference* between A_2 and A_1 the quantity

$$S^A_2 - S^A_1 = -\frac{(\Delta E^{R})^{\rm swrev}_{A_1A_2}}{T_R} \tag{17}$$

where $(\Delta E^{R})^{\rm swrev}_{A_1A_2}$ is the energy change of R in any reversible standard weight process for AR from A_1 to A_2, and T_R is the temperature of R. On account of Corollary 3, the right hand side of Eq. (17) is determined uniquely by states A_1 and A_2.

Let A_0 be a reference state in which A is separable and uncorrelated from its environment, to which we assign an arbitrarily chosen value of entropy S^A_0. Then, the value of the entropy of A in any other state A_1 in which A is separable and uncorrelated from its environment, is determined uniquely by the equation

$$S^A_1 = S^A_0 - \frac{(\Delta E^{R})^{\rm swrev}_{A_1A_0}}{T_R} \,, \tag{18}$$

where $(\Delta E^{R})^{\rm swrev}_{A_1A_0}$ is the energy change of R in any reversible standard weight process for AR from A_0 to A_1, and T_R is the temperature of R. Such a process exists for every state A_1, on account of Assumption 2. Therefore, entropy is a property of A and is defined for every state of A in which A is separable and uncorrelated from its environment.

Theorem 4. Additivity of entropy differences for uncorrelated states. Consider the pairs of states $(C_1 = A_1B_1, C_2 = A_2B_2)$ in which the composite system $C = AB$ is separable and uncorrelated from its environment, and systems A and B are separable and uncorrelated from each other. Then,

$$S^{AB}_{A_2B_2} - S^{AB}_{A_1B_1} = S^A_2 - S^A_1 + S^B_2 - S^B_1 \ . \tag{19}$$

Proof. Let us choose a thermal reservoir R, with temperature T_R, and consider the composite process (Π_{AR}, Π_{BR}) where Π_{AR} is a reversible standard weight process for AR from A_1 to A_2, while Π_{BR} is a reversible standard weight process for BR from B_1 to B_2. The composite process (Π_{AR}, Π_{BR}) is a reversible standard weight process for CR from C_1 to C_2, in which the energy change of R is the sum of the energy changes in the constituent processes Π_{AR} and Π_{BR}, *i.e.*, $(\Delta E^R)^{\text{swrev}}_{C_1C_2} = (\Delta E^R)^{\text{swrev}}_{A_1A_2} + (\Delta E^R)^{\text{swrev}}_{B_1B_2}$. Therefore:

$$\frac{(\Delta E^R)^{\text{swrev}}_{C_1C_2}}{T_R} = \frac{(\Delta E^R)^{\text{swrev}}_{A_1A_2}}{T_R} + \frac{(\Delta E^R)^{\text{swrev}}_{B_1B_2}}{T_R} \ . \tag{20}$$

Equation (20) and the definition of entropy (17) yield Eq. (19).

Comment. As a consequence of Theorem 4, if the values of entropy are chosen so that they are additive in the reference states, entropy results as an additive property. Note, however, that the proof of additivity requires that (A_1, B_1) and (A_2, B_2) are pairs of states such that the subsystems A and B are uncorrelated from each other.

Theorem 5. Let (A_1, A_2) be any pair of states in which a closed system A is separable and uncorrelated from its environment and let R be a thermal reservoir with temperature T_R. Let $\Pi_{AR\text{irr}}$ be any irreversible standard weight process for AR from A_1 to A_2 and let $(\Delta E^R)^{\text{swirr}}_{A_1A_2}$ be the energy change of R in this process. Then

$$-\frac{(\Delta E^R)^{\text{swirr}}_{A_1A_2}}{T_R} < S^A_2 - S^A_1 \ . \tag{21}$$

Proof. Let $\Pi_{AR\text{rev}}$ be any reversible standard weight process for AR from A_1 to A_2 and let $(\Delta E^R)^{\text{swrev}}_{A_1A_2}$ be the energy change of R in this process. On account of Theorem 2,

$$(\Delta E^R)^{\text{swrev}}_{A_1A_2} < (\Delta E^R)^{\text{swirr}}_{A_1A_2} \ . \tag{22}$$

Since T_R is positive, from Eqs. (22) and (17) one obtains

$$-\frac{(\Delta E^R)^{\text{swirr}}_{A_1A_2}}{T_R} < -\frac{(\Delta E^R)^{\text{swrev}}_{A_1A_2}}{T_R} = S^A_2 - S^A_1 \ . \tag{23}$$

Theorem 6. Principle of entropy nondecrease. Let (A_1, A_2) be a pair of states in which a closed system A is separable and uncorrelated from its environment and let $(A_1 \to A_2)_W$ be any weight process for A from A_1 to A_2. Then, the entropy difference $S^A_2 - S^A_1$ is equal to zero if and only if the weight process is reversible; it is strictly positive if and only if the weight process is irreversible.

Proof. If $(A_1 \to A_2)_W$ is reversible, then it is a special case of a reversible standard weight process for AR in which the initial stable equilibrium state of R does not change. Therefore, $(\Delta E^R)^{\text{swrev}}_{A_1A_2} = 0$ and by applying the definition of entropy, Eq. (17), one obtains

$$S^A_2 - S^A_1 = -\frac{(\Delta E^R)^{\text{swrev}}_{A_1A_2}}{T_R} = 0 \ . \tag{24}$$

If $(A_1 \to A_2)_W$ is irreversible, then it is a special case of an irreversible standard weight process for AR in which the initial stable equilibrium state of R does not change. Therefore, $(\Delta E^R)^{\text{swirr}}_{A_1A_2} = 0$ and Equation (21) yields

$$S_2^A - S_1^A > -\frac{(\Delta E^R)^{\text{swirr}}_{A_1A_2}}{T_R} = 0 \ . \tag{25}$$

Moreover: if a weight process $(A_1 \to A_2)_W$ for A is such that $S_2^A - S_1^A = 0$, then the process must be reversible, because we just proved that for any irreversible weight process $S_2^A - S_1^A > 0$; if a weight process $(A_1 \to A_2)_W$ for A is such that $S_2^A - S_1^A > 0$, then the process must be irreversible, because we just proved that for any reversible weight process $S_2^A - S_1^A = 0$.

Corollary 4. If states A_1 and A_2 can be interconnected by means of a reversible weight process for A, they have the same entropy. If states A_1 and A_2 can be interconnected by means of a zero-work reversible weight process for A, they have the same energy and the same entropy.

Proof. These are straightforward consequences of Theorem 6 together with the definition of energy.

Theorem 7. Highest-entropy principle. Among all the states of a closed system A such that A is separable and uncorrelated from its environment, the constituents of A are contained in a given set of regions of space $\boldsymbol{R}^A$ and the value of the energy E^A of A is fixed, the entropy of A has the highest value only in the unique stable equilibrium state A_{se} determined by $\boldsymbol{R}^A$ and E^A.

Proof. Let A_g be any other state of A in the set of states considered here. On account of the first law and of the definition of energy, A_g and A_{se} can be interconnected by a zero work weight process for A, either $(A_g \to A_{se})_W$ or $(A_{se} \to A_g)_W$. However, the existence of a zero work weight process $(A_{se} \to A_g)_W$ would violate the definition of stable equilibrium state. Therefore, a zero work weight process $(A_g \to A_{se})_W$ exists and is irreversible, so that Theorem 6 implies $S_{se}^A > S_g^A$.

Assumption 3. Existence of spontaneous decorrelations and impossibility of spontaneous creation of correlations. Consider a system AB composed of two closed subsystems A and B. Let $(AB)_1$ be a state in which AB is separable and uncorrelated from its environment and such that in the corresponding states A_1 and B_1, systems A and B are separable but correlated; let A_1B_1 be the state of AB such that the corresponding states A_1 and B_1 of A and B are the same as for state $(AB)_1$, but A and B are uncorrelated. Then, a zero work weight process $((AB)_1 \to A_1B_1)_W$ for AB is possible, while a weight process $(A_1B_1 \to (AB)_1)_W$ for AB is impossible.

Corollary 5. Energy difference between states of a composite system in which subsystems are correlated with each other. Let $(AB)_1$ and $(AB)_2$ be states of a composite system AB in which AB is separable and uncorrelated from its environment, while systems A and B are separable but correlated with each other. We have

$$E^{AB}_{(AB)_2} - E^{AB}_{(AB)_1} = E^{AB}_{A_2B_2} - E^{AB}_{A_1B_1} = E_2^A - E_1^A + E_2^B - E_1^B \ . \tag{26}$$

Proof. Since a zero work weight process $((AB)_1 \to A_1B_1)_W$ for AB exists on account of Assumption 3, states $(AB)_1$ and A_1B_1 have the same energy. In other words, the energy of a composite system in state $(AB)_1$ with separable but correlated subsystems coincides with the energy of the composite system in state A_1B_1 where its separable subsystems are uncorrelated in the corresponding states A_1 and A_2.

Definition of energy for a state in which a system is correlated with its environment. On account of Eq. (26), we will say that the energy of a system A in a state A_1 in which A is correlated with its environment is equal to the energy of system A in the corresponding state A_1 in which A is uncorrelated from its environment.
Comment. Equation (26) and the definition of energy for a state in which a system is correlated with its environment extend the definition of energy and the proof of the additivity of energy differences presented in (Gyftopoulos & Beretta, 2005; Zanchini, 1986) to the case in which systems A and B are separable but correlated with each other.
To our knowledge, Assumption 3 (never made explicit) underlies all reasonable models of relaxation and decoherence.
Corollary 6. De-correlation entropy. Given a pair of (different) states $(AB)_1$ and A_1B_1 as defined in Assumption 3, then we have

$$\sigma^{AB}_{(AB)_1} = S^{AB}_{A_1B_1} - S^{AB}_{(AB)_1} > 0 \ , \tag{27}$$

where the positive quantity σ_1^{AB} is called the *de-correlation entropy*[1] of state $(AB)_1$. Clearly, if the subsystems are uncorrelated, *i.e.*, if $(AB)_1 = A_1B_1$, then $\sigma^{AB}_{(AB)_1} = \sigma^{AB}_{A_1B_1} = 0$.
Proof. On account of Assumption 3, a zero work weight process $\Pi_{AB} = ((AB)_1 \to A_1B_1)_W$ for AB exists. Process Π_{AB} is irreversible, because the reversibility of Π_{AB} would require the existence of a zero work weight process for AB from A_1B_1 to $(AB)_1$, which is excluded by Assumption 3. Since Π_{AB} is irreversible, Theorem 6 yields the conclusion.
Comment. Let $(AB)_1$ and $(AB)_2$ be a pair of states of a composite system AB such that AB is separable and uncorrelated from its environment, while subsystems A and B are separable but correlated with each other. Let A_1B_1 and A_2B_2 be the corresponding pairs of states of AB, in which the subsystems A and B are in the same states as before, but are uncorrelated from each other. Then, the entropy difference between $(AB)_2$ and $(AB)_1$ is not equal to the entropy difference between A_2B_2 and A_1B_1 and therefore, on account of Eq. (19), it is not equal to the sum of the entropy difference between A_2 and A_1 and the entropy difference between B_2 and B_1, evaluated in the corresponding states in which subsystems A and B are uncorrelated from each other. In fact, combining Eq. (19) with Eq. (27), we have

$$S^{AB}_{(AB)_2} - S^{AB}_{(AB)_1} = (S^A_2 - S^A_1) + (S^B_2 - S^B_1) - (\sigma^{AB}_{(AB)_2} - \sigma^{AB}_{(AB)_1}) \ . \tag{28}$$

6. Fundamental relation, temperature, and Gibbs relation for closed systems

Set of equivalent stable equilibrium states. We will call *set of equivalent stable equilibrium states* of a closed system A, denoted ESE^A, a subset of its stable equilibrium states such that any pair of states in the set:

- differ from one another by some geometrical features of the regions of space $\boldsymbol{R}^A$;
- have the same composition;
- can be interconnected by a zero-work reversible weight process for A and, hence, by Corollary 4, have the same energy and the same entropy.

Comment. Let us recall that, for all the stable equilibrium states of a closed system A in a scenario AB, system A is separable and the external force field $\mathbf{F}_e^A = \mathbf{F}_e^{AB}$ is the same; moreover, all the compositions of A belong to the same set of compatible compositions $(\boldsymbol{n}^{0A}, \nu^A)$.

[1]Explicit expressions of this property in the quantum formalism are given, *e.g.*, in Wehrl (1978); Beretta et al. (1985); Lloyd (1989).

Parameters of a closed system. We will call *parameters* of a closed system A, denoted by $\boldsymbol{\beta}^A = \beta_1^A, \ldots, \beta_s^A$, a minimal set of real variables sufficient to fully and uniquely parametrize all the different sets of equivalent stable equilibrium states ESE^A of A. In the following, we will consider systems with a finite number s of parameters.
Examples. Consider a system A consisting of a single particle confined in spherical region of space of volume V; the box is centered at position $\mathbf{r}$ which can move in a larger region where there are no external fields. Then, it is clear that any rotation or translation of the spherical box within the larger region can be effected in a zero-work weight process that does not alter the rest of the state. Therefore, the position of the center of the box is *not* a parameter of the system. The volume instead is a parameter. The same holds if the box is cubic. If it is a parallelepiped, instead, the parameters are the sides ℓ_1, ℓ_2, ℓ_3 but not its position and orientation. For a more complex geometry of the box, the parameters are any minimal set of geometrical features sufficient to fully describe its shape, regardless of its position and orientation. The same if instead of one, the box contains many particles.
Suppose now we have a spherical box, with one or many particles, that can be moved in a larger region where there are k subregions, each much larger than the box and each with an external electric field everywhere parallel to the x axis and with uniform magnitude E_{ek}. As part of the definition of the system, let us restrict it only to the states such that the box is fully contained in one of these regions. For this system, the magnitude of E_e can be changed in a weight process by moving A from one uniform field subregion to another, but this in general will vary the energy. Therefore, in addition to the volume of the sphere, this system will have k as a parameter identifying the subregion where the box is located. Equivalently, the subregion can be identified by the parameter E_e taking values in the set $\{E_{ek}\}$. For each value of the energy E, system A has a set ESE^A for every pair of values of the parameters (V, E_e) with E_e in $\{E_{ek}\}$.
Corollary 7. Fundamental relation for the stable equilibrium states of a closed system. On the set of all the stable equilibrium states of a closed system A (in scenario AB, for given initial composition $\boldsymbol{n}^{0A}$, stoichiometric coefficients $\boldsymbol{\nu}^{\mathbf{A}}$ and external force field $\mathbf{F}_e^A$), the entropy is given by a single valued function

$$S_{\rm se}^A = S_{\rm se}^A(E^A, \boldsymbol{\beta}^A) \;, \tag{29}$$

which is called *fundamental relation* for the stable equilibrium states of A. Moreover, also the reaction coordinates are given by a single valued function

$$\boldsymbol{\varepsilon}_{\rm se}^A = \boldsymbol{\varepsilon}_{\rm se}^A(E^A, \boldsymbol{\beta}^A) \;, \tag{30}$$

which specifies the unique composition compatible with the initial composition $\boldsymbol{n}^{0A}$, called the *chemical equilibrium composition*.
Proof. On account of the Second Law and Lemma 1, among all the states of a closed system A with energy E^A, the regions of space $\boldsymbol{R}^A$ identify a unique stable equilibrium state. This implies the existence of a single valued function $A_{\rm se} = A_{\rm se}(E^A, \boldsymbol{R}^A)$, where $A_{\rm se}$ denotes the state, in the sense of Eq. (3). By definition, for each value of the energy E^A, the values of the parameters $\boldsymbol{\beta}^A$ fully identify all the regions of space $\boldsymbol{R}^A$ that correspond to a set of equivalent stable equilibrium states ESE^A, which have the same value of the entropy and the same composition. Therefore, the values of E^A and $\boldsymbol{\beta}^A$ fix uniquely the values of $S_{\rm se}^A$ and of $\boldsymbol{\varepsilon}_{\rm se}^A$. This implies the existence of the single valued functions written in Eqs. (29) and (30).
Comment. Clearly, for a non-reactive closed system, the composition is fixed and equal to the initial, *i.e.*, $\boldsymbol{\varepsilon}_{\rm se}^A(E^A, \boldsymbol{\beta}^A) = 0$.

Usually (Hatsopoulos & Keenan, 1965; Gyftopoulos & Beretta, 2005), in view of the equivalence that defines them, each set ESE^A is thought of as a single state called "a stable equilibrium state" of A. Thus, for a given closed system A (and, hence, given initial amounts of constituents), it is commonly stated that the energy and the parameters of A determine "a unique stable equilibrium state" of A, which is called "the chemical equilibrium state" of A if the system is reactive according to a given set of stoichiometric coefficients. For a discussion of the implications of Eq. (30) and its reduction to more familiar chemical equilibrium criteria in terms of chemical potentials see, *e.g.*, (Beretta & Gyftopoulos, 2004).

Assumption 4. The fundamental relation (29) is continuous and differentiable with respect to each of the variables E^A and $\boldsymbol{\beta}^A$.

Theorem 8. For any closed system, for fixed values of the parameters the fundamental relation (29) is a strictly increasing function of the energy.

Proof. Consider two stable equilibrium states $A_{\rm se1}$ and $A_{\rm se2}$ of a closed system A, with energies E_1^A and E_2^A, entropies $S_{\rm se1}^A$ and $S_{\rm se2}^A$, and with the same regions of space occupied by the constituents of A (and therefore the same values of the parameters). Assume $E_2^A > E_1^A$. By Assumption 1, we can start from state $A_{\rm se1}$ and, by a weight process for A in which the regions of space occupied by the constituents of A have no net changes, add work so that the system ends in a non-equilibrium state A_2 with energy E_2^A. By Theorem 6, we must have $S_2^A \geq S_{\rm se1}^A$. Now, on account of Lemma 2, we can go from state A_2 to $A_{\rm se2}$ with a zero-work irreversible weight process for A. By Theorem 6, we must have $S_{\rm se2}^A > S_2^A$. Combining the two inequalities, we find that $E_2^A > E_1^A$ implies $S_{\rm se2}^A > S_{\rm se1}^A$.

Corollary 8. The fundamental relation for any closed system A can be rewritten in the form

$$E_{\rm se}^A = E_{\rm se}^A(S^A, \boldsymbol{\beta}^A) \ . \tag{31}$$

Proof. By Theorem 8, for fixed $\boldsymbol{\beta}^A$, Eq. (29) is a strictly increasing function of E^A. Therefore, it is invertible with respect to E^A and, as a consequence, can be written in the form (31).

Temperature of a closed system in a stable equilibrium state. Consider a stable equilibrium state $A_{\rm se}$ of a closed system A identified by the values of E^A and $\boldsymbol{\beta}^A$. The partial derivative of the fundamental relation (31) with respect to S^A, is denoted by

$$T^A = \left(\frac{\partial E_{\rm se}^A}{\partial S^A}\right)_{\boldsymbol{\beta}^A} \ . \tag{32}$$

Such derivative is always defined on account of Assumption 4. When evaluated at the values of E^A and $\boldsymbol{\beta}^A$ that identify state $A_{\rm se}$, it yields a value that we call the *temperature* of state $A_{\rm se}$.

Comment. One can prove (Gyftopoulos & Beretta, 2005, p.127) that two stable equilibrium states A_1 and A_2 of a closed system A are mutual stable equilibrium states if and only if they have the same temperature, *i.e.*, if $T_1^A = T_2^A$. Moreover, it is easily proved (Gyftopoulos & Beretta, 2005, p.136) that, when applied to a thermal reservoir R, Eq. (32) yields that all the stable equilibrium states of a thermal reservoir have the same temperature which is equal to the temperature T_R of R defined by Eq. (12).

Corollary 9. For any stable equilibrium state of any (normal) closed system, the temperature is non-negative.

Proof. The thesis follows immediately from the definition of temperature, Eq. (32), and Theorem 8.

Gibbs equation for a non-reactive closed system. By differentiating Eq. (31), one obtains (omitting the superscript "A" and the subscript "se" for simplicity)

$$dE = T\,dS + \sum_{j=1}^{s} F_j d\beta_j \ , \tag{33}$$

where F_j is called generalized force conjugated to the j-th parameter of A, $F_j = \left(\partial E_{\mathrm{se}}/\partial\beta_j\right)_{S,\boldsymbol{\beta}'}$. If all the regions of space $\boldsymbol{R}^A$ coincide and the volume V of any of them is a parameter, the negative of the conjugated generalized force is called *pressure*, denoted by p, $p = -\left(\partial E_{\mathrm{se}}/\partial V\right)_{S,\boldsymbol{\beta}'}$.

Fundamental relation in the quantum formalism. Let us recall that the measurement procedures that define energy and entropy must be applied, in general, to a (homogeneous) ensemble of identically prepared replicas of the system of interest. Because the numerical outcomes may vary (fluctuate) from replica to replica, the values of the energy and the entropy defined by these procedures are arithmetic means. Therefore, what we have denoted so far, for simplicity, by the symbols E^A and S^A should be understood as $\langle E^A\rangle$ and $\langle S^A\rangle$. Where appropriate, like in the quantum formalism implementation, this more precise notation should be preferred. Then, written in full notation, the fundamental relation (29) for a closed system is

$$\langle S^A\rangle_{\mathrm{se}} = S^A_{\mathrm{se}}(\langle E^A\rangle, \boldsymbol{\beta}^A) \ , \tag{34}$$

and the corresponding Gibbs relation

$$d\langle E\rangle = T\,d\langle S\rangle + \sum_{j=1}^{s} F_j d\beta_j \ . \tag{35}$$

7. Definitions of energy and entropy for an open system

Our definition of energy is based on the First Law, by which a weight process is possible between any pair of states A_1 and A_2 in which a closed system A is separable and uncorrelated from its environment. Our definition of entropy is based on Assumption 2, by which a reversible standard weight process for AR is possible between any pair of states A_1 and A_2 in which a closed system A is separable and uncorrelated from its environment. In both cases, A_1 and A_2 have compatible compositions. In this section, we extend the definitions of energy and entropy to a set of states in which an open system O is separable and uncorrelated from its environment; two such states of O have, in general, non-compatible compositions.

Separable open system uncorrelated from its environment. Consider an open system O that has Q as its (open) environment, *i.e.*, the composite system OQ is isolated in $\mathbf{F}_e^{OQ}$. We say that system O is *separable* from Q at time t if the state $(OQ)_t$ of OQ can be reproduced as (*i.e.*, coincides with) a state $(AB)_t$ of an isolated system AB in $\mathbf{F}_e^{AB} = \mathbf{F}_e^{OQ}$ such that A and B are closed and separable at time t. If the state $(AB)_t = A_t B_t$, *i.e.*, is such that A and B are uncorrelated from each other, then we say that the open system O is *uncorrelated from its environment* at time t, and we have $O_t = A_t$, $Q_t = B_t$, and $(OQ)_t = O_t Q_t$.

Set of elemental species. Following (Gyftopoulos & Beretta, 2005, p.545), we will call *set of elemental species* a *complete* set of *independent* constituents with the following features: (1) (*completeness*) there exist reaction mechanisms by which all other constituents can be formed starting only from constituents in the set; and (2) (*independence*) there exist no reaction mechanisms that involve only constituents in the set.

For example, in chemical thermodynamics we form a set of elemental species by selecting among all the chemical species formed by atomic nuclei of a single kind those that have the most stable molecular structure and form of aggregation at standard temperature and pressure.

Energy and entropy of a separable open system uncorrelated from its environment. Let OQ be an isolated system in $\mathbf{F}_e^{OQ}$, with O and Q open systems, and let us choose scenario OQ, so that Q is the environment of O. Let us suppose that O has r single-constituent regions of space and a set of allowed reaction mechanisms with stoichiometric coefficients $\boldsymbol{\nu}^O$. Let us consider a state O_1 in which O is separable and uncorrelated from its environment and has composition $\boldsymbol{n}_1^O = (n_1^O, \ldots, n_i^O, \ldots, n_r^O)_1$. Let $A^{\boldsymbol{n}_1^O}B$ be an isolated system in $\mathbf{F}_e^{A^{\boldsymbol{n}_1^O}B} = \mathbf{F}_e^{OQ}$, such that $A^{\boldsymbol{n}_1^O}$ is closed, has the same allowed reaction mechanisms as O and compositions compatible with $\boldsymbol{n}_1^O$. Let $A_1^{\boldsymbol{n}_1^O}$ be a state of $A^{\boldsymbol{n}_1^O}$ such that, in that state, system $A^{\boldsymbol{n}_1^O}$ is a separable system in $\mathbf{F}_e^{A^{\boldsymbol{n}_1^O}} = \mathbf{F}_e^{A^{\boldsymbol{n}_1^O}B}$ and is uncorrelated from its environment; moreover, the state $A_1^{\boldsymbol{n}_1^O}$ coincides with O_1, *i.e.*, has the same values of all the properties. We will define as energy and entropy of O, in state O_1, the energy and the entropy of $A^{\boldsymbol{n}_1^O}$ in state $A_1^{\boldsymbol{n}_1^O}$, namely $E_1^O = E_1^{A^{\boldsymbol{n}_1^O}}$ and $S_1^O = S_1^{A^{\boldsymbol{n}_1^O}}$. The existence of system $A^{\boldsymbol{n}_1^O}$ and of state $A_1^{\boldsymbol{n}_1^O}$ is granted by the definition of separability for O in state O_1.

The values of the energy and of the entropy of $A^{\boldsymbol{n}_1^O}$, in state $A_1^{\boldsymbol{n}_1^O}$, are determined by choosing a reference state $A_0^{\boldsymbol{n}_1^O}$ of $A^{\boldsymbol{n}_1^O}$ and by applying Eqs. (7) and (18). The reference state $A_0^{\boldsymbol{n}_1^O}$ and the reference values $E_0^{A^{\boldsymbol{n}_1^O}}$ and $S_0^{A^{\boldsymbol{n}_1^O}}$ are selected as defined below.

We choose $A^{\boldsymbol{n}_1^O}$ as the composite of q closed subsystems, $A^{\boldsymbol{n}_1^O} = A^1 A^2 \cdots A^i \cdots A^q$, each one containing an elemental species, chosen so that the composition of $A^{\boldsymbol{n}_1^O}$ is compatible with that of O in state O_1. Each subsystem, A^i, contains n_i particles of the i-th elemental species and is constrained by a wall in a spherical box with a variable volume V^{A^i}; each box is very far from the others and is placed in a position where the external force field $\mathbf{F}_e^{A^{\boldsymbol{n}_1^O}}$ is vanishing. We choose the reference state $A_0^{\boldsymbol{n}_1^O}$ to be such that each subsystem A^i is in a stable equilibrium state A_0^i with a prescribed temperature, T_0, and a volume $V_0^{A^i}$ such that the pressure has a prescribed value p_0.

We fix the reference values of the energy and the entropy of the reference state $A_0^{\boldsymbol{n}_1^O}$ as follows:

$$E_0^{A^{\boldsymbol{n}_1^O}} = \sum_{i=1}^{q} E_0^{A^i} , \tag{36}$$

$$S_0^{A^{\boldsymbol{n}_1^O}} = \sum_{i=1}^{q} S_0^{A^i} , \tag{37}$$

with the values of $E_0^{A^i}$ and $S_0^{A^i}$ fixed arbitrarily. Notice that by construction $V_0^{A^{\boldsymbol{n}_1^O}} = \sum_{i=1}^q V_0^{A^i}$ and, therefore, we also have $E_0^{A^{\boldsymbol{n}_1^O}} + p_0 V_0^{A^{\boldsymbol{n}_1^O}} = \sum_{i=1}^q (E_0^{A^i} + p_0 V_0^{A^i})$. In chemical thermodynamics, it is customary to set $E_0^{A^i} + p_0 V_0^{A^i} = 0$ and $S_0^{A^i} = 0$ for each elemental species. Similarly to what seen for a closed system, the definition of energy for O can be extended to the states of O in which O is separable but correlated with its environment.

8. Fundamental relation for an open system

Stable equilibrium state of an open system. A state of an open system O in which O is a separable open system in $\mathbf{F}_e^O$ and is uncorrelated from its environment Q is called a stable equilibrium state if it can be reproduced as a stable equilibrium state of a closed system A in $\mathbf{F}_e^A = \mathbf{F}_e^O$.
We will consider separately the two different cases:
a) the constituents of O are non-reactive, *i.e.*, no reaction mechanism is allowed for O;
b) reactions with stoichiometric coefficients $\boldsymbol{\nu}^O$ are allowed for O.
Fundamental relation for the stable equilibrium states of an open system with non-reactive constituents. Let SE^O be the set of all the stable equilibrium states of an open system O with r non-reactive constituents and s parameters, $\boldsymbol{\beta}^O = \beta_1^O, \ldots, \beta_s^O$. Let us consider the subset $SE^O_{\boldsymbol{n}_1^O}$ of all the states of SE^O that have the composition $\boldsymbol{n}_1^O$, and let $A^{\boldsymbol{n}_1^O}$ be a closed system with composition $\boldsymbol{n}_1^O$, such that its stable equilibrium states coincide with those of the subset $SE^O_{\boldsymbol{n}_1^O}$ and therefore also the parameters coincide, *i.e.*, $\boldsymbol{\beta}^{A^{n_1^O}} = \boldsymbol{\beta}^O$. Then, every subset $ESE^{A^{n_1^O}}$ of equivalent stable equilibrium states of $A^{\boldsymbol{n}_1^O}$, which is determined by the energy $E^{A^{n_1^O}}$ and the parameters $\boldsymbol{\beta}^{A^{n_1^O}}$, coincides with a subset of equivalent stable equilibrium states of O with composition $\boldsymbol{n}_1^O$. The same argument can be repeated for every composition of O. Therefore, on the whole set SE^O, a relation with the form

$$S_{se}^O = S_{se}^O(E^O, \boldsymbol{n}^O, \boldsymbol{\beta}^O) \tag{38}$$

is defined and is called fundamental relation for O. Since the relation $S_{se}^O = S_{se}^O(E^O)$, for fixed values of $\boldsymbol{n}^O$ and $\boldsymbol{\beta}^O$, is strictly increasing, Eq. (38) can be rewritten as

$$E_{se}^O = E_{se}^O(S^O, \boldsymbol{n}^O, \boldsymbol{\beta}^O) \ . \tag{39}$$

Gibbs equation for a non-reactive open system. If the system has non-reactive constituents, the fundamental relation given by Eq. (39) applies. By differentiating Eq. (39), one obtains (omitting the superscript "O" and the subscript "se" for simplicity)

$$dE = TdS + \sum_{i=1}^{r} \mu_i \, \mathrm{d}n_i + \sum_{j=1}^{s} F_j \, \mathrm{d}\beta_j \ , \tag{40}$$

where μ_i is called the *total potential* of i-th constituent of O.
In Eq. (40), it is assumed that Eq. (39) is continuous and differentiable also with respect to $\boldsymbol{n}$. For systems with very large values of the amounts of constituents this condition is fulfilled. However, for very few particle closed systems, the variable $\boldsymbol{n}$ takes on only discrete values, and, according to our definition, a separable state of an open system must be reproduced as a separable state of a closed system. Thus, the extension of Eq. (40) to few particles open systems requires an *extended definition* of a separable state of an open system, which includes states with non integer numbers of particles. This extension will not be presented here.
Fundamental relation for the stable equilibrium states of an open system with reactive constituents. Let SE^O be the set of all the stable equilibrium states of an open system O with parameters $\boldsymbol{\beta}^O$ and constituents which can react according to a set of reaction mechanisms defined by the stoichiometric coefficients $\boldsymbol{\nu}^O$. Let $(\boldsymbol{n}_1^{0O}, \boldsymbol{\nu}^O)$ be the set of the compositions of O which are compatible with the initial composition $\boldsymbol{n}_1^{0O} = (n_1^{0O}, \ldots, n_r^{0O})_1$. Let $SE^{\boldsymbol{n}_1^{0O}}$ be the

subset of SE^O with compositions compatible with $(\boldsymbol{n}_1^{0O}, \boldsymbol{\nu}^O)$ and let $A^{\boldsymbol{n}_1^{0O}}$ be a closed system with compositions compatible with $(\boldsymbol{n}_1^{0O}, \boldsymbol{\nu}^O)$ and stable equilibrium states that coincide with those of the subset $SE^{\boldsymbol{n}_1^{0O}}$ so that also the parameters coincide, *i.e.*, $\boldsymbol{\beta}^{A^{\boldsymbol{n}_1^{0O}}} = \boldsymbol{\beta}^O$.
Then, every subset $ESE^{A^{\boldsymbol{n}_1^{0O}}}$ of equivalent stable equilibrium states of $A^{\boldsymbol{n}_1^{0O}}$, which is determined by the energy $E^{A^{\boldsymbol{n}_1^{0O}}}$ and the parameters $\boldsymbol{\beta}^{A^{\boldsymbol{n}_1^{0O}}}$, coincides with a subset of equivalent stable equilibrium states in the set $SE^{\boldsymbol{n}_1^{0O}}$. The same argument can be repeated for every set of compatible compositions of O, $(\boldsymbol{n}_2^{0O}, \boldsymbol{\nu}^O)$, $(\boldsymbol{n}_3^{0O}, \boldsymbol{\nu}^O)$, etc. Therefore, on the whole set SE^O, the following single-valued relation is defined

$$S_{se}^O = S_{se}^O(E^O, \boldsymbol{n}^{0O}, \boldsymbol{\beta}^O) \tag{41}$$

which is called fundamental relation for O. Since the relation $S_{se}^O = S_{se}^O(E^O)$, for fixed values of $\boldsymbol{n}^{0O}$ and $\boldsymbol{\beta}^O$, is strictly increasing, Eq. (41) can be rewritten as

$$E_{se}^O = E_{se}^O(S^O, \boldsymbol{n}^{0O}, \boldsymbol{\beta}^O) \ . \tag{42}$$

Comment. On the set SE^O of the stable equilibrium states of O, also the reaction coordinates are given by a single valued function

$$\boldsymbol{\varepsilon}_{se}^O = \boldsymbol{\varepsilon}_{se}^O(E^O, \boldsymbol{n}^{0O}, \boldsymbol{\beta}^O) \ , \tag{43}$$

which defines the chemical equilibrium composition. The existence of Eq. (43) is a consequence of the existence of a single valued function such as Eq. (30) for each of the closed systems $A^{\boldsymbol{n}_1^{0O}}$, $A^{\boldsymbol{n}_2^{0O}}$, ... used to reproduce the stable equilibrium states of O with sets of amounts of constituents compatible with the initial compositions, $\boldsymbol{n}_1^{0O}$, $\boldsymbol{n}_2^{0O}$, etc.

9. Conclusions

In this paper, a general definition of entropy is presented, based on operative definitions of all the concepts employed in the treatment, designed to provide a clarifying and useful, complete and coherent, minimal but general, rigorous logical framework suitable for unambiguous fundamental discussions on Second Law implications.
Operative definitions of system, state, isolated system, environment of a system, process, separable system, system uncorrelated from its environment and parameters of a system are stated, which are valid also in the presence of internal semipermeable walls and reaction mechanisms. The concepts of heat and of quasistatic process are never mentioned, so that the treatment holds also for nonequilibrium states, both for macroscopic and few particles systems.
The role of correlations on the domain of definition and on the additivity of energy and entropy is discussed: it is proved that energy is defined for any separable system, even if correlated with its environment, and is additive for separable subsystems even if correlated with each other; entropy is defined only for a separable system uncorrelated from its environment and is additive only for separable subsystems uncorrelated from each other; the concept of decorrelation entropy is defined.
A definition of thermal reservoir less restrictive than in previous treatments is adopted: it is fulfilled, with an excellent approximation, by any single-constituent simple system contained in a fixed region of space, provided that the energy values are restricted to a suitable finite range. The proof that entropy is a property of the system is completed by a new explicit proof

that the entropy difference between two states of a system is independent of the initial state of the auxiliary thermal reservoir chosen to measure it.
The definition of a reversible process is given with reference to a given *scenario*, *i.e.*, the largest isolated system whose subsystems are available for interaction; thus, the operativity of the definition is improved and the treatment becomes compatible also with recent interpretations of irreversibility in the quantum mechanical framework.
Rigorous extensions of the definitions of energy and entropy to open systems are stated. The existence of a fundamental relation for the stable equilibrium states of an open system with reactive constituents is proved rigorously; it is shown that the amounts of constituents which correspond to given fixed values of the reaction coordinates should appear in this equation.

10. Acknowledgments

G.P. Beretta gratefully acknowledges the Cariplo–UniBS–MIT-MechE faculty exchange program co-sponsored by UniBS and the CARIPLO Foundation, Italy under grant 2008-2290.

11. References

Bennett, C.H. (2008). The Second Law and Quantum Physics, in *Meeting the Entropy Challenge, AIP Conf. Proc. Series* 1033: 66-79.

Beretta, G.P. (1984). On the Relation Between Classical and Quantum Thermodynamic Entropy, *J. Math. Phys.* 25: 1507 (1984).

Beretta, G.P. (1987). Quantum Thermodynamics of Nonequilibrium. Onsager Reciprocity and Dispersion-Dissipation Relations, *Found. Phys.* 17: 365-381.

Beretta, G.P. (2006). Nonlinear Model Dynamics for Closed-System, Constrained, Maximal-Entropy-Generation Relaxation by Energy Redistribution, *Phys. Rev. E* 73: 026113.

Beretta, G.P. (2009). Nonlinear Quantum Evolution Equations to Model Irreversible Adiabatic Relaxation With Maximal Entropy Production and Other Nonunitary Processes, *Reports on Mathematical Physics* 64: 139-168.

Beretta, G.P.; Ghoniem, A.F. & Hatsopoulos, G.N., Editors (2008). *Meeting the Entropy Challenge, AIP Conf. Proc. Series* 1033.

Beretta, G.P. & Gyftopoulos, E.P. (2004). Thermodynamic Derivations of Conditions for Chemical Equilibrium and of Onsager Reciprocal Relations for Chemical Reactors, *J. Chem. Phys.* 121: 2718-2728.

Beretta, G.P.; Gyftopoulos, E.P.; Park, J.L. & Hatsopoulos, G.N. (1984). Quantum Thermodynamics: a New Equation of Motion for a Single Constituent of Matter, *Nuovo Cimento B* 82: 169-191.

Beretta, G.P.; Gyftopoulos, E.P. & Park, J.L. (1985). Quantum Thermodynamics: a New Equation of Motion for a General Constituent of Matter, *Nuovo Cimento B* 87: 77-97.

Callen, H.B. (1985). *Thermodynamics and Introduction to Thermostatics*, 2nd Ed., Wiley.

Clausius, R. (1865). Üer Verschiedene für die Anwendungen Bequeme Formen der Hauptgleichungen der Mechanischen Warmetheorie. Memoir read at the Philos. Soc. Zürich on April 24, *Pogg. Ann.* 125: 353-400. English translation in: Kestin, J. Ed. (1976). *The Second Law of Thermodynamics*, Dowden, Hutchinson and Ros: Stroudsburg: 162-193.

Fermi, E. (1937). *Thermodynamics*, Prentice-Hall.

Feynman, R.P. (1963). *Lectures on Physics*, Vol. 1, Addison-Welsey.

Giovannetti, V.; Lloyd, S. & Maccone, L. (2003). Quantum Limits to Dynamical Evolution, *Phys. Rev. A*, 67: 052109.

Goldstein, S.; Lebowitz, J.L.; Tumulka, R. & Zanghí, N. (2006). Canonical Typicality, *Phys. Rev. Lett.* 96: 050403.

Guggenheim, E.A. (1967). *Thermodynamics*, North-Holland, 7th Ed., p. 10.

Gyftopoulos, E.P. & Beretta, G.P. (2005). *Thermodynamics. Foundations and Applications*, Dover, Mineola (first edition, Macmillan, 1991).

Harremöes, P., Editor (2007). Facets of Entropy, in *Entropy*, Special Issue: Proceedings of the workshop in Copenhagen, 24-26 October 2007.

Hatsopoulos, G.N. & Beretta, G.P. (2008). Where is the entropy challenge?, in *Meeting the Entropy Challenge*, *AIP Conf. Proc. Series* 1033: 34-54.

Hatsopoulos, G.N. & Gyftopoulos, E.P. (1976). A Unified Quantum Theory of Mechanics and Thermodynamics, *Found. Phys.* 6: 15-31, 127-141, 439-455, 561-570.

Hatsopoulos, G.N. & Keenan, J.H. (1965). *Principles of General Thermodynamics*, Wiley, p. xxiii.

Horodecki, R.; Horodecki, M. & Horodecki, P. (2001). Balance of Information in Bipartite Quantum-Communication Systems: Entanglement-Energy Analogy, *Phys. Rev. A* 63: 022310.

Keenan, J.H. (1941). *Thermodynamics*, Wiley, p. 6.

Kim, Y.; Yu, R.; Kulik, S.P.; Shih, Y. & Scully, M.O. (2000). Delayed Choice Quantum Eraser, *Phys. Rev. Lett.* 84: 1-5.

Landau, L.D. & Lifshitz, E.M. (1980). *Statistical Physics*, Part I, 3rd Ed., Revised by Lifshitz, E.M. & Pitaevskii, L.P., Translated by Sykes J.B. & Kearsley, M.J., Pergamon Press, p. 45.

Lieb, E.H. & Yngvason, J. (1999). *Phys. Reps.* 310: 1-96.

Lloyd, S. (1989). Use of Mutual Information to Decrease Entropy: Implications for the Second Law of Thermodynamics, *Phys. Rev. A* 39: 5378-5386.

Lloyd, S. (1997). Quantum-Mechanical Maxwell's Demon, *Phys. Rev. A* 56: 3374-3382.

Lloyd, S. (2008). The Once and Future Second Law of Thermodynamics, in *Meeting the Entropy Challenge*, *AIP Conf. Proc. Series* 1033: 143-157.

Maccone, L. (2009). Quantum Solution to the Arrow-of-Time Dilemma, *Phys. Rev. Lett.* 103: 080401.

Scully, M.O. (2001). Extracting Work from a Single Thermal Bath via Quantum Negentropy, *Phys. Rev. Lett.* 87: 220601.

Scully, M.O. (2002). Quantum Afterburner: Improving the Efficiency of an Ideal Heat Engine, *Phys. Rev. Lett.* 88: 050602.

Scully, M.O. & Drühl, K. (1982). Quantum Eraser: a Proposed Photon Correlation Experiment Concerning Observation and Delayed Choice in Quantum Mechanics, *Phys. Rev. A* 25: 2208-2213.

Tisza, L. (1966). *Generalized Thermodynamics*, MIT Press, p. 16.

Wehrl, A. (1978). General Properties of Entropy, *Rev. Mod. Phys.* 50: 221-260.

Zanchini, E. (1986). On the Definition of Extensive Property Energy by the First Postulate of Thermodynamics, *Found. Phys.* 16: 923-935.

Zanchini, E. (1988). Thermodynamics: Energy of Closed and Open Systems, *Il Nuovo Cimento B* 101: 453-465.

Zanchini, E. (1992). Thermodynamics: Energy of Nonsimple Systems and Second Postulate, *Il Nuovo Cimento B* 107: 123-139.

3

Heat Flow, Work Energy, Chemical Reactions, and Thermodynamics: A Dynamical Systems Perspective

Wassim M. Haddad[1], Sergey G. Nersesov[2] and VijaySekhar Chellaboina[3]
[1]*Georgia Institute of Technology*
[2]*Villanova University*
[3]*Tata Consultancy Services*
[1,2]*USA*
[3]*INDIA*

1. Introduction

There is no doubt that thermodynamics is a theory of universal proportions whose laws reign supreme among the laws of nature and are capable of addressing some of science's most intriguing questions about the origins and fabric of our universe. The laws of thermodynamics are among the most firmly established laws of nature and play a critical role in the understanding of our expanding universe. In addition, thermodynamics forms the underpinning of several fundamental life science and engineering disciplines, including biological systems, physiological systems, chemical reaction systems, ecological systems, information systems, and network systems, to cite but a few examples. While from its inception its speculations about the universe have been grandiose, its mathematical foundation has been amazingly obscure and imprecise (Truesdell (1969; 1980); Arnold (1990); Haddad et al. (2005)). This is largely due to the fact that classical thermodynamics is a physical theory concerned mainly with equilibrium states and does not possess equations of motion. The absence of a state space formalism in classical thermodynamics, and physics in general, is quite disturbing and in our view largely responsible for the monomeric state of classical thermodynamics.

In recent research, Haddad et al. (2005; 2008) combined the two universalisms of thermodynamics and dynamical systems theory under a single umbrella to develop a dynamical system formalism for classical thermodynamics so as to harmonize it with classical mechanics. While it seems impossible to reduce thermodynamics to a mechanistic world picture due to microscopic reversibility and Poincaré recurrence, the system thermodynamic formulation of Haddad et al. (2005) provides a harmonization of classical thermodynamics with classical mechanics. In particular, our dynamical system formalism captures all of the key aspects of thermodynamics, including its fundamental laws, while providing a mathematically rigorous formulation for thermodynamical systems out of equilibrium by unifying the theory of heat transfer with that of classical thermodynamics. In addition, the concept of entropy for a nonequilibrium state of a dynamical process is defined, and its global existence and uniqueness is established. This state space formalism of thermodynamics shows that the behavior of heat, as described by the conservation equations of thermal transport and as described by classical thermodynamics, can be derived from the same basic principles and is part of the same scientific discipline. Connections between irreversibility, the second law of thermodynamics, and the entropic arrow of time are also established in Haddad et al. (2005). Specifically, we show a state irrecoverability and, hence, a state irreversibility nature of thermodynamics. State irreversibility reflects time-reversal non-invariance, wherein time-reversal is not meant literally; that is, we consider dynamical systems whose trajectory reversal is or is not allowed and not a reversal of time itself. In addition, we show that for every nonequilibrium system state and corresponding system trajectory of our thermodynamically consistent dynamical system, there does not exist a state such that the corresponding system trajectory completely recovers the initial system state of the dynamical system and at the same time restores the energy supplied by the environment back to its original condition. This, along with the existence of a global strictly increasing entropy function on every nontrivial system trajectory, establishes the existence of a completely ordered time set having a topological structure involving a closed set homeomorphic to the real line giving a clear time-reversal asymmetry characterization of thermodynamics and establishing an emergence of the direction of time flow.

In this paper, we reformulate and extend some of the results of Haddad et al. (2005). In particular, unlike the framework in Haddad et al. (2005) wherein we establish the existence and uniqueness of a global entropy function of a specific form for our thermodynamically consistent system model, in this paper we assume the existence of a continuously differentiable, strictly concave function that leads to an entropy inequality that can be identified with the second law of thermodynamics as a statement about entropy increase. We then turn our attention to stability and convergence. Specifically, using Lyapunov stability theory and the Krasovskii-LaSalle invariance principle, we show that for an adiabatically isolated system the proposed interconnected dynamical system model is Lyapunov stable with convergent trajectories to equilibrium states where the temperatures of all subsystems are equal. Finally, we present a state-space dynamical system model for chemical thermodynamics. In particular, we use the law of mass-action to obtain the dynamics of chemical reaction networks. Furthermore, using the notion of the chemical potential (Gibbs (1875; 1878)), we unify our state space mass-action kinetics model with our thermodynamic dynamical system model involving energy exchange. In addition, we show that entropy production during chemical reactions is nonnegative and the dynamical system states of our chemical thermodynamic state space model converge to a state of temperature equipartition and zero affinity (i.e., the difference between the chemical potential of the reactants and the chemical potential of the products in a chemical reaction).

2. Mathematical preliminaries

In this section, we establish notation, definitions, and provide some key results necessary for developing the main results of this paper. Specifically, $\mathbb{R}$ denotes the set of real numbers, $\overline{\mathbb{Z}}_+$ (respectively, $\mathbb{Z}_+$) denotes the set of nonnegative (respectively, positive) integers, $\mathbb{R}^q$ denotes the set of $q \times 1$ column vectors, $\mathbb{R}^{n\times m}$ denotes the set of $n \times m$ real matrices, $\mathbb{P}^n$ (respectively, $\mathbb{N}^n$) denotes the set of positive (respectively, nonnegative) definite matrices, $(\cdot)^{\mathrm{T}}$ denotes transpose, I_q or I denotes the $q \times q$ identity matrix, $\mathbf{e}$ denotes the ones vector of order q, that is, $\mathbf{e} \triangleq [1,\ldots,1]^{\mathrm{T}} \in \mathbb{R}^q$, and $\mathbf{e}_i \in \mathbb{R}^q$ denotes a vector with unity in the ith component and zeros elsewhere. For $x \in \mathbb{R}^q$ we write $x \geq\geq 0$ (respectively, $x >> 0$) to indicate that every component of x is nonnegative (respectively, positive). In this case, we say that x is *nonnegative* or *positive*, respectively. Furthermore, $\overline{\mathbb{R}}_+^q$ and $\mathbb{R}_+^q$ denote the nonnegative and positive orthants of $\mathbb{R}^q$, that is, if $x \in \mathbb{R}^q$, then $x \in \overline{\mathbb{R}}_+^q$ and $x \in \mathbb{R}_+^q$ are equivalent, respectively, to $x \geq\geq 0$ and $x >> 0$. Analogously, $\overline{\mathbb{R}}_+^{n\times m}$ (respectively, $\mathbb{R}_+^{n\times m}$) denotes the set of $n \times m$ real matrices whose entries are nonnegative (respectively, positive). For vectors $x,y \in \mathbb{R}^q$, with components x_i and y_i, $i = 1,\ldots,q$, we use $x \circ y$ to denote component-by-component multiplication, that is, $x \circ y \triangleq [x_1y_1,\ldots,x_qy_q]^{\mathrm{T}}$. Finally, we write $\partial\mathcal{S}$, $\overset{\circ}{\mathcal{S}}$, and $\overline{\mathcal{S}}$ to denote the boundary, the interior, and the closure of the set $\mathcal{S}$, respectively.

We write $\|\cdot\|$ for the Euclidean vector norm, $V'(x) \triangleq \frac{\partial V(x)}{\partial x}$ for the Fréchet derivative of V at x, $\mathcal{B}_\varepsilon(\alpha)$, $\alpha \in \mathbb{R}^q$, $\varepsilon > 0$, for the *open ball centered* at α with *radius* ε, and $x(t) \to \mathcal{M}$ as $t \to \infty$ to denote that $x(t)$ approaches the set $\mathcal{M}$ (that is, for every $\varepsilon > 0$ there exists $T > 0$ such that $\mathrm{dist}(x(t),\mathcal{M}) < \varepsilon$ for all $t > T$, where $\mathrm{dist}(p,\mathcal{M}) \triangleq \inf_{x\in\mathcal{M}} \|p - x\|$). The notions of openness, convergence, continuity, and compactness that we use throughout the paper refer to the topology generated on $\mathcal{D} \subseteq \mathbb{R}^q$ by the norm $\|\cdot\|$. A subset $\mathcal{N}$ of $\mathcal{D}$ is *relatively open* in $\mathcal{D}$ if $\mathcal{N}$ is open in the subspace topology induced on $\mathcal{D}$ by the norm $\|\cdot\|$. A point $x \in \mathbb{R}^q$ is a *subsequential limit* of the sequence $\{x_i\}_{i=0}^\infty$ in $\mathbb{R}^q$ if there exists a subsequence of $\{x_i\}_{i=0}^\infty$ that converges to x in the norm $\|\cdot\|$. Recall that every bounded sequence has at least one subsequential limit. A *divergent sequence* is a sequence having no convergent subsequence.

Consider the nonlinear autonomous dynamical system

$$\dot{x}(t) = f(x(t)), \qquad x(0) = x_0, \qquad t \in \mathcal{I}_{x_0}, \tag{1}$$

where $x(t) \in \mathcal{D} \subseteq \mathbb{R}^n$, $t \in \mathcal{I}_{x_0}$, is the system state vector, $\mathcal{D}$ is a relatively open set, $f : \mathcal{D} \to \mathbb{R}^n$ is continuous on $\mathcal{D}$, and $\mathcal{I}_{x_0} = [0, \tau_{x_0})$, $0 \leq \tau_{x_0} \leq \infty$, is the *maximal interval of existence* for the solution $x(\cdot)$ of (1). We assume that, for every initial condition $x(0) \in \mathcal{D}$, the differential equation (1) possesses a unique right-maximally defined continuously differentiable solution which is defined on $[0,\infty)$. Letting $s(\cdot,x)$ denote the right-maximally defined solution of (1) that satisfies the initial condition $x(0) = x$, the above assumptions imply that the map $s : [0,\infty) \times \mathcal{D} \to \mathcal{D}$ is continuous (Hartman, 1982, Theorem V.2.1), satisfies the *consistency* property $s(0,x) = x$, and possesses the *semigroup* property $s(t,s(\tau,x)) = s(t+\tau,x)$ for all $t,\tau \geq 0$ and $x \in \mathcal{D}$. Given $t \geq 0$ and $x \in \mathcal{D}$, we denote the map $s(t,\cdot) : \mathcal{D} \to \mathcal{D}$ by s_t and the map $s(\cdot,x) : [0,\infty) \to \mathcal{D}$ by s^x. For every $t \in \mathbb{R}$, the map s_t is a homeomorphism and has the inverse s_{-t}.

The *orbit* $\mathcal{O}_x$ of a point $x \in \mathcal{D}$ is the set $s^x([0,\infty))$. A set $\mathcal{D}_c \subseteq \mathcal{D}$ is *positively invariant* relative to (1) if $s_t(\mathcal{D}_c) \subseteq \mathcal{D}_c$ for all $t \geq 0$ or, equivalently, $\mathcal{D}_c$ contains the orbits of all its points. The set $\mathcal{D}_c$ is *invariant* relative to (1) if $s_t(\mathcal{D}_c) = \mathcal{D}_c$ for all $t \geq 0$. The *positive limit set* of $x \in \mathbb{R}^q$ is the set $\omega(x)$ of all subsequential limits of sequences of the form $\{s(t_i,x)\}_{i=0}^\infty$, where $\{t_i\}_{i=0}^\infty$ is an increasing divergent sequence in $[0,\infty)$. $\omega(x)$ is closed and invariant, and $\overline{\mathcal{O}}_x = \mathcal{O}_x \cup \omega(x)$ (Haddad & Chellaboina (2008)). In addition, for every $x \in \mathbb{R}^q$ that has bounded positive orbits, $\omega(x)$ is nonempty and compact, and, for every neighborhood $\mathcal{N}$ of $\omega(x)$, there exists $T > 0$ such that $s_t(x) \in \mathcal{N}$ for every $t > T$ (Haddad & Chellaboina (2008)). Furthermore, $x_e \in \mathcal{D}$ is an *equilibrium point* of (1) if and only if $f(x_e) = 0$ or, equivalently, $s(t,x_e) = x_e$ for all $t \geq 0$. Finally, recall that if all solutions to (1) are bounded, then it follows from the Peano-Cauchy theorem (Haddad & Chellaboina, 2008, p. 76) that $\mathcal{I}_{x_0} = \mathbb{R}$.

Definition 21 *(Haddad et al., 2010, pp. 9,10) Let $f = [f_1,\ldots,f_n]^{\mathrm{T}} : \mathcal{D} \subseteq \overline{\mathbb{R}}_+^n \to \mathbb{R}^n$. Then f is* essentially nonnegative *if $f_i(x) \geq 0$, for all $i = 1,\ldots,n$, and $x \in \overline{\mathbb{R}}_+^n$ such that $x_i = 0$, where x_i denotes the ith component of x.*

Proposition 2.1 *(Haddad et al., 2010, p. 12) Suppose* $\overline{\mathbb{R}}_+^n \subset \mathcal{D}$*. Then* $\overline{\mathbb{R}}_+^n$ *is an invariant set with respect to (1) if and only if* $f : \mathcal{D} \to \mathbb{R}^n$ *is essentially nonnegative.*

Definition 2.2 *(Haddad et al., 2010, pp. 13, 23) An equilibrium solution* $x(t) \equiv x_e \in \overline{\mathbb{R}}_+^n$ *to (1) is* Lyapunov stable with respect to $\overline{\mathbb{R}}_+^n$ *if, for all* $\varepsilon > 0$*, there exists* $\delta = \delta(\varepsilon) > 0$ *such that if* $x \in \mathcal{B}_\delta(x_e) \cap \overline{\mathbb{R}}_+^n$*, then* $x(t) \in \mathcal{B}_\varepsilon(x_e) \cap \overline{\mathbb{R}}_+^n$*,* $t \geq 0$*. An equilibrium solution* $x(t) \equiv x_e \in \overline{\mathbb{R}}_+^n$ *to (1) is* semistable with respect to $\overline{\mathbb{R}}_+^n$ *if it is Lyapunov stable with respect to* $\overline{\mathbb{R}}_+^n$ *and there exists* $\delta > 0$ *such that if* $x_0 \in \mathcal{B}_\delta(x_e) \cap \overline{\mathbb{R}}_+^n$*, then* $\lim_{t\to\infty} x(t)$ *exists and corresponds to a Lyapunov stable equilibrium point with respect to* $\overline{\mathbb{R}}_+^n$*. The system (1) is said to be* semistable with respect to $\overline{\mathbb{R}}_+^n$ *if every equilibrium point of (1) is semistable with respect to* $\overline{\mathbb{R}}_+^n$*. The system (1) is said to be* globally semistable with respect to $\overline{\mathbb{R}}_+^n$ *if (1) is semistable with respect to* $\overline{\mathbb{R}}_+^n$ *and, for every* $x_0 \in \overline{\mathbb{R}}_+^n$*,* $\lim_{t\to\infty} x(t)$ *exists.*

Proposition 2.2 *(Haddad et al., 2010, p. 22) Consider the nonlinear dynamical system (1) where* f *is essentially nonnegative and let* $x \in \overline{\mathbb{R}}_+^n$*. If the positive limit set of (1) contains a Lyapunov stable (with respect to* $\overline{\mathbb{R}}_+^n$*) equilibrium point* y*, then* $y = \lim_{t\to\infty} s(t, x)$*.*

3. Interconnected thermodynamic systems: A state space energy flow perspective

The fundamental and unifying concept in the analysis of thermodynamic systems is the concept of energy. The energy of a state of a dynamical system is the measure of its ability to produce changes (motion) in its own system state as well as changes in the system states of its surroundings. These changes occur as a direct consequence of the energy flow between different subsystems within the dynamical system. Heat (energy) is a fundamental concept of thermodynamics involving the capacity of hot bodies (more energetic subsystems) to produce work. As in thermodynamic systems, dynamical systems can exhibit energy (due to friction) that becomes unavailable to do useful work. This in turn contributes to an increase in system entropy, a measure of the tendency of a system to lose the ability to do useful work. In this section, we use the state space formalism to construct a mathematical model of a thermodynamic system that is consistent with basic thermodynamic principles.

Specifically, we consider a large-scale system model with a combination of subsystems (compartments or parts) that is perceived as a single entity. For each subsystem (compartment) making up the system, we postulate the existence of an *energy* state variable such that the knowledge of these subsystem state variables at any given time $t = t_0$, together with the knowledge of any inputs (heat fluxes) to each of the subsystems for time $t \geq t_0$, completely determines the behavior of the system for any given time $t \geq t_0$. Hence, the (energy) state of our dynamical system at time t is uniquely determined by the state at time t_0 and any external inputs for time $t \geq t_0$ and is independent of the state and inputs before time t_0.

More precisely, we consider a large-scale interconnected dynamical system composed of a large number of units with aggregated (or lumped) energy variables representing homogenous groups of these units. If all the units comprising the system are identical (that is, the system is perfectly homogeneous), then the behavior of the dynamical system can be captured by that of a single plenipotentiary unit. Alternatively, if every interacting system unit is distinct, then the resulting model constitutes a microscopic system. To develop a middle-ground thermodynamic model placed between complete aggregation (classical thermodynamics) and complete disaggregation (statistical thermodynamics), we subdivide

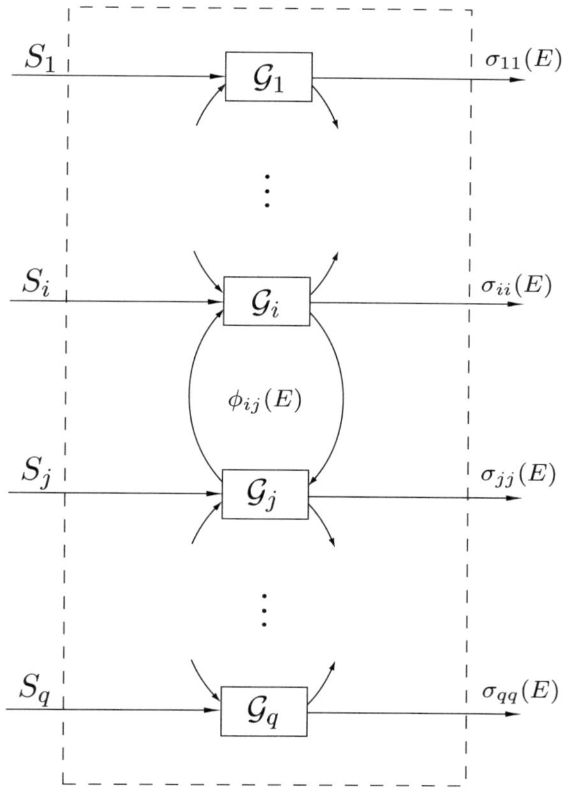

Fig. 1. Interconnected dynamical system $\mathcal{G}$.

the large-scale dynamical system into a finite number of compartments, each formed by a large number of homogeneous units. Each compartment represents the energy content of the different parts of the dynamical system, and different compartments interact by exchanging heat. Thus, our compartmental thermodynamic model utilizes subsystems or compartments to describe the energy distribution among distinct regions in space with intercompartmental flows representing the heat transfer between these regions. Decreasing the number of compartments results in a more aggregated or homogeneous model, whereas increasing the number of compartments leads to a higher degree of disaggregation resulting in a heterogeneous model.

To formulate our state space thermodynamic model, consider the interconnected dynamical system $\mathcal{G}$ shown in Figure 1 involving energy exchange between q interconnected subsystems. Let $E_i : [0,\infty) \to \overline{\mathbb{R}}_+$ denote the energy (and hence a nonnegative quantity) of the ith subsystem, let $S_i : [0,\infty) \to \mathbb{R}$ denote the external power (heat flux) supplied to (or extracted from) the ith subsystem, let $\phi_{ij} : \overline{\mathbb{R}}^q_+ \to \mathbb{R}$, $i \neq j$, $i,j = 1,\ldots,q$, denote the net instantaneous rate of energy (heat) flow from the jth subsystem to the ith subsystem, and let $\sigma_{ii} : \overline{\mathbb{R}}^q_+ \to \overline{\mathbb{R}}_+$, $i = 1,\ldots,q$, denote the instantaneous rate of energy (heat) dissipation from the ith subsystem to the environment. Here, we assume that $\phi_{ij} : \overline{\mathbb{R}}^q_+ \to \mathbb{R}$, $i \neq j$, $i,j = 1,\ldots,q$, and $\sigma_{ii} : \overline{\mathbb{R}}^q_+ \to \overline{\mathbb{R}}_+$, $i = 1,\ldots,q$, are locally Lipschitz continuous on $\overline{\mathbb{R}}^q_+$ and $S_i : [0,\infty) \to \mathbb{R}$, $i = 1,\ldots,q$, are bounded piecewise continuous functions of time.

An *energy balance* for the ith subsystem yields

$$E_i(T) = E_i(t_0) + \left[\sum_{j=1, j\neq i}^{q} \int_{t_0}^{T} \phi_{ij}(E(t))\mathrm{d}t\right] - \int_{t_0}^{T} \sigma_{ii}(E(t))\mathrm{d}t + \int_{t_0}^{T} S_i(t)\mathrm{d}t, \quad T \geq t_0, \tag{2}$$

or, equivalently, in vector form,

$$E(T) = E(t_0) + \int_{t_0}^{T} w(E(t))\mathrm{d}t - \int_{t_0}^{T} d(E(t))\mathrm{d}t + \int_{t_0}^{T} S(t)\mathrm{d}t, \quad T \geq t_0, \tag{3}$$

where $E(t) \triangleq [E_1(t),\ldots,E_q(t)]^{\mathrm{T}}$, $t \geq t_0$, is the system energy state, $d(E(t)) \triangleq [\sigma_{11}(E(t)),\ldots, \sigma_{qq}(E(t))]^{\mathrm{T}}$, $t \geq t_0$, is the system dissipation, $S(t) \triangleq [S_1(t),\ldots,S_q(t)]^{\mathrm{T}}$, $t \geq t_0$, is the system heat flux, and $w = [w_1,\ldots,w_q]^{\mathrm{T}} : \overline{\mathbb{R}}_{+}^{q} \to \mathbb{R}^q$ is such that

$$w_i(E) = \sum_{j=1, j\neq i}^{q} \phi_{ij}(E), \quad E \in \overline{\mathbb{R}}_{+}^{q}. \tag{4}$$

Since $\phi_{ij} : \overline{\mathbb{R}}_{+}^{q} \to \mathbb{R}$, $i \neq j$, $i,j = 1,\ldots,q$, denotes the net instantaneous rate of energy flow from the jth subsystem to the ith subsystem, it is clear that $\phi_{ij}(E) = -\phi_{ji}(E)$, $E \in \overline{\mathbb{R}}_{+}^{q}$, $i \neq j$, $i,j = 1,\ldots,q$, which further implies that $\mathbf{e}^{\mathrm{T}} w(E) = 0$, $E \in \overline{\mathbb{R}}_{+}^{q}$.
Note that (2) yields a conservation of energy equation and implies that the energy stored in the ith subsystem is equal to the external energy supplied to (or extracted from) the ith subsystem plus the energy gained by the ith subsystem from all other subsystems due to subsystem coupling minus the energy dissipated from the ith subsystem to the environment. Equivalently, (2) can be rewritten as

$$\dot{E}_i(t) = \left[\sum_{j=1, j\neq i}^{q} \phi_{ij}(E(t))\right] - \sigma_{ii}(E(t)) + S_i(t), \quad E_i(t_0) = E_{i0}, \quad t \geq t_0, \tag{5}$$

or, in vector form,

$$\dot{E}(t) = w(E(t)) - d(E(t)) + S(t), \quad E(t_0) = E_0, \quad t \geq t_0, \tag{6}$$

where $E_0 \triangleq [E_{10},\ldots,E_{q0}]^{\mathrm{T}}$, yielding a *power balance* equation that characterizes energy flow between subsystems of the interconnected dynamical system $\mathcal{G}$. We assume that $\phi_{ij}(E) \geq 0$, $E \in \overline{\mathbb{R}}_{+}^{q}$, whenever $E_i = 0$, $i \neq j$, $i,j = 1,\ldots,q$, and $\sigma_{ii}(E) = 0$, whenever $E_i = 0$, $i = 1,\ldots,q$. The above constraint implies that if the energy of the ith subsystem of $\mathcal{G}$ is zero, then this subsystem cannot supply any energy to its surroundings nor can it dissipate energy to the environment. In this case, $w(E) - d(E)$, $E \in \overline{\mathbb{R}}_{+}^{q}$, is essentially nonnegative (Haddad & Chellaboina (2005)). Thus, if $S(t) \equiv 0$, then, by Proposition 2.1, the solutions to (6) are nonnegative for all nonnegative initial conditions. See Haddad & Chellaboina (2005); Haddad et al. (2005; 2010) for further details.
Since our thermodynamic compartmental model involves intercompartmental flows representing energy transfer between compartments, we can use graph-theoretic notions with *undirected graph topologies* (i.e., bidirectional energy flows) to capture the compartmental system interconnections. Graph theory (Diestel (1997); Godsil & Royle (2001)) can be useful

in the analysis of the connectivity properties of compartmental systems. In particular, an undirected graph can be constructed to capture a compartmental model in which the compartments are represented by nodes and the flows are represented by edges or arcs. In this case, the environment must also be considered as an additional node.

For the interconnected dynamical system $\mathcal{G}$ with the power balance equation (6), we define a *connectivity matrix*[1] $\mathcal{C} \in \mathbb{R}^{q\times q}$ such that for $i \neq j$, $i,j = 1,\ldots,q$, $\mathcal{C}_{(i,j)} \triangleq 1$ if $\phi_{ij}(E) \not\equiv 0$ and $\mathcal{C}_{(i,j)} \triangleq 0$ otherwise, and $\mathcal{C}_{(i,i)} \triangleq -\sum_{k=1,k\neq i}^{q} \mathcal{C}_{(k,i)}$, $i = 1,\ldots,q$. Recall that if $\operatorname{rank}\mathcal{C} = q-1$, then $\mathcal{G}$ is strongly connected (Haddad et al. (2005)) and energy exchange is possible between any two subsystems of $\mathcal{G}$. The next definition introduces a notion of entropy for the interconnected dynamical system $\mathcal{G}$.

Definition 3.1 *Consider the interconnected dynamical system $\mathcal{G}$ with the power balance equation (6). A continuously differentiable, strictly concave function $\mathcal{S} : \overline{\mathbb{R}}_+^q \to \mathbb{R}$ is called the* entropy *function of $\mathcal{G}$ if*

$$\left(\frac{\partial \mathcal{S}(E)}{\partial E_i} - \frac{\partial \mathcal{S}(E)}{\partial E_j}\right)\phi_{ij}(E) \geq 0, \quad E \in \overline{\mathbb{R}}_+^q, \quad i \neq j, \quad i,j = 1,\ldots,q, \tag{7}$$

and $\frac{\partial \mathcal{S}(E)}{\partial E_i} = \frac{\partial \mathcal{S}(E)}{\partial E_j}$ *if and only if* $\phi_{ij}(E) = 0$ *with* $\mathcal{C}_{(i,j)} = 1$, $i \neq j$, $i,j = 1,\ldots,q$.

It follows from Definition 3.1 that for an *isolated system* $\mathcal{G}$, that is, $S(t) \equiv 0$ and $d(E) \equiv 0$, the entropy function of $\mathcal{G}$ is a nondecreasing function of time. To see this, note that

$$\begin{aligned}
\dot{\mathcal{S}}(E) &= \frac{\partial \mathcal{S}(E)}{\partial E}\dot{E} \\
&= \sum_{i=1}^{q} \frac{\partial \mathcal{S}(E)}{\partial E_i} \sum_{j=1, j\neq i}^{q} \phi_{ij}(E) \\
&= \sum_{i=1}^{q} \sum_{j=i+1}^{q} \left(\frac{\partial \mathcal{S}(E)}{\partial E_i} - \frac{\partial \mathcal{S}(E)}{\partial E_j}\right)\phi_{ij}(E) \\
&\geq 0, \quad E \in \overline{\mathbb{R}}_+^q,
\end{aligned} \tag{8}$$

where $\frac{\partial \mathcal{S}(E)}{\partial E} \triangleq \left[\frac{\partial \mathcal{S}(E)}{\partial E_1},\ldots,\frac{\partial \mathcal{S}(E)}{\partial E_q}\right]$ and where we used the fact that $\phi_{ij}(E) = -\phi_{ji}(E)$, $E \in \overline{\mathbb{R}}_+^q$, $i \neq j$, $i,j = 1,\ldots,q$.

Proposition 3.1 *Consider the isolated (i.e., $S(t) \equiv 0$ and $d(E) \equiv 0$) interconnected dynamical system $\mathcal{G}$ with the power balance equation (6). Assume that $\operatorname{rank}\mathcal{C} = q-1$ and there exists an entropy function $\mathcal{S} : \overline{\mathbb{R}}_+^q \to \mathbb{R}$ of $\mathcal{G}$. Then, $\sum_{j=1}^{q} \phi_{ij}(E) = 0$ for all $i = 1,\ldots,q$ if and only if $\frac{\partial \mathcal{S}(E)}{\partial E_1} = \cdots = \frac{\partial \mathcal{S}(E)}{\partial E_q}$. Furthermore, the set of nonnegative equilibrium states of (6) is given by $\mathcal{E}_0 \triangleq \left\{E \in \overline{\mathbb{R}}_+^q : \frac{\partial \mathcal{S}(E)}{\partial E_1} = \cdots = \frac{\partial \mathcal{S}(E)}{\partial E_q}\right\}$.*

[1]The negative of the connectivity matrix, that is, $-\mathcal{C}$, is known as the graph Laplacian in the literature.

Proof. If $\frac{\partial \mathcal{S}(E)}{\partial E_i} = \frac{\partial \mathcal{S}(E)}{\partial E_j}$, then $\phi_{ij}(E) = 0$ for all $i,j = 1,\ldots,q$, which implies that $\sum_{j=1}^{q}\phi_{ij}(E) = 0$ for all $i = 1,\ldots,q$. Conversely, assume that $\sum_{j=1}^{q}\phi_{ij}(E) = 0$ for all $i = 1,\ldots,q$, and, since $\mathcal{S}$ is an entropy function of $\mathcal{G}$, it follows that

$$
\begin{aligned}
0 &= \sum_{i=1}^{q}\sum_{j=1}^{q} \frac{\partial \mathcal{S}(E)}{\partial E_i}\phi_{ij}(E) \\
&= \sum_{i=1}^{q-1}\sum_{j=i+1}^{q}\left(\frac{\partial \mathcal{S}(E)}{\partial E_i} - \frac{\partial \mathcal{S}(E)}{\partial E_j}\right)\phi_{ij}(E) \\
&\geq 0,
\end{aligned}
$$

where we have used the fact that $\phi_{ij}(E) = -\phi_{ji}(E)$ for all $i,j = 1,\ldots,q$. Hence,

$$
\left(\frac{\partial \mathcal{S}(E)}{\partial E_i} - \frac{\partial \mathcal{S}(E)}{\partial E_j}\right)\phi_{ij}(E) = 0
$$

for all $i,j = 1,\ldots,q$. Now, the result follows from the fact that rank $\mathcal{C} = q-1$. □

Theorem 3.1 *Consider the isolated (i.e., $S(t) \equiv 0$ and $d(E) \equiv 0$) interconnected dynamical system $\mathcal{G}$ with the power balance equation (6). Assume that $\operatorname{rank}\mathcal{C} = q-1$ and there exists an entropy function $\mathcal{S} : \overline{\mathbb{R}}_+^q \to \mathbb{R}$ of $\mathcal{G}$. Then the isolated system $\mathcal{G}$ is globally semistable with respect to $\overline{\mathbb{R}}_+^q$.*

Proof. Since $w(\cdot)$ is essentially nonnegative, it follows from Proposition 2.1 that $E(t) \in \overline{\mathbb{R}}_+^q$, $t \geq t_0$, for all $E_0 \in \overline{\mathbb{R}}_+^q$. Furthermore, note that since $\mathbf{e}^{\mathrm{T}}w(E) = 0$, $E \in \overline{\mathbb{R}}_+^q$, it follows that $\mathbf{e}^{\mathrm{T}}\dot{E}(t) = 0$, $t \geq t_0$. In this case, $\mathbf{e}^{\mathrm{T}}E(t) = \mathbf{e}^{\mathrm{T}}E_0$, $t \geq t_0$, which implies that $E(t)$, $t \geq t_0$, is bounded for all $E_0 \in \overline{\mathbb{R}}_+^q$. Now, it follows from (8) that $\mathcal{S}(E(t))$, $t \geq t_0$, is a nondecreasing function of time, and hence, by the Krasovskii-LaSalle theorem (Haddad & Chellaboina (2008)), $E(t) \to \mathcal{R} \triangleq \{E \in \overline{\mathbb{R}}_+^q : \dot{\mathcal{S}}(E) = 0\}$ as $t \to \infty$. Next, it follows from (8), Definition 3.1, and the fact that $\operatorname{rank}\mathcal{C} = q-1$, that $\mathcal{R} = \left\{E \in \overline{\mathbb{R}}_+^q : \frac{\partial \mathcal{S}(E)}{\partial E_1} = \cdots = \frac{\partial \mathcal{S}(E)}{\partial E_q}\right\} = \mathcal{E}_0$.
Now, let $E_{\mathrm{e}} \in \mathcal{E}_0$ and consider the continuously differentiable function $V : \mathbb{R}^q \to \mathbb{R}$ defined by

$$
V(E) \triangleq \mathcal{S}(E_{\mathrm{e}}) - \mathcal{S}(E) - \lambda_{\mathrm{e}}(\mathbf{e}^{\mathrm{T}}E_{\mathrm{e}} - \mathbf{e}^{\mathrm{T}}E),
$$

where $\lambda_{\mathrm{e}} \triangleq \frac{\partial \mathcal{S}}{\partial E_1}(E_{\mathrm{e}})$. Next, note that $V(E_{\mathrm{e}}) = 0$, $\frac{\partial V}{\partial E}(E_{\mathrm{e}}) = -\frac{\partial \mathcal{S}}{\partial E}(E_{\mathrm{e}}) + \lambda_{\mathrm{e}}\mathbf{e}^{\mathrm{T}} = 0$, and, since $\mathcal{S}(\cdot)$ is a strictly concave function, $\frac{\partial^2 V}{\partial E^2}(E_{\mathrm{e}}) = -\frac{\partial^2 \mathcal{S}}{\partial E^2}(E_{\mathrm{e}}) > 0$, which implies that $V(\cdot)$ admits a local minimum at E_{e}. Thus, $V(E_{\mathrm{e}}) = 0$, there exists $\delta > 0$ such that $V(E) > 0$, $E \in \mathcal{B}_\delta(E_{\mathrm{e}})\backslash\{E_{\mathrm{e}}\}$, and $\dot{V}(E) = -\dot{\mathcal{S}}(E) \leq 0$ for all $E \in \mathcal{B}_\delta(E_{\mathrm{e}})\backslash\{E_{\mathrm{e}}\}$, which shows that $V(\cdot)$ is a Lyapunov function for $\mathcal{G}$ and E_{e} is a Lyapunov stable equilibrium of $\mathcal{G}$. Finally, since, for every $E_0 \in \overline{\mathbb{R}}_+^n$, $E(t) \to \mathcal{E}_0$ as $t \to \infty$ and every equilibrium point of $\mathcal{G}$ is Lyapunov stable, it follows from Proposition 2.2 that $\mathcal{G}$ is globally semistable with respect to $\overline{\mathbb{R}}_+^q$. □

In classical thermodynamics, the partial derivative of the system entropy with respect to the system energy defines the reciprocal of the system temperature. Thus, for the interconnected dynamical system $\mathcal{G}$,

$$
T_i \triangleq \left(\frac{\partial \mathcal{S}(E)}{\partial E_i}\right)^{-1}, \quad i = 1,\ldots,q, \tag{9}
$$

represents the temperature of the ith subsystem. Condition (7) is a manifestation of the *second law of thermodynamics* and implies that if the temperature of the jth subsystem is greater than the temperature of the ith subsystem, then energy (heat) flows from the jth subsystem to the ith subsystem. Furthermore, $\frac{\partial \mathcal{S}(E)}{\partial E_i} = \frac{\partial \mathcal{S}(E)}{\partial E_j}$ if and only if $\phi_{ij}(E) = 0$ with $\mathcal{C}_{(i,j)} = 1, i \neq j, i,j = 1,\ldots,q$, implies that temperature equality is a necessary and sufficient condition for thermal equilibrium. This is a statement of the *zeroth law of thermodynamics*. As a result, Theorem 3.1 shows that, for a strongly connected system $\mathcal{G}$, the subsystem energies converge to the set of equilibrium states where the temperatures of all subsystems are equal. This phenomenon is known as *equipartition of temperature* (Haddad et al. (2010)) and is an emergent behavior in thermodynamic systems. In particular, all the system energy is eventually transferred into heat at a uniform temperature, and hence, all dynamical processes in $\mathcal{G}$ (system motions) would cease.

The following result presents a sufficient condition for energy equipartition of the system, that is, the energies of all subsystems are equal. And this state of energy equipartition is uniquely determined by the initial energy in the system.

Theorem 3.2 *Consider the isolated (i.e., $S(t) \equiv 0$ and $d(E) \equiv 0$) interconnected dynamical system $\mathcal{G}$ with the power balance equation (6). Assume that rank $\mathcal{C} = q-1$ and there exists a continuously differentiable, strictly concave function $f : \overline{\mathbb{R}}_+ \to \mathbb{R}$ such that the entropy function $\mathcal{S} : \overline{\mathbb{R}}_+^q \to \mathbb{R}$ of $\mathcal{G}$ is given by $\mathcal{S}(E) = \sum_{i=1}^q f(E_i)$. Then, the set of nonnegative equilibrium states of (6) is given by $\mathcal{E}_0 = \{\alpha\mathbf{e} : \alpha \geq 0\}$ and $\mathcal{G}$ is semistable with respect to $\overline{\mathbb{R}}_+^q$. Furthermore, $E(t) \to \frac{1}{q}\mathbf{e}\mathbf{e}^{\mathrm{T}}E(t_0)$ as $t \to \infty$ and $\frac{1}{q}\mathbf{e}\mathbf{e}^{\mathrm{T}}E(t_0)$ is a semistable equilibrium state of $\mathcal{G}$.*

Proof. First, note that since $f(\cdot)$ is a continuously differentiable, strictly concave function it follows that

$$\left(\frac{\mathrm{d}f}{\mathrm{d}E_i} - \frac{\mathrm{d}f}{\mathrm{d}E_j}\right)(E_i - E_j) \leq 0, \quad E \in \overline{\mathbb{R}}_+^q, \quad i,j = 1,\ldots,q,$$

which implies that (7) is equivalent to

$$\left(E_i - E_j\right)\phi_{ij}(E) \leq 0, \quad E \in \overline{\mathbb{R}}_+^q, \quad i \neq j, \quad i,j = 1,\ldots,q,$$

and $E_i = E_j$ if and only if $\phi_{ij}(E) = 0$ with $\mathcal{C}_{(i,j)} = 1$, $i \neq j$, $i,j = 1,\ldots,q$. Hence, $-E^{\mathrm{T}}E$ is an entropy function of $\mathcal{G}$. Next, with $\mathcal{S}(E) = -\frac{1}{2}E^{\mathrm{T}}E$, it follows from Proposition 3.1 that $\mathcal{E}_0 = \{\alpha\mathbf{e} \in \overline{\mathbb{R}}_+^q, \alpha \geq 0\}$. Now, it follows from Theorem 3.1 that $\mathcal{G}$ is globally semistable with respect to $\overline{\mathbb{R}}_+^q$. Finally, since $\mathbf{e}^{\mathrm{T}}E(t) = \mathbf{e}^{\mathrm{T}}E(t_0)$ and $E(t) \to \mathcal{M}$ as $t \to \infty$, it follows that $E(t) \to \frac{1}{q}\mathbf{e}\mathbf{e}^{\mathrm{T}}E(t_0)$ as $t \to \infty$. Hence, with $\alpha = \frac{1}{q}\mathbf{e}^{\mathrm{T}}E(t_0)$, $\alpha\mathbf{e} = \frac{1}{q}\mathbf{e}\mathbf{e}^{\mathrm{T}}E(t_0)$ is a semistable equilibrium state of (6). □

If $f(E_i) = \log_e(c + E_i)$, where $c > 0$, so that $\mathcal{S}(E) = \sum_{i=1}^q \log_e(c + E_i)$, then it follows from Theorem 3.2 that $\mathcal{E}_0 = \{\alpha\mathbf{e} : \alpha \geq 0\}$ and the isolated (i.e., $S(t) \equiv 0$ and $d(E) \equiv 0$) interconnected dynamical system $\mathcal{G}$ with the power balance equation (6) is semistable. In this case, the absolute temperature of the ith compartment is given by $c + E_i$. Similarly, if $\mathcal{S}(E) = -\frac{1}{2}E^{\mathrm{T}}E$, then it follows from Theorem 3.2 that $\mathcal{E}_0 = \{\alpha\mathbf{e} : \alpha \geq 0\}$ and the isolated (i.e., $S(t) \equiv 0$ and $d(E) \equiv 0$) interconnected dynamical system $\mathcal{G}$ with the power balance equation (6) is semistable. In both these cases, $E(t) \to \frac{1}{q}\mathbf{e}\mathbf{e}^{\mathrm{T}}E(t_0)$ as $t \to \infty$. This shows that the steady-state energy of the isolated interconnected dynamical system $\mathcal{G}$ is given by

$\frac{1}{q}\mathbf{e}\mathbf{e}^{\mathrm{T}}E(t_0) = \frac{1}{q}\sum_{i=1}^{q} E_i(t_0)\mathbf{e}$, and hence, is uniformly distributed over all subsystems of $\mathcal{G}$. This phenomenon is known as *energy equipartition* (Haddad et al. (2005)). The aforementioned forms of $\mathcal{S}(E)$ were extensively discussed in the recent book by Haddad et al. (2005) where $\mathcal{S}(E) = \sum_{i=1}^{q} \log_e(c + E_i)$ and $-\mathcal{S}(E) = \frac{1}{2}E^{\mathrm{T}}E$ are referred to, respectively, as the entropy and the ectropy functions of the interconnected dynamical system $\mathcal{G}$.

4. Work energy, free energy, heat flow, and Clausius' inequality

In this section, we augment our thermodynamic energy flow model $\mathcal{G}$ with an additional (deformation) state representing subsystem volumes in order to introduce the notion of work into our thermodynamically consistent state space energy flow model. Specifically, we assume that each subsystem can perform (positive) work on the environment as well as the environment can perform (negative) work on the subsystems. The rate of work done by the ith subsystem on the environment is denoted by $d_{\mathrm{w}i} : \overline{\mathbb{R}}_+^q \times \mathbb{R}_+^q \to \overline{\mathbb{R}}_+, i = 1,\ldots,q$, the rate of work done by the environment on the ith subsystem is denoted by $S_{\mathrm{w}i} : [0,\infty) \to \overline{\mathbb{R}}_+$, $i = 1,\ldots,q$, and the volume of the ith subsystem is denoted by $V_i : [0,\infty) \to \mathbb{R}_+, i = 1,\ldots,q$. The net work done by each subsystem on the environment satisfies

$$p_i(E,V)\mathrm{d}V_i = (d_{\mathrm{w}i}(E,V) - S_{\mathrm{w}i}(t))\mathrm{d}t, \tag{10}$$

where $p_i(E,V), i = 1,\ldots,q$, denotes the *pressure* in the ith subsystem and $V \triangleq [V_1,\ldots,V_q]^{\mathrm{T}}$. Furthermore, in the presence of work, the energy balance equation (5) for each subsystem can be rewritten as

$$\mathrm{d}E_i = w_i(E,V)\mathrm{d}t - (d_{\mathrm{w}i}(E,V) - S_{\mathrm{w}i}(t))\mathrm{d}t - \sigma_{ii}(E,V)\mathrm{d}t + S_i(t)\mathrm{d}t, \tag{11}$$

where $w_i(E,V) \triangleq \sum_{j=1, j\neq i}^{q} \phi_{ij}(E,V)$, $\phi_{ij} : \overline{\mathbb{R}}_+^q \times \mathbb{R}_+^q \to \mathbb{R}$, $i \neq j, i,j = 1,\ldots,q$, denotes the net instantaneous rate of energy (heat) flow from the jth subsystem to the ith subsystem, σ_{ii} : $\overline{\mathbb{R}}_+^q \times \mathbb{R}_+^q \to \overline{\mathbb{R}}_+, i = 1,\ldots,q$, denotes the instantaneous rate of energy dissipation from the ith subsystem to the environment, and, as in Section 3, $S_i : [0,\infty) \to \mathbb{R}$, $i = 1,\ldots,q$, denotes the external power supplied to (or extracted from) the ith subsystem. It follows from (10) and (11) that positive work done by a subsystem on the environment leads to a decrease in internal energy of the subsystem and an increase in the subsystem volume, which is consistent with the first law of thermodynamics.

The definition of entropy for $\mathcal{G}$ in the presence of work remains the same as in Definition 3.1 with $\mathcal{S}(E)$ replaced by $\mathcal{S}(E,V)$ and with all other conditions in the definition holding for every $V >> 0$. Next, consider the ith subsystem of $\mathcal{G}$ and assume that E_j and $V_j, j \neq i, i = 1,\ldots,q$, are constant. In this case, note that

$$\frac{\mathrm{d}\mathcal{S}}{\mathrm{d}t} = \frac{\partial \mathcal{S}}{\partial E_i}\frac{\mathrm{d}E_i}{\mathrm{d}t} + \frac{\partial \mathcal{S}}{\partial V_i}\frac{\mathrm{d}V_i}{\mathrm{d}t} \tag{12}$$

and define

$$p_i(E,V) \triangleq \left(\frac{\partial \mathcal{S}}{\partial E_i}\right)^{-1}\left(\frac{\partial \mathcal{S}}{\partial V_i}\right), \quad i = 1,\ldots,q. \tag{13}$$

It follows from (10) and (11) that, in the presence of work energy, the power balance equation (6) takes the new form involving energy and deformation states

$$\dot{E}(t) = w(E(t),V(t)) - d_{\rm w}(E(t),V(t)) + S_{\rm w}(t) - d(E(t),V(t)) + S(t), \quad E(t_0) = E_0, \quad t \geq t_0, \tag{14}$$

$$\dot{V}(t) = D(E(t),V(t))(d_{\rm w}(E(t),V(t)) - S_{\rm w}(t)), \quad V(t_0) = V_0, \tag{15}$$

where $w(E,V) \triangleq [w_1(E,V),\ldots,w_q(E,V)]^{\rm T}$, $d_{\rm w}(E,V) \triangleq [d_{\rm w1}(E,V),\ldots, d_{{\rm w}q}(E,V)]^{\rm T}$, $S_{\rm w}(t) \triangleq [S_{\rm w1}(t),\ldots,S_{{\rm w}q}(t)]^{\rm T}$, $d(E,V) \triangleq [\sigma_{11}(E,V),\ldots,\sigma_{qq}(E,V)]^{\rm T}$, $S(t) \triangleq [S_1(t),\ldots,S_q(t)]^{\rm T}$, and

$$D(E,V) \triangleq \mathrm{diag}\left[\left(\frac{\partial \mathcal{S}}{\partial E_1}\right)\left(\frac{\partial \mathcal{S}}{\partial V_1}\right)^{-1},\ldots,\left(\frac{\partial \mathcal{S}}{\partial E_q}\right)\left(\frac{\partial \mathcal{S}}{\partial V_q}\right)^{-1}\right]. \tag{16}$$

Note that

$$\left(\frac{\partial \mathcal{S}(E,V)}{\partial V}\right) D(E,V) = \frac{\partial \mathcal{S}(E,V)}{\partial E}. \tag{17}$$

The power balance and deformation equations (14) and (15) represent a statement of the first law of thermodynamics. To see this, define the work L done by the interconnected dynamical system $\mathcal{G}$ over the time interval $[t_1,t_2]$ by

$$L \triangleq \int_{t_1}^{t_2} \mathbf{e}^{\rm T}[d_{\rm w}(E(t),V(t)) - S_{\rm w}(t)]\mathrm{d}t, \tag{18}$$

where $[E^{\rm T}(t),V^{\rm T}(t)]^{\rm T}$, $t \geq t_0$, is the solution to (14) and (15). Now, premultiplying (14) by $\mathbf{e}^{\rm T}$ and using the fact that $\mathbf{e}^{\rm T}w(E,V) = 0$, it follows that

$$\Delta U = -L + Q, \tag{19}$$

where $\Delta U = U(t_2) - U(t_1) \triangleq \mathbf{e}^{\rm T}E(t_2) - \mathbf{e}^{\rm T}E(t_1)$ denotes the variation in the total energy of the interconnected system $\mathcal{G}$ over the time interval $[t_1,t_2]$ and

$$Q \triangleq \int_{t_1}^{t_2} \mathbf{e}^{\rm T}[S(t) - d(E(t),V(t))]\mathrm{d}t \tag{20}$$

denotes the net energy received by $\mathcal{G}$ in forms other than work.

This is a statement of the *first law of thermodynamics* for the interconnected dynamical system $\mathcal{G}$ and gives a precise formulation of the equivalence between work and heat. This establishes that heat and mechanical work are two different aspects of energy. Finally, note that (15) is consistent with the classical thermodynamic equation for the rate of work done by the system $\mathcal{G}$ on the environment. To see this, note that (15) can be equivalently written as $\mathrm{d}L = \mathbf{e}^{\rm T}D^{-1}(E,V)\mathrm{d}V$, which, for a single subsystem with volume V and pressure p, has the classical form

$$\mathrm{d}L = p\mathrm{d}V. \tag{21}$$

It follows from Definition 3.1 and (14)–(17) that the time derivative of the entropy function satisfies

$$\begin{aligned}
\dot{\mathcal{S}}(E,V) &= \frac{\partial \mathcal{S}(E,V)}{\partial E}\dot{E} + \frac{\partial \mathcal{S}(E,V)}{\partial V}\dot{V} \\
&= \frac{\partial \mathcal{S}(E,V)}{\partial E} w(E,V) - \frac{\partial \mathcal{S}(E,V)}{\partial E}(d_{\mathrm{w}}(E,V) - S_{\mathrm{w}}(t)) \\
&\quad - \frac{\partial \mathcal{S}(E,V)}{\partial E}(d(E,V) - S(t)) + \frac{\partial \mathcal{S}(E,V)}{\partial V} D(E,V)(d_{\mathrm{w}}(E,V) - S_{\mathrm{w}}(t)) \\
&= \sum_{i=1}^{q} \frac{\partial \mathcal{S}(E,V)}{\partial E_i} \sum_{j=1, j\neq i}^{q} \phi_{ij}(E,V) + \sum_{i=1}^{q} \frac{\partial \mathcal{S}(E,V)}{\partial E_i}(S_i(t) - d_i(E,V)) \\
&= \sum_{i=1}^{q} \sum_{j=i+1}^{q} \left(\frac{\partial \mathcal{S}(E,V)}{\partial E_i} - \frac{\partial \mathcal{S}(E,V)}{\partial E_j} \right) \phi_{ij}(E,V) \\
&\quad + \sum_{i=1}^{q} \frac{\partial \mathcal{S}(E,V)}{\partial E_i}(S_i(t) - d_i(E,V)) \\
&\geq \sum_{i=1}^{q} \frac{\partial \mathcal{S}(E,V)}{\partial E_i}(S_i(t) - d_i(E,V)), \quad (E,V) \in \overline{\mathbb{R}}_+^q \times \overline{\mathbb{R}}_+^q. \qquad (22)
\end{aligned}$$

Noting that $\mathrm{d}Q_i \triangleq [S_i - \sigma_{ii}(E)]\mathrm{d}t$, $i = 1, \ldots, q$, is the infinitesimal amount of the net heat received or dissipated by the ith subsystem of $\mathcal{G}$ over the infinitesimal time interval $\mathrm{d}t$, it follows from (22) that

$$\mathrm{d}\mathcal{S}(E) \geq \sum_{i=1}^{q} \frac{\mathrm{d}Q_i}{T_i}. \qquad (23)$$

Inequality (23) is the classical *Clausius inequality* for the variation of entropy during an infinitesimal irreversible transformation.

Note that for an *adiabatically isolated* interconnected dynamical system (i.e., no heat exchange with the environment), (22) yields the universal inequality

$$\mathcal{S}(E(t_2), V(t_2)) \geq \mathcal{S}(E(t_1), V(t_1)), \quad t_2 \geq t_1, \qquad (24)$$

which implies that, for any dynamical change in an adiabatically isolated interconnected system $\mathcal{G}$, the entropy of the final system state can never be less than the entropy of the initial system state. In addition, in the case where $(E(t), V(t)) \in \mathcal{M}_{\mathrm{e}}$, $t \geq t_0$, where $\mathcal{M}_{\mathrm{e}} \triangleq \{(E,V) \in \overline{\mathbb{R}}_+^q \times \overline{\mathbb{R}}_+^q : E = \alpha \mathbf{e}, \alpha \geq 0, V \in \overline{\mathbb{R}}_+^q\}$, it follows from Definition 3.1 and (22) that inequality (24) is satisfied as a strict inequality for all $(E,V) \in (\overline{\mathbb{R}}_+^q \times \overline{\mathbb{R}}_+^q) \backslash \mathcal{M}_{\mathrm{e}}$. Hence, it follows from Theorem 2.15 of Haddad et al. (2005) that the adiabatically isolated interconnected system $\mathcal{G}$ does not exhibit Poincaré recurrence in $(\overline{\mathbb{R}}_+^q \times \overline{\mathbb{R}}_+^q) \backslash \mathcal{M}_{\mathrm{e}}$.

Next, we define the *Gibbs free energy*, the *Helmholtz free energy*, and the *enthalpy* functions for the interconnected dynamical system $\mathcal{G}$. For this exposition, we assume that the entropy of $\mathcal{G}$ is a sum of individual entropies of subsystems of $\mathcal{G}$, that is, $\mathcal{S}(E,V) = \sum_{i=1}^{q} \mathcal{S}_i(E_i, V_i)$, $(E,V) \in$

$\overline{\mathbb{R}}_+^q \times \mathbb{R}_+^q$. In this case, the Gibbs free energy of $\mathcal{G}$ is defined by

$$G(E,V) \triangleq \mathbf{e}^{\mathrm{T}}E - \sum_{i=1}^{q}\left(\frac{\partial\mathcal{S}(E,V)}{\partial E_i}\right)^{-1}\mathcal{S}_i(E_i,V_i) + \sum_{i=1}^{q}\left(\frac{\partial\mathcal{S}(E,V)}{\partial E_i}\right)^{-1}\left(\frac{\partial\mathcal{S}(E,V)}{\partial V_i}\right)V_i, \quad (E,V)\in\overline{\mathbb{R}}_+^q\times\mathbb{R}_+^q, \quad (25)$$

the Helmholtz free energy of $\mathcal{G}$ is defined by

$$F(E,V) \triangleq \mathbf{e}^{\mathrm{T}}E - \sum_{i=1}^{q}\left(\frac{\partial\mathcal{S}(E,V)}{\partial E_i}\right)^{-1}\mathcal{S}_i(E_i,V_i), \quad (E,V)\in\overline{\mathbb{R}}_+^q\times\mathbb{R}_+^q, \quad (26)$$

and the enthalpy of $\mathcal{G}$ is defined by

$$H(E,V) \triangleq \mathbf{e}^{\mathrm{T}}E + \sum_{i=1}^{q}\left(\frac{\partial\mathcal{S}(E,V)}{\partial E_i}\right)^{-1}\left(\frac{\partial\mathcal{S}(E,V)}{\partial V_i}\right)V_i, \quad (E,V)\in\overline{\mathbb{R}}_+^q\times\mathbb{R}_+^q. \quad (27)$$

Note that the above definitions for the Gibbs free energy, Helmholtz free energy, and enthalpy are consistent with the classical thermodynamic definitions given by $G(E,V) = U + pV - TS$, $F(E,V) = U - TS$, and $H(E,V) = U + pV$, respectively. Furthermore, note that if the interconnected system $\mathcal{G}$ is *isothermal* and *isobaric*, that is, the temperatures of subsystems of $\mathcal{G}$ are equal and remain constant with

$$\left(\frac{\partial\mathcal{S}(E,V)}{\partial E_1}\right)^{-1} = \cdots = \left(\frac{\partial\mathcal{S}(E,V)}{\partial E_q}\right)^{-1} = T > 0, \quad (28)$$

and the pressure $p_i(E,V)$ in each subsystem of $\mathcal{G}$ remains constant, respectively, then any transformation in $\mathcal{G}$ is reversible.
The time derivative of $G(E,V)$ along the trajectories of (14) and (15) is given by

$$\begin{aligned}\dot{G}(E,V) &= \mathbf{e}^{\mathrm{T}}\dot{E} - \sum_{i=1}^{q}\left(\frac{\partial\mathcal{S}(E,V)}{\partial E_i}\right)^{-1}\left[\frac{\partial\mathcal{S}(E,V)}{\partial E_i}\dot{E}_i + \frac{\partial\mathcal{S}(E,V)}{\partial V_i}\dot{V}_i\right]\\ &\quad + \sum_{i=1}^{q}\left(\frac{\partial\mathcal{S}(E,V)}{\partial E_i}\right)^{-1}\left(\frac{\partial\mathcal{S}(E,V)}{\partial V_i}\right)\dot{V}_i\\ &= 0, \end{aligned} \quad (29)$$

which is consistent with classical thermodynamics in the absence of chemical reactions.
For an isothermal interconnected dynamical system $\mathcal{G}$, the time derivative of $F(E,V)$ along the trajectories of (14) and (15) is given by

$$\begin{aligned}\dot{F}(E,V) &= \mathbf{e}^{\mathrm{T}}\dot{E} - \sum_{i=1}^{q}\left(\frac{\partial\mathcal{S}(E,V)}{\partial E_i}\right)^{-1}\left[\frac{\partial\mathcal{S}(E,V)}{\partial E_i}\dot{E}_i + \frac{\partial\mathcal{S}(E,V)}{\partial V_i}\dot{V}_i\right]\\ &= -\sum_{i=1}^{q}\left(\frac{\partial\mathcal{S}(E,V)}{\partial E_i}\right)^{-1}\left(\frac{\partial\mathcal{S}(E,V)}{\partial V_i}\right)\dot{V}_i\\ &= -\sum_{i=1}^{q}(d_{\mathrm{w}i}(E,V) - S_{\mathrm{w}i}(t))\\ &= -L, \end{aligned} \quad (30)$$

where L is the net amount of work done by the subsystems of $\mathcal{G}$ on the environment. Furthermore, note that if, in addition, the interconnected system $\mathcal{G}$ is *isochoric*, that is, the volumes of each of the subsystems of $\mathcal{G}$ remain constant, then $\dot{F}(E,V) = 0$. As we see in the next section, in the presence of chemical reactions the interconnected system $\mathcal{G}$ evolves such that the Helmholtz free energy is minimized.
Finally, for the isolated ($S(t) \equiv 0$ and $d(E,V) \equiv 0$) interconnected dynamical system $\mathcal{G}$, the time derivative of $H(E,V)$ along the trajectories of (14) and (15) is given by

$$\begin{aligned}\dot{H}(E,V) &= \mathbf{e}^{\mathrm{T}}\dot{E} + \sum_{i=1}^{q}\left(\frac{\partial\mathcal{S}(E,V)}{\partial E_i}\right)^{-1}\left(\frac{\partial\mathcal{S}(E,V)}{\partial V_i}\right)\dot{V}_i \\ &= \mathbf{e}^{\mathrm{T}}\dot{E} + \sum_{i=1}^{q}(d_{\mathrm{w}i}(E,V) - S_{\mathrm{w}i}(t)) \\ &= \mathbf{e}^{\mathrm{T}}w(E,V) \\ &= 0. \end{aligned} \tag{31}$$

5. Chemical equilibria, entropy production, and chemical thermodynamics

In its most general form thermodynamics can also involve reacting mixtures and combustion. When a chemical reaction occurs, the bonds within molecules of the *reactant* are broken, and atoms and electrons rearrange to form *products*. The thermodynamic analysis of reactive systems can be addressed as an extension of the compartmental thermodynamic model described in Sections 3 and 4. Specifically, in this case the compartments would qualitatively represent different quantities in the same space, and the intercompartmental flows would represent transformation rates in addition to transfer rates. In particular, the compartments would additionally represent quantities of different chemical substances contained within the compartment, and the compartmental flows would additionally characterize transformation rates of reactants into products. In this case, an additional mass balance equation is included for addressing conservation of energy as well as conservation of mass. This additional mass conservation equation would involve the law of mass-action enforcing proportionality between a particular reaction rate and the concentrations of the reactants, and the law of superposition of elementary reactions assuring that the resultant rates for a particular species is the sum of the elementary reaction rates for the species.
In this section, we consider the interconnected dynamical system $\mathcal{G}$ where each subsystem represents a substance or species that can exchange energy with other substances as well as undergo chemical reactions with other substances forming products. Thus, the reactants and products of chemical reactions represent subsystems of $\mathcal{G}$ with the mechanisms of heat exchange between subsystems remaining the same as delineated in Section 3. Here, for simplicity of exposition, we do not consider work done by the subsystem on the environment nor work done by the environment on the system. This extension can be easily addressed using the formulation in Section 4.
To develop a dynamical systems framework for thermodynamics with chemical reaction networks, let q be the total number of species (i.e., reactants and products), that is, the number of subsystems in $\mathcal{G}$, and let X_j, $j = 1,\ldots,q$, denote the jth species. Consider a single chemical reaction described by

$$\sum_{j=1}^{q} A_j X_j \xrightarrow{k} \sum_{j=1}^{q} B_j X_j, \tag{32}$$

where A_j, B_j, $j = 1,\ldots,q$, are the *stoichiometric coefficients* and k denotes the *reaction rate*. Note that the values of A_j corresponding to the products and the values of B_j corresponding to the reactants are zero. For example, for the familiar reaction

$$2\mathrm{H}_2 + \mathrm{O}_2 \xrightarrow{k} 2\mathrm{H}_2\mathrm{O}, \tag{33}$$

X_1, X_2, and X_3 denote the species H_2, O_2, and H_2O, respectively, and $A_1 = 2$, $A_2 = 1$, $A_3 = 0$, $B_1 = 0$, $B_2 = 0$, and $B_3 = 2$.

In general, for a reaction network consisting of $r \geq 1$ reactions, the ith reaction is written as

$$\sum_{j=1}^{q} A_{ij} X_j \xrightarrow{k_i} \sum_{j=1}^{q} B_{ij} X_j, \quad i = 1,\ldots,r, \tag{34}$$

where, for $i = 1,\ldots,r$, $k_i > 0$ is the reaction rate of the ith reaction, $\sum_{j=1}^{q} A_{ij} X_j$ is the reactant of the ith reaction, and $\sum_{j=1}^{q} B_{ij} X_j$ is the product of the ith reaction. Each stoichiometric coefficient A_{ij} and B_{ij} is a nonnegative integer. Note that each reaction in the reaction network (34) is represented as being irreversible.[2] Reversible reactions can be modeled by including the reverse reaction as a separate reaction. The reaction network (34) can be written compactly in matrix-vector form as

$$AX \xrightarrow{k} BX, \tag{35}$$

where $X = [X_1,\ldots,X_q]^{\mathrm{T}}$ is a column vector of species, $k = [k_1,\ldots,k_r]^{\mathrm{T}} \in \mathbb{R}_+^r$ is a positive vector of reaction rates, and $A \in \mathbb{R}^{r\times q}$ and $B \in \mathbb{R}^{r\times q}$ are nonnegative matrices such that $A_{(i,j)} = A_{ij}$ and $B_{(i,j)} = B_{ij}$, $i = 1,\ldots,r$, $j = 1,\ldots,q$.

Let $n_j : [0,\infty) \to \overline{\mathbb{R}}_+$, $j = 1,\ldots,q$, denote the *mole number* of the jth species and define $n \triangleq [n_1,\ldots,n_q]^{\mathrm{T}}$. Invoking the *law of mass-action* (Steinfeld et al. (1989)), which states that, for an *elementary reaction*, that is, a reaction in which all of the stoichiometric coefficients of the reactants are one, the rate of reaction is proportional to the product of the concentrations of the reactants, the species quantities change according to the dynamics (Haddad et al. (2010); Chellaboina et al. (2009))

$$\dot{n}(t) = (B - A)^{\mathrm{T}} K n^A(t), \quad n(0) = n_0, \quad t \geq t_0, \tag{36}$$

where $K \triangleq \mathrm{diag}[k_1,\ldots,k_r] \in \mathbb{P}^r$ and

$$n^A \triangleq \begin{bmatrix} \prod_{j=1}^{q} n_j^{A_{1j}} \\ \vdots \\ \prod_{j=1}^{q} n_j^{A_{rj}} \end{bmatrix} = \begin{bmatrix} n_1^{A_{11}} \cdots n_q^{A_{1q}} \\ \vdots \\ n_1^{A_{r1}} \cdots n_q^{A_{rq}} \end{bmatrix} \in \overline{\mathbb{R}}_+^r. \tag{37}$$

For details regarding the law of mass-action and Equation (36), see Erdi & Toth (1988); Haddad et al. (2010); Steinfeld et al. (1989); Chellaboina et al. (2009). Furthermore, let $M_j > 0$,

[2] *Irreversibility* here refers to the fact that part of the chemical reaction involves generation of products from the original reactants. Reversible chemical reactions that involve generation of products from the reactants and vice versa can be modeled as two irreversible reactions; one of which involves generation of products from the reactants and the other involving generation of the original reactants from the products.

$j = 1,\ldots,q$, denote the *molar mass* (i.e., the mass of one mole of a substance) of the jth species, let $m_j : [0,\infty) \to \overline{\mathbb{R}}_+$, $j = 1,\ldots,q$, denote the mass of the jth species so that $m_j(t) = M_j n_j(t)$, $t \geq t_0$, $j = 1,\ldots,q$, and let $m \triangleq [m_1,\ldots,m_q]^{\mathrm{T}}$. Then, using the transformation $m(t) = Mn(t)$, where $M \triangleq \mathrm{diag}[M_1,\ldots,M_q] \in \mathbb{P}^q$, (36) can be rewritten as the *mass balance equation*

$$\dot{m}(t) = M(B-A)^{\mathrm{T}}\tilde{K}m^A(t), \quad m(0) = m_0, \quad t \geq t_0, \tag{38}$$

where $\tilde{K} \triangleq \mathrm{diag}\left[\frac{k_1}{\prod_{j=1}^q M_j^{A_{1j}}},\ldots,\frac{k_r}{\prod_{j=1}^q M_j^{A_{rj}}}\right] \in \mathbb{P}^r$.

In the absence of nuclear reactions, the total mass of the species during each reaction in (35) is conserved. Specifically, consider the ith reaction in (35) given by (34) where the mass of the reactants is $\sum_{j=1}^q A_{ij}M_j$ and the mass of the products is $\sum_{j=1}^q B_{ij}M_j$. Hence, conservation of mass in the ith reaction is characterized as

$$\sum_{j=1}^{q}(B_{ij} - A_{ij})M_j = 0, \quad i = 1,\ldots,r, \tag{39}$$

or, in general for (35), as

$$\mathbf{e}^{\mathrm{T}}M(B-A)^{\mathrm{T}} = 0. \tag{40}$$

Note that it follows from (38) and (40) that $\mathbf{e}^{\mathrm{T}}\dot{m}(t) \equiv 0$.

Equation (38) characterizes the change in masses of substances in the interconnected dynamical system $\mathcal{G}$ due to chemical reactions. In addition to the change of mass due to chemical reactions, each substance can exchange energy with other substances according to the energy flow mechanism described in Section 3; that is, energy flows from substances at a higher temperature to substances at a lower temperature. Furthermore, in the presence of chemical reactions, the exchange of matter affects the change of energy of each substance through the quantity known as the *chemical potential*.

The notion of the chemical potential was introduced by Gibbs in 1875–1878 (Gibbs (1875; 1878)) and goes far beyond the scope of chemistry effecting virtually every process in nature (Baierlein (2001); Fuchs (1996); Job & Herrmann (2006)). The chemical potential has a strong connection with the second law of thermodynamics in that *every process in nature evolves from a state of higher chemical potential towards a state of lower chemical potential*. It was postulated by Gibbs (1875; 1878) that the change in energy of a homogeneous substance is proportional to the change in mass of this substance with the coefficient of proportionality given by the chemical potential of the substance.

To elucidate this, assume the jth substance corresponds to the jth compartment and consider the rate of energy change of the jth substance of $\mathcal{G}$ in the presence of matter exchange. In this case, it follows from (5) and Gibbs' postulate that the rate of energy change of the jth substance is given by

$$\dot{E}_j(t) = \left[\sum_{k=1,k\neq j}^{q}\phi_{jk}(E(t))\right] - \sigma_{jj}(E(t)) + S_j(t) + \mu_j(E(t),m(t))\dot{m}_j(t), \quad E_j(t_0) = E_{j0}, \quad t \geq t_0, \tag{41}$$

where $\mu_j : \overline{\mathbb{R}}_+^q \times \overline{\mathbb{R}}_+^q \to \mathbb{R}$, $j = 1,\ldots,q$, is the chemical potential of the jth substance. It follows from (41) that $\mu_j(\cdot,\cdot)$ is the chemical potential of a unit mass of the jth substance. We assume that if $E_j = 0$, then $\mu_j(E,m) = 0$, $j = 1,\ldots,q$, which implies that if the energy of the jth substance is zero, then its chemical potential is also zero.

Next, using (38) and (41), the energy and mass balance equations for the interconnected dynamical system $\mathcal{G}$ can be written as

$$\dot{E}(t) = w(E(t)) + P(E(t),m(t))M(B-A)^{\mathrm{T}}\tilde{K}m^{A}(t) - d(E(t)) + S(t), \quad E(t_0) = E_0, \quad t \geq t_0, \tag{42}$$

$$\dot{m}(t) = M(B-A)^{\mathrm{T}}\tilde{K}m^{A}(t), \quad m(0) = m_0, \tag{43}$$

where $P(E,m) \triangleq \mathrm{diag}[\mu_1(E,m),\ldots,\mu_q(E,m)] \in \mathbb{R}^{q\times q}$ and where $w(\cdot)$, $d(\cdot)$, and $S(\cdot)$ are defined as in Section 3. It follows from Proposition 1 of Chellaboina et al. (2009) that the dynamics of (43) are essentially nonnegative and, since $\mu_j(E,m) = 0$ if $E_j = 0$, $j = 1,\ldots,q$, it also follows that, for the isolated dynamical system $\mathcal{G}$ (i.e., $S(t) \equiv 0$ and $d(E) \equiv 0$), the dynamics of (42) and (43) are essentially nonnegative.

Note that, for the ith reaction in the reaction network (35), the chemical potentials of the reactants and the products are $\sum_{j=1}^{q} A_{ij}M_j\mu_j(E,m)$ and $\sum_{j=1}^{q} B_{ij}M_j\mu_j(E,m)$, respectively. Thus,

$$\sum_{j=1}^{q} B_{ij}M_j\mu_j(E,m) - \sum_{j=1}^{q} A_{ij}M_j\mu_j(E,m) \leq 0, \quad (E,m) \in \overline{\mathbb{R}}_+^q \times \overline{\mathbb{R}}_+^q, \tag{44}$$

is a restatement of the principle that a chemical reaction evolves from a state of a greater chemical potential to that of a lower chemical potential, which is consistent with the second law of thermodynamics. The difference between the chemical potential of the reactants and the chemical potential of the products is called *affinity* (DeDonder (1927); DeDonder & Rysselberghe (1936)) and is given by

$$\nu_i(E,m) = \sum_{j=1}^{q} A_{ij}M_j\mu_j(E,m) - \sum_{j=1}^{q} B_{ij}M_j\mu_j(E,m) \geq 0, \quad i = 1,\ldots,r. \tag{45}$$

Affinity is a driving force for chemical reactions and is equal to zero at the state of *chemical equilibrium*. A nonzero affinity implies that the system in not in equilibrium and that chemical reactions will continue to occur until the system reaches an equilibrium characterized by zero affinity. The next assumption provides a general form for the inequalities (44) and (45).

Assumption 5.1 *For the chemical reaction network (35) with the mass balance equation (43), assume that $\mu(E,m) >> 0$ for all $E \neq 0$ and*

$$(B-A)M\mu(E,m) \leq\leq 0, \quad (E,m) \in \overline{\mathbb{R}}_+^q \times \overline{\mathbb{R}}_+^q, \tag{46}$$

or, equivalently,

$$\nu(E,m) = (A-B)M\mu(E,m) \geq\geq 0, \quad (E,m) \in \overline{\mathbb{R}}_+^q \times \overline{\mathbb{R}}_+^q, \tag{47}$$

where $\mu(E,m) \triangleq [\mu_1(E,m),\ldots,\mu_q(E,m)]^{\mathrm{T}}$ is the vector of chemical potentials of the substances of $\mathcal{G}$ and $\nu(E,m) \triangleq [\nu_1(E,m),\ldots,\nu_r(E,m)]^{\mathrm{T}}$ is the affinity vector for the reaction network (35).

Note that equality in (46) or, equivalently, in (47) characterizes the state of chemical equilibrium when the chemical potentials of the products and reactants are equal or, equivalently, when the affinity of each reaction is equal to zero. In this case, no reaction occurs and $\dot{m}(t) = 0$, $t \geq t_0$.

Next, we characterize the entropy function for the interconnected dynamical system $\mathcal{G}$ with the energy and mass balance equations (42) and (43). The definition of entropy for $\mathcal{G}$ in the presence of chemical reactions remains the same as in Definition 3.1 with $\mathcal{S}(E)$ replaced by $\mathcal{S}(E,m)$ and with all other conditions in the definition holding for every $m >> 0$. Consider the jth subsystem of $\mathcal{G}$ and assume that E_k and m_k, $k \neq j$, $k = 1,\ldots,q$, are constant. In this case, note that

$$\frac{\mathrm{d}\mathcal{S}}{\mathrm{d}t} = \frac{\partial\mathcal{S}}{\partial E_j}\frac{\mathrm{d}E_j}{\mathrm{d}t} + \frac{\partial\mathcal{S}}{\partial m_j}\frac{\mathrm{d}m_j}{\mathrm{d}t} \tag{48}$$

and recall that

$$\frac{\partial\mathcal{S}}{\partial E}P(E,m) + \frac{\partial\mathcal{S}}{\partial m} = 0. \tag{49}$$

Next, it follows from (49) that the time derivative of the entropy function $\mathcal{S}(E,m)$ along the trajectories of (42) and (43) is given by

$$\begin{aligned}
\dot{\mathcal{S}}(E,m) &= \frac{\partial\mathcal{S}(E,m)}{\partial E}\dot{E} + \frac{\partial\mathcal{S}(E,m)}{\partial m}\dot{m} \\
&= \frac{\partial\mathcal{S}(E,m)}{\partial E}w(E) + \left(\frac{\partial\mathcal{S}(E,m)}{\partial E}P(E,m) + \frac{\partial\mathcal{S}(E,m)}{\partial m}\right)M(B-A)^{\mathrm{T}}\tilde{K}m^{A} \\
&\quad + \frac{\partial\mathcal{S}(E,m)}{\partial E}S(t) - \frac{\partial\mathcal{S}(E,m)}{\partial E}d(E) \\
&= \frac{\partial\mathcal{S}(E,m)}{\partial E}w(E) + \frac{\partial\mathcal{S}(E,m)}{\partial E}S(t) - \frac{\partial\mathcal{S}(E,m)}{\partial E}d(E) \\
&= \sum_{i=1}^{q}\sum_{j=i+1}^{q}\left(\frac{\partial\mathcal{S}(E,m)}{\partial E_i} - \frac{\partial\mathcal{S}(E,m)}{\partial E_j}\right)\phi_{ij}(E) + \frac{\partial\mathcal{S}(E,m)}{\partial E}S(t) - \frac{\partial\mathcal{S}(E,m)}{\partial E}d(E), \\
&\qquad (E,m) \in \overline{\mathbb{R}}_+^q \times \overline{\mathbb{R}}_+^q.
\end{aligned} \tag{50}$$

For the isolated system $\mathcal{G}$ (i.e., $S(t) \equiv 0$ and $d(E) \equiv 0$), the entropy function of $\mathcal{G}$ is a nondecreasing function of time and, using identical arguments as in the proof of Theorem 3.1, it can be shown that $(E(t), m(t)) \to \mathcal{R} \triangleq \left\{(E,m) \in \overline{\mathbb{R}}_+^q \times \overline{\mathbb{R}}_+^q : \frac{\partial\mathcal{S}(E,m)}{\partial E_1} = \cdots = \frac{\partial\mathcal{S}(E,m)}{\partial E_q}\right\}$ as $t \to \infty$ for all $(E_0, m_0) \in \overline{\mathbb{R}}_+^q \times \overline{\mathbb{R}}_+^q$.

The entropy production in the interconnected system $\mathcal{G}$ due to chemical reactions is given by

$$\begin{aligned}
\mathrm{d}\mathcal{S}_{\mathrm{i}}(E,m) &= \frac{\partial\mathcal{S}(E,m)}{\partial m}\mathrm{d}m \\
&= -\frac{\partial\mathcal{S}(E,m)}{\partial E}P(E,m)M(B-A)^{\mathrm{T}}\tilde{K}m^{A}\mathrm{d}t, \quad (E,m) \in \overline{\mathbb{R}}_+^q \times \overline{\mathbb{R}}_+^q.
\end{aligned} \tag{51}$$

If the interconnected dynamical system $\mathcal{G}$ is isothermal, that is, all subsystems of $\mathcal{G}$ are at the same temperature

$$\left(\frac{\partial\mathcal{S}(E,m)}{\partial E_1}\right)^{-1} = \cdots = \left(\frac{\partial\mathcal{S}(E,m)}{\partial E_q}\right)^{-1} = T, \tag{52}$$

where $T > 0$ is the system temperature, then it follows from Assumption 5.1 that

$$\begin{aligned} \mathrm{d}\mathcal{S}_{\mathrm{i}}(E,m) &= -\frac{1}{T}\mathbf{e}^{\mathrm{T}}P(E,m)M(B-A)^{\mathrm{T}}\tilde{K}m^{A}\mathrm{d}t \\ &= -\frac{1}{T}\mu^{\mathrm{T}}(E,m)M(B-A)^{\mathrm{T}}\tilde{K}m^{A}\mathrm{d}t \\ &= \frac{1}{T}\nu^{\mathrm{T}}(E,m)\tilde{K}m^{A}\mathrm{d}t \\ &\geq 0, \quad (E,m)\in\overline{\mathbb{R}}_{+}^{q}\times\overline{\mathbb{R}}_{+}^{q}. \end{aligned} \tag{53}$$

Note that since the affinity of a reaction is equal to zero at the state of a chemical equilibrium, it follows that equality in (53) holds if and only if $\nu(E,m) = 0$ for some $E \in \overline{\mathbb{R}}_{+}^{q}$ and $m \in \overline{\mathbb{R}}_{+}^{q}$.

Theorem 5.1 *Consider the isolated (i.e., $S(t) \equiv 0$ and $d(E) \equiv 0$) interconnected dynamical system $\mathcal{G}$ with the power and mass balance equations (42) and (43). Assume that rank $\mathcal{C} = q-1$, Assumption 5.1 holds, and there exists an entropy function $\mathcal{S} : \overline{\mathbb{R}}_{+}^{q} \times \overline{\mathbb{R}}_{+}^{q} \to \mathbb{R}$ of $\mathcal{G}$. Then $(E(t),m(t)) \to \mathcal{R}$ as $t \to \infty$, where $(E(t),m(t))$, $t \geq t_0$, is the solution to (42) and (43) with the initial condition $(E_0,m_0) \in \overline{\mathbb{R}}_{+}^{q} \times \overline{\mathbb{R}}_{+}^{q}$ and*

$$\mathcal{R} = \left\{(E,m)\in\overline{\mathbb{R}}_{+}^{q}\times\overline{\mathbb{R}}_{+}^{q} : \frac{\partial\mathcal{S}(E,m)}{\partial E_1} = \cdots = \frac{\partial\mathcal{S}(E,m)}{\partial E_q} \text{ and } \nu(E,m)=0\right\}, \tag{54}$$

where $\nu(\cdot,\cdot)$ is the affinity vector of $\mathcal{G}$.

Proof. Since the dynamics of the isolated system $\mathcal{G}$ are essentially nonnegative, it follows from Proposition 2.1 that $(E(t),m(t)) \in \overline{\mathbb{R}}_{+}^{q} \times \overline{\mathbb{R}}_{+}^{q}$, $t \geq t_0$, for all $(E_0,m_0) \in \overline{\mathbb{R}}_{+}^{q} \times \overline{\mathbb{R}}_{+}^{q}$. Consider a scalar function $v(E,m) = \mathbf{e}^{\mathrm{T}}E + \mathbf{e}^{\mathrm{T}}m$, $(E,m) \in \overline{\mathbb{R}}_{+}^{q} \times \overline{\mathbb{R}}_{+}^{q}$, and note that $v(0,0) = 0$ and $v(E,m) > 0$, $(E,m) \in \overline{\mathbb{R}}_{+}^{q} \times \overline{\mathbb{R}}_{+}^{q}$, $(E,m) \neq (0,0)$. It follows from (40), Assumption 5.1, and $\mathbf{e}^{\mathrm{T}}w(E) \equiv 0$ that the time derivative of $v(\cdot,\cdot)$ along the trajectories of (42) and (43) satisfies

$$\begin{aligned} \dot{v}(E,m) &= \mathbf{e}^{\mathrm{T}}\dot{E} + \mathbf{e}^{\mathrm{T}}\dot{m} \\ &= \mathbf{e}^{\mathrm{T}}P(E,m)M(B-A)^{\mathrm{T}}\tilde{K}m^{A} \\ &= \mu^{\mathrm{T}}(E,m)M(B-A)^{\mathrm{T}}\tilde{K}m^{A} \\ &= -\nu^{\mathrm{T}}(E,m)\tilde{K}m^{A} \\ &\leq 0, \quad (E,m)\in\overline{\mathbb{R}}_{+}^{q}\times\overline{\mathbb{R}}_{+}^{q}, \end{aligned} \tag{55}$$

which implies that the solution $(E(t),m(t))$, $t \geq t_0$, to (42) and (43) is bounded for all initial conditions $(E_0,m_0) \in \overline{\mathbb{R}}_{+}^{q} \times \overline{\mathbb{R}}_{+}^{q}$.

Next, consider the function $\tilde{v}(E,m) = \mathbf{e}^{\mathrm{T}}E + \mathbf{e}^{\mathrm{T}}m - \mathcal{S}(E,m)$, $(E,m) \in \overline{\mathbb{R}}_{+}^{q} \times \overline{\mathbb{R}}_{+}^{q}$. Then it follows from (50) and (55) that the time derivative of $\tilde{v}(\cdot,\cdot)$ along the trajectories of (42) and (43) satisfies

$$\begin{aligned} \dot{\tilde{v}}(E,m) &= \mathbf{e}^{\mathrm{T}}\dot{E} + \mathbf{e}^{\mathrm{T}}\dot{m} - \dot{\mathcal{S}}(E,m) \\ &= -\nu^{\mathrm{T}}(E,m)\tilde{K}m^{A} - \sum_{i=1}^{q}\sum_{j=i+1}^{q}\left(\frac{\partial\mathcal{S}(E,m)}{\partial E_i} - \frac{\partial\mathcal{S}(E,m)}{\partial E_j}\right)\phi_{ij}(E) \\ &\leq 0, \quad (E,m)\in\overline{\mathbb{R}}_{+}^{q}\times\overline{\mathbb{R}}_{+}^{q}, \end{aligned} \tag{56}$$

which implies that $\tilde{v}(\cdot,\cdot)$ is a nonincreasing function of time, and hence, by the Krasovskii-LaSalle theorem (Haddad & Chellaboina (2008)), $(E(t),m(t)) \to \mathcal{R} \triangleq \{(E,m) \in \overline{\mathbb{R}}_+^q \times \overline{\mathbb{R}}_+^q : \dot{\tilde{v}}(E,m) = 0\}$ as $t \to \infty$. Now, it follows from Definition 3.1, Assumption 5.1, and the fact that $\operatorname{rank} \mathcal{C} = q-1$ that

$$\begin{aligned}\mathcal{R} &= \left\{(E,m) \in \overline{\mathbb{R}}_+^q \times \overline{\mathbb{R}}_+^q : \frac{\partial \mathcal{S}(E,m)}{\partial E_1} = \cdots = \frac{\partial \mathcal{S}(E,m)}{\partial E_q}\right\} \\ &\quad \cap \{(E,m) \in \overline{\mathbb{R}}_+^q \times \overline{\mathbb{R}}_+^q : \nu(E,m) = 0\},\end{aligned} \tag{57}$$

which proves the result. □

Theorem 5.1 implies that the state of the interconnected dynamical system $\mathcal{G}$ converges to the state of thermal and chemical equilibrium when the temperatures of all substances of $\mathcal{G}$ are equal and the masses of all substances reach a state where all reaction affinities are zero corresponding to a halting of all chemical reactions.

Next, we assume that the entropy of the interconnected dynamical system $\mathcal{G}$ is a sum of individual entropies of subsystems of $\mathcal{G}$, that is, $\mathcal{S}(E,m) = \sum_{j=1}^q \mathcal{S}_j(E_j,m_j)$, $(E,m) \in \overline{\mathbb{R}}_+^q \times \overline{\mathbb{R}}_+^q$. In this case, the Helmholtz free energy of $\mathcal{G}$ is given by

$$F(E,m) = \mathbf{e}^{\mathrm{T}}E - \sum_{j=1}^q \left(\frac{\partial \mathcal{S}(E,m)}{\partial E_j}\right)^{-1} \mathcal{S}_j(E_j,m_j), \quad (E,m) \in \overline{\mathbb{R}}_+^q \times \overline{\mathbb{R}}_+^q. \tag{58}$$

If the interconnected dynamical system $\mathcal{G}$ is isothermal, then the derivative of $F(\cdot,\cdot)$ along the trajectories of (42) and (43) is given by

$$\begin{aligned}\dot{F}(E,m) &= \mathbf{e}^{\mathrm{T}}\dot{E} - \sum_{j=1}^q \left(\frac{\partial \mathcal{S}(E,m)}{\partial E_j}\right)^{-1} \dot{\mathcal{S}}_j(E_j,m_j) \\ &= \mathbf{e}^{\mathrm{T}}\dot{E} - \sum_{j=1}^q \left(\frac{\partial \mathcal{S}(E,m)}{\partial E_j}\right)^{-1} \left[\frac{\partial \mathcal{S}_j(E_j,m_j)}{\partial E_j}\dot{E}_j + \frac{\partial \mathcal{S}_j(E_j,m_j)}{\partial m_j}\dot{m}_j\right] \\ &= \mu^{\mathrm{T}}(E,m)M(B-A)^{\mathrm{T}}\tilde{K}m^A \\ &= -\nu^{\mathrm{T}}(E,m)\tilde{K}m^A \\ &\le 0, \quad (E,m) \in \overline{\mathbb{R}}_+^q \times \overline{\mathbb{R}}_+^q,\end{aligned} \tag{59}$$

with equality in (59) holding if and only if $\nu(E,m) = 0$ for some $E \in \overline{\mathbb{R}}_+^q$ and $m \in \overline{\mathbb{R}}_+^q$, which determines the state of chemical equilibrium. Hence, the Helmholtz free energy of $\mathcal{G}$ evolves to a minimum when the pressure and temperature of each subsystem of $\mathcal{G}$ are maintained constant, which is consistent with classical thermodynamics. A similar conclusion can be arrived at for the Gibbs free energy if work energy considerations to and by the system are addressed. Thus, the Gibbs and Helmholtz free energies are a measure of the tendency for a reaction to take place in the interconnected system $\mathcal{G}$, and hence, provide a measure of the work done by the interconnected system $\mathcal{G}$.

6. Conclusion

In this paper, we developed a system-theoretic perspective for classical thermodynamics and chemical reaction processes. In particular, we developed a nonlinear compartmental model involving heat flow, work energy, and chemical reactions that captures all of the key aspects of thermodynamics, including its fundamental laws. In addition, we showed that the interconnected compartmental model gives rise to globally semistable equilibria involving states of temperature equipartition. Finally, using the notion of the chemical potential, we combined our heat flow compartmental model with a state space mass-action kinetics model to capture energy and mass exchange in interconnected large-scale systems in the presence of chemical reactions. In this case, it was shown that the system states converge to a state of temperature equipartition and zero affinity.

7. References

Arnold, V. (1990). Contact geometry: The geometrical method of Gibbs' thermodynamics, *in* D. Caldi & G. Mostow (eds), *Proceedings of the Gibbs Symposium*, American Mathematical Society, Providence, RI, pp. 163–179.

Baierlein, R. (2001). The elusive chemical potential, *Amer. J. Phys.* 69(4): 423–434.

Chellaboina, V., Bhat, S. P., Haddad, W. M. & Bernstein, D. S. (2009). Modeling and analysis of mass action kinetics: Nonnegativity, realizability, reducibility, and semistability, Contr. Syst. Mag. 29(4): 60–78.

DeDonder, T. (1927). *L'Affinité*, Gauthiers-Villars, Paris.

DeDonder, T. & Rysselberghe, P. V. (1936). *Affinity*, Stanford University Press, Menlo Park, CA.

Diestel, R. (1997). *Graph Theory*, Springer-Verlag, New York, NY.

Erdi, P. & Toth, J. (1988). *Mathematical Models of Chemical Reactions: Theory and Applications of Deterministic and Stochastic Models*, Princeton University Press, Princeton, NJ.

Fuchs, H. U. (1996). *The Dynamics of Heat*, Springer-Verlag, New York, NY.

Gibbs, J. W. (1875). On the equilibrium of heterogeneous substances, *Tras. Conn. Acad. Sci.* III: 108–248.

Gibbs, J. W. (1878). On the equilibrium of heterogeneous substances, *Trans. Conn. Acad. Sci.* III: 343–524.

Godsil, C. & Royle, G. (2001). *Algebraic Graph Theory*, Springer-Verlag, New York.

Haddad, W. M. & Chellaboina, V. (2005). Stability and dissipativity theory for nonnegative dynamical systems: A unified analysis framework for biological and physiological systems, *Nonlinear Analysis: Real World Applications* 6: 35–65.

Haddad, W. M. & Chellaboina, V. (2008). *Nonlinear Dynamical Systems and Control. A Lyapunov-Based Approach*, Princeton University Press, Princeton, NJ.

Haddad, W. M., Chellaboina, V. & Hui, Q. (2010). *Nonnegative and Compartmental Dynamical Systems*, Princeton University Press, Princeton, NJ.

Haddad, W. M., Chellaboina, V. & Nersesov, S. G. (2005). *Thermodynamics. A Dynamical Systems Approach*, Princeton University Press, Princeton, NJ.

Haddad, W. M., Chellaboina, V. & Nersesov, S. G. (2008). Time-reversal symmetry, Poincaré recurrence, irreversibility, and the entropic arrow of time: From mechanics to system thermodynamics, *Nonlinear Analysis: Real World Applications* 9: 250–271.

Hartman, P. (1982). *Ordinary Differential Equations*, Birkhäuser, Boston.

Job, G. & Herrmann, F. (2006). Chemical potential – a quantity in search of recognition, *Eur. J. Phys.* 27: 353–371.

Steinfeld, J. I., Francisco, J. S. & Hase, W. L. (1989). *Chemical Kinetics and Dynamics*, Prentice-Hall, Upper Saddle River, NJ.

Truesdell, C. (1969). *Rational Thermodynamics*, McGraw-Hill, New York, NY.

Truesdell, C. (1980). *The Tragicomical History of Thermodynamics 1822-1854*, Springer-Verlag, New York, NY.

4

Modern Stochastic Thermodynamics

A. D. Sukhanov and O. N. Golubjeva
Joint Institute for Nuclear Research
Russia

For our beloved son Eugene

1. Introduction
Limitations of thermodynamics based on the quantum statistical mechanics

An increased interest in using equilibrium thermodynamics as an independent macrotheory can be observed in recent years. From a fundamental standpoint, thermodynamics gives an universal macrodescription of nature in which using specific micromodels of objects is unnecessary. From a pragmatic standpoint, there is obviously a demand for using thermodynamics both to describe the behavior of relatively small objects (nanoparticles, etc.) at low temperatures and to study high-energy physics (including the quark–gluon plasma).
As is well known, phenomenological thermodynamics is based on four laws. Among them, the zero law is basic. It relates the fundamental idea of thermal equilibrium of an object to its environment, called a *heat bath*. In this theory, in which all macroparameters are exactly defined, the zero law is a strict condition determining the concept of temperature:

$$T \equiv T_0, \tag{1}$$

where T is the object temperature and T_0 is the heat bath temperature.
In the same time there exists also statistical thermodynamics (ST). In its nonquantum version founders of which were Gibbs and Einstein {LaLi68},{Su05} all macroparameters are considered random values fluctuating about their means. It is assumed here that the concept of thermal equilibrium is preserved, but its content is generalized. It is now admitted that the object temperature experiences also fluctuations δT because of the thermal stochastic influence of the heat bath characterized by the Boltzmann constant k_B. As a result, the zero law takes the form of a soft condition, namely,

$$T = T_0 \pm \delta T = \langle T \rangle \pm \delta T. \tag{2}$$

Here the average object temperature $\langle T \rangle$ coincides with T_0 and $\langle (\delta T)^2 \rangle \equiv (\Delta T)^2$ has the meaning of the object temperature dispersion.
To preserve the thermodynamic character of this description, it is simultaneously assumed that the values of the dispersion of any macroparameter A_i is bounded by the condition $(\Delta A_i)^2 / \langle A_i \rangle^2 \leq 1$. This means that for the dispersion $(\Delta T)^2$ there is the requirement

$$\frac{(\Delta T)^2}{T_0^2} \leq 1. \tag{3}$$

In other words, the zero law of the nonquantum version of statistical thermodynamics is not just one condition (2) but the set of conditions (2) and (3).
We stress that nonquantum version of statistical thermodynamics (see chap. 12 in {LaLi68}) absolutely does not take the quantum stochastic influence characterized by the Planck constant $\hbar$ into account. At the same time, it is well known from quantum dynamics that the characteristics of an object can experience purely quantum fluctuations when there are no thermal effects. In the general case, both quantum and thermal types of environment stochastic influences determining macroparameters and their fluctuations are simultaneously observed in experiments. In this regard, it is necessary to develop a theory such that the approaches of quantum mechanics and nonquantum version of statistical thermodynamics can be combined.
Today, there exists a sufficiently widespread opinion that thermodynamics based on quantum statistical mechanics (QSM-based thermodynamics) has long played the role of such a theory quite effectively. But this theoretical model is probably inadequate for solving a number of new problems. In our opinion, this is due to the following significant factors.
First, QSM-based thermodynamics is not a consistent quantum theory because it plays the role of a quasiclassical approximation in which the nonzero energy of the ground state is not taken into account. Second, the theory is not a consistent statistical theory because it does not initially contain fluctuations of intensive macroparameters (primarily, of temperature). However, the temperature fluctuations in low-temperature experiments are sufficiently noticeable for small objects, including nanoparticles and also for critical phenomena. Third, the assertion that the minimal entropy is zero in it, is currently very doubtful. Fourth, in this theory, the expression $\Theta = k_B T$ is used as a modulus of the distribution for any objects at any temperature. This corresponds to choosing the classical model of the heat bath {Bog67} as a set of weakly coupled classical oscillators. Then a microobject with quantized energy is placed in it. Thus, quantum and thermal influences are considered as additive. Fifth, in this theory at enough low temperatures the condition (3) is invalid for relative fluctuations of temperature. As a result, in QSM-based thermodynamics, it is possible to calculate the means of the majority of extensive macroparameters with the account of quantum stochastic influence. However, using the corresponding apparatus to calculate fluctuations of the same macroparameters leads to the violation of condition that is analogical one (3). This means that full value statistical thermodynamics as a macrotheory cannot be based completely on QSM as a microtheory.
To obtain a consistent quantum-thermal description of natural objects, or modern stochastic thermodynamics (MST), in our opinion, it is possible to use two approaches. Nevertheless, they are both based on one general idea, namely, replacing the classical model of the heat bath with an adequate quantum model, or a *quantum heat bath* (QHB) {Su99}.
The first of these approaches is described in the Sect. 1 {Su08}. We modify the macrodescription of objects in the heat bath by taking quantum effects into account in the framework of nonquantum version of statistical thermodynamics with an inclusion of temperature fluctuations but without using the operator formalism. In this case, based on intuitive considerations, we obtain a theory of effective macroparameters (TEM) as a macrotheory.
In the Sect. 2 we modify standard quantum mechanics taking thermal effects into account {SuGo09}. As a result, we formulate a quantum-thermal dynamics or, briefly, $(\hbar, k)$-dynamics ($\hbar kD$) as a microtheory. The principal distinction from QSM is that in such a theory, the state of a microobject under the conditions of contact with the QHB is generally described not by

the density matrix but by a temperature-dependent complex wave function.
In the Sect. 3 we overcome the main paradox appearing in QSM-based thermodynamics at calculation of macroparameters fluctuations. It is that at account of quantum effects its results fall outside the scope of the thermodynamics. We develop the theory of the effective macroparameters fluctuations (TEMF) combining TEM and $\hbar kD$. We also investigate effective macroparameters obeying the uncertainties relations (URs) and offer a criterion for the choice of conjugate quantities.

2. Theory of effective macroparameters as a macroscopic ground of modern stochastic thermodynamics

At first we construct MST in the form of a macrotheory or TEM. That is a generalization of nonquantum version of statistical thermodynamics. The development of this theory is based on a main MST postulate reduced to statements:

A. Stochastic influences of quantum and thermal types are realized by an environment to which the QHB model is assigned.

B. The state of thermal equilibrium between the object and the QHB is described by an effective temperature.

C. The physical characteristics of objects of any complexity at any temperature are described by effective macroparameters to which random *c*-number quantities are assigned.

D. The main thermodynamic relations are formulated for the corresponding effective macroparameters; moreover, their standard forms are preserved, including zero law (2)-(3).

2.1 Effective temperature

We note that by changing the form of the zero law from (1) to (2) - (3), we take into account that the object temperature can fluctuate. Therefore, the only possibility (probably still remaining) is to modify the model of the heat bath, which is a source of stochastic influences, by organically including a quantum-type influence in it.
Because an explicit attempt to modify the heat bath model is made by as for the first time, it is useful first to make clear what is tacitly taken for such a model in the nonquantum version of statistical thermodynamics. As follows from Chap. 9 in the Gibbs's monograph {Gi60}, it is based on the canonical distribution

$$dw(\mathcal{E}) = e^{(F-\mathcal{E})/\Theta}\, d\mathcal{E} \tag{4}$$

in the macroparameters space [1]. The object energy $\mathcal{E} = \mathcal{E}(V,T)$ in it is a random quantity whose fluctuations (for $V = \text{const}$) depend on object temperature fluctuations according to zero law (2)-(3); F is the free energy determined by the normalization condition. The distribution modulus

$$\Theta \equiv k_B T_0 \tag{5}$$

has a sense of the energy typical of a definite heat bath model.
Up to now, according to the ideas of Bogoliubov {Bog67}, a heat bath is customarily modeled by an infinite set of normal modes each of which can be treated as an excitation of a chain

[1] We emphasize that distribution (4) is similar to the canonical distribution in classical statistical mechanics (CSM) only in appearance. The energy $\varepsilon = \varepsilon(p,q)$ in the latter distribution is also a random quantity, but its fluctuations depend on the fluctuations of the microparameters p and q at the object temperature defined by the formula (1).

of weakly coupled oscillators. As follows from experiments, the quantity $k_B T_0$ in relatively narrow ranges of frequencies and temperatures has the meaning of the average energy $\langle \varepsilon_{cl} \rangle$ of the classical normal mode. It can therefore be concluded from formulae (4) and (5) that the heat bath model that can be naturally called classical is used in the nonquantum version of statistical thermodynamics. From a modern standpoint, the experimental data in some cases cannot be interpreted using such a model, on which, we stress, QSM is also based.

In what follows, we propose an alternative method for simultaneously including quantum- and thermal-type stochastic influences. According to the main MST postulate, we pass from the classical heat bath model to a more general quantum model, or QHB. As a result, all effects related to both types of environment stochastic influences on the objects can be attributed to the generalized heat bath. However, the thermodynamic language used to describe thermal equilibrium can be preserved, i.e., we can explicitly use no the operator formalism in this language. For this, as the QHB model, we propose to choose the set consisting of an infinite number of quantum normal modes, each with the average energy

$$\langle \varepsilon_{qu} \rangle = \frac{\hbar\omega}{2} + \hbar\omega\left(e^{\hbar\omega/(k_B T_0)} - 1\right)^{-1} = \frac{\hbar\omega}{2}\coth\frac{\hbar\omega}{2k_B T_0} \tag{6}$$

over the entire ranges of frequencies and temperatures, which agrees with experiments. This means that in the QHB, we determine the expression for the distribution modulus Θ by the more general condition $\Theta = \langle \varepsilon_{qu} \rangle$, instead of the condition $\Theta = \langle \varepsilon_{cl} \rangle$ typical of the classical model.

Further, according to the main MST postulate, we propose to write the quantity Θ as

$$\Theta \equiv k_B (T_{ef})_0. \tag{7}$$

It is significant that the introduced quantity

$$(T_{ef})_0 \equiv \frac{\langle \varepsilon_{qu} \rangle}{k_B} = \frac{\hbar\omega}{2k_B}\coth\frac{\hbar\omega}{2k_B T_0} \tag{8}$$

has the meaning of the effective QHB temperature. It fixes the thermal equilibrium condition in the case when stochastic influences of both types are taken into account on equal terms. It depends on both fundamental constants $\hbar$ and k_B.

We could now formulate a zero law similar to (2)-(3) as the interrelation condition for the effective object and QHB temperatures T_{ef} and $(T_{ef})_0$. But we restrict ourself here to the consideration of problems in which it is not necessary to take the fluctuations of the effective object temperature into account. We therefore set

$$(T_{ef})_0 \equiv T_{ef} \quad \text{and} \quad T_0 \equiv T$$

in all formulae of Sections 1 and 2. The quantum generalization of macroparameters fluctuations theory (TEMF) is the subject of the Sect 3.

We call attention to the fact that the effective object temperature T_{ef} is a function of two object characteristics ω and T. In this case, equilibrium thermal radiation with a continuous spectrum is manifested as a QHB with a temperature T on the Kelvin scale. Under these conditions, we have not only $T = T_0$ but also $\omega = \omega_0$ as the thermal equilibrium state is reached, i.e., it is as if the object made a resonance choice of one of the QHB modes whose frequency ω_0 coincides with its characteristic frequency ω. It is necessary to choose the corresponding frequency from either the experiment or some intuitive considerations. In this regard, we can assume that the MST can at least be applied to a wide class of objects whose periodic or conditionally periodic motions can be assigned to independent degrees of freedom.

We note that the frequency ω and the temperature T (and consequently the effective temperature T_{ef}) are intensive quantities, which stresses that they are conceptually close to each other but makes them qualitatively different from extensive quantities (for example, energy), for which additivity is typical. They are holistic characteristics of the state of the object–environment system and have the transitivity property. It is significant that the characteristics ω, T, and T_{ef}, being c-number quantities, are not initially related to the number of observables to which operators are assigned in quantum theory.

We can elucidate the physical meaning of the effective temperature T_{ef} by considering its behavior in the limiting cases. Thus, as the temperature (on the Kelvin scale) $T \to 0$, the effective temperature T_{ef} becomes nonzero,

$$T_{ef} \to \frac{\hbar\omega}{2k_B} \equiv T^0_{ef}, \tag{9}$$

where T^0_{ef} has the meaning of the minimal effective temperature of the object with the characteristic frequency ω.

The effective temperature in turn becomes

$$T_{ef} \equiv T^0_{ef} \coth \frac{T^0_{ef}}{T} \to T\left[1 + \frac{1}{3}\left(\frac{T^0_{ef}}{T}\right)^2 + \cdots\right] \tag{10}$$

in the limit of high temperatures T. Of course, the concepts of low and high temperatures for each object essentially depend on the ratio T^0_{ef}/T.

2.2 Effective entropy

To calculate the effective macroparameters in terms of the corresponding distribution function, we must generalize canonical distribution (4) introduced by Gibbs in the nonquantum version of statistical thermodynamics. As above, according to the main MST postulate, this generalization reduces to replacing expression (5) for the distribution modulus Θ with expression (7), i.e., to replacing $T = T_0$ with $T_{ef} = (T_{ef})_0$. The desired distribution thus becomes

$$dw(\mathcal{E}) = \rho(\mathcal{E})\, d\mathcal{E} = \frac{1}{k_B T_{ef}} e^{-\mathcal{E}/(k_B T_{ef})}\, d\mathcal{E}, \tag{11}$$

where $\mathcal{E}$ is the random energy of the object's independent degree of freedom to which the model of the oscillator with the frequency ω is assigned.

Based on distribution (11), we can calculate the internal energy of the object as a macroparameter:

$$\mathcal{E}_{ef} = \int \mathcal{E}\rho(\mathcal{E})\, d\mathcal{E} = k_B T_{ef}. \tag{12}$$

Because $\mathcal{E}_{ef}$ with account (8) coincides with $\langle \varepsilon_{qu} \rangle$ of form (6), this means that the quantum oscillator in the heat bath is chosen as an object model.

To calculate the effective entropy S_{ef} of such an object, it is convenient to write formula (11) in the form in which the distribution density $\tilde{\rho}(\mathcal{E}) = \rho(\mathcal{E})\hbar\omega/2$ is dimensionless,

$$dw(\mathcal{E}) = \tilde{\rho}(\mathcal{E}) \left(\frac{\hbar\omega}{2}\right)^{-1} d\mathcal{E} = e^{(F_{ef} - \mathcal{E})/(k_B T_{ef})} \left(\frac{\hbar\omega}{2}\right)^{-1} d\mathcal{E}, \tag{13}$$

where the effective free energy is given by

$$F_{ef} = -k_{\rm B} T_{ef} \log \left(\frac{T_{ef}}{T^0_{ef}} \right). \tag{14}$$

We then obtain

$$S_{ef} = -k_{\rm B} \int \tilde{\rho}(\mathcal{E}) \log \tilde{\rho}(\mathcal{E}) \left(\frac{\hbar\omega}{2} \right)^{-1} d\mathcal{E} =$$

$$= k_{\rm B} \left[1 + \log \left(\coth \frac{\hbar\omega}{2k_{\rm B}T} \right) \right] = k_{\rm B} \left[1 + \log \left(\coth \frac{T^0_{ef}}{T} \right) \right]. \tag{15}$$

It follows from formula (15) that in the high-temperature limit $T \gg T^0_{ef}$, the effective entropy is written as

$$S_{ef} \to k_{\rm B} \log T + \text{const}, \tag{16}$$

which coincides with the expression for the oscillator entropy in thermodynamics based on classical statistical mechanics (CSM-based thermodynamics). At the same time, in the low-temperature limit $T \ll T^0_{ef}$, the effective entropy is determined by the world constant $k_{\rm B}$:

$$S_{ef} \to S^0_{ef} = k_{\rm B}. \tag{17}$$

Thus, in TEM, the behavior of the effective entropy of the degree of freedom of the object for which the periodic motion is typical corresponds to the initial formulation of Nernst's theorem, in which the minimum entropy is nonzero. Moreover, the range of temperatures T where we have $S_{ef} \approx S^0_{ef}$ can be very considerable, depending on the ratio T^0_{ef}/T.

It is obvious that using the model of the QHB, we can combine the quantum- and thermal-type influences (traditionally considered as specific influences only for the respective micro- and macrolevels) to form a holistic stochastic influence in the TEM framework. But using such an approach, we need not restrict ourself to generalizing only the traditional macroparameters, such as temperature and entropy. It becomes possible to give a meaning to the concept of effective action, as a new macroparameter which is significantly related to the quantum-type stochastic influence on the microlevel.

2.3 Effective action as a new macroparameter

The problem of introducing the concept of action into thermodynamics and of establishing the interrelation between the two widespread (but used in different areas of physics) quantities (entropy and action) has attracted the attention of many the most outstanding physicists, including Boltzmann {Bol22}, Boguslavskii, de Broglie. But the results obtained up to now were mainly related to CSM-based thermodynamics, and quantum effects were taken into account only in the quasiclassical approximation. Our aim is to extend them to the TEM. To do this, we choose the harmonic oscillator as an initial model of a periodically moving object. If we pass from the variables p and q to the action–angle variables when analyzing it in the framework of classical mechanics, then we can express the action j (having the meaning of a generalized momentum) in terms of the oscillator energy ε as

$$j = \frac{\varepsilon}{\omega}. \tag{18}$$

In passing to thermodynamics, we should preserve interrelation (18) between the action and the energy. In the same time, in place of the microparameters j and ε we use the corresponding macroparameters in this case, namely, the average quantities $\langle j\rangle$ and $\langle\varepsilon\rangle$. Their specific expressions depend on the choice of the heat bath model (classical or non-classical) used for averaging.

It is quite natural that in the framework of CSM-based thermodynamics, Boltzmann assumed that

$$\mathcal{J}_T \equiv \langle j\rangle = \frac{\langle\varepsilon_{\rm cl}\rangle}{\omega} = \frac{k_{\rm B}T}{\omega}. \tag{19}$$

Following the same idea, we determine the effective action $\mathcal{J}_{ef}$ in the TEM framework by the formula

$$\mathcal{J}_{ef} = \langle j\rangle = \frac{\langle\varepsilon_{\rm qu}\rangle}{\omega} = \frac{\mathcal{E}_{ef}}{\omega} = \frac{k_{\rm B}T_{ef}}{\omega}. \tag{20}$$

This means that in the TEM, we start from the fact that the effective action for all objects to which the model of the quantum oscillator in the QHB is applicable has the form

$$\mathcal{J}_{ef} = \frac{\hbar}{2}\coth\frac{\hbar\omega}{2k_{\rm B}T} = \mathcal{J}^0_{ef}\coth\frac{T^0_{ef}}{T}, \tag{21}$$

where accordingly (20)

$$\mathcal{J}^0_{ef} = \frac{k_B T^0_{ef}}{\omega} = \frac{\hbar}{2} \tag{22}$$

is the minimal effective action for $T\to 0$. Of course, in the limit $T \gg T^0_{ef}$, we have the effective action

$$\mathcal{J}_{ef} \to \mathcal{J}_T\left[1+\frac{1}{3}\left(\frac{T^0_{ef}}{T}\right)^2+\cdots\right],$$

i.e., it goes to expression (19) obtained in the CSM-based thermodynamics. Thus, both at low and high temperatures the formulae (22) and (19) for the effective action $\mathcal{J}_{ef}$ are written by the minimal effective temperature T^0_{ef}. This means that even purely quantum influence (at $T = 0$) can be interpreted as a peculiar thermal influence. Thus, one cannot assume that quantum and thermal influences can be considered separately. In other words, they are non-additive notions.

As is well-known, the original Planck formula for the average energy of the quantum oscillator in QSM

$$\mathcal{E}_{\rm quasi} = \frac{\hbar\omega}{e^{\hbar\omega/(k_{\rm B}T)}-1} \equiv \mathcal{E}_{ef} - \frac{\hbar\omega}{2} \tag{23}$$

is only applicable in the quasiclassical approximation framework. Substituting the expression $\mathcal{E}_{\rm quasi}$ of form (23) in formula (20) instead of $\mathcal{E}_{ef}$, we also obtain the effective action in the quasiclassical approximation:

$$\mathcal{J}_{\rm quasi} = \frac{\mathcal{E}_{\rm quasi}}{\omega} = \frac{\hbar}{e^{\hbar\omega/(k_{\rm B}T)}-1} \equiv \mathcal{J}_{ef} - \frac{\hbar}{2}. \tag{24}$$

The quasiclassical nature of expressions (23) and (24) is manifested, in particular, in the fact that these both quantities tend to zero as $T\to 0$.

2.4 The interrelation between the effective action and the effective entropy

To establish the interrelation between the action and the entropy, Boltzmann assumed that the isocyclic motions of the oscillator in mechanics for which $\omega = \text{const}$ correspond to the isothermal processes in thermodynamics. In this case, the oscillator energy can be changed under external influence that can be treated as the work δA^{dis} of dissipative forces equivalent to the heat δQ.

Generalizing this idea, we assume that every energy transferred at stochastic influence (quantum and thermal) in the TEM can be treated as the effective work $\delta A_{ef}^{\text{dis}}$ of dissipative forces equivalent to the effective heat δQ_{ef}. This means that for isothermal processes, the same change in the effective energy $d\mathcal{E}_{ef}$ of the macroobject to which the model of the quantum oscillator in the QHB can be assigned can be represented in two forms

$$d\mathcal{E}_{ef} = \delta A_{ef}^{\text{dis}} = \omega\, d\mathcal{J}_{ef} \quad \text{or} \quad d\mathcal{E}_{ef} = \delta Q_{ef} = T_{ef}\, dS_{ef}. \tag{25}$$

Furthermore, following Boltzmann, we choose the ratio $d\mathcal{E}_{ef}/\mathcal{E}_{ef}$ as a measure of energy transfer from the QHB to the object in such processes. The numerator and denominator in this ratio can be expressed in terms of either the effective action $\mathcal{J}_{ef}$ or the effective entropy S_{ef} and effective temperature T_{ef} using formulas (25) and (20). Equating the obtained expressions for the ratio $d\mathcal{E}_{ef}/\mathcal{E}_{ef}$, we obtain the differential equation

$$\frac{d\mathcal{E}_{ef}}{\mathcal{E}_{ef}} = \frac{\omega\, d\mathcal{J}_{ef}}{\omega \mathcal{J}_{ef}} = \frac{T_{ef}\, dS_{ef}}{k_{\text{B}} T_{ef}} \tag{26}$$

relating the effective entropy to the effective action. Its solution has the form

$$S_{ef} = k_{\text{B}} \int \frac{d\mathcal{J}_{ef}}{\mathcal{J}_{ef}} = k_{\text{B}} \log \frac{\mathcal{J}_{ef}}{J_0} = k_{\text{B}} \log\left(\frac{\hbar}{2J_0} \coth \frac{\hbar\omega}{2k_{\text{B}}T}\right), \tag{27}$$

where J_0 is the arbitrary constant of action dimensionality.

Choosing the quantity $\hbar/2e$ as J_0, where e is the base of the natural logarithms, we can make expression (27) coincides with the expression for the effective entropy S_{ef} of form (15). Taking into account that $S_{ef}^0 = k_{\text{B}}$, we have

$$S_{ef} = S_{ef}^0 \left[1 + \log \frac{\mathcal{J}_{ef}}{\mathcal{J}_{ef}^0}\right] = S_{ef}^0 \left[1 + \log \coth \frac{T_{ef}^0}{T}\right]. \tag{28}$$

For the entropy of the quantum oscillator in QSM-based thermodynamics, i.e. in the quasiclassical approximation, the well-known expression

$$S_{\text{quasi}} = -k_{\text{B}} \left\{ \frac{\hbar\omega}{k_{\text{B}}T} \left(1 - e^{\hbar\omega/(k_{\text{B}}T)}\right)^{-1} + \log\left(1 - e^{-\hbar\omega/(k_{\text{B}}T)}\right) \right\} \tag{29}$$

is applicable. It will be interesting to compare (28) with the analogical expression from QSM-based thermodynamics. For this goal we rewrite the formula (29), taking into account (24) in the form

$$S_{quasi} = \frac{\omega}{T} \mathcal{J}_{quasi} + k_B \log\left(1 + \frac{\mathcal{J}_{quasi}}{\hbar}\right). \tag{30}$$

In contrast to S_{ef} in the form (28) the quantity S_{quasi} tends to zero as $T \to 0$.

2.5 The first holistic stochastic-action constant

We note that according to formulas (21) and (28), the ratio of the effective action to the effective entropy is given by

$$\frac{\mathcal{J}_{ef}}{S_{ef}} = \frac{\mathcal{J}_{ef}^0}{S_{ef}^0} \cdot \frac{\coth(T_{ef}^0/T)}{1+\log\coth(T_{ef}^0/T)} = \varkappa \frac{\coth(\varkappa\omega/T)}{1+\log\coth(\varkappa\omega/T)}. \tag{31}$$

In this expression,

$$\varkappa \equiv \frac{\mathcal{J}_{ef}^0}{S_{ef}^0} = \frac{\hbar}{2k_\mathrm{B}} \tag{32}$$

is the minimal ratio (31) for $T \ll T_{ef}^0$.

In our opinion, the quantity

$$\varkappa = 3.82 \cdot 10^{-12}\,\mathrm{K}\cdot\mathrm{s} \tag{33}$$

is not only the notation for one of the possible combinations of the world constants $\hbar$ and k_B. It also has its intrinsic physical meaning. In addition to the fact that the ratio $\mathcal{J}_{ef}/S_{ef}$ of form (31) at any temperature can be expressed in terms of this quantity, it is contained in definition (2.5) of the effective temperature

$$T_{ef} = \varkappa\omega\coth\frac{\varkappa\omega}{T} \tag{34}$$

and also in the Wien's displacement law $T/\omega_{\max} = 0.7\varkappa$ for equilibrium thermal radiation. Starting from the preceding, we can formulate the hypothesis according to which the quantity $\varkappa$ plays the role of the *first* constant essentially characterizing the holistic stochastic action of environment on the object.

Hence, the minimal ratio of the action to the entropy in QSM-based thermodynamics is reached as $T \to 0$ and is determined by the formula

$$\frac{\mathcal{J}_{quasi}}{S_{quasi}} = \frac{T}{\omega}\left[1+\frac{k_B T}{\omega\mathcal{J}_{quasi}}\log\left(1+\frac{\mathcal{J}_{quasi}}{\hbar}\right)\right]^{-1} \to \frac{T}{\omega}\left[1+\frac{k_B T}{\hbar\omega}\right]^{-1} \to 0. \tag{35}$$

We have thus shown that not only $\mathcal{J}_\mathrm{quasi} \to 0$ and $S_\mathrm{quasi} \to 0$ but the ratio $\mathcal{J}_\mathrm{quasi}/S_\mathrm{quasi} \to 0$ in this microtheory too. This result differs sharply from the limit $\mathcal{J}_{ef}/S_{ef} \to \varkappa \neq 0$ for the corresponding effective quantities in the TEM. Therefore, it is now possible to compare the two theories (TEM and QSM) experimentally by measuring the limiting value of this ratio.

The main ideas on which the QST as a macrotheory is based were presented in the foregoing. The stochastic influences of quantum and thermal types over the entire temperature range are taken into account simultaneously and on equal terms in this theory. As a result, the main macroparameters of this theory are expressed in terms of the single macroparameter $\mathcal{J}_{ef}$ and combined fundamental constant $\varkappa = \hbar/2k_\mathrm{B}$. The experimental detection $\varkappa$ as the minimal nonzero ratio $\mathcal{J}_{ef}/S_{ef}$ can confirm that the TEM is valid in the range of sufficiently low temperatures. The first indications that the quantity $\varkappa$ plays an important role were probably obtained else in Andronikashvili's experiments (1948) on the viscosity of liquid helium below the λ point.

3. $(\hbar, k)$-dynamics as a microscopic ground of modern stochastic thermodynamics

In this section, following ideas of paper {Su06}, where we introduced the original notions of $\hbar k$D, we develop this theory further as a microdescription of an object under thermal equilibrium conditions {SuGo09}. We construct a model of the object environment, namely, QHB at zero and finite temperatures. We introduce a new microparameter, namely, the stochastic action operator, or Schrödingerian. On this ground we introduce the corresponding macroparameter, the effective action, and establish that the most important effective macroparameters —internal energy, temperature, and entropy—are expressed in terms of this macroparameter. They have the physical meaning of the standard macroparameters for a macrodescription in the frame of TEM describing in the Sect.1.

3.1 The model of the quantum heat bath: the "cold" vacuum

In constructing the $\hbar k$D, we proceed from the fact that no objects are isolated in nature. In other words, we follow the Feynman idea, according to which any system can be represented as a set of the object under study and its environment (the "rest of the Universe"). The environment can exert both regular and stochastic influences on the object. Here, we study only the stochastic influence. Two types of influence, namely, quantum and thermal influences characterized by the respective Planck and Boltzmann constants, can be assigned to it.

To describe the environment with the holistic stochastic influence we introduce a concrete model of environment, the QHB. It is a natural generalization of the classical thermal bath model used in the standard theories of thermal phenomena {Bog67}, {LaLi68}. According to this, the QHB is a set of weakly coupled quantum oscillators with all possible frequencies. The equilibrium thermal radiation can serve as a preimage of such a model in nature.

The specific feature of our understanding of this model is that we assume that we must apply it to both the "thermal" ($T \neq 0$) and the "cold" ($T = 0$) vacua. Thus, in the sense of Einstein, we proceed from a more general understanding of the thermal equilibrium, which can, in principle, be established for any type of environmental stochastic influence (purely quantum, quantum-thermal, and purely thermal).

We begin our presentation by studying the "cold" vacuum and discussing the description of a single quantum oscillator from the number of oscillators forming the QHB model for $T = 0$ from a new standpoint. For the purpose of the subsequent generalization to the case $T \neq 0$, not its well-known eigenstates $\Psi_n(q)$ in the q representation but the coherent states (CS) turn out to be most suitable.

But we recall that the lowest state in the sets of both types is the same. In the occupation number representation, the "cold" vacuum in which the number of particles is $n = 0$ corresponds to this state. In the q representation, the same ground state of the quantum oscillator is in turn described by the real wave function

$$\Psi_0(q) = [2\pi(\Delta q_0)^2]^{-1/4} e^{-q^2/4(\Delta q_0)^2}. \tag{36}$$

In view of the properties of the Gauss distribution, the Fourier transform $\Psi_0(p)$ of this function has a similar form (with q replaced with p); in this case, the respective momentum and coordinate dispersions are

$$(\Delta p_0)^2 = \frac{\hbar m\omega}{2}, \qquad (\Delta q_0)^2 = \frac{\hbar}{2m\omega}. \tag{37}$$

As is well known, CS are the eigenstates of the non-Hermitian particle annihilation operator $\hat{a}$ with complex eigenvalues. But they include one isolated state $|0_a\rangle$ of the particle vacuum in which the eigenvalue of $\hat{a}$ is zero

$$\hat{a}|0_a\rangle = 0|0_a\rangle, \quad \text{or} \quad \hat{a}\Psi_0(q) = 0. \tag{38}$$

In what follows, it is convenient to describe the QHB in the q representation. Therefore, we express the annihilation operator $\hat{a}$ and the creation operator $\hat{a}^\dagger$ in terms of the operators $\hat{p}$ and $\hat{q}$ using the traditional method. We have

$$\hat{a} = \frac{1}{2}\left(\frac{\hat{p}}{\sqrt{\Delta p_0^2}} - i\frac{\hat{q}}{\sqrt{\Delta q_0^2}}\right), \qquad \hat{a}^\dagger = \frac{1}{2}\left(\frac{\hat{p}}{\sqrt{\Delta p_0^2}} + i\frac{\hat{q}}{\sqrt{\Delta q_0^2}}\right). \tag{39}$$

The particle number operator then becomes

$$\hat{N}_a = \hat{a}^\dagger \hat{a} = \frac{1}{\hbar\omega}\left(\frac{\hat{p}^2}{2m} + \frac{m\omega^2\hat{q}^2}{2} - \frac{\hbar\omega}{2}\hat{I}\right), \tag{40}$$

where $\hat{I}$ is the unit operator. The sum of the first two terms in the parentheses forms the Hamiltonian $\hat{\mathcal{H}}$ of the quantum oscillator, and after multiplying relations (40) by $\hbar\omega$ on the left and on the right, we obtain the standard interrelation between the expressions for the Hamiltonian in the q and n representations:

$$\hat{\mathcal{H}} = \frac{\hat{p}^2}{2m} + \frac{m\omega^2\hat{q}^2}{2} = \hbar\omega\left(\hat{N}_a + \frac{1}{2}\hat{I}\right). \tag{41}$$

From the thermodynamics standpoint, we are concerned with the effective internal energy of the quantum oscillator in equilibrium with the "cold" QHB. Its value is equal to the mean of the Hamiltonian calculated over the state $|0_a\rangle \equiv |\Psi_0(q)\rangle$:

$$\mathcal{E}_{ef}^0 = \langle\Psi_0(q)|\hat{\mathcal{H}}|\Psi_0(q)\rangle = \hbar\omega\langle\Psi_0(q)|\hat{N}_a|\Psi_0(q)\rangle + \frac{\hbar\omega}{2} = \frac{\hbar\omega}{2} = \varepsilon_0. \tag{42}$$

It follows from formula (42) that in the given case, the state without particles coincides with the state of the Hamiltonian with the minimal energy ε_0. The quantity ε_0, traditionally treated as the zero point energy, takes the physical meaning of a macroparameter, or the effective internal energy $\mathcal{E}_{ef}^0$ of the quantum oscillator in equilibrium with the "cold" vacuum.

3.2 The model of the quantum heat bath: passage to the "thermal" vacuum

We can pass from the "cold" to the "thermal" vacuum using the Bogoliubov (u,v) transformation with the complex temperature-dependent coefficients { SuGo09}

$$u = \left(\frac{1}{2}\coth\frac{\hbar\omega}{2k_BT} + \frac{1}{2}\right)^{1/2} e^{i\pi/4}, \qquad v = \left(\frac{1}{2}\coth\frac{\hbar\omega}{2k_BT} - \frac{1}{2}\right)^{1/2} e^{-i\pi/4}. \tag{43}$$

In the given case, this transformation is canonical but leads to a unitarily nonequivalent representation because the QHB at any temperature is a system with an infinitely large number of degrees of freedom.
In the end, such a transformation reduces to passing from the set of quantum oscillator CS to a more general set of states called the thermal correlated CS (TCCS) {Su06}. They are selected because they ensure that the Schrödinger coordinate–momentum uncertainties relation is saturated at any temperature.

From the of the second-quantization apparatus standpoint, the Bogoliubov (u,v) transformation ensures the passage from the original system of particles with the "cold" vacuum $|0_a\rangle$ to the system of quasiparticles described by the annihilation operator $\hat{b}$ and the creation operator $\hat{b}^\dagger$ with the "thermal" vacuum $|0_b\rangle$. In this case, the choice of transformation coefficients (43) is fixed by the requirement that for any method of description, the expression for the mean energy of the quantum oscillator in thermal equilibrium be defined by the Planck formula (6)

$$\mathcal{E}_{Pl} = [\langle\Psi_T(q)|\hat{\mathcal{H}}|\Psi_T(q)\rangle] = \langle\varepsilon_{qu}\rangle = \frac{\hbar\omega}{2}\coth\frac{\hbar\omega}{2k_BT}, \tag{44}$$

which can be obtained from experiments. As shown in {Su06}, the state of the "thermal" vacuum $|0_b\rangle \equiv |\Psi_T(q)\rangle$ in the q representation corresponds to the complex wave function

$$\Psi_T(q) = [2\pi(\Delta q_{ef})^2]^{-1/4}\exp\left\{-\frac{q^2}{4(\Delta q_{ef})^2}(1-i\alpha)\right\}, \tag{45}$$

where

$$(\Delta q_{ef})^2 = \frac{\hbar}{2m\omega}\coth\frac{\hbar\omega}{2k_BT}, \qquad \alpha = \left[\sinh\frac{\hbar\omega}{2k_BT}\right]^{-1}. \tag{46}$$

For its Fourier transform $\Psi_T(p)$, a similar expression with the same coefficient α and

$$(\Delta p_{ef})^2 = \frac{\hbar m\omega}{2}\coth\frac{\hbar\omega}{2k_BT} \tag{47}$$

holds. We note that the expressions for the probability densities $\rho_T(q)$ and $\rho_T(p)$ have already been obtained by Bloch (1932), but the expressions for the phases that depend on the parameter α play a very significant role and were not previously known. It is also easy to see that as $T \to 0$, the parameter $\alpha \to 0$ and the function $\Psi_T(q)$ from TCCS passes to the function $\Psi_0(q)$ from CS.

Of course, the states from TCCS are the eigenstates of the non-Hermitian quasiparticle annihilation operator $\hat{b}$ with complex eigenvalues. They also include one isolated state of the quasiparticle vacuum in which the eigenvalue of b is zero,

$$\hat{b}|0_b\rangle = 0|0_b\rangle, \text{ or } \hat{b}\Psi_T(q) = 0. \tag{48}$$

Using condition (48) and expression (45) for the wave function of the "thermal" vacuum, we obtain the expression for the operator $\hat{b}$ in the q representation:

$$\hat{b} = \frac{1}{2}\left(\coth\frac{\hbar\omega}{2k_BT}\right)^{\frac{1}{2}}\left[\frac{\hat{p}}{\sqrt{\Delta p_0^2}} - i\frac{\hat{q}}{\sqrt{\Delta q_0^2}}\left(\coth\frac{\hbar\omega}{2k_BT}\right)^{-1}(1-i\alpha)\right]. \tag{49}$$

The corresponding quasiparticle creation operator has the form

$$\hat{b}^\dagger = \frac{1}{2}\left(\coth\frac{\hbar\omega}{2k_BT}\right)^{\frac{1}{2}}\left[\frac{\hat{p}}{\sqrt{\Delta p_0^2}} + i\frac{\hat{q}}{\sqrt{\Delta q_0^2}}\left(\coth\frac{\hbar\omega}{2k_BT}\right)^{-1}(1+i\alpha)\right]. \tag{50}$$

We can verify that as $T \to 0$, the operators $\hat{b}^\dagger$ and $\hat{b}$ for quasiparticles pass to the operators $\hat{a}^\dagger$ and $\hat{a}$ for particles.

Acting just as above, we obtain the expression for the effective Hamiltonian, which is proportional to the quasiparticle number operator in the q representation

$$\hat{\mathcal{H}}_{ef} = \hbar\omega\hat{N}_b = \coth\frac{\hbar\omega}{2k_BT}\left(\frac{\hat{p}^2}{2m} + \frac{m\omega^2 q^2}{2}\right) - \frac{\hbar\omega}{2}\left(\hat{I} + \frac{\alpha}{\hbar}\{\hat{p},\hat{q}\}\right), \tag{51}$$

where we take $1+\alpha^2 = \coth^2(\hbar\omega/2k_BT)$ into account. Obviously, $\hat{\mathcal{H}}_{ef}\Psi_T(q) = 0$, i.e. $\Psi_T(q)$- an eigenfunction of $\hat{\mathcal{H}}_{ef}$.
Passing to the original Hamiltonian, we obtain

$$\hat{\mathcal{H}} = \hbar\omega\left(\coth\frac{\hbar\omega}{2k_BT}\right)^{-1}\left[\hat{N}_b + \frac{1}{2}\left(\hat{I} + \frac{\alpha}{\hbar}\{\hat{p},\hat{q}\}\right)\right]. \tag{52}$$

We stress that the operator $\{\hat{p},\hat{q}\}$ in formula (52) can also be expressed in terms of bilinear combinations of the operators $\hat{b}^{\dagger}$ and $\hat{b}$, but they differ from the quasiparticle number operator. This means that the operators $\hat{\mathcal{H}}$ and $\hat{N}_b$ do not commute and that the wave function of form (45) characterizing the state of the "thermal" vacuum is therefore not the eigenfunction of the Hamiltonian $\hat{\mathcal{H}}$.
As before, we are interested in the macroparameter, namely, the effective internal energy $\mathcal{E}_{ef}$ of the quantum oscillator now in thermal equilibrium with the "thermal" QHB. Calculating it just as in Sec. 3.1, we obtain

$$\mathcal{E}_{ef} = \hbar\omega\left[\langle\Psi_T(q)|\hat{N}_b|\Psi_T(q)\rangle\right] + \frac{\hbar\omega}{2\coth(\hbar\omega/2k_BT)}\left(1 + \frac{\alpha}{\hbar}\langle\Psi_T(q)|\{p,q\}|\Psi_T(q)\rangle\right) \tag{53}$$

in the q representation. Because we average over the quasiparticle vacuum in formula (53), the first term in it vanishes. At the same time, it was shown by us {Su06} that

$$\langle\Psi_T(q)|\{\hat{p},\hat{q}\}|\Psi_T(q)\rangle = \hbar\alpha. \tag{54}$$

As a result, we obtain the expression for the effective internal energy of the quantum oscillator in the "thermal" QHB in the $\hbar k$D framework:

$$\mathcal{E}_{ef} = \frac{\hbar\omega}{2\coth(\hbar\omega/2k_BT)}(1+\alpha^2) = \frac{\hbar\omega}{2}\coth\frac{\hbar\omega}{2k_BT} = \mathcal{E}_{Pl}, \tag{55}$$

that coincides with the formula (44). This means that the average energy of the quantum oscillator at $T \neq 0$ has the meaning of effective internal energy as a macroparameter in the case of equilibrium with the "thermal" QHB. As $T \to 0$, it passes to a similar quantity corresponding to equilibrium with the "cold" QHB.
Although final result (55) was totally expected, several significant conclusions follow from it.
1. In the $\hbar k$D, in contrast to calculating the internal energy in QSM, where all is defined by the probability density $\rho_T(q)$, the squared parameter α determining the phase of the wave function contributes significantly to the same expression, which indicates that the quantum ideology is used more consistently.
2. In the $\hbar k$D, the expression for $\coth(\hbar\omega/2k_BT)$ in formula (55) appears as an holistic quantity, while the contribution $\varepsilon_0 = \hbar\omega/2$ to the same formula (6) in QSM usually arises separately as an additional quantity without a thermodynamic meaning and is therefore often neglected.
3. In the $\hbar k$D, the operators $\hat{\mathcal{H}}$ and $\hat{N}_b$ do not commute. It demonstrates that the number of quasiparticles is not preserved, which is typical of the case of spontaneous

symmetry breaking. In our opinion, the proposed model of the QHB is a universal model of the environment with a stochastic influence on an object. Therefore, the manifestations of spontaneous symmetry breaking in nature must not be limited to superfluidity and superconductivity phenomena.

3.3 Schrödingerian as a stochastic action operator

The effective action as a macroparameter was postulated in the Section 1 in the framework of TEM by generalizing concepts of adiabatic invariants. In the $\hbar k$D framework, we base our consistent microdescription of an object in thermal equilibrium on the model of the QHB described by a wave function of form (45).

Because the original statement of the $\hbar k$D is the idea of the holistic stochastic influence of the QHB on the object, we introduce a new operator in the Hilbert space of microobject states to implement it. As leading considerations, we use an analysis of the right-hand side of the Schrödinger coordinate–momentum uncertainties relation in the saturated form {Su06}:

$$(\Delta p)^2(\Delta q)^2 = |\widetilde{R}_{pq}|^2. \tag{56}$$

For not only a quantum oscillator in a heat bath but also any object, the complex quantity in the right-hand side of (56)

$$\widetilde{R}_{pq} = \langle \Delta p | \Delta q \rangle = \langle | \Delta \hat{p} \Delta \hat{q} | \rangle \tag{57}$$

has a double meaning. On one hand, it is the amplitude of the transition from the state $|\Delta q\rangle$ to the state $|\Delta p\rangle$; on the other hand, it can be treated as the mean of the Schrödinger quantum correlator calculated over an arbitrary state $|\ \rangle$ of some operator.

As is well known, the nonzero value of quantity (57) is the fundamental attribute of nonclassical theory in which the environmental stochastic influence on an object plays a significant role. Therefore, it is quite natural to assume that the averaged operator in the formula has a fundamental meaning. In view of dimensional considerations, we call it the stochastic action operator, or Schrödingerian

$$\hat{j} \equiv \Delta \hat{p} \Delta \hat{q}. \tag{58}$$

Of course, it should be remembered that the operators $\Delta \hat{q}$ and $\Delta \hat{p}$ do not commute and their product is a non-Hermitian operator.

To analyze further, following Schrödinger (1930) {DoMa87}, we can express the given operator in the form

$$\hat{j} = \frac{1}{2}\{\Delta \hat{p}, \Delta \hat{q}\} + \frac{1}{2}[\hat{p}, \hat{q}] = \hat{\sigma} - i\hat{j}_0, \tag{59}$$

which allows separating the Hermitian part (the operator $\hat{\sigma}$) in it from the anti-Hermitian one, in which the Hermitian operator is

$$\hat{j}_0 = \frac{i}{2}[\hat{p}, \hat{q}] \equiv \frac{\hbar}{2}\hat{I}. \tag{60}$$

It is easy to see that the mean $\sigma = \langle |\hat{\sigma}| \rangle$ of the operator $\hat{\sigma}$ resembles the expression for the standard correlator of coordinate and momentum fluctuations in classical probability theory; it transforms into this expression if the operators $\Delta \hat{q}$ and $\Delta \hat{p}$ are replaced with c-numbers. It reflects the contribution to the transition amplitude $\widetilde{R}_{pq}$ of the environmental stochastic influence. Therefore, we call the operator $\hat{\sigma}$ the external stochastic action operator in what follows. Previously, the possibility of using a similar operator was discussed by Bogoliubov

and Krylov (1939) as a quantum analogue of the classical action variable in the set of action–angle variables.
At the same time, the operators $\hat{j}_0$ and $\hat{j}$ were not previously introduced. The operator of form (60) reflects a specific peculiarity of the objects to be "sensitive" to the minimal stochastic influence of the "cold" vacuum and to respond to it adequately regardless of their states. Therefore, it should be treated as a minimal stochastic action operator. Its mean $\mathcal{J}_0 = \langle|\hat{j}_0|\rangle = \hbar/2$ is independent of the choice of the state over which the averaging is performed, and it hence has the meaning of the invariant eigenvalue of the operator $\hat{j}_0$.
This implies that in the given case, we deal with the universal quantity $\mathcal{J}_0$, which we call the minimal action. Its fundamental character is already defined by its relation to the Planck world constant $\hbar$. But the problem is not settled yet. Indeed, according to the tradition dating back to Planck, the quantity $\hbar$ is assumed to be called the elementary quantum of the action. At the same time, the factor $1/2$ in the quantity $\mathcal{J}_0$ plays a significant role, while half the quantum of the action is not observed in nature. Therefore, the quantities $\hbar$ and $\hbar/2$, whose dimensions coincide, have different physical meanings and must hence be named differently, in our opinion. From this standpoint, it would be more natural to call the quantity $\hbar$ the external quantum of the action.
Hence, the quantity $\hbar$ is the minimal portion of the action transferred to the object from the environment or from another object. Therefore, photons and other quanta of fields being carriers of fundamental interactions are first the carriers of the minimal action equal to $\hbar$. The same is also certainly related to phonons.
Finally, we note that only the quantity $\hbar$ is related to the discreteness of the spectrum of the quantum oscillator energy in the absence of the heat bath. At the same time, the quantity $\hbar/2$ has an independent physical meaning. It reflects the minimal value of stochastic influence of environment at $T = 0$, specifying by formula (42) the minimal value of the effective internal energy $\mathcal{E}^0_{ef}$ of the quantum oscillator.

3.4 Effective action in $(\hbar,k)$-dynamics

Now we can turn to the macrodescription of objects using their microdescription in the $\hbar k$D framework. It is easy to see that the mean $\widetilde{\mathcal{J}}$ of the operator $\hat{j}$ of form (59) coincides with the complex transition amplitude $\widetilde{R}_{pq}$ and, in thermal equilibrium, can be expressed as

$$\widetilde{\mathcal{J}} = \langle\Psi_T(q)|\hat{j}|\Psi_T(q)\rangle = \sigma - i\mathcal{J}_0 = (\widetilde{R}_{pq})_{ef}. \tag{61}$$

In what follows, we regard the modulus of the complex quantity $\widetilde{\mathcal{J}}$,

$$|\widetilde{\mathcal{J}}| = \sqrt{\sigma^2 + \mathcal{J}_0^2} = \sqrt{\sigma^2 + \frac{\hbar^2}{4}} \equiv \mathcal{J}_{ef} \tag{62}$$

as a new macroparameter and call it the effective action. It has the form

$$\mathcal{J}_{ef} = \frac{\hbar}{2} \coth \frac{\hbar\omega}{2k_B T}, \tag{63}$$

that coincides with a similar quantity $\mathcal{J}_{ef}$ postulated as a fundamental macroparameter in TEM framework (see the Sect.1.) from intuitive considerations.
We now establish the interrelation between the effective action and traditional macroparameters. Comparing expression (63) for $|\widetilde{\mathcal{J}}|$ with (55) for the effective internal

energy $\mathcal{E}_{ef}$, we can easily see that

$$\mathcal{E}_{ef} = \omega|\tilde{\mathcal{J}}| = \omega\mathcal{J}_{ef}. \tag{64}$$

In the high-temperature limit, where

$$\sigma \to \mathcal{J}_T = \frac{k_B T}{\omega} \gg \frac{\hbar}{2}, \tag{65}$$

relation (64) becomes

$$\mathcal{E} = \omega\mathcal{J}_T. \tag{66}$$

Boltzmann {Bol22} previously obtained this formula for macroparameters in CSM-based thermodynamics by generalizing the concept of adiabatic invariants used in classical mechanics.

Relation (64) also allows expressing the interrelation between the effective action and the effective temperature T_{ef} (8) in explicit form:

$$T_{ef} = \frac{\omega}{k_B}\mathcal{J}_{ef}. \tag{67}$$

This implies that

$$T^0_{ef} = \frac{\omega}{k_B}\mathcal{J}^0_{ef} = \frac{\hbar\omega}{2k_B} \neq 0, \tag{68}$$

where $\mathcal{J}^0_{ef} \equiv \mathcal{J}_0$. Finally, we note that using formulas (56), (61)– (64), (46), and (47), we can rewrite the saturated Schrödinger uncertainties relation for the quantum oscillator for $T \neq 0$ as

$$\Delta p_{ef} \cdot \Delta q_{ef} = \mathcal{J}_{ef} = \frac{\mathcal{E}_{ef}}{\omega} = \frac{\hbar}{2}\coth\frac{\hbar\omega}{2k_B T}. \tag{69}$$

3.5 Effective entropy in the $(\hbar, k)$-dynamics

The possibility of introducing entropy in the $\hbar k$D is also based on using the wave function $\Psi_T(q)$ instead of the density operator. To define the entropy as the initial quantity, we take the formal expression

$$-k_B\left\{\int \rho(q)\log\rho(q)\,dq + \int \rho(p)\log\rho(p)\,dp\right\} \tag{70}$$

described in {DoMa87}. Here, $\rho(q) = |\Psi(q)|^2$ and $\rho(p) = |\Psi(p)|^2$ are the dimensional densities of probabilities in the respective coordinate and momentum representations.

Using expression (45) for the wave function of the quantum oscillator, we reduce $\rho(q)$ to the dimensionless form:

$$\tilde{\rho}(\tilde{q}) = \left[\frac{2\pi}{\delta}\coth\frac{\hbar\omega}{2k_B T}\right]^{-1} e^{-\tilde{q}^2/2}, \qquad \tilde{q}^2 = \frac{q^2}{(\Delta q_{ef})^2}, \tag{71}$$

where δ is an arbitrary constant. A similar expression for its Fourier transform $\tilde{\rho}(\tilde{p})$ differs by only replacing q with p.

Using the dimensionless expressions, we propose to define entropy in the $\hbar k$D framework by the equality

$$S_{qp} = -k_B\left\{\int \tilde{\rho}(\tilde{q})\log\tilde{\rho}(\tilde{q})\,d\tilde{q} + \int \tilde{\rho}(\tilde{p})\log\tilde{\rho}(\tilde{p})\,d\tilde{p}\right\}. \tag{72}$$

Substituting the corresponding expressions for $\tilde{\rho}(\tilde{q})$ and $\tilde{\rho}(\tilde{p})$ in (72), we obtain

$$S_{qp} = k_B\left\{\left(1+\log\frac{2\pi}{\delta}\right)+\log\coth\frac{\hbar\omega}{2k_BT}\right\}. \tag{73}$$

Obviously, the final result depends on the choice of the constant δ.
Choosing $\delta = 2\pi$, we can interpret expression (73) as the quantum-thermal entropy or, briefly, the QT entropy S_{QT} because it coincides exactly with the effective entropy S_{ef} (15). This ensures the consistency between the main results of our proposed micro- and macrodescriptions, i.e. $\hbar kD$ and TEM, and their correspondence to experiments.
We can approach the modification of original formal expression (70) in another way. Combining both terms in it, we can represent it in the form

$$-k_B\int d\varepsilon\, W(\varepsilon)\log W(\varepsilon). \tag{74}$$

It is easy to see that $W(\varepsilon)$ is the Wigner function for the quantum oscillator in the QHB:

$$W(\varepsilon) = \{2\pi\Delta q\Delta p\}^{-1}\exp\left\{-\frac{p^2}{2(\Delta p)^2}-\frac{q^2}{2(\Delta q)^2}\right\} = \frac{\omega}{2\pi k_B T_{ef}}e^{-\varepsilon/k_BT_{ef}}. \tag{75}$$

After some simple transformations the expression (74) takes also the form $S_{ef} = S_{QT}$.
Modifying expressions (70) (for $\delta = 2\pi$) or (74) in the $\hbar k$D framework thus leads to the expression for the QT, or effective, entropy of form (15). From the microscopic standpoint, they justify the expression for the effective entropy as a macroparameter in MST. We note that the traditional expression for entropy in QSM-based thermodynamics turns out to be only a quasiclassical approximation of the QT, or effective entropy.

3.6 Some thermodynamics relations in terms of the effective action

The above presentation shows that using the $\hbar k$D developed here, we can introduce the effective action $\mathcal{J}_{ef}$ as a new fundamental macroparameter. The advantage of this macroparameter is that in the given case, it has a microscopic preimage, namely, the stochastic action operator $\hat{j}$, or Schrödingerian. Moreover, we can in principle express the main macroparameters of objects in thermal equilibrium in terms of it. As is well known, temperature and entropy are the most fundamental of them. It is commonly accepted that they have no microscopic preimages but take the environment stochastic influence on the object generally into account. In the traditional presentation, the temperature is treated as a "degree of heating," and entropy is treated as a "measure of system chaos."
If the notion of effective action is used, these heuristic considerations about T_{ef} and S_{ef} can acquire an obvious meaning. For this, we turn to expression (67) for T_{ef}, whence it follows that the effective action is also an *intensive* macroparameter characterizing the stochastic influence of the QHB. In view of this, the zero law of MST can be rewritten as

$$\mathcal{J}_{ef} = (\mathcal{J}_{ef})_0 \pm \delta\mathcal{J}_{ef}, \tag{76}$$

where $(\mathcal{J}_{ef})_0$ is the effective action of a QHB and $\mathcal{J}_{ef}$ and $\delta\mathcal{J}_{ef}$ are the means of the effective reaction of an object and its fluctuation. The state of thermal equilibrium can actually be described in the sense of Newton, assuming that "the stochastic action is equal to the stochastic counteraction" in such cases.

We now turn to the effective entropy S_{ef}. In the absence of a mechanical contact, its differential in MST is

$$dS_{ef} = \frac{\delta Q_{ef}}{T_{ef}} = \frac{d\mathcal{E}_{ef}}{T_{ef}}. \tag{77}$$

Substituting the expressions for effective internal energy (64) and effective temperature (67) in this relation, we obtain

$$dS_{ef} = k_B \frac{\omega\, d\mathcal{J}_{ef}}{\omega \mathcal{J}_{ef}} = k_B \cdot d\left(\log \frac{\mathcal{J}_{ef}}{\mathcal{J}_{ef}^0}\right) = dS_{QT}. \tag{78}$$

It follows from this relation that the effective or QT entropy, being an *extensive* macroparameter, can be also expressed in terms of $\mathcal{J}_{ef}$.
As a result, it turns out that two qualitatively different characteristics of thermal phenomena on the macrolevel, namely, the effective temperature and effective entropy, embody the presence of two sides of stochastization the characteristics of an object in nature in view of the contact with the QHB. At any temperature, they can be expressed in terms of the same macroparameter, namely, the effective action $\mathcal{J}_{ef}$. This macroparameter has the stochastic action operator, or Schrödingerian simultaneously dependent on the Planck and Boltzmann constants as a microscopic preimage in the $\hbar k$D.

4. Theory of effective macroparameters fluctuations and their correlation

In the preceding sections we considered effective macroparameters as random quantities but the subject of interest were only problems in which the fluctuations of the effective temperature and other effective object macroparameters can be not taken into account. In given section we consistently formulate a noncontradictory theory of quantum-thermal fluctuations of effective macroparameters (TEMF) and their correlation. We use the apparatus of two approaches developed in sections 2 and 3 for this purpose.
This theory is based on the rejection of the classical thermostat model in favor of the quantum one with the distribution modulus $\Theta_{qu} = k_B T_{ef}$. This allows simultaneously taking into account the quantum and thermal stochastic influences of environment describing by effective action. In addition, it is assumed that some of macroparameters fluctuations are obeyed the nontrivial uncertainties relations. It appears that correlators of corresponding fluctuations are proportional to effective action $\mathcal{J}_{ef}$.

4.1 Inapplicability QSM-based thermodynamics for calculation of the macroparameters fluctuations

As well known, the main condition of applicability of thermodynamic description is the following inequality for relative dispersion of macroparameter A_i :

$$\frac{(\Delta A_i)^2}{\langle A_i \rangle^2} \leqslant 1, \tag{79}$$

where

$$(\Delta A_i)^2 \equiv \langle (\delta A_i)^2 \rangle = \langle A_i^2 \rangle - \langle A_i \rangle^2$$

is the dispersion of the quantity A_i.
In the non-quantum version of statistical thermodynamics, the expressions for macroparameters dispersions can be obtained. So, for dispersions of the temperature

and the internal energy of the object for $V = const$ we have according to Einstein {LaLi68}

$$(\Delta T)^2 = \frac{1}{k_B C_V}\Theta_{cl}^2 = \frac{k_B}{C_V}T^2 \text{ and } (\Delta\mathcal{E})^2 = \frac{C_V}{k_B}\Theta_{cl}^2 = k_B C_V T^2, \tag{80}$$

where $C_V = \frac{\partial\langle\mathcal{E}\rangle}{\partial T}\Big|_V$ is the heat capacity for the constant volume. At high temperatures the condition (79) is satisfies for any macroparameters and any objects including the classical oscillator.
For its internal energy $\mathcal{E} = \langle\varepsilon\rangle = k_B T$ with account $C_V = k_B$ we obtain its dispersion

$$(\Delta\mathcal{E})^2 = k_B C_V T^2 = k_B^2 T^2 = \mathcal{E}^2. \tag{81}$$

So, the condition (79) is valid for $\mathcal{E}$ and this object can also be described in the framework of thermodynamics.
For the account of quantum effects in QSM-based thermodynamics instead of (80) are used the following formulae

$$(\Delta T)^2 = 0 \text{ and } (\Delta\mathcal{E}_{qu})^2 = k_B (C_V)_{qu} T^2. \tag{82}$$

The difference is that instead of C_V, it contains

$$(C_V)_{qu} = \frac{\partial\mathcal{E}_{qu}}{\partial T}\Big|_V,$$

where $\mathcal{E}_{qu} = \langle\varepsilon_{qu}\rangle$ is the internal energy of the object calculated in the QSM framework.
For a quantum oscillator in this case we have

$$\langle\mathcal{E}_{qu}\rangle = \frac{\hbar\omega}{\exp\{2\varkappa\frac{\omega}{T}\} - 1} = \frac{\hbar\omega}{2}\cdot\frac{\exp\{-\varkappa\frac{\omega}{T}\}}{\sinh(\varkappa\frac{\omega}{T})}, \tag{83}$$

and its heat capacity is

$$(C_V)_{qu} = k_B\left(\frac{\hbar\omega}{k_B T}\right)^2 \frac{\exp\{2\varkappa\frac{\omega}{T}\}}{(\exp\{2\varkappa\frac{\omega}{T}\} - 1)^2} = k_B\left(\varkappa\frac{\omega}{T}\right)^2 \frac{1}{\sinh^2(\varkappa\frac{\omega}{T})}. \tag{84}$$

According to general formula (82), the dispersion of the quantum oscillator internal energy has the form

$$(\Delta\mathcal{E}_{qu})^2 = k_B (C_V)_{qu} T^2 = \left(\frac{\hbar\omega}{2}\right)^2 \cdot \frac{1}{\sinh^2(\varkappa\frac{\omega}{T})} =$$
$$= \hbar\omega\langle\mathcal{E}_{qu}\rangle + \langle\mathcal{E}_{qu}\rangle^2 = \exp\{2\varkappa\frac{\omega}{T}\}\langle\mathcal{E}_{qu}\rangle^2, \tag{85}$$

and the relative dispersion of its energy is

$$\frac{(\Delta\mathcal{E}_{qu})^2}{\langle\mathcal{E}_{qu}\rangle^2} = \frac{\hbar\omega}{\langle\mathcal{E}_{qu}\rangle} + 1 = \exp\{2\varkappa\frac{\omega}{T}\}. \tag{86}$$

We note that in expression (83) the zero-point energy $\varepsilon_0 = \hbar\omega/2$ is absent. It means that the relative dispersion of internal energy stimulating by thermal stochastic influence are only the subject of interest. So, we can interpret this calculation as a quasiclassical approximation.

A similar result exists for the relative dispersion of the energy of thermal radiation in the spectral interval $(\omega,\omega+\Delta\omega)$ for the volume V :

$$\frac{(\Delta\mathcal{E}_{\omega})^2}{\langle\mathcal{E}_{\omega}\rangle^2}=\frac{\hbar\omega}{\langle\mathcal{E}_{\omega}\rangle}+\frac{\pi^2c^3}{V\omega^2\Delta\omega}=\frac{\pi^2c^3}{V\omega^2\Delta\omega}\exp\{2\varkappa\frac{\omega}{T}\}. \tag{87}$$

We can see that at $T\to 0$ expressions (86) and (87) tend to infinity. However, few people paid attention to the fact that thereby the condition (79) of the applicability of the thermodynamic description does not satisfy. A.I. Anselm {An73} was the only one who has noticed that ordinary thermodynamics is inapplicable as the temperature descreases. We suppose that in this case instead of QSM-based thermodynamics can be fruitful MST based on $\hbar k$D.

4.2 Fluctuations of the effective internal energy and effective temperature

To calculate dispersions of macroparameters in the quantum domain, we use MST instead of QSM-based thermodynamics in 4.2 and 4.3, i.e., we use the macrotheory described in Sect.1. It is based on the Gibbs distribution in the effective macroparameters space {Gi60}

$$dW(\mathcal{E})=\rho(\mathcal{E})d\mathcal{E}=\frac{1}{k_BT_{ef}}\exp\{-\frac{\mathcal{E}}{k_BT_{ef}}\}d\mathcal{E}. \tag{88}$$

Here, T_{ef} is the effective temperature of form (8), simultaneously taking the quantum–thermal effect of the QHB into account and $\mathcal{E}$ is the random object energy to which the conditional frequency ω can be assigned at least approximately.

Using distribution (88), we find the expression for the effective internal energy of the object coinciding with the Planck formula

$$\mathcal{E}_{ef}=\langle\varepsilon_{qu}\rangle=\int\mathcal{E}\rho(\mathcal{E})d\mathcal{E}=k_BT_{ef}\equiv\mathcal{E}_{Pl}=\frac{\hbar\omega}{2}\coth\varkappa\frac{\omega}{T}, \tag{89}$$

the average squared effective internal energy

$$\langle\mathcal{E}_{ef}^2\rangle=\int\mathcal{E}^2\rho(\mathcal{E})d\mathcal{E}=2\langle\mathcal{E}_{ef}\rangle^2, \tag{90}$$

and the dispersion of the effective internal energy

$$(\Delta\mathcal{E}_{ef})^2=\langle\mathcal{E}_{ef}^2\rangle-\langle\mathcal{E}_{ef}\rangle^2=\langle\mathcal{E}_{ef}\rangle^2. \tag{91}$$

It is easy to see that its relative dispersion is unity, so that condition (79) holds in this case.

For the convenience of the comparison of the obtained formulae with the non quantum version of ST {LaLi68}, we generalize the concept of heat capacity, introducing the effective heat capacity of the object

$$(C_V)_{ef}\equiv\frac{\partial\langle\mathcal{E}_{ef}\rangle}{\partial T_{ef}}=k_B. \tag{92}$$

This allows writing formula (91) for the dispersion of the internal energy in a form that is similar to formula (83), but the macroparameters are replaced with their effective analogs in this case:

$$(\Delta\mathcal{E}_{ef})^2=k_B(C_V)_{ef}T_{ef}^2=k_B^2T_{ef}^2. \tag{93}$$

It should be emphasized that we assumed in all above-mentioned formulae in Sect.4 that $T_{ef}=(T_{ef})_0$ and $T=T_0$, where $(T_{ef})_0$ and T_0 are the effective and Kelvin temperature of the QHB correspondingly.

Indeed, in the macrotheory under consideration, we start from the fact that the effective object temperature T_{ef} also experiences fluctuations. Therefore, the zero law according to (67) and (76) becomes

$$T_{ef} = T^0_{ef} \pm \delta T_{ef}, \tag{94}$$

where δT_{ef} is the fluctuation of the effective object temperature. According to the main MST postulate, the form of the expression for the dispersion of the effective object temperature is similar to that of expression (80):

$$(\Delta T_{ef})^2 = \frac{k_B}{(C_V)_{ef}} T^2_{ef} = T^2_{ef}, \tag{95}$$

so that the relative dispersion of the effective temperature also obeys condition (79).
To compare the obtained formulae with those in QSM-based thermodynamics, we represent dispersion of the effective internal energy (93) in the form

$$(\Delta \mathcal{E}_{ef})^2 = \left(\frac{\hbar\omega}{2}\right)^2 (\coth \varkappa \frac{\omega}{T})^2 = \left(\frac{\hbar\omega}{2}\right)^2 \cdot [1 + \sinh^{-2}(\varkappa \frac{\omega}{T})]. \tag{96}$$

The comparison of formula (96) with expression (85), where the heat capacity has form (84), allows writing the second term in (96) in the form resembling initial form (81)

$$(\Delta \mathcal{E}_{ef})^2 = (\frac{\hbar\omega}{2})^2 + k_B (C_V)_{qu} T^2. \tag{97}$$

However, in contrast to formula (85), the sum in it is divided into two terms differently. Indeed, the first term in formula (97) can be written in the form

$$(\frac{\hbar\omega}{2})^2 = \frac{\hbar}{2} \rho_\omega(\omega,0)\omega^2, \tag{98}$$

where

$$\rho_\omega(\omega,0) \equiv \left.\frac{\partial \langle \mathcal{E}_{ef} \rangle}{\partial \omega}\right|_{T=0} = \frac{\hbar}{2}$$

is the spectral density of the effective internal energy at $T = 0$. Then formula (97) for the dispersion of the effective internal energy acquires the form generalizing formula (85):

$$(\Delta \mathcal{E}_{ef})^2 = \frac{\hbar}{2} \rho_\omega(\omega,0)\omega^2 + k_B [C_V(\omega,T)]_{qu} T^2. \tag{99}$$

It is of interest to note that in contrast to formula (85) for the quantum oscillator or a similar formula for thermal radiation, an additional term appears in formula (99) and is also manifested in the cold vacuum. The symmetric form of this formula demonstrates that the concepts of characteristics, such as frequency and temperature, are similar, which is manifested in the expression for the minimal effective temperature $T^0_{ef} = \varkappa\omega$. The corresponding analogies between the world constant $\hbar/2$ and k_B and also between the characteristic energy "densities" ρ_ω and $(C_V)_{qu}$ also exist.
In the limit $T \to 0$, only the first term remains in formula (99), and, as a result,

$$(\Delta \mathcal{E}^0_{ef})^2 = (\mathcal{E}^0_{ef})^2 = (\frac{\hbar\omega}{2})^2 \neq 0. \tag{100}$$

In our opinion, we have a very important result. This means that zero-point energy is "smeared", i.e. it has a non-zero width. It is natural that the question arises as to what is the reason for the fluctuations of the effective internal energy in the state with $T = 0$. This is because the peculiar stochastic thermal influence exists even at zero Kelvin temperature due to $T_{ef} \neq 0$. In this case the influence of "cold" vacuum in the form (100) is equivalent to $k_B T^0_{ef}/\omega$. In contrast to this, $(\Delta\mathcal{E}_{qu})^2 \to 0$, as $T \to 0$ in QSM-based thermodynamics, because the presence of the zero point energies is taken into account not at all in this theory.

4.3 Correlation between fluctuations and the uncertainties relations for effective macroparameters

Not only the fluctuations of macroparameters, but also the correlation between them under thermal equilibrium play an important role in MST. This correlation is reflected in correlators contained in the uncertainties relations (UR) of macroparameters {Su05}

$$\Delta A_i \Delta A_j \geqslant \langle \delta A_i, \delta A_j \rangle, \tag{101}$$

where the uncertainties ΔA_i and ΔA_j on the left and the correlator on the right must be calculated independently. If the right side of (101) is not equal to zero restriction on the uncertainties arise.

We now pass to analyzing the correlation between the fluctuations of the effective macroparameters in thermal equilibrium. We recall that according to main MST postulate, the formulae for dispersions and correlators remain unchanged, but all macroparameters contained in them are replaced with the effective ones: $A_i \to (A_{ef})_i$.

a). Independent effective macroparameters

Let us consider a macrosystem in the thermal equilibrium characterizing in the space of effective macroparameters by the pair of variables T_{ef} and V_{ef}.Then the probability density of fluctuations of the effective macroparameters becomes {LaLi68}, {An73}

$$W(\delta T_{ef}, \delta V_{ef}) = C \exp\left\{ -\frac{1}{2}\left(\frac{\delta T_{ef}}{\Delta T_{ef}}\right)^2 - \frac{1}{2}\left(\frac{\delta V_{ef}}{\Delta V_{ef}}\right)^2 \right\}. \tag{102}$$

Here, C is the normalization constant, the dispersion of the effective temperature $(\Delta T_{ef})^2$ has form (95), and the dispersion of the effective volume δV_{ef} is

$$(\Delta V_{ef})^2 = -k_B T_{ef} \left.\frac{\partial V_{ef}}{\partial P_{ef}}\right|_{T_{ef}} . \tag{103}$$

We note that both these dispersions are nonzero for any T.

Accordingly to formula (102) the correlator of these macroparameters $\langle \delta T_{ef}, \delta V_{ef} \rangle = 0$. This equality confirms the independence of the fluctuations of the effective temperature and volume. Hence it follows that the UR for these quantities has the form $\Delta T_{ef} \Delta V_{ef} \geqslant 0$, i.e., no additional restrictions on the uncertainties ΔT_{ef} and ΔV_{ef} arise from this relation.

b). Conjugate effective macroparameters

As is well known, the concept of conjugate quantities is one of the key concepts in quantum mechanics. Nevertheless, it is also used in thermodynamics but usually on the basis of

heuristic considerations. Without analyzing the physical meaning of this concept in MST (which will be done in 4.4), we consider the specific features of correlators and URs for similar pairs of effective macroparameters.
Based on the first law of thermodynamics, Sommerfeld emphasized {So52}] that entropy is a macroparameter conjugate to temperature. To obtain the corresponding correlator, we calculate the fluctuation of the effective entropy S_{ef} :

$$\delta S_{ef} = \left.\frac{\partial S_{ef}}{\partial T_{ef}}\right|_{V_{ef}} \delta T_{ef} + \left.\frac{\partial S_{ef}}{\partial V_{ef}}\right|_{T_{ef}} \delta V_{ef} = \frac{(C_V)_{ef}}{T_{ef}} \delta T_{ef} + \left.\frac{\partial P_{ef}}{\partial T_{ef}}\right|_{V_{ef}} \delta V_{ef} \tag{104}$$

In the calculation of the correlator of fluctuations of the macroparameters δS_{ef} and δT_{ef} using distribution (102), the cross terms vanish because of the independence of the quantities δV_{ef} and δT_{ef}. As a result, the correlator contains only one term proportional to $(\Delta T_{ef})^2$ so that $\langle \delta S_{ef}, \delta T_{ef} \rangle$ becomes

$$\langle \delta S_{ef}, \delta T_{ef} \rangle = \frac{(C_V)_{ef}}{T_{ef}} (\Delta T_{ef})^2 = k_B T_{ef}. \tag{105}$$

We note that, the obtained expression depends linearly on T_{ef}.
To analyze the desired UR, we find the dispersion $(\Delta S_{ef})^2$, using distribution (102):

$$(\Delta S_{ef})^2 = \left(\frac{(C_V)_{ef}}{T_{ef}}\right)^2 (\Delta T_{ef})^2 + \left(\left.\frac{\partial P_{ef}}{\partial T_{ef}}\right|_{V_{ef}}\right)^2 (\Delta V_{ef})^2, \tag{106}$$

where ΔT_{ef} and ΔV_{ef} are defined by formulas (95) and (103). This expression can be simplified for $V_{ef} = const$. Thus, if (92) and (95) are taken into account, the uncertainty ΔS_{ef} becomes

$$\Delta S_{ef} = \frac{(C_V)_{ef}}{T_{ef}} (\Delta T_{ef}) = k_B. \tag{107}$$

As a result, the uncertainties product in the left-hand side of the UR has the form

$$(\Delta S_{ef})(\Delta T_{ef}) = k_B T_{ef}. \tag{108}$$

Combining formulas (108) and (105), we finally obtain the "effective entropy–effective temperature" UR in the form of an equality

$$(\Delta S_{ef})(\Delta T_{ef}) = k_B T_{ef} = \langle \delta S_{ef}, \delta T_{ef} \rangle. \tag{109}$$

In the general case, for $V_{ef} \neq const$, the discussed UR implies the inequality

$$\Delta S_{ef} \Delta T_{ef} \geqslant k_B T_{ef}. \tag{110}$$

In other words, the uncertainties product in this case is restricted to the characteristic of the QHB, namely, its effective temperature, which does not vanish in principle. This is equivalent to the statement that the mutual restrictions imposed on the uncertainties ΔS_{ef} and ΔT_{ef} are governed by the state of thermal equilibrium with the environment. Analogical result is valid for conjugate effective macroparameters the pressure P_{ef} and V_{ef}.

4.4 Interrelation between the correlators of conjugate effective macroparameters fluctuations and the stochastic action. The second holistic stochastic-action constant

To clarify the physical meaning of the correlation of macroparameters fluctuations we turn to results of the sections 2 and 3. In this case, we proceed from the Bogoliubov idea, according to which only the environmental stochastic influence can be the reason for the appearance of a nontrivial correlation between fluctuations of both micro and macroparameters.

We recall that the effective action $\mathcal{J}_{ef}$ in MST which is connected with the Schrödingerian in $\hbar kD$ is a characteristic of stochastic influence. Its definition in formula (62) was related to the quantum correlator of the canonically conjugate quantities, namely, the coordinate and momentum in the thermal equilibrium state. In this state, the corresponding UR is saturated {Su06}:

$$\Delta p_{ef} \Delta q_{ef} \equiv \mathcal{J}_{ef}, \tag{111}$$

where uncertainties are

$$\Delta p_{ef} = \sqrt{m\omega}\sqrt{\mathcal{J}_{ef}} \quad \text{and} \quad \Delta q_{ef} = \frac{1}{\sqrt{m\omega}}\sqrt{\mathcal{J}_{ef}}.$$

We stress that in this context, the quantities p_{ef} and q_{ef} also have the meaning of the effective macroparameters, which play an important role in the theory of Brownian motion.

We show that correlator of the effective macroparameters (105) introduced above also depend on $\mathcal{J}_{ef}$. We begin our consideration with the correlator of "effective entropy–effective temperature" fluctuations. Using (110), we can write relation (105) in the form

$$\langle \delta S_{ef}, \delta T_{ef} \rangle = \omega \mathcal{J}_{ef} \quad \text{or} \quad \langle \delta S_{ef}, \delta \mathcal{J}_{ef} \rangle = k_B \mathcal{J}_{ef}. \tag{112}$$

Thus, we obtain two correlators of different quantities. They depend linearly on the effective action $\mathcal{J}_{ef}$; so, they are equivalent formally.

However, the pair of correlators in formula (112) is of interest from the physical point of view because their external identity is deceptive. In our opinion, the second correlator is more important because it reflects the interrelation between the environmental stochastic influence in the form $\delta\mathcal{J}_{ef}$ and the response of the object in the form of entropy fluctuation δS_{ef} to it.

To verify this, we consider the limiting value of this correlator as $T \to 0$ that is equal to the production

$$k_B \mathcal{J}^0_{ef} = k_B \frac{\hbar}{2} \equiv \widetilde{\varkappa}, \tag{113}$$

where $\widetilde{\varkappa}$ is the *second* holistic stochastic action constant differing from the *first* one $\varkappa = \hbar/2k_B$. In the macrotheory, it is a minimal restriction on the uncertainties product of the effective entropy and the effective action:

$$\Delta S^0_{ef} \Delta \mathcal{J}^0_{ef} = k_B \frac{\hbar}{2} = \widetilde{\varkappa} \neq 0. \tag{114}$$

The right-hand side of this expression contains the combination of the world constants k_B and $\frac{\hbar}{2}$, which was not published previously.

We compare expression (114) with the limiting value of the Schrödinger quantum correlator for the "coordinate–momentum" microparameters {Su06}, which are unconditionally assumed to be conjugate. In the microtheory, it is a minimum restriction on the product of the uncertainties Δp and Δq and is equal to

$$\Delta p_0 \Delta q_0 = \mathcal{J}^0_{ef} = \frac{\hbar}{2},$$

i.e., it also depends only on the world constant. Accordingly, convincing arguments used to admit that $\mathcal{J}_{ef}$ and S_{ef} are conjugate macroparameters appear.
Summarizing the above considerations, we formulate the criterion that allows us to independently estimate, what pair of the macroparameters can be considered conjugate.In our opinion, it reduces to the following conditions: a). the correlator of their fluctuations depends on $\mathcal{J}_{ef}$ linearly, and b). the minimum restriction on the uncertainties product is fixed by either one of the world constants $\frac{1}{2}\hbar$ and k_B or their product.
We note that the correlators of conjugate macroparameters fluctuations vanish in the case of the classical limit where environmental stochastic influence of quantum and thermal types are not taken into account. In this case, the corresponding quantities can be considered independent, the URs for them become trivial, and any restrictions on the values of their uncertainties vanish.

4.5 Transport coefficients and their interrelation with the effective action

We now turn to the analysis of transport coefficients. It follows from the simplest considerations of kinetic theory that all these coefficients are proportional to each other. We show below, what is the role of the effective action $\mathcal{J}_{ef}$ in this interrelation.
As we established {Su06}, "coordinate–momentum" UR (111) for the quantum oscillator in a thermostat can be written in the form

$$\Delta p_{ef} \Delta q_{ef} = m D_{ef}. \tag{115}$$

Then, for the effective self-diffusion coefficient with account (111), we have the expression

$$D_{ef} = \frac{\mathcal{J}_{ef}}{m}. \tag{116}$$

We now take into account the relation between the effective shear viscosity coefficient η_{ef} and the coefficient D_{ef}. We then obtain

$$\eta_{ef} = D_{ef} \rho_m = \frac{\mathcal{J}_{ef}}{V}, \tag{117}$$

where ρ_m is the mass density.
In our opinion, the ratio of the heat conductivity to the electroconductivity contained in the Wiedemann–Franz law is also of interest:

$$\frac{\lambda}{\sigma} = \gamma (\frac{k_B}{e})^2 T = \gamma \frac{k_B}{e^2} (k_B T), \tag{118}$$

where γ is a numerical coefficient. Obviously, the presence of the factor $k_B T$ in it implies that the classical heat bath model is used.
According to the main MST postulate, the generalization of this law to the QHB model must have the form

$$\frac{\lambda_{ef}}{\sigma_{ef}} = \gamma (\frac{k_B}{e})^2 T_{ef} = \gamma \frac{k_B}{e^2} (k_B T_{ef}) = \gamma \frac{k_B}{e^2} (k_B T^0_{ef}) \coth \varkappa \frac{\omega}{T}. \tag{119}$$

It is probable that this formula, which is also valid at low temperatures, has not been considered in the literature yet. As $T \to 0$, from (119), we obtain

$$\left(\frac{\lambda^0_{ef}}{\sigma^0_{ef}} \right) = \gamma (\frac{k_B}{e})^2 T^0_{ef} = \gamma \frac{\omega}{e^2} (k_B \frac{\hbar}{2}) = \gamma \frac{\omega}{e^2} \widetilde{\varkappa}, \tag{120}$$

where $T^0_{ef} = \varkappa\omega$, and the constant $\widetilde{\varkappa}$ coincides with the correlator $\langle \delta S^0_{ef}, \delta \mathcal{J}^0_{ef} \rangle$ according to (114). We assume that the confirmation of this result by experiments is of interest.

5. Conclusion

So, we think that QSM and non-quantum version of ST as before keep their concernment as the leading theories in the region of their standard applications.
But as it was shown above, MST allows filling gaps in domains that are beyond of these frameworks. MST is able to be a ground theory at calculation of effective macroparameters and, their dispersions and correlators at low temperatures.
In the same time, MST can be also called for explanation of experimental phenomena connected with behavior of the ratio "shift viscosity to the volume density of entropy" in different mediums. This is an urgent question now for describing of nearly perfect fluids features.
In additional, the problem of zero-point energy smearing is not solved in quantum mechanics. In this respect MST can demonstrate its appreciable advantage because it from very beginning takes the stochastic influence of cold vacuum into account. This work was supported by the Russian Foundation for Basic Research (project No. 10-01-90408).

6. References

Anselm, A.I. (1973). *The Principles of Statistical Physics and Thermodynamics*, Nauka, ISBN 5-354-00079-3 Moscow

Bogoliubov, N.N. (1967). *Lectures on Quantum Statistics*, Gordon & Breach, Sci.Publ.,Inc. New York V.1 Quantum Statistics. 250 p.

Boltzmann, L. (1922). *Vorlesungen über die Prinzipien der Mechanik*, Bd.2, Barth, Leipzig

Dodonov, V.V. & Man'ko, V.I. (1987). Generalizations of Unsertainties Relations in Quantum Mechanics. *Trudy Lebedev Fiz. Inst.*, Vol.183, (September 1987) (5-70), ISSN 0203-5820

Gibbs, J.W. (1960). *Elementary Principles in Statistical Mechanics*, Dover, ISBN 1-881987-17-5 New York

Landau, L.D. & Lifshits E.M. (1968). *Course of Theoretical Physics*, V.5, Pergamon Press, ISBN 5-9221-00055-6 Oxford

Sommerfeld, A. (1952). *Thermodynamics and Statistical Physics*, Cambridge Univ. Press, ISBN 0-521-28796-0 Cambridge

Sukhanov A.D. (1999). On the Global Interrelation between Quantum Dynamics and Thermodynamics, *Proceedings of 11-th Int. Conf "Problems of Quantum Field Theory"*, pp. 232-236, ISBN 5-85165-523-2, Dubna, July 1998, JINR, Dubna

Sukhanov, A.D. (2005). Einstein's Statistical-Thermodynamic Ideas in a Modern Physical Picture of the World. *Phys. Part. Nucl.*, Vol.36, No.6, (December 2005) (667-723), ISSN 0367-2026

Sukhanov, A.D. (2006). Schrödinger Uncertainties Relation for Quantum Oscillator in a Heat Bath. *Theor. Math. Phys*, Vol.148, No.2, (August 2006) (1123-1136), ISSN 0564-6162

Sukhanov, A.D. (2008). Towards a Quantum Generalization of Equilibrium Statistical Thermodynamics: Effective Macroparameters *Theor. Math. Phys*, Vol.153, No.1, (January 2008) (153-164), ISSN 0564-6162

Sukhanov, A.D. & Golubjeva O.N. (2009). Towards a Quantum Generalization of Equilibrium Statistical Thermodynamics: $(\hbar - k)$– Dynamics *Theor. Math. Phys*, Vol.160, No.2, (August 2009) (1177-1189), ISSN 0564-6162

5

On the Two Main Laws of Thermodynamics

Martina Costa Reis and Adalberto Bono Maurizio Sacchi Bassi
Universidade Estadual de Campinas
Brazil

1. Introduction

The origins of thermodynamics date back to the first half of the nineteenth century, when the industrial revolution occurred in Europe. Initially developed for engineers only, thermodynamics focused its attention on studying the functioning of thermal machines. Years after the divulgation of results obtained by Carnot on the operation of thermal machines, Clausius, Kelvin, Rankine, and others, re-discussed some of the ideas proposed by Carnot, so creating classical thermodynamics. The conceptual developments introduced by them, in the mid of XIX century, have allowed two new lines of thought: the kinetic theory of gases and equilibrium thermodynamics. Thus, thermodynamics was analyzed on a microscopic scale and with a mathematical precision that, until then, had not been possible (Truesdell, 1980). However, since mathematical rigor had been applied to thermodynamics through the artifice of timelessness, it has become a science restricted to the study of systems whose states are in thermodynamic equilibrium, distancing itself from the other natural sciences.

The temporal approach was resumed in the mid-twentieth century only, by the works of Onsager (Onsager, 1931a, b), Eckart (Eckart, 1940) and Casimir (Casimir, 1945), resulting in the thermodynamics of irreversible processes (De Groot & Mazur, 1984). Later in 1960, Toupin & Truesdell (Toupin & Truesdell, 1960) started the modern thermodynamics of continuous media, or continuum mechanics, today the most comprehensive thermodynamic theory. This theory uses a rigorous mathematical treatment, is extensively applied in computer modeling of various materials and eliminates the artificial separation between thermodynamics and chemical kinetics, allowing a more consistent approach to chemical processes.

In this chapter, a radical simplification of thermodynamics of continuous media is obtained by imposing the homogeneous restriction on the process, that is, all the extensive and intensive properties of the system are functions of time, but are not functions of space. Improved physical understanding of some of the fundamental concepts of thermodynamics, such as internal energy, enthalpy, entropy, and the Helmholtz and Gibbs energies is presented. Further, the temporal view is applied to the first and second laws of thermodynamics. The conservation of linear and angular momenta, together with the rigid body concept, stresses the union with mechanics for the first law. For the second law, intrinsic characteristics of the system are central for understanding dissipation in thermally homogeneous processes. Moreover, including the definitions for non equilibrium states, the basic intensive properties of temperature, pressure and chemical potential are re-discussed.

This is accomplished without making use of statistical methods and by selecting a mathematically coherent, but simplified temporal theory.

2. Some basic concepts

2.1 Continuous media and thermodynamic properties

The concept of continuous medium is derived from mathematics. The set of real numbers is continuous, since between any two real numbers there is infinity of numbers and it will always be possible to find a real number between the pair of original numbers, no matter how close they are (Mase & Mase, 1999). Similarly, the physical space occupied by a body is continuous, although the matter is not continuous, because it is made up of atoms, which are composed of even smaller particles. Clearly, a material body does not fill the space it occupies, because the space occupied by its mass is smaller than the space occupied by its volume. But, according to the continuity of matter assumption, any chemical homogeneous body can be divided into ever-smaller portions retaining all the chemical properties of the original body, so one can assume that bodies completely fill the space they occupy. Moreover, this approach provides a solid mathematical treatment on the behavior of the body, which is correctly described by continuous real functions of time (Bassi, 2005a; Nery & Bassi, 2009a).

With continuity imposed on matter, the body is called a system and, obviously, the mass and the volume of any system occupy the same space. If the outside boundary of the system is impermeable to energy and matter, the system is considered isolated. Otherwise, the system will be considered closed if the boundary that separates it from the outside is impermeable to mass only. The amount of any thermodynamic quantity is indirectly or directly perceived by an observer located within the system. A thermodynamic quantity whose amount cannot be verified by an observer located within an isolated system is not a property. The value assigned to any property is relative to some well established referential (m, mole etc.), but a referential does not need to be numerically well defined (the concept of mole is well established, but it is not numerically well defined).

Properties are further classified into intensive, additive extensive and non-additive extensive properties. Intensive properties are those that, at time t, may present real values at each point $<n_1, n_2, n_3>$ of the system. Thus, if a is an intensive property, there is a specific temporal function $a = a\ (t, n_1, n_2, n_3)$ defining the values of a. Examples of intensive properties are pressure, density, concentration, temperature and their inverses. In turn, extensive properties are those that have null value only (additive) or cannot present a real value (non-additive) at all points of the system. Examples of additive extensive properties are volume, mass, internal energy, Helmholtz and Gibbs energies, entropy and amount of substance. Inverses of additive extensive properties are non-additive extensive properties, but the most useful of these are products of additive extensive properties by inverses of additive extensive properties, such as the mean density of a system (Bassi, 2006a).

2.2 Mathematical formalism

Let a continuous function $y= f(x)$ be defined in an open interval of real numbers (a, b). If a fixed real number x within this range is chosen, there is a quotient,

$$\frac{f(x+h)-f(x)}{h}, \tag{1}$$

where h is a positive or negative real and $x+h$ is a real within the interval (a, b). If h approaches zero and the limit of the quotient tends to some well defined real value, then that limit defines the derivative of the function $y= f(x)$ at x (Apostol, 1967),

$$f\,'(x)=\frac{\mathrm{d}y}{\mathrm{d}x}=\lim_{h\to 0}\frac{f(x+h)-f(x)}{h}. \quad (2)$$

The first equality of Equation 2 could still be represented by $\mathrm{d}y= f\,'(x)\mathrm{d}x$, but not by multiplication of both its sides by the inverse of $\mathrm{d}y$, because the values of $\mathrm{d}y$ and $\mathrm{d}x$ may be null and their inverses may diverge, thus the integrity of Equation 2 would not be maintained. It is fundamental to remember that the $\mathrm{d}y$ and $\mathrm{d}x$ values include not only finite quantities but necessarily zero, because there is a qualitative difference between null and finite quantities, no matter how small the finite quantities become. Thus, as well as Equation 2 cannot be multiplied by the inverse of $\mathrm{d}y$, the equation $\mathrm{d}y= f\,'(x)\mathrm{d}x$ does not refer to an interval $y_2 - y_1 = f(x_2) - f(x_1)$, no matter how small the finite interval becomes, but uniquely to the fixed real value x (as well as Equation 2).
Certainly, both the mathematical function and its derivative should maintain consistency with physical reality. For example, the $\mathrm{w}= \mathrm{w}(t)$ and $\mathrm{q}= \mathrm{q}(t)$ functions and their derivatives should express the intrinsic characteristics of work and heat and should retain their characteristics for any theory where these quantities are defined. Thus, because the Fourier equation for heat conduction defines $\frac{\mathrm{dq}}{\mathrm{d}t}$, acceptance of its validity implies accepting the existence of a differentiable temporal function $\mathrm{q}= \mathrm{q}(t)$ in any natural science. However, evidently the acceptance of the Fourier equation do not force all existing theories to include the equality $\mathrm{q}= \mathrm{q}(t)$. Surely, it will not be considered by timeless thermodynamics, but that is a constraint imposed on this theory.
Differential equations mathematically relate different quantities that an observer would be able to measure in the system. Some of these relations arise from specific properties of the material (constitutive functions), while others follow the physical laws that are independent of the nature of the material (thermodynamic functions). If the process is not specified, the differentiable function of state $z= u(x, y)$, and the process functions z, respectively correspond to an exact and inexact differential equations. Indeed, one has

$$M(x,y)\mathrm{d}x+N(x,y)\mathrm{d}y=\mathrm{d}z\,, \quad (3)$$

where $\frac{\partial M(x,y)}{\partial y}=\frac{\partial}{\partial y}\left(\frac{\partial u(x,y)}{\partial x}\right)$ and $\frac{\partial N(x,y)}{\partial x}=\frac{\partial}{\partial x}\left(\frac{\partial u(x,y)}{\partial y}\right)$ for $z= u(x, y)$. Because $\frac{\partial^2 u(x,y)}{\partial y\partial x}=\frac{\partial^2 u(x,y)}{\partial x\partial y}$, if

$$\frac{\partial N(x,y)}{\partial x}=\frac{\partial M(x,y)}{\partial y}, \quad (4)$$

then $z= u(x, y)$ and the differential equation (Equation 3) is said to be exact. Otherwise, it is inexact. Thus, for an exact differential equation the function $z= u(x, y)$ can be found, but for solving an inexact differential equation the process must be specified. An important mathematical corollary indicates that the integral of an exact differential equation is independent of the path that leads from state 1 to state 2 (Bassi, 2005b; Agarwal & O'Regan, 2008), because it equals $z_2 - z_1 = u(x_2, y_2) - u(x_1, y_1)$, while this is not true for integrals of inexact differential equations.

Mathematically, the concept of state comprises the smallest set of measurements of system properties, at time *t*, enough to ensure that all measures of properties of the system are known, at that very moment. The definition of state implies that if X is the value of any property of the system at instant *t* and Ξ is the state of the system at that same time, there must be a constitutive or thermodynamic function X =X(Ξ). On the other hand, if Y does not correspond to the value of a property of the system at time *t*, the existence of a function Y=Y(Ξ) is not guaranteed. This shows that all integrals of exact differential equations are function of state differences between two states, while differential equations involving the differentials of properties included in Ξ generally are inexact (Nery & Bassi, 2009b). Thus, all properties are functions of state and, if the process is not specified, all functions of state are properties.

2.3 Relative and absolute scales

Consider a sequence of systems ordered according to the continuous increment of a specific property of them, as for example their volume. This ordering may be represented by a continuous sequence of real numbers named a dimensionless scale. Dimensionless scales can be related each other by choosing functions whose derivatives are always positive. Linear functions do not alter the physical content of the chosen property, but non-linear ones do not expand or contract proportionally all scale intervals. Thus, dimensionless scales related by non-linear functions attribute different physical characteristics to the considered property. For instance, because the dimensionless scales corresponding to empirical and absolute temperatures are related by a non-linear function, empirical temperatures cannot be substituted for absolute temperatures in thermodynamic equations.

The entire real axis is a possible dimensionless scale. Because the real axis does not have a real number as a lower bound neither as an upper bound, it is not sufficient to choose a value in the scale and relate this value to a particular system, in order to convert the dimensionless scale to a dimensional one (Truesdell, 1984). To do this, it is essential to employ at least two values, as for empirical temperature scales. But only one value is needed if a pre-defined unit is used, as in the case of the Pascal unit for pressure ($Pa=Kg\ m^{-1}s^{-2}$, where Kg, m and s are, respectively, the pre-defined units for mass, distance and time). The dimensional scales for empirical temperatures and for pressure are examples of relative scales.

So, if X belongs to the real axis, for $-\infty < X < \infty$ one may propose the new dimensionless scale

$$Y = \exp(X). \tag{5}$$

This new scale, contrasting with the previous one, only includes the positive semi-axis of real numbers with the zero lower bound being as unattainable as the lower bound of the real axis, $-\infty$. By imposing X=0, Equation 5 gives Y=1, where the dimensionless 1 can be related to any system for defining the scale unit. Any scale containing only the positive semi-axis of real numbers that assigns a well defined physical meaning to Y=1 is a dimensional scale called absolute. The physical contents of some properties, as for example the volume, require absolute scales for measuring their amounts in the system (for the volume, Y=1 may be assigned to 1 m^3 and there is not a null volume system).

3. First law of thermodynamics

3.1 Internal energy

According to the thermodynamics of continuous media, the mathematical expression for the first law of thermodynamics is a balance of energy that, along with the balance equations of mass and linear and angular momenta, applies to phenomena that involve the production or absorption of heat. In this approach, conservation of linear and angular momenta is explicit in the energy balance, while in classical thermodynamics conservation of linear and angular momenta are implicitly assumed. Actually, because classical thermodynamics focuses its attention on systems which are macroscopically stationary, linear and angular momenta are arbitrarily zero, restricting the study of several physical systems (Liu, 2002).

The principle of conservation of energy was first enunciated by Joule, near the mid of XIX century, who demonstrated through numerous experiments that heat and work are uniformly and universally inter-convertible. Moreover, the principle of conservation of energy requires that for any positive change in the energy content of the system, there must be an inflow of energy of equal value. Similarly, for any negative change of the energy content of the system, there must be liberation of the same energy value.

Consider a body whose composition is fixed. Moreover, suppose that the positions and relative orientations of the constituent particles of the body are unchanged, but the body can move in space. This body is defined as rigid body and its energy content is the body's energy E_R. Now, consider that the restrictions on the number of particles, their positions and orientations are abolished, so the body's energy is E. Thus, the energy content of the body can be separated into two additive parts

$$E= U+E_R, \tag{6}$$

where U is the internal energy and represents the sum of the energies of the motions, of the constituent particles and into them, which do not change the total linear and angular momenta of the body (internal motions).

While the energy of the rigid body is well defined by the laws of mechanics, the comprehension of internal energy values depends on the microscopic model adopted to describe material bodies. The difference $\Delta U_{a\to b} = U(t_b)-U(t_a)$, between the internal energy at two instants t_a and t_b of a gas supposed ideal, can be experimentally determined. However, it is not possible to experimentally determine the internal energy of any body at instant t.

Similarly, the energy exchange between a body and its exterior is divided into two additive portions named heat and work. Heat, q, is an exchange of energy in which total linear and angular momenta of the body, as well as total linear and angular momenta of its exterior, are not changed. Thus, heat involves only the internal energies of the body and its exterior and cannot be absorbed or emitted by the energy of a rigid body (Moreira & Bassi, 2001; Bassi, 2006b). In turn, work, w, involves both the internal and rigid body energies. Hence, there is no restriction on the rigid body absorption or emission of work (Williams, 1971). Equation 6, as well as the concepts of rigid body energy, internal energy, heat and work is valid not only for bodies, but also for systems.

Considering the time of existence of a process in a closed system, the heat exchanged from the initial instant $t_\#$ until the instant t is denoted by $\Delta q(t)= q(t)-q_\#$, where $q_\#$ represents the heat exchanged from a referential moment until the initial instant $t_\#$ of the process and $q(t)$ indicates the heat exchanged from the referential moment until instant t. Likewise, one has $\Delta w(t)= w(t)-w_\#$, $\Delta w_R(t)= w_R(t)-w_{R\#}$ and, by imposing $q_\#= 0$, $w_\#= 0$ and $w_{R\#}= 0$, respectively $\Delta q(t)= q(t)$, $\Delta w(t)= w(t)$ and $\Delta w_R(t)= w_R(t)$. Assuming $\Delta E_R(t)= E_R(t)-E_{R\#}$ and $\Delta U(t)= U(t)-U_\#$, energy conservation implies that

$$\Delta E_R(t)+ \Delta U(t)= \Delta q(t)+\Delta w_R(t)+\Delta w(t)\,, \tag{7}$$

where $\Delta w(t)$ indicates the portion of the work that is transformed into internal energy or comes from it.

The more general statement of the first law of thermodynamics is:

"The internal energy and the energy of rigid body do not interconvert (Šilhavý, 1989)."

Therefore, according to the statement on the first law and Equation 7,

$$\Delta E_R(t)= \Delta w_R(t), \tag{8}$$

and, subtracting Equation 8 from Equation 7,

$$\Delta U(t)= \Delta q(t)+\Delta w(t). \tag{9}$$

Equation 9 is the mathematical expression of the first law of thermodynamics for closed systems. For the range from t_a to t_b, where $t_\#< t_a \leq t \leq t_b < t^\#$, Equation 9 may be written

$$\int_{t_a}^{t_b} \frac{dU}{dt} dt = \int_{t_a}^{t_b} \frac{dq}{dt} dt + \int_{t_a}^{t_b} \frac{dw}{dt} dt\,, \tag{10}$$

and, by making $t_a \to t$ and $t_b \to t$, the limit of Equation 10 is

$$\frac{dU}{dt} = \frac{dq}{dt} + \frac{dw}{dt}\,, \tag{11}$$

where $\frac{dU}{dt}$ is the rate of change of internal energy of the system at time t, and $\frac{dq}{dt}$ and $\frac{dw}{dt}$ are respectively the thermal and the non thermal powers that the system exchanges with the outside at that instant. Defining the differentials

$$dU= \frac{dU}{dt}dt\,,\; dq= \frac{dq}{dt}dt \text{ and } dw= \frac{dw}{dt}dt\,, \tag{12}$$

Equation 11 may be written

$$dU= dq+dw. \tag{13}$$

Considering the entire range of existence of a process $t_\# < t < t^\#$ and imposing $q_\#= 0$ and $w_\#= 0$, Equation 9 can be rewritten

$$\Delta U= q+w, \tag{14}$$

which is the most usual form of the first law. Equations 9 to 14 reflect the conservation of energy in the absence of changes of total linear and angular momenta.

Because differentials are not extremely small finite intervals, it should be noted that Equation 9 cannot be extrapolated to Equation 13. But in some textbooks Equation 13 is proposed considering that: (a) dU is an exact differential, but dq and dw are inexact differentials or (b) dq and dw are finite intervals, while dU is a differential. Such considerations arise from a mistaken view of the differential concept. Indeed: (1) in order to a differential equation to have mathematical meaning, its differentials must be defined using derivatives, as in Equations 3 (by using the process specifications if needed) and 12; (2) the subtraction of two different well-defined real values corresponds to a well-defined finite interval and produces a well-defined real, no matter how small, but never a differential, which is an undetermined real and (3) there are exact and inexact differential equations, but there is not such classification for differentials. In short, Equation 13 is a consequence of Equation 9 if and only if the differentials dU, dq and dw are defined using derivatives, while the validity of Equation 14 does not require this (Gurtin, 1971; Nery & Bassi, 2009b).

3.2 Enthalpy

Suppose a closed system whose outside homogeneously exerts, on the system boundary, a well defined constant pressure p' during the entire existence of a process occurring in the system, including the initial and final instants of the process. Additionally, consider homogeneous the system pressure at the initial, $p_\#$, and final, $p^\#$, process instants, that is, consider that, at the initial and final process instants, the system is in mechanical equilibrium with outside, so that $p_\# = p^\# = p'$. Therefore, for a process under constant pressure it is necessary that the system be in mechanical equilibrium at $t_\#$ and $t^\#$, but it is not necessary that this also occurs during the existence interval of the process, $t_\# < t < t^\#$. If, excluding the volumetric work performed by p' or against p', $\Delta w_{nv}(t)$ is the work exchanged by the system from the initial instant to an instant t such that $t_\# < t < t^\#$, thus, for a process under constant pressure occurring in a closed system,

$$\Delta w_{nv}(t)+\Delta q(t)= \Delta U(t)+\Delta(pV)(t)= \Delta H(t), \tag{15}$$

because the enthalpy at instant t, $H(t)$, is defined by

$$H(t)= U(t)+(pV)(t). \tag{16}$$

If $\Delta w_{nv}(t)= 0$, Equation 15 indicates that the heat exchanged with the outside during a process under constant pressure is the enthalpy change $\Delta H(t)$ (Planck, 1945). This result is of fundamental importance for thermo-chemistry, because in this system the enthalpy behaves similarly to the internal energy in a closed system limited by rigid walls. In analogy to the mathematical expression of the first law of thermodynamics for closed systems (Equation 9), ΔH indicates the module and the direction of the exchange of energy $\Delta w_{nv}(t)+\Delta q(t)$ between the system and its surroundings. Considering $\Delta w_{nv}(t)= 0$, if $\Delta H < 0$ the process is said to be exothermic and, if $\Delta H > 0$, the process is endothermic.

4. Second law of thermodynamics

4.1 Statement for the second law

The first law of thermodynamics is not sufficient to determine the occurrence of physical or chemical processes. Whereas the first law addresses just the energetic content of system, the second law demands further conditions for the existence of a process. Treatises on classical thermodynamics contain several statements about the second law, which are frequently associated with the works of Clausius, Kelvin, Carnot and Planck. Despite some differences among the various statements, all of them claim that to produce an amount of work in a cyclic process, the system must not only absorb heat, but it must also emit some amount of it (Kestin, 1976).

This is equivalent to the establishment that, for any closed system at a homogeneous temperature, work and internal energy may always be converted into heat according to the first law, but there is a limit for the rate of absorbing heat,

$$\frac{dq}{dt} \leq T\frac{dS}{dt}, \tag{17}$$

and for the rate of producing work (Šilhavý, 1997),

$$-\frac{dw}{dt} \leq T\frac{dS}{dt} - \frac{dU}{dt}, \tag{18}$$

where T is the homogeneous absolute temperature and S is the entropy. The variables T, S and U correspond to properties of the closed system but, because time derivatives of state functions are not state functions (Nery & Bassi, 2009b), Equation 17 does not necessarily impose a constraint on the rate of heat absorption. On the contrary, given this rate, Equation 17 causes an entropy rate increase at least equal to $\frac{1}{T}\frac{dq}{dt}$ and, given $\frac{dU}{dt}$, Equation 18 shows that a larger entropy rate increase corresponds to a larger rate of producing work. But the system must return to the same state for cyclic processes, thus in such processes restrictions are imposed to the variations of state functions. Indeed, Equation 17 indicates that, for thermally homogeneous cyclic processes that occur in closed systems (Serrin, 1979),

$$\oint \frac{dq}{T} \leq 0. \tag{19}$$

Equation 17 also introduces the idea of dissipation (Šilhavý, 1983). If the dissipation is defined by

$$\frac{d\delta}{dt} \equiv T\frac{dS}{dt} - \frac{dq}{dt}, \tag{20}$$

Equation 17 shows that

$$\frac{d\delta}{dt} \geq 0. \tag{21}$$

Hence, the second law of thermodynamics asserts that there exists in nature an amount which, by changes on a closed system at homogeneous temperature, either remains constant (non-dissipative processes) or increases (dissipative processes). The concept of dissipation presented here is analogous to friction (Truesdell, 1984). However, it is an internal friction in the system and not between the system and its outside. Dissipation always occurs when, in the state considered, there is a tendency to change the internal motions of the system.

So far, the second law of thermodynamics has been defined for thermally homogeneous closed systems. If additional restrictions are imposed on the system such as system isolation, according to Equations 17 and 21 the second law states that

$$\frac{dS}{dt} \geq 0 \text{ and } \frac{d\delta}{dt} \geq 0 . \tag{22}$$

Equation 22 confirms that dissipation may occur even for a system which do not exchange energy with its outside, reinforcing the fact that dissipation is an internal phenomenon of the system. Isolated systems are not the only special thermally homogeneous closed systems of interest. Thus, some specific situations are detailed in the following text. But, first, note that thermodynamic reservoirs are not considered in this approach because, by definition, reservoirs are systems which do not obey the same physical laws of the system under study. However, it is possible to make experimentally confirmed deductions by imposing that the environment obeys the same physical laws as the body (Hutter, 1977; Serrin, 1979; Nery & Bassi, 2009b). Now, consider a thermally homogeneous closed system under:

Adiabatic process: If no heat exchange with the outside is imposed, $\frac{dq}{dt} = 0$ and, according to Equation 17,

$$\frac{dS}{dt} \geq 0 . \tag{23}$$

An important consequence obtained from Equation 23 is that, in a non-dissipative process, the words "adiabatic" and "isentropic" have the same meaning, but for a dissipative adiabatic process clearly $\frac{dS}{dt} > 0$ (Truesdell, 1991).

Isoenergetic process: If an isoenergetic process is considered, but interactions between the system and its outside are allowed, Equation 18 shows that

$$-\frac{dw}{dt} \leq T\frac{dS}{dt} . \tag{24}$$

Thus, for a non-dissipative process the entropy of a system can be decreased by doing work on the system. This is a very interesting assertion, because it eliminates the wrong idea that in any process whatever the entropy of a system always remains either constant or increases. Obviously, if the absence of both volumetric and non-volumetric work is imposed, changes of $\frac{dS}{dt}$ coincide with those of an adiabatic process (Day, 1987).

Isentropic process: If entropy S is maintained constant, from Equation 17

$$\frac{dq}{dt} \leq 0 . \tag{25}$$

Thus, heat cannot be absorbed in an isentropic process. A non-dissipative isentropic process is adiabatic and the change of internal energy coincides with the work exchanged. In the absence of both volumetric and non-volumetric work, Equation 25 becomes

$$\frac{dU}{dt} \leq 0 . \tag{26}$$

Thus, if no work is exchanged during the process, the internal energy does not increase in an isentropic process.

Isothermal process: If the homogeneous temperature is kept constant in time, the time derivative of the Helmholtz energy,

$$A(t)= U(t)-(TS)(t), \tag{27}$$

is

$$\frac{dA}{dt} = \frac{dU}{dt} - T\frac{dS}{dt}, \tag{28}$$

and, using Equation 28, Equation 18 can be described by

$$\frac{dA}{dt} \leq \frac{dw}{dt} . \tag{29}$$

Hence, in an isothermal process, the increase of Helmholtz energy is not greater than the work done on the system. In addition, if no work is exchanged during the process,

$$\frac{dA}{dt} \leq 0 . \tag{30}$$

Equation 30 implies that the Helmholtz energy does not increase. Note that all these conclusions are restricted to isothermal processes. If the process is thermally homogeneous, but not isothermal, instead of Equation 29 the correct relation is $\frac{dA}{dt} \leq \frac{dw}{dt} - S\frac{dT}{dt}$, which is far more complicated.

Isothermal-isobaric process: If both temperature and pressure are homogeneous and constant in time, the time derivative of the Gibbs energy,

$$G(t)= H(t)-(TS)(t), \tag{31}$$

is

$$\frac{dG}{dt} = \frac{dH}{dt} - T\frac{dS}{dt}, \tag{32}$$

and Equation 18 can be replaced by

$$\frac{dU}{dt} - T\frac{dS}{dt} + p\frac{dV}{dt} \leq \frac{dw_{nv}}{dt} . \tag{33}$$

Using Equation 16, Equation 33 may be written

$$\frac{dH}{dt} - T\frac{dS}{dt} \leq \frac{dw_{nv}}{dt}, \tag{34}$$

or, using Equation 32,

$$\frac{dG}{dt} \leq \frac{dw_{nv}}{dt}. \tag{35}$$

For example, for a spontaneous process $\frac{dw_{nv}}{dt}$ may be zero, thus $\frac{dG}{dt} \leq 0$. If an isothermal-isobaric endothermic reaction is spontaneous, $T\frac{dS}{dt}$ is positive and large enough to surpass the positive value $\frac{dH}{dt}$. Therefore, isothermal-isobaric endothermic reactions are driven by the increase of entropy. On the other hand, for a spontaneous isothermal-isobaric exothermic reaction, entropy may decrease but $\frac{dH}{dt}$ must be negative enough to surpass the negative value $T\frac{dS}{dt}$.

Although thermally homogeneous processes must be studied, natural (heterogeneous) processes must also be mentioned. All natural processes will approach thermal homogeneity as the forward process rate decreases, in relation to a finite thermal homogenization rate considered constant. If this happens in a closed system, the process will approach obedience to Equation 17. Nevertheless, because the process tends to a stationary state, the dissipation $\frac{d\delta}{dt}$ tends to zero faster than Equation 17 becomes obeyed. This means that when the forward process rate of the process tends to zero, both a thermally homogeneous dissipative process and a natural process tend towards a thermally homogeneous non-dissipative process. On the other hand, a natural process will approach a thermally homogeneous dissipative process when its thermal homogenization rate is increased, in relation to a finite and constant forward rate of the natural process. Thus, Equation 17 is a limiting equation for natural processes.

4.2 Maximization of missing information

A possible statistical way for expressing the second law of thermodynamics is:

"A system may change over time until the state with the highest density of possible microstates is reached. Once this state is achieved, the system cannot alter it anymore, unless the conditions imposed on the system are modified."

To illustrate this statement of the second law, consider a sphere divided by an imaginary diametrical plane into two compartments I and II. Also, consider two indistinguishable mathematical points moving at random, so the probability of occurrence of any of the following microstates is equal to 1/4: "x in I, y in II", "y in I, x in II", "x and y in I" and "x and y in II". However, since the x and y points are indistinguishable, the probability of the state "one point in I, one point in II" is twice the probability of occurrence for each one of the states "two points in I" and "two points in II" (Bassi, 2005c). Because probability theory rests upon set theory, it is reasonable to introduce states as sets of equally probable microstates.

Now, suppose a gas consisting of only 10 molecules occupying the entire volume of a closed vessel. The probability that all molecules are in the left half of the vessel at the same time t is $1/2^{10} = 1/1024$, that is, for every 1024 seconds this configuration could be observed, on average, during one second. However, thermodynamics deals only with macroscopic systems, where the number of constituents is of the order of the Avogadro constant. So, for one mole of molecules in a gaseous state, the probability that all they are in the left or right half of the vessel is, for all purposes, zero and then one can consider that such state does not exist. But, because the thermodynamic state varies continuously, the concept of the number of microstates corresponding to each state must be replaced by the continuously varying non-dimensional density of microstates, $\gamma \geq 1$, related to each state (Fermi, 1956).
In general, for a macroscopic system the density of possible microstates may be considered null for all states, except for the state with the highest density of possible microstates, which is called the stable state. But, because potential barriers can restrain changes of state, the system may remain in an unstable state until a perturbation suddenly alters the system state. This is the reason for not imposing that the system will change over time until the state with the highest density of possible microstates is reached, in the statistical statement of the second law. Note that, as the density of possible microstates corresponding to the state increases, the partial knowledge about the state of the system decreases. Thus, in the stable state the ignorance (missing information) about the characteristics of the system is maximized.

4.3 Missing structural information and other missing information

In the previous section 4.1 the existence of a thermodynamic property called entropy was introduced, which helps in understanding how a thermodynamic process will evolve. In the present section, an interpretation of entropy is presented, based on the structural characteristics of the system. First, by supposing that the values for all properties that cannot change in an isolated system (such as mass, volume, and internal energy) are already known, for any system define structural information as additional information. Then, for any system, entropy is proportional to the quantity φ of missing structural information (Brillouin, 1962; Gray, 1990).
In an isolated system the missing structural information is associated with the density of microstates by

$$\varphi = c\ln(\gamma), \tag{36}$$

where c is an arbitrary constant of proportionality that defines the unit for measurement of missing structural information. By considering $c = k_B$, where k_B is the Boltzmann constant, Equation 36 is written

$$S = k_B\ln(\gamma), \tag{37}$$

which is the familiar relationship between entropy and the density of microstates of the isolated system (Boltzmann, 1964). Note that the entropy is proportional to the missing structural information for whatever system but, only for an isolated system, entropy is proportional to the logarithm of the density of microstates. Using the statistical statement for the second law, Equation 37 indicates that:

"In an isolated system, the entropy never decreases as time increases."

Therefore, the combined effect of the first and second laws of thermodynamics states that, as time progresses, the internal energy of an isolated system may redistribute without altering its total amount, in order to increase the entropy until the latter reaches a maximum, at the stable state. This statement coincides with the known extreme principles (Šilhavý, 1997).

The interpretation of entropy as a measure of the well defined missing structural information allows a more precise comprehension of this important property, without employing subjective adjectives such as organized and unorganized. For example, consider a gaseous isolated system consisting of one mole of molecules and suppose that all the molecules occupy the left or right half of the vessel. The entropy of this state is lower than the entropy of the stable state because, for an isolated system, the entropy is related to the density of microstates (which, for this state, is lower than the density for the stable state) and, for any system, the entropy is related to the ignorance about the structural conditions of the system (which, for this state, is lower than the ignorance for the stable state). Thus, the entropy does not furnish any information about whether this state is ordered or not (Michaelides, 2008).

Because $\gamma \geq 1$, according to Equation 37 entropy is an additive extensive property whose maximum lower bound value is zero, so that $S \geq 0$. But it is not assured that, for all systems, S can in fact be zero or very close zero. For instance, unlike crystals in which each atom has a fixed mean position in time, in glassy states the positions of the atoms do not cyclically vary. That is, even if the temperature should go to absolute zero, the entropies of glassy systems would not disappear completely, so that they present the residual entropy

$$S_{RES} = k_B \ln(\gamma_G), \tag{38}$$

where $\gamma_G > 1$ represents the density of microstates at 0 K. This result does not contradict Nernst's heat theorem. Indeed, in 1905 Walther Nernst stated that the variation of entropy for any chemical or physical transformation will tend to zero as the temperature approaches indefinitely absolute zero, that is,

$$\lim_{T \to 0} (\Delta S) = 0 . \tag{39}$$

But there is no doubt that the value of S_{RES}, for any substance, is negligible when compared with the entropy value of the same substance at 298.15 K. Therefore, at absolute zero the entropy is considered to be zero. This assertion is equivalent to the statement made by Planck in 1910 that, as the temperature decreases indefinitely, the entropy of a chemical homogeneous body of finite density tends to zero (Planck, 1945), that is,

$$\lim_{T \to 0} (S) = 0 . \tag{40}$$

This assertion allows the establishment of a criterion to distinguish stable states from steady states, because stable states are characterized by a null limiting entropy, whereas for steady states the limiting entropy is not null (Šilhavý, 1997).

Although it is known that γ is not directly associated with the entropy for a non-isolated system, γ still exists and is related to some additive extensive property of the system denoted by ζ (Tolman, 1938; Mcquarrie, 2000). By requiring that the unit for ζ is the same as for S, the generalized Boltzmann equation is written

$$\zeta = k_B \ln(\gamma), \tag{41}$$

where ζ is proportional to some kind of missing information. Considering the special processes discussed in the previous section 4.1, in some cases the property denoted by ζ (Equation 41) can be easily found. For instance, since $\frac{dA}{dt} \leq 0$ for an isothermal process in a closed system which does not exchange work with its surroundings, then $\zeta = -\frac{A}{T} = S - \frac{U}{T}$ for thermally homogeneous closed systems that cannot exchange work with the outside. Analogously, if both the temperature and the pressure of a closed system are homogeneous and the system can only exchange volumetric work with the outside, then $\zeta = -\frac{G}{T} = S - \frac{H}{T}$.

5. Homogeneous processes

5.1 Fundamental equation for homogeneous processes

During the time of existence of a homogeneous process, the value of each one of the intensive properties of the system may vary over time, but at any moment the value is the same for all geometric points of the system. The state of a homogeneous system consisting of J chemical species is characterized by the values of entropy, volume and amount of substance for each one of the J chemical species, that is, the state is specified by the set of values $\Phi = \langle S, V, n_1, \ldots, n_J \rangle$. Obviously, this assertion implies that all other independent properties of the system, as for instance its electric or magnetic polarization, are considered material characteristics which are held constant during the time of existence of the process. Should some of them vary, the set of values Φ would not be enough for specifying the state of the system, but such variations are not allowed in the usual theory. This assertion also implies that S exists, independently of satisfying the equality dq= TdS. This approach was proposed by Planck and is very important, since it allows introducing the entropy without employing concepts such as Carnot cycles (Planck, 1945).
Thus, at every moment t the value of the internal energy U is a state function $U(t) = U(S(t), V(t), n_1(t), \ldots, n_J(t))$. Moreover, since this function is differentiable for any set of values $\Phi = \langle S, V, n_1, \ldots, n_J \rangle$, the equation defining the relationship between dU, dS, dV, and $dn_1, \ldots, dn_J$, is the exact differential equation

$$dU = \frac{\partial U}{\partial S}(\Phi)dS + \frac{\partial U}{\partial V}(\Phi)dV + \sum_{j=1}^{J} \frac{\partial U}{\partial n_j}(\Phi)dn_j \,. \tag{42}$$

The internal energy, the entropy, the volume and the amounts of substance are called the phase (homogeneous system) primitive properties, that is, all other phase properties can be derived from them. For instance, the temperature, the pressure and the chemical potential of any chemical species are phase intensive properties respectively defined by $T = \frac{\partial U}{\partial S}(\Phi)$, $p = -\frac{\partial U}{\partial V}(\Phi)$ and $\mu_j = \frac{\partial U}{\partial n_j}(\Phi)$ for $j = 1, \ldots, J$. Thus, by substituting T, p and μ_j for their corresponding derivatives in Equation 42, the fundamental equation of homogeneous processes is obtained,

$$dU = TdS - pdV + \sum_{j=1}^{J} \mu_j dn_j \, . \qquad (43)$$

Equation 43 cannot be deduced from both Equation 13 and the equalities dq= TdS and dw= -pdV (Nery & Bassi, 2009b). Since the phase can exchange types of work other than the volumetric one, these obviously should be included in the expression of first law, but the fundamental equation of homogeneous processes might not be altered. For instance, an electrochemical cell exchanges electric work, while the electric charge of the cell does not change, thus it is not included in the variables defining the system state, and a piston expanding against a null external pressure produces no work, but the cylinder volume is not held constant, thus the volume is included in the variables defining the system state. Moreover, there is not a "chemical work", because chemical reactions may occur inside isolated systems, but work is a non-thermal energy exchanged with the system outside (section 3.1).

Equations 13 and 43 only coincide for non-dissipative homogeneous processes in closed systems that do not alter the system composition and exchange only volumetric work with the outside. But neither Equation 13, nor Equation 43 is restricted to non-dissipative processes, and a differential equation for dissipative processes cannot be inferred from a differential equation restricted to non-dissipative ones, because differential equations do not refer to intervals, but to unique values of the variables (section 2.2), so invalidating an argument often found in textbooks. Indeed, homogeneous processes in closed systems that do not alter the system composition and exchange only volumetric work with the outside cannot be dissipative processes. Moreover, Equation 13 is restricted to closed systems, while Equation 43 is not. In short, Equation 43, as well as the corresponding equation in terms of time derivatives,

$$\frac{dU}{dt} = T\frac{dS}{dt} - p\frac{dV}{dt} + \sum_{j=1}^{J} \mu_j \frac{dn_j}{dt} \, , \qquad (44)$$

refer to a single instant and a single state of a homogeneous process, which needs not to be a stable state (a state in thermodynamic equilibrium).

The Equations 43 and 44 just demand that the state of the system presents thermal, baric and chemical homogeneity. Because each phase in a multi-phase system has its own characteristics (for instance, its own density), Φ separately describes the state of each phase in the system. But, because the internal energy, the entropy, the volume and the amounts of substance are additive extensive properties, their differentials for the multi-phase system can be obtained by adding the corresponding differentials for a finite number of phases. Thus, the thermal, baric and chemical homogeneities guarantee the validity of Equations 43 and 44 for multi-phase systems containing a finite number of phases.

Further, if an interior part of the system is separated from the remaining part by an imaginary boundary, this open subsystem will still be governed by Equations 43 and 44. Because any additive extensive property will approach zero when the subsystem under study tends to a point, sometimes it is convenient to substitute $u = u(s, v, c_1, \ldots, c_J)$, where $u = \frac{U}{M}$, $s = \frac{S}{M}$, $v = \frac{V}{M}$, $c_j = \frac{n_j}{M}$ for $j = 1, \ldots, J$, and M is the subsystem mass at instant t, for $U = U(S, V, n_1, \ldots, n_J)$. Hence, the equation

$$\mathrm{du} = \mathrm{Tds} - \mathrm{pdv} + \sum_{j=1}^{J} \mu_j \mathrm{dc_j}\,, \tag{45}$$

may substitute Equation 43. Indeed, Equation 45 is a fundamental equation of continuum mechanics.

5.2 Thermodynamic potentials

Not only is the function $U = U(\Phi)$ differentiable for all values of the set Φ, but also the functions $\frac{\partial U}{\partial S}(\Phi)$, $\frac{\partial U}{\partial V}(\Phi)$, and $\frac{\partial U}{\partial n_j}(\Phi)$ for $j = 1,\ldots,J$ are differentiable. Moreover, because $\frac{\partial^2 U}{\partial S^2}(\Phi) \neq 0$, $\frac{\partial^2 U}{\partial V^2}(\Phi) \neq 0$, and $\frac{\partial^2 U}{\partial {n_j}^2}(\Phi) \neq 0$ for $j = 1,\ldots,J$ at any instant t, the state of any phase, besides being described by the set of values Φ, can also be described by any of the following sets

$$\Phi_V(t) \equiv \left\langle S(t), \frac{\partial U}{\partial V}(\Phi(t)), n_1(t), \ldots, n_J(t) \right\rangle, \tag{46}$$

$$\Phi_S(t) \equiv \left\langle \frac{\partial U}{\partial S}(\Phi(t)), V(t), n_1(t), \ldots, n_J(t) \right\rangle, \tag{47}$$

$$\Phi_{nj}(t) \equiv \left\langle S(t), V(t), n_1(t), \ldots, \frac{\partial U}{\partial n_j}(\Phi(t)), \ldots, n_J(t) \right\rangle, \tag{48}$$

$$\Phi_{SV}(t) \equiv \left\langle \frac{\partial U}{\partial S}(\Phi(t)), \frac{\partial U}{\partial V}(\Phi(t)), n_1(t), \ldots, n_J(t) \right\rangle, \tag{49}$$

among others. Actually, the phase state is described by any one of a family of 2^{J+2} possible sets of values and, for each set, there is an additive extensive property which is named the thermodynamic potential of the set (Truesdell, 1984). For instance, the thermodynamic potential corresponding to $\Phi_S(t)$ is the Helmholtz energy A and, from Equation 43 and the definition $A = U - TS$,

$$\mathrm{dA} = -\mathrm{SdT} - \mathrm{pdV} + \sum_{j=1}^{J} \mu_j \mathrm{dn_j}\,, \tag{50}$$

where $\frac{\partial^2 A}{\partial T^2}(\Phi_S) \neq 0$, $\frac{\partial^2 A}{\partial V^2}(\Phi_S) \neq 0$, and $\frac{\partial^2 A}{\partial {n_j}^2}(\Phi_S) \neq 0$ for $j = 1,\ldots,J$ at any instant t.

Analogously, the thermodynamic potential corresponding to $\Phi_V(t)$ is the enthalpy $H = U + pV$,

$$\mathrm{dH} = \mathrm{TdS} + \mathrm{Vdp} + \sum_{j=1}^{J} \mu_j \mathrm{dn_j}\,, \tag{51}$$

and $\frac{\partial^2 H}{\partial S^2}(\Phi_V) \neq 0$, $\frac{\partial^2 H}{\partial p^2}(\Phi_V) \neq 0$, and $\frac{\partial^2 H}{\partial n_j^2}(\Phi_V) \neq 0$ for *j*= 1,…,*J* at any instant *t*. The thermodynamic potential referring to the set $\Phi_{nj}(t)$ is $Y_j = U - \mu_j n_j$. By substituting Equation 43 in the expression for dY_j it follows that

$$dY_j = TdS - pdV + \mu_1 dn_1 + ... - n_j d\mu_j + ... + \mu_J dn_J\ , \tag{52}$$

and $\frac{\partial^2 Y_j}{\partial S^2}(\Phi_{nj}) \neq 0$, $\frac{\partial^2 Y_j}{\partial V^2}(\Phi_{nj}) \neq 0$, $\frac{\partial^2 Y_j}{\partial n_i^2}(\Phi_{nj}) \neq 0$ for *i*= 1,…,*J* but *i*≠*j*, and $\frac{\partial^2 Y_j}{\partial \mu_j^2}(\Phi_{nj}) \neq 0$ at any instant *t*. Finally, the thermodynamic potential corresponding to $\Phi_{SV}(t)$ is the Gibbs energy G= *U*-TS+pV,

$$dG = -SdT + Vdp + \sum_{j=1}^{J} \mu_j dn_j\ , \tag{53}$$

and $\frac{\partial^2 G}{\partial T^2}(\Phi_{SV}) \neq 0$, $\frac{\partial^2 G}{\partial p^2}(\Phi_{SV}) \neq 0$, and $\frac{\partial^2 G}{\partial n_j^2}(\Phi_{SV}) \neq 0$ for *j*= 1,…,*J* at any instant *t*. Note that *U* is the thermodynamic potential corresponding to Φ= <S, V, n_1, ..., n_J>, but S is not a thermodynamic potential for the set $< U, V, n_1, \ldots, n_J >$, since it is not possible to ensure that the derivative $\frac{\partial^2 S}{\partial V^2}(U, V, n_1, \ldots, n_J)$ is not zero. Thus, the maximization of S for the stable states of isolated systems does not guarantee that S is a thermodynamic potential.

5.3 Temperature

When the volume and the amount of all substances in the phase do not vary, *U* is a monotonically increasing function of S, and then the partial derivative $\frac{\partial U}{\partial S}(\Phi)$ is a positive quantity. Thus, because this partial derivative is the definition of temperature,

$$T = \frac{\partial U}{\partial S}(\Phi) > 0\ . \tag{54}$$

Because the internal energy is the thermodynamic potential corresponding to the set of values Φ, $\frac{\partial^2 U}{\partial S^2} \neq 0$ and, to complete the temperature definition, the sign of this second derivative must be stated. In fact, $\frac{\partial^2 U}{\partial S^2}(\Phi) = \frac{\partial T}{\partial S}(\Phi) > 0$. Thus, temperature is a concept closely related to the second law of thermodynamics but the first scale of temperature proposed by Kelvin in 1848 emerged as a logical consequence of Carnot's work, without even mentioning the concepts of internal energy and entropy.

Kelvin's first scale includes the entire real axis of dimensionless real numbers and is independent of the choice of the body employed as a thermometer (Truesdell & Baratha, 1988). The corresponding dimensional scales of temperature are called empirical. In 1854, Kelvin proposed a dimensionless scale including only the positive semi-axis of the real numbers. For the corresponding absolute scale (section 2.3), the dimensionless 1 may stand for a phase at $\frac{1}{273.15}$ of the temperature value of water at its triple point. The second scale proposed by Kelvin is completely consistent with the gas thermometer experimental results known in 1854. Moreover, it is consistent with the heat theorem proposed by Nernst in 1905, half a century later.

Because, according to the expression $\frac{\partial T}{\partial S}(\Phi) > 0$, the variations of temperature and entropy have the same sign, when temperature tends to its maximum lower bound, the same must occur for entropy. But, if the maximum lower bound of entropy is zero as proposed by Planck in 1910, when this value is reached a full knowledge about a state of an isolated homogeneous system should be obtained. Then, because the null absolute temperature is not attainable, another statement could have been made by Planck on Nernst's heat theorem:

"It is impossible to obtain full knowledge about an isolated homogeneous system."

5.4 Pressure

In analogy to temperature, pressure is defined by a partial derivative of $U = U(S, V, n_1, \ldots n_J)$,

$$p = -\frac{\partial U}{\partial V}(\Phi), \tag{55}$$

or, alternatively, by

$$p = -\frac{\partial A}{\partial V}(\Phi_S). \tag{56}$$

But, for completing the pressure definition, the signs of the second derivatives of U and A must be established. Actually, it is easily proved that these second derivatives must have the same sign, so that it is sufficient to state that $\frac{\partial p}{\partial V}(\Phi_S) < 0$, in agreement with the mechanical concept of pressure. Equation 55 demonstrates that, when $p>0$, U increases owing to the contraction of phase volume. Hence, according to the principle of conservation of energy, for a closed phase with constant composition and entropy, $p>0$ indicates that the absorption of energy from the outside is followed by volumetric contraction, while $p<0$ implies that absorption of energy from outside is accompanied by volumetric expansion. The former corresponds to an expansive phase tendency, while the latter corresponds to a contractive phase tendency. Evidently, when $p = 0$ no energy exchange between the system and the outside follows volumetric changes. So, the latter corresponds to a non-expansive and non-contractive tendency.

It is clear that p can assume any value, in contrast to temperature. Hence, the scale for pressure is analogous to Kelvin's first scale, that is, p can take any real number. For gases, p is always positive, but for liquids and solids p can be positive or negative. A stable state of a solid at negative pressure is a solid under tension, but a liquid at negative pressure is in a meta-stable state (Debenedetti, 1996). Thermodynamics imposes no unexpected restriction on the value of $\frac{\partial p}{\partial T}(\Phi_S)$ but, because in most cases this derivative is positive, several textbooks consider any stable state presenting a negative value for this derivative as being anomalous. The most well known "anomaly" is related to water, even though there are many others.

5.5 Chemical potential

In analogy to pressure, the chemical potential is defined by a partial derivative of $U = U(S, V, n_1, ..., n_J)$,

$$\mu_j = \frac{\partial U}{\partial n_j}(\Phi), \tag{57}$$

or, alternatively, by

$$\mu_j = \frac{\partial G}{\partial n_j}(\Phi_{SV}). \tag{58}$$

Moreover, to complete the chemical potential definition the signs of the second derivatives of U and G must be established. Because these derivatives must have the same sign, it is enough to state that $\frac{\partial \mu_j}{\partial n_j}(\Phi_{SV}) > 0$, which illustrates that both μ_j and n_j must have variations with the same sign when temperature, pressure and all the other J-1 amounts of substance remain unchanged. Remembering that, for the $j^{\underline{th}}$ chemical species the partial molar value z_j of an additive extensive property z is, by definition,

$$z_j = \frac{\partial z}{\partial n_j}(\Phi_{SV}), \tag{59}$$

Equation 58 shows that $\mu_j = G_j$, that is, the chemical potential of the $j^{\underline{th}}$ chemical species is its partial molar Gibbs energy in the phase.

Although μ_j is called a chemical potential, in fact μ_j is not a thermodynamic potential like U, H, A, Y_j and G. This denomination is derived from an analogy with physical potentials that control the movement of charges or masses. In this case, the chemical potential controls the diffusive flux of a certain chemical substance, that is, μ_j controls the movement of the particles of a certain chemical substance when their displacement is only due to random motion. In order to demonstrate this physical interpretation, let two distinct but otherwise closed phases with the same homogeneous temperature and pressure be in contact by means of a wall that is only permeable to the $j^{\underline{th}}$ species. Considering that both phases can only perform volumetric work and are maintained at fixed temperature and pressure, according to Equations 35 and 53

$$dG = \mu_{j1} dn_{j1} + \mu_{j2} dn_{j2} \leq 0, \tag{60}$$

where the subscripts "1" and "2" describe the phases in contact. But, because $dn_{j2} = -dn_{j1}$, it follows that

$$(\mu_{j1} - \mu_{j2})dn_{j1} \le 0 . \quad (61)$$

Thus, $dn_{j1} > 0$ implies $\mu_{j1} - \mu_{j2} \le 0$, that is, the substance *j* flows from the phase in which it has a larger potential to the phase in which its chemical potential is smaller.

6. Conclusion

By using elementary notions of differential and integral calculus, the fundamental concepts of thermodynamics were re-discussed according to the thermodynamics of homogeneous processes, which may be considered an introductory theory to the mechanics of continuum media. For the first law, the importance of knowing the defining equations of the differentials dq, dw and d*U* was stressed. Moreover, the physical meaning of q, w and *U* was emphasized and the fundamental equation for homogeneous processes was clearly separated from the first law expression.

In addition, for the second law, a thermally homogeneous closed system was used. This approach was employed to derive the significance of Helmholtz and Gibbs energies. Further, entropy was defined by using generic concepts such as the correspondence between states and microstates and the missing structural information. Thus, it was shown that the concept of entropy, which had been defined only for systems in equilibrium, can be extended to other systems much more complex than the thermal machines. The purpose of this chapter was to expand the understanding and the applicability of thermodynamics.

7. Acknowledgement

The authors would like to acknowledge Professor Roy Bruns for an English revision of this manuscript and CNPQ.

8. References

Agarwal, R.P. & O'Regan, D. (2008). *Ordinary and Partial Differential Equations: With Special Functions, Fourier Series, and Boundary Value Problems*, Springer-Verlag, 978-0-387-79145-6, New York

Apostol, T.M. (1967). *Calculus. One-Variable Calculus, with an Introduction to Linear Algebra*, John-Wiley & Sons, 0-471-00005-1, New York

Bassi, A.B.M.S. (2005, a). Quantidade de substância. *Chemkeys*, (September, 2005) pp. 1-3

Bassi, A.B.M.S. (2005, b). Matemática e termodinâmica. *Chemkeys*, (September, 2005) pp. 1-9

Bassi, A.B.M.S. (2005, c). Entropia e energias de Helmholtz e de Gibbs. *Chemkeys*, (November, 2005) pp.1-14

Bassi, A.B.M.S. (2006, a). O conceito de propriedade termodinâmica. *Chemkeys*, (May, 2006) pp.1-10

Bassi, A.B.M.S. (2006, b). As duas expressões fundamentais da termodinâmica. *Chemkeys*, (April, 2006) pp.1-8

Boltzmann, L.E. (1964). *Lectures on Gas Theory*, University of California Press, 0-486-68455-5, Ontario

Brillouin, L. (1962). *Science and Information Theory*, Academic Press, 0121349500, New York

Casimir, H.B.G. (1945). On Onsager's principle of microscopic reversibility. *Review of Modern Physics*, Vol. 12, No. 2-3 (April-June, 1945) pp. 343-350, 0034-6861

Day, W.A. (1987). A comment on a formulation of the second law of thermodynamics. *Archive of Rational Mechanics and Analysis*, Vol. 98, No. 3, (September, 1987) pp. 211-227, 0003-9527

De Groot, S.R. & Mazur, P. (1984). *Non-equilibrium Thermodynamics*, Dover Publications, 0-486-64741-2, Toronto

Debenedetti, P.G. (1996). *Metastable liquids: Concepts and Principles*, Princeton University Press, 0691085951, Princeton

Eckart, C. (1940). The thermodynamics of irreversible processes I. The simple fluid. Physical Review, Vol. 58, No. 3, (August, 1940) pp. 267-269, 1050-2947

Fermi, E. (1956). *Thermodynamics*, Dover Publications, 486-60361-X, Mineola

Gray, R.M. (1990). *Entropy and Theory Information*, Springer-Verlag, 0387973710, New York

Gurtin, M.E. (1971). On the first law of thermodynamics. *Archive of Rational Mechanics and Analysis*, Vol. 42, No. 2, (January, 1971) pp.77-92, 0003-9527

Hutter, K. (1977). The foundations of thermodynamics, its basic postulates and implications. A review of modern thermodynamics. *Acta Mechanica*, Vol. 27, No.1-4, (March, 1977) pp. 1-54, 0001-5970

Kestin, J. (1976). *The Second Law of Thermodynamics*, Halsted Press, 0879332425, New York

Liu, I-Shih. (2002). *Continuum Mechanics*, Springer-Verlag, 3-540-43019-9, Berlin

Mase, G.T. & Mase, G.E. (1999). *Continuum Mechanics for Engineers*, CRC Press, 0-8493-1885-6, Boca Raton

Mcquarrie, D.A. (2000). *Statistical Mechanics*, University Science Books, 1891389157, New York.

Michaelides, E.E. (2008). Entropy, order and disorder. *The Open Thermodynamics Journal*, Vol. 2, No. 2, (March, 2008) pp. 7-11, 1874-396X

Moreira, N.H. & Bassi, A.B.M.S. (2001). Sobre a primeira lei da termodinâmica. *Química Nova*, Vol. 24, No. 4, (July-August, 2001) pp. 563-567, 3-540-43019-9

Nery, A.R.L. & Bassi, A.B.M.S. (2009, a). A quantidade de matéria nas ciências clássicas. *Química Nova*, Vol. 32, No. 7, (August, 2009) pp. 1961-1964, 1678-7064

Nery, A.R.L. & Bassi, A.B.M.S. (2009, b). A primeira lei da termodinâmica dos processos homogêneos. *Química Nova*, Vol. 32, No. 2, (February, 2009) pp. 522-529, 1678-7064

Onsager, L. (1931, a). Reciprocal relations in irreversible processes I. *Physical Review*, Vol.37, No.4, (February, 1931) pp. 405-426, 1050-2947

Onsager, L. (1931, b). Reciprocal relations in irreversible processes II. *Physical Review*, Vol. 38, No. 12, (December, 1931) pp. 2265-2279, 1050-2947

Planck, M. (1945). *Treatise on Thermodynamics*, Dover Publications, 048666371X, New York

Serrin, J. (1979). Conceptual analysis of the classical second law of thermodynamics. *Archive of Rational Mechanics and Analysis*, Vol. 70, No. 4, (December, 1979) pp. 355-371, 0003-9527

Šilhavý, M. (1983). On the Clausius inequality. *Archive of Rational Mechanics and Analysis*, Vol. 81, No. 3, (September, 1983) pp. 221-243, 0003-9527

Šilhavý, M. (1989). Mass, internal energy and Cauchy's equations of motions in frame-indifferent thermodynamics. Archive of Rational Mechanics and Analysis, Vol. 107, No. 1, (March, 1989) pp. 1-22, 0003-9527

Šilhavý, M. (1997). *The Mechanics and Thermodynamics of Continuous Media*, Springer-Verlag, 3-540-58378-5, Berlin

Tolman, R.C. (1938). *The Principles of Statistical Mechanics*, Oxford Press, 0-486-63896-0, New York

Toupin, R. & Truesdell, C.A. (1960). The classical field of theories, In: *Handbuch der Physik*, S. Flügge (Ed.), pp. 226-858, Springer-Verlag, 0085-140X, Berlin

Truesdell, C.A. & Baratha, S. (1988). *The Concepts and Logic of Classical Thermodynamics as a Theory of Heat Engines: Rigorously Constructed upon the Foundation Laid by S. Carnot and F. Reech*, Springer-Verlag, 3540079718, New York

Truesdell, C.A. (1980). *The Tragicomical History of Thermodynamics*, 1822-1854, Springer-Verlag, 0-387904034, New York

Truesdell, C.A. (1984). *Rational Thermodynamics*, Springer-Verlag, 0-387-90874-9, New York

Truesdell, C.A. (1991). *A First Course in Rational Continuum Mechanics*, Academic Press, 0127013008, Boston

Williams, W.O. (1971). Axioms for work and energy in general continua. *Archive of Rational Mechanics and Analysis*, Vol. 42, No. 2, (January, 1972) pp. 93-114, 0003-9527

6

Non-extensive Thermodynamics of Algorithmic Processing – the Case of Insertion Sort Algorithm

Dominik Strzałka and Franciszek Grabowski
Rzeszów University of Technology
Poland

1. Introduction

In this chapter it will be shown that there can exist possible connections of Tsallis non-extensive definition of entropy (Tsallis, 1988) with the statistical analysis of simple insertion sort algorithm behaviour. This will be done basing on the connections between the idea of Turing machines (Turing, 1936) as a basis of considerations in computer science and especially in algorithmic processing and the proposal of non-equilibrium thermodynamics given by Constatino Tsallis (Tsallis, 1988; Tsallis, 2004) for indication of the possible existence of non-equilibrium states in the case of one sorting algorithm behaviour. Moreover, it will be also underlined that a some kind of paradigm change (Kuhn, 1962) is needed in the case of computer systems analysis because if one considers the computers as physical implementations of Turing machines should take into account that such implementations always need energy for their work (Strzałka, 2010) - Turing machine as a mathematical model of processing does not need energy. Because there is no (computer) machine that have the efficiency η = 100%, thus the problem of entropy production appears during their work. If we note that the process of sorting is also the introduction of order (obviously, according to a given appropriate relation) into the processed set (sometimes sorting is considered as an ordering (Knuth, 1997)), thus if one orders it must decrease the entropy in sorted set and increase it somewhere else (outside the Turing machine - in physical world outside its implementation). The connections mentioned above will be given basing on the analysis of insertion sorting, which behaviour for some cases can lead to the levels of entropy production that can be considered in terms of non-extensivity. The presented deliberations can be also related to the try of finding a new thermodymical basis for important part of authors' interest, i.e., the physics of computer processing.

2. Importance of physical approach

The understanding of concept of entropy is intimately linked with the concept of energy that is omnipresent in our lives. The principle of conservation of energy says that the difference of internal energy in the system must be equal to the amount of energy delivered to the system during the conversion, minus the energy dissipated during the transformation. The principle allows to write an appropriate equation but does not impose any restrictions on the quantities used in this equation. What's more, it does not give any indications of how the energy should be supplied or drained from the system, or what laws (if any exist) should govern the transformations of energy from one form to another. Only the differences of transformed energy are important. However, there are the rules governing the energy transformations. A concept of entropy and other related notions create a space of those rules.

Let's note that Turing machine is a basis of many considerations in computer science. It was introduced by Alan Mathison Turing in the years 1935–1936 as a response to the problem posed in 1900 by David Hilbert known as the *Entscheidungsproblem* (Penrose, 1989). The conception of Turing machine is powerful enough to model the algorithmic processing and so far it haven't been invented its any real improvements, which would increase the area of decidable languages or which will improve more than polynomial its time of action (Papadimitriou, 1994). For this reason, it is a model which can be used to implement any algorithm. This can be followed directly from Alonso Church's thesis, which states that (Penrose, 1989; Wegner & Goldin, 2003):

"Any reasonable attempt to create a mathematical model of algorithmic computation and to define its time of action must lead to the model of calculations and the associated measure of time cost, which are polynomial equivalent to the Turing machines."

Note also that the Turing machine is, in fact, the concept of mathematics, not a physical device. The traditional and widely acceptable definition of machine is connected with physics. It assumes that it is a physical system operating in a deterministic way in a well-defined cycles, built by a man, whose main goal is focusing energy dispersion for the execution of a some physical work (Horákowá et al., 2003). Such a machine works almost in accordance with the concept of the mechanism specified by Deutsch – as a perfect machinery for moving in a cyclical manner according to the well-known and described laws of physics, acting as a simple (maybe sometimes complicated) system (Deutsch, 1951; Grabowski & Strzałka, 2009; Amral & Ottino, 2004).

On the other hand the technological advances have led to a situation in which there is a huge number of different types of implementations of Turing machines and each such an implementation is a physical system. Analysis of the elementary properties of Turing machines as a mathematical concept tells us, that this is a model based on unlimited resources: for example, in the Turing machine tape length is unlimited and the consumption of energy for processing is 0 (Stepney et al., 2006). This means that between the mathematical model and its physical implementation there are at least two quite subtle but crucial differences: first, a mathematical model that could work, does not need any Joule of energy, while its physical implementation so, and secondly, the resources of (surrounding) environment are always limited: in reality the length of Turing machine tape is limited (Stepney et al., 2006).

Because in the mathematical model of algorithmic computations there is no consumption of energy, i.e., the problem of physical efficiency of the model (understood as the ratio of energy supplied to it for work, which the machine will perform) does not exist. Moreover, it seems that since the machine does not consume energy the possible connections between thermodynamics and problems of entropy production aren't interesting and don't exist. However, this problem is not so obvious, not only due to the fact that the implementations of Turing machines are physical systems, but also because the use of a Turing machine for the solution of algorithmic problems can be also associated with the conception such as the order, which is (roughly speaking) *anti-entropic*. A classic example of this type of problem is sorting. It is usually one of the first problems discussed at the courses of algorithms to show what is the algorithmic processing and to explain the idea of computational complexity (see for example the first chapter in famous book (Cormen et al., 2001)).

Generally, the main objective of sorting is in fact find such a permutation (ordering change) $< a_1^{'}, a_2^{'}, ..., a_N^{'} >$ of the input consisting of N numbers (or in general N keys) $< a_1, a_2, ..., a_N >$ to ensure that $a_1^{'} \leq a_2^{'} \leq ... \leq a_N^{'}$. As one can see the search for an appropriate permutation is carried out using the ordering relation < defined on the values (keys) and the following conditions for three values a, b, c are satisfied:

- there is true exactly one of the possibilities $a < b$, $a = b$, $b < a$;
- if $a < b$ and $b < c$, then $a < c$.

In this chapter, basing on the context of so far presented considerations, it will be discussed a simple algorithm for sorting based on the idea of insertion sort. This is one of the easiest and most intuitive sorting algorithms (based on the behaviour of bridge player who sorts his cards before the game) and its detailed description can be found in the literature (Cormen et al., 2001). It is not too fast algorithm (for the worst-case it belongs to a class of algorithms of complexity $O(n^2)$, however for the optimistic case it has the complexity $\Omega(n)$), but it is very simple, because it only consists of two loops: the outer guarantees sorting of all elements and the internal one, which finds the right place for each key in the sorted set. This loop is a key-point of our analysis because it will represent a very interesting behaviour in the context of analysis of algorithm dynamics for all possible input set instances. This follows from the fact that the number of this inner loop executions, which can also be identified with the duration of this loop, depends on (Strzałka & Grabowski, 2008):

- the number of sorted keys (the size n of the task). If, for example, the pessimistic case is sorted for long input sets and elements of small key values, the duration of this loop can be very long especially for the data contained at the end of the input sorted set;
- currently sorted value of the key. If the sorting is done in accordance with the relation "<", then for large values of data keys finding the right place in output set should last a very short period of time, while for small values of keys it should take a lot of inside loop executions. Thus, all parts of the input set close to the optimistic case, i.e., the parts with preliminary, rough sort of data (e.g., as a result of the local growing trend in input), will result in fewer executions of inner loop, while the parts of input set closer to the worst-case (that is, for example, those with falling local trends) will mean the need of many executions of inside loop.

The third condition is visible when the algorithm will be viewed as a some kind of *black box* (system), in which the input set is the system INPUT and the sorted data is the system OUTPUT (this approach is consistent with the considerations, which are given by Knuth in (Knuth, 1997) where in his definition of algorithm there are 5 key features among which are the *input* and *output* or with the approach presented by Cormen in (Cormen et al., 2001)). Then it can be seen that there is a third additional condition for the number of inner loop executions: the so far sorted values of processed set contained in this part of the output, where the sorting was already done, influence on the number of this loop executions. Thus we have an elementary feedback. The position of each new sorted element depends not only on its numerical value (understood here as the input IN), but also on the values of the items already sorted (that is, *de facto* output OUT). If it were not so, each new element in the sorted input would be put on pre-defined place in already sorted sequence (for example, it would be always included at the beginning, end or elsewhere within the output – such a situation is for example in the case of sorting by selection).

The above presented observations will influence the dynamics of analysed algorithm and its analysis will be conducted in the context of thermodynamic conditions. Let's note once again that the sorting is an operation that introduces the order into the processed set and in other words it is an operation that reduces the level of entropy considered as the measure of disorder. In the case of the classical approach, which is based on a mathematical model of Turing machines the processing will cause the entropy reduction in the input set but will not cause its growth in the surroundings of the machine (it doesn't consume the energy). But in the case of the physical implementation of Turing machine, the processing of input set must result in an increase of entropy in the surroundings of the machine. This follows from the fact that even if the sorting operation is done by the machine that has the efficiency $\eta = 100\%$ it still will require the energy consumption – this energy should be produced at the source and this lead to the increase of the entropy "*somewhere*" near the source.

3. Levels of entropy production in insertion-sort algorithm

The presented analysis will be based on the following approach (Strzałka & Grabowski, 2008). If the sorted data set is of size n, then it can occur $n!$ of possible key arrangements (input instances). One of them will relate to the case of the proper arrangement of elements in the set (i.e., the set is already sorted – the case is the optimistic one), while the second one will relate to the worst-case (in the set there will be arrangement, but different from that required). For both of these situations it can be given the exact number of dominant operations that should be done by the algorithm, while for the most of other $n! - 2$ cases this is not necessary so simple. However, the analysis of insertion sorting can be performed basing on the conception of inversions (Knuth, 1997). The number of inversions can be used to calculate how many times the dominant operation in insertion sort algorithm should be done, but it is also an indication of the level of entropy in the processed set, since the number of inversions is information about how many elements of the set are not ordered. Of course, the arrangement will reduce the entropy in the set, but it will increase the entropy in the environment.

Therefore, we can consider the levels of entropy production during insertion sorting. If we denote by M the total number of executions of inside and outside loops needed for successive n_i elements processed from the input set of size n, then for each key $M = n_i$. Let M_1 will be the number of outer loop requests for each sorted key – always it will be $M_1 = 1$. If by M_2 we will denote the number of inner loop calls, then it may vary from 0 to $n_i - 1$, and if by M_3 we determine the number of such inside loop executions that may have occurred but not occurred due to the some properties of sorted set, we will have $M = M_1 + M_2 + M_3$. For the numbers M_1, M_2 and M_3 one can specify the number of possible configurations of inner and outer loop executions in the following cases: optimistic, pessimistic and others. By the analogy, this approach can be interpreted as a try to determine the number of allowed microstates (configurations), which will be used to the analysis of entropy levels production in the context of the number of necessary internal loop executions.

This number will be equal to the number of possible combinations $C_M^{M_1}$ multiplied by $C_{M-M_1}^{M_2}$ (this number is multiplicative):

$$W = C_M^{M_1} \cdot C_{M-M_1}^{M_2} = \frac{M!}{M_1!(M-M_1)!} \cdot \frac{(M-M_1)!}{M_2!(M-M_1-M_2)} = \frac{M!}{M_1!M_2!M_3!}, \quad (1)$$

i.e., the number C of M_1 combinations of necessary outer loop calls from M executions multiplied by C combinations of M_2 necessary executions of inner loop from the rest possible $M - M_1$ calls.

An optimistic case is characterized by the need of a single execution of outer loop ($M_1 = 1$) for each sorted key, the lack of inside loop calls ($M_2 = 0$) and $n_i - 1$ no executions of this loop ($M_3 = n_i - 1$), which means that the number of possible W_O configurations of these two loops will be equal

$$W_O = \frac{n_i!}{1!0!(n_i-1)!} = n_i \,. \quad (2)$$

For the pessimistic case it will be: $M_1 = 1$ and $M_2 = n_i - 1$ – one need to use this loop a maximal available times – $M_3 = 0$, thus W_P (P – pessimistic) will be equal

$$W_P = \frac{n_i!}{1!(n_i-1)!0!} = n_i \,. \quad (3)$$

Thus, the number of microstate configurations in both cases is the same ($W_O = W_P$). It might seem a little surprising, but it is worth to note that although in the worst case the elements are arranged in reverse order than the assumed in sorting process, it is still the order. From the perspective of thermodynamics the optimistic and pessimistic cases are the same because they are characterised by the entropy production at the lowest possible level; in any other cases W will be greater. For example let's consider the case when one needs only one excess dominant operation for key n_i, i.e., : $M_1 = 1$, $M_2 = 1$, $M_3 = n_i - 2$, so W_D (D – dynamical) will be equal

$$W_D = \frac{n_i!}{1!1!(n_i-2)!} = \frac{(n_i-2)!(n_i-1)n_i}{(n_i-2)!} = n_i(n_i-1) \,. \quad (4)$$

The lowest possible levels of entropy production for the optimistic or pessimistic cases correspond to the relationship given by Onsager (Prigogine & Stengers, 1984). They show that if a system is in a state close to thermodynamic equilibrium, the entropy production is at the lowest possible level. Thus, while sorting by the insertion-sort algorithm the optimistic or pessimistic cases, Turing machine is in (quasi)equilibrium state.

It can be seen that in the optimistic and pessimistic cases the process of sorting (or entropy production) is extensive, but it is not known if these considerations are entitled to the other instances. However, one can see this by doing a description of the micro scale, examining the behavior of the algorithm for input data sets with certain properties (let's note that this is in contradiction to the commonly accepted approach in computer science where one of the most important assumptions in computational complexity assumes that this measure should be independent on specific instances properties, thus usually the worst case is considered (Mertens, 2002)). Moreover, to avoid problems associated with determining the number of inversions, one can analyze the behavior of this algorithm by recording for each sorted key the number of executed dominant operations (it will be labeled as $Y(n)$) and then examine the process of increments of number of dominant operations (i.e., $Y'(n)$); in other words – to consider how the process

$$Y'(n) = Y(n+1) - Y(n) \tag{5}$$

can behave.

The equation (1) shows the entropy production for each sorted key. In the classical analysis of algorithms computational complexity two similar ways can be taken: one can consider a total number of dominant operations executions or a number of dominant operations for each processed element of input set (the increments of the first number). The second approach will be more interesting one and it will show the properties of distribution of all possible increments of number of dominant operations: for each sorted key n_i the number of dominant operations is a random variable and its values can appear with changing probabilities for each n_i. We would like to know how the distribution of increments looks like when $n_i \to n$ and of course when $n \to \infty$. It is not hard to see that in the optimistic case, the expression (5) will always be zero, and for the worst-case is always equal to one. If there will be sorted the instances "similar" to the cases that are optimistic or pessimistic, the deviations from the above number of increments will be small and their probability distributions should be characterized by quickly vanishing tails, therefore, it will belong to the Gaussian basin of attraction. As we know this is a distribution, which is a natural consequence of the assumptions underlying the classical definition of Boltzmann-Gibbs entropy. However, it may also turn-out that certain properties of sorted input sets (e.g., long raising and falling trends) will cause that the probability distributions of the number of dominant operations increments will have a different character, and then the concept of non-extensive entropy will be useful.

4. Non-equilibrium states of insertion sort algorithm behaviour

In order to visualize the so far presented deliberations a simple experiment involving the sorting of input data sets by insertion-sort algorithm was done. Sorted sets were the trajectories of one-dimensional Brownian motion (random walk) – denoted by $X(t)$; see Fig. 1. Each sorted set has 10^6 elements. One of the most characteristic feature of these sets is the presence of local increasing or decreasing trends, which for the sorting algorithm can be regarded as a local optimistic or pessimistic cases. These trends and their changes should affect the dynamics of algorithm behavior (considered as the changing number of executed dominant operations) – see for example Fig. 1. If sorting is done according to the non-decreasing order (i.e., by the relation $\leq$), any raising trend would be the case of initially correct order of keys in input data (in other words it can be very roughly treated as a case similar to the optimistic one) – in mathematical analysis, this situation would be described by a small number of inversions. However, any falling trend will be the case of improper order of data (i.e., very roughly – similar to the worst-case) – in mathematical analysis, this situation involves a large number of inversion. Any raising trend in input set will cause the decline of the number of dominant operations, while the falling trend its rapid growth (Fig. 1).

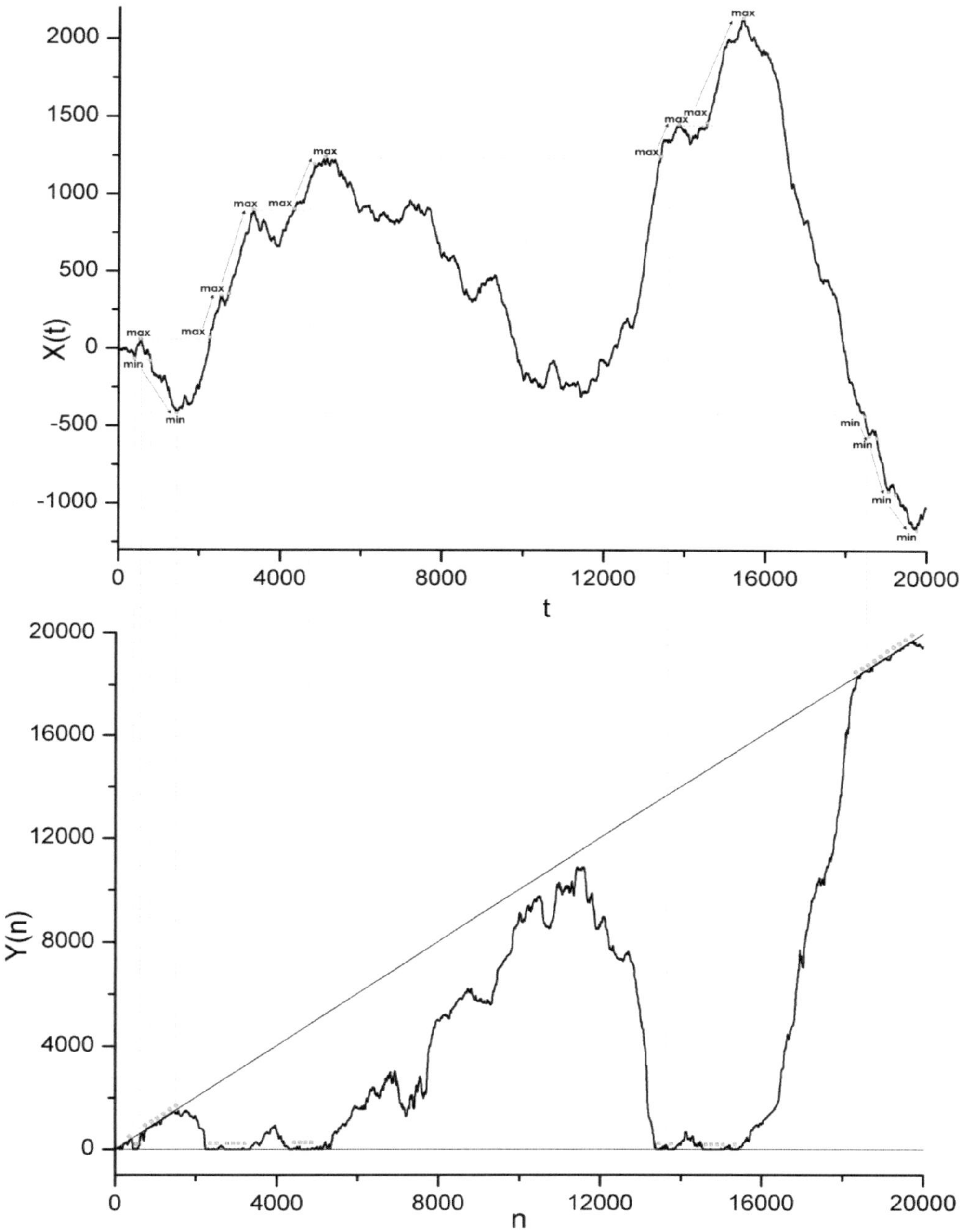

Fig. 1. The example of one input set (top) with the behaviour of algorithm (down), i.e., the recorded set of dominant operations needed for sorting n = 20000 keys. As it can be seen when in input set there is a local minimum, there is a need to execute the maximal number of dominant operations, and when in input set there is a local maximum there is a need to execute only one dominant operation

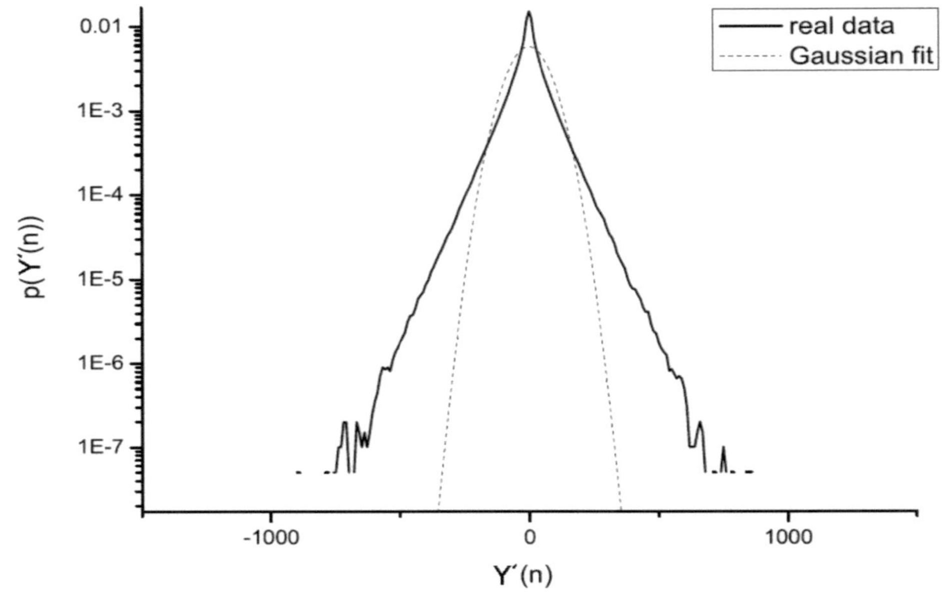

Fig. 2. The example of one empirical probability distribution of increments $Y'(n)$ of dominant operations used by insertion-sort algorithm. Line with dots stands for a normal distribution fitted by calculated process mean and variance; continuous line represents the empirical distribution obtained by a kernel estimator

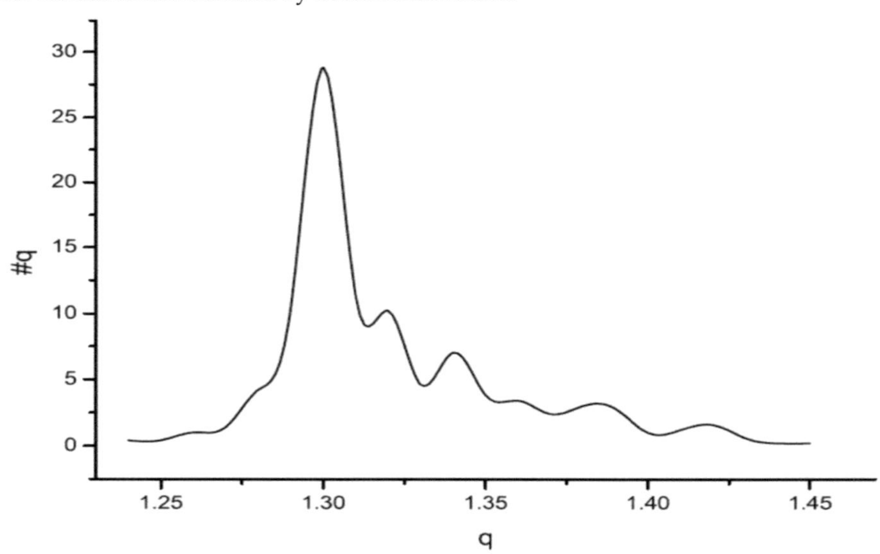

Fig. 3. The distribution of q values for 500 sorted sets; as it can bee seen in most cases there is $q \approx 1.3$, but its range lies between 1.25 and 1.45

During the experiment, the numbers of dominant (needed for each sorted key) operations $Y(n)$ were recorded. 500 sets of input data have been sorted and as a result we received the set of sorting processes realisations. Next, primarily the empirical probability density distributions of increments $Y'(n)$ of the number of dominant operations were examined. As one can see (Fig. 2), the empirical distribution of dominant operations increments has slowly vanishing tails than fitted by process mean and variance normal distribution. This is a first example, which shows that the process of sorting by insertion input sets that are the trajectories of random walk can be related to the idea of Tsallis approach with his proposal of q-Gaussian distributions (Tsallis et al., 1995; Alemany & Zanette, 1994). In this case the estimated value of q parameter equals ≈ 1.3.

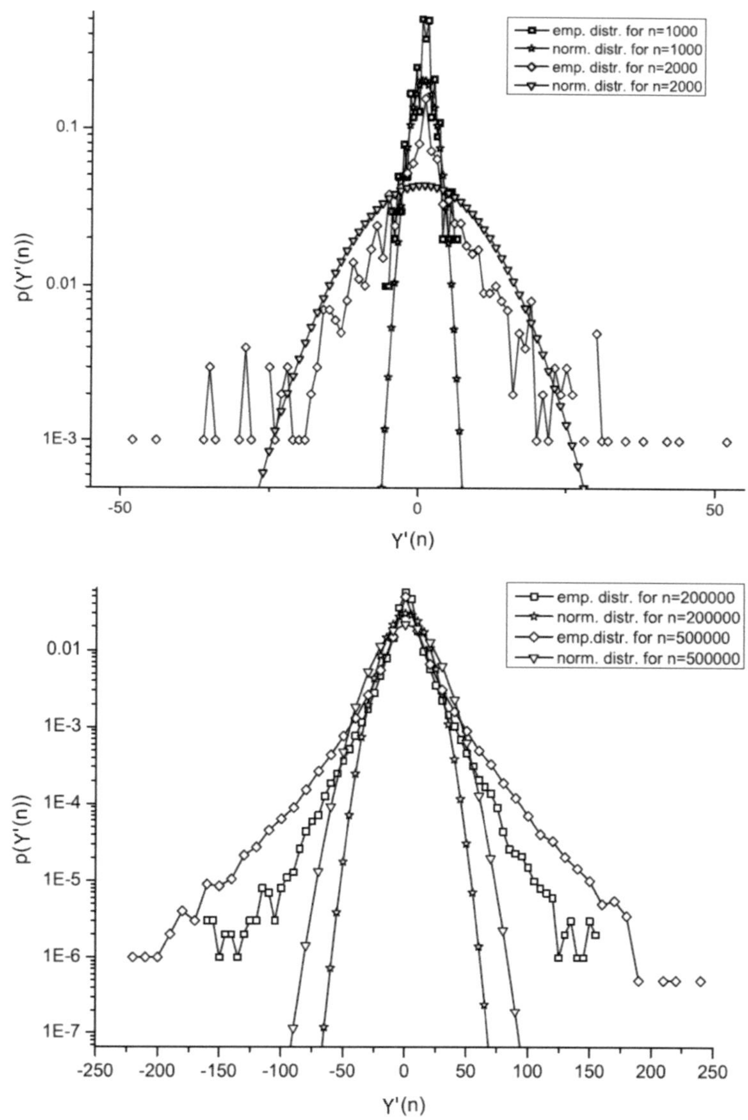

Fig. 5. The comparison of empirical and fitted normal distributions for increment processes $Y'(n)$, when the input size n = 1000 and n = 2000 (top figure) and when n = 200000 and n = 500000 (bottom figure)

The presented case allows us to ask some interesting questions. First one is obvious: is it only the one case when such a situation appears, or this situation is a "normal" one. The answer to the following question can be given immediately because as it was written above in the experiment 500 sets were sorted and for each sorted set we can see similar behaviour of empirical distributions. But this obviously allows to ask another important question: how the values of q parameter change for different input sets. To answer this we perform an analysis for all sorted sets an its results can be seen on Fig. 3 – the values of q parameter are between 1.25 and 1.45.

The second important problem can be the size n of input set: it can be very interesting for what n we can see the first symptoms of slowly vanishing tails for distributions of increments. If the number of keys in the sorted set is small (Fig. 4 shows the empirical and fitted Gaussian distributions for sets with different input sizes n), for example less than 1000 keys the q-Gaussian probability distribution of increments for sorting process is not clearly visible, but if the input set has quit large n the considered effect is definitely more visible. One of the conclusions drawn from these observations can be the suggestion that the classical (mathematical) analysis of this algorithm behavior (for example, shown in (Cormen et al., 2001)), in which the behavior of the algorithm for small data sets also shows the possible behavior for any size of input data, i.e., in a some sense it is *extended* to sets of input data of any size n, not quite well shows the nature of all possible behaviors of the analyzed algorithm. It seems that the problem considered in this chapter (non-extensive behavior of entropy) is rather the emergent one – it can appear when the size n of input is numerous.

Of course, the insertion sort algorithm isn't a very efficient one (we know this basing on its computational complexity) and sorting of large data sets is rather done for example by Quick-sort, but this simple algorithm can open a new field of discussion concerning the analysis of algorithms in the context of not only the determinants of the nature of mathematics, but also the conditions related to the characteristics of a physical nature – this can be viewed as a some kind of paradigm change in the so far presented approaches.

5. Conclusions

The chapter presents considerations concerning the existence of possible connections between non-extensive definition of entropy proposed by C. Tsallis and algorithmic processing basing on the example of sorting by insertion-sort. The figures presenting empirical distributions in log-lin scale show the existence of a slowly vanishing tails of probability distributions indicating that the thermodynamic conditions of analysed algorithm work emerge for a suitably large data sets – in a some sense, it seems that this is a feature of an emergent character; in the classical analysis of algorithms it can't be taken into account. In the chapter it was also shown that some features of input sets can also be transferred on the level of dynamic behaviour of the algorithm and the number of necessary dominant operations that are performed during its work. This differs from the assumption in the classical computational complexity analysis where the most interesting and authoritative case is the pessimistic (worst) one even if one takes into account also the analysis of the average case. Meanwhile, the analysis of algorithm combined with knowledge about the properties of processed data shows that there can appear interesting phenomena that may be of fundamental importance for the analysis of Turing machines that are not treated as a mathematical models, but considered in the context of their physical properties of the implementations.

6. References

Alemany, P. A. & Zanette, D. H. (1994). Fractal random walks from a variational formalism for Tsallis entropies, *Physical Reviev E*, Vol. 49, (1994) p. R956, ISSN 1539-3755

Amaral, L. A. N. & Ottino, J. M. (2004). Complex Systems and Networks: Challenges and Opportunities for Chemical and Biological Engineers, *Chemical Engineering Science*, Vol. 59, No. 8-9, (2004) pp. 1653–1666, ISSN 0009-2509

Cormen, T. H.; Leiserson, Ch. E.; Rivest, R. L. & Stein, C. (2001). *Introduction to Algorithms* (second ed.), MIT Press and McGraw-Hill, ISBN 978-0-262-53196-2, New York

Deutsch, K. (1951). Mechanism, Organism, and Society, *Philosophy of Science*, Vol. 18, (1951) pp. 230–252, ISSN 00318248

Grabowski, F. & Strzałka, D. (2009). Conception of paradigms evolution in science - towards the complex systems approach, *Egitania Sciencа*, Vol. 4, (2009) pp. 197–210, ISSN 1646-8848

Horákowá, J.; Kelemen, J, &, Čapek, J. (2003). Turing, von Neumann, and the 20th Century Evolution of the Concept of Machine. *International Conference in Memoriam John von Neumann*, John von Neumann Computer Society, pp. 121–135, Budapešť, 2003.

Knuth, D. E. (1997). *The art of computer programming, vol. 1*, Addison-Wesley, ISBN 0-201-89683-4, Massachusetts

Kuhn, T. S. (1962). *The Structure of Scientific Revolutions*, University of Chicago Press, ISBN 0-226-45808-3, Chicago

Mertens, S. (2002). Computational Complexity for Physicists, Computing in Science and Engineering, Vol. 4, No. 3, (May 2002) pp. 31–47, ISSN 1521-9615

Papadimitriou, Ch. H. (1994). *Computational Complexity*, Addison Wesley, ISBN 0201530821, Massachusetts

Penrose, R. (1989). *The Emperor's New Mind: Concerning Computers, Minds and The Laws of Physics*, Oxford University Press, ISBN 0-198-51973-7, New York

Prigogine, I. & Stengers, I. (1984). *Order out of Chaos: Man's new dialogue with nature*, Flamingo, ISBN 0006541151,

Turing, A. M. (1936). On computable numbers, with an application to the Entscheidungsproblem, *Proceedings of the London Mathematical Society*, Series 2(42), (1936) pp. 230–265. Errata appeared in Series 2(43), (1937) pp. 544–546

Tsallis, C. (1988). Possible generalization of Boltzmann-Gibbs statistics, *Journal Statistical Physics*, Vol. 52, (1988) p. 479, ISSN 0022-4715

Tsallis, C.; Levy, S. V. F.; Souza, A. M . C. and Mayanard, R. (1995). Statistical-Machanical Foundation of the Ubiquity of Lèvy Distributions in Nature, *Physical Review Letters*, Vol. 75, (1995) p. 3589, ISSN 0031-9007

Tsallis, C. (2004). What should a statistical mechanics satisfy to reflect nature?, *Physica D*, 193, (2004) pp. 3-34, ISSN 0167-2789

Stepney, S.; Braunstein, S. L.; Clark, J. A.; Tyrrell, A.; Adamatzky, A.; Smith, R. E.; Addis, T.; Johnson, C.; Timmis, J.; Welch, P.; Milner, R. & Partridge, D. (2006). Journeys in non-classical computation II: Initial journeys and waypoints, *International Journal of Parallel, Emergent and Distributed Systems*, Vol. 21, No. 2, (2006) pp. 97–125, ISSN 1744-5760

Strzałka, D. & Grabowski, F. (2008). Towards possible non-extensive thermodynamics of algorithmic processing – statistical mechanics of insertion sort algorithm,

International Journal of Modern Physics C, Vol. 19, No. 9, (2008) pp. 1443–1458, ISSN 0129-1831

Strzałka, D. (2010). Paradigms evolution in computer science, *Egitania Scienca*, Vol. 6, (2010) p. 203, ISSN 1646-8848

Wegner, P. & Goldin, D. (2003). Computation Beyond Turing Machines, *Communications of the ACM*, Vol. 46, No. 4, (Apr. 2003), pp. 100–102, ISSN 0001-0782

7

Lorentzian Wormholes Thermodynamics

Prado Martín-Moruno and Pedro F. González-Díaz
Instituto de Física Fundamental,
Consejo Superior de Investigaciones Científicas
Spain

1. Introduction

The term wormhole was initially coined by Misner and Wheeler (Misner & Wheeler, 1957) (see also Wheeler (Wheeler, 1955)) in order to describe the extra connections which could exist in a spacetime, composed by two mouths and a throat, denoting therefore more general structures than that was initially considered by Einstein and Rosen (Einstein & Rosen, 1935). Nevertheless, the study of macroscopic wormholes in general relativity was left in some way behind when Fuller and Wheeler (Fuller & Wheeler, 1962) showed the instability of the Einstein-Rosen bridge.

Although other solutions of the wormhole type, stable and traversable, were studied in those years (Ellis, 1973; Bronnikov, 1973; Kodama, 1978), it was in 1988 when the physics of wormhole was revived by the work of Morris and Thorne (Morris & Thorne, 1988). These authors considered the characteristics that should have a spacetime in order to describe a wormhole, which could be used by a intrepid traveler either as a short-cut between two regions of the same universe or as a gate to another universe. They found that such structure must be generated by a stuff not only with a negative radial pressure, but with a radial pressure so negative that this exotic material violates the null energy condition. Such pathological characteristic could have been suspected through the mentioned previous studies (Ellis, 1973; Bronnikov, 1973; Kodama, 1978), that pointed out the necessity to change the sign of the kinetic term of the scalar field which supports the geometry in order to maintain the stability of the wormhole. Moreover, in the work of Morris and Thorne (Morris & Thorne, 1988) it is included a comment by Page indicating that the exoticity of the material would be not only needed in the static and spherically symmetric case, but in more general cases. Although Morris and Thorne was aware of violations of the null energy condition, both in theoretical examples and in the laboratory, they also studied the possibility to minimize the use of this odd stuff, which they called exotic matter.

Nevertheless, it seems that exotic matter should not longer be minimized since the universe itself could be an inexhaustible source of this stuff. Recent astronomical data (Mortlock & Webster, 2000) indicate that the Universe could be dominated by a fluid which violates the null energy condition, dubbed phantom energy (Caldwell, 2002). In fact, Sushkov and Lobo (Sushkov, 2005; Lobo, 2005), independently, have shown that phantom energy could well be the class of exotic matter which is required to support traversable wormholes. This result could be regarded to be one of the most powerful arguments in favor of the idea that wormholes should no longer be regarded as just describing purely mathematical toy spacetime models with interest only for science fictions writers, but also as plausible physical realities that could exist in the very spacetemporal fabric of our Universe. The realization of this fact has motivated a renaissance of the study of wormhole spacetimes, the special interest being the consideration of the possible accretion of phantom energy onto wormholes, which may actually cause the growth of their mouths (Gonzalez-Diaz, 2004)

Due to the status at which wormholes have been promoted, it seems that the following step should be the search of these objects in our Universe. Cramer et al. (Cramer et al., 1995) were the first in noticing that a negative mass, like a wormhole, would deflect the rays coming from a luminous source, similarly to a positive mass but taking the term deflection its proper meaning in this case. Considering a wormhole between the source and the observer, the observer would either measure an increase of the intensity or receive no signal if he/she is in a certain umbral region. Following this line of thinking other works have studied the effects of microlensing (Torres et al., 1998a;b; Safonova et al., 2002) or macrolensing (Safonova et al., 2001) that wormholes could originate. Nevertheless, wormholes would affect the trajectory not only of light rays passing at some distance of them, but of rays going through them coming from other universe or other region of the same universe (Shatskiy, 2007; n.d.). In this case, as it could be expected (Morris et al., 1988), the wormhole would cause the divergence of the light rays, which would form an image of a disk with an intensity reaching several relative maxima and minima, and an absolute maximum in the edge (Shatskiy, n.d.). However, if for any reason the intensity in the edge could be much higher than in the interior region (Shatskiy, 2007), then this image may be confused with an Einstein ring, like one generated by the deformation of light due to a massive astronomical object with positive mass situated on the axis formed by the source and the observer, between them[1] (Gonzalez-Diaz, n.d.). In summary, whereas the deformation in the trajectory of light rays passing close to the wormhole could be due to any other astronomical object with negative mass, if it might exist, the observational trace produced by the light rays coming through the hole could be confused with the deformation produced by massive object with positive mass. Therefore, if it would be possible to measure in the future both effects together, then we might find a wormhole.

On the other hand, it is well known that the thermodynamical description of black holes (Bardeen et al., 1973) and other vacuum solutions, as the de Sitter model (Gibbons & Hawking, 1977), has provided these spacetimes with quite a more robust consistency, allowing moreover for a deeper understanding of their structure and properties. Following this spirit, a possible thermodynamical representation of wormholes could lead to a deeper understanding of both, these objects and the exotic material which generates them, which could perhaps be the largest energy source in the universe. Therefore, in the present chapter, we consider the potential thermodynamical properties of Lorentzian traversable wormholes. Such study should necessarily be considered in terms of local concepts, as trapping horizons, since in the considered spacetime the definition of an event horizon is no longer possible.

The importance of the use of trapping horizons in order to characterize the black holes themselves in terms of local quantities has been emphasized by Hayward (Hayward, 1994a; 1996; 1998; 2004), since global properties can not be measured by a real observer with a finite life. In this way, the mentioned author has developed a formalism able to describe the thermodynamical properties of dynamical and spherically symmetric black holes, based

[1]It must be pointed out that whereas an structure of this kind is obtained in Ref. (Shatskiy, 2007), in Ref. (Shatskiy, n.d.) it is claimed that a correct interpretation of the results would indicate an image in the form of a luminous spot in the case that the number of stars in the other universe would be infinite, which would tend to a situation in which the maxima and minima could be distinguished when the real case tends to separate of the mentioned idealization.

in the existence of trapping horizons. Therefore, the presence of trapping horizons in the wormhole spacetime would also make possible the study of these objects. Moreover, since both objects, black holes and wormholes, can be characterized by outer trapping horizons, which are spacelike or null and timelike, respectively, they could show certain similar properties (Hayward, 1999), in particular, an analogous thermodynamics.

As we will show, the key point in this study will not lie only in applying the formalism developed by Hayward (Hayward, 1994a; 1996; 1998; 2004) to the wormhole spacetime, but in noticing that the results coming from the accretion method, (Babichev et al., 2004; Martin-Moruno et al., 2006) and (Gonzalez-Diaz, 2004; Gonzalez-Diaz & Martin-Moruno, 2008), must be equivalent to those which will be obtained by the mentioned formalism; this fact will allow a univocal characterization of wormholes. Such a characterization, together with some results about phantom thermodynamics (Gonzalez-Diaz & Siguenza, 2004; Saridakis et al., 2009), which concluded that phantom energy would possess a negative temperature, would provide any possible Hawking-like radiation from wormholes with a well defined physical meaning.

In this chapter, we will start by summarizing some previous concepts on the Morris and Thorne solution 2.1 and the Hayward formalism 2.1. This formalism will be applied to Morris-Thorne wormholes in Sec. 3. In Sec. 4 we will introduce a consistent characterization of dynamical wormholes, which will allow us to derive a thermal radiation and formulate a whole thermodynamics in Sec. 5. Finally, in Sec.6, the conclusions are summarized and further comments are added. Throughout this chapter, we use the signature convention $(-,+,+,+)$.

2. Preliminaries

2.1 The Morris-Thorne wormholes

Morris and Thorne (Morris & Thorne, 1988) considered the most general static and spherically symmetric metric able to describe a stable and traversable wormhole. That solution describes a throat connecting two asymptotically flat regions of the spacetime, without any event horizon. This metric is

$$ds^2 = -e^{2\Phi(l)}dt^2 + dl^2 + r^2(l)\left[d\theta^2 + \sin^2 d\varphi^2\right], \tag{1}$$

where the coordinate $-\infty < l < \infty$ and the function $\Phi(l)$ should be positive definite for any value of l. In order to recover the asymptotic limit, $r(l)/|l| \to 1$ and $\Phi(l) \to$ constant, when $l \to \pm\infty$. On the other hand, the wormhole throat is the minimum of the function $r(l)$, r_0, which we can suppose, without loss of generality, placed at $l = 0$; therefore $l < 0$ and $l > 0$ respectively cover the two asymptotically flat regions connected through the throat at $l = 0$.

It is useful to express metric (1) in terms of the Schwarzschild coordinates, which yields

$$ds^2 = -e^{2\Phi(r)}dt^2 + \frac{dr^2}{1 - K(r)/r} + r^2\left[d\theta^2 + \sin^2 d\varphi^2\right], \tag{2}$$

where $\Phi(r)$ and $K(r)$ are the redshift function and the shape function, respectively, and it must be pointed out that now two sets of coordinates are needed in order to cover both spacetime regions, both with $r_0 \leq r \leq \infty$. For preserving asymptotic flatness, both such functions[2], $\Phi(r)$ and $K(r)$, must tend to a constant value when the radial coordinate goes to infinity. On the other hand, the minimum radius, r_0, corresponds to the throat, where $K(r_0) = r_0$. Although

[2]In general, there could be different functions $\Phi(r)$ and $K(r)$ in each region, (Visser, 1995), although, for our present purposes, this freedom is not of interest.

the metric coefficient g_{rr} diverges at the throat, this surface is only an apparent singularity, since the proper radial distance

$$l(r) = \pm \int_{r_0}^{r} \frac{dr^*}{\sqrt{1 - K(r^*)/r^*}}, \tag{3}$$

must be finite everywhere.
In order to interpret this spacetime, we can use an embedding diagram (Morris & Thorne, 1988) (see also (Visser, 1995) or (Lobo, n.d.))). This embedding diagram, Fig. 1, can be obtained by using the spherical symmetry of this spacetime, which allow us to consider, without lost of generality, a slice defined by $\theta = \pi/2$. Such a slice is described at constant time by

$$ds^2 = \frac{dr^2}{1 - K(r)/r} + r^2 d\varphi^2. \tag{4}$$

Now, we consider the Euclidean three-dimensional spacetime in cylindrical coordinates, i. e.

$$ds^2 = dz^2 + dr^2 + r^2 d\varphi^2. \tag{5}$$

In this spacetime the slice is an embedded surface described by an equation $z = z(r)$. Therefore, Eq. (5) evaluated at the surface yields

$$ds^2 = \left[1 + \left(\frac{dz}{dr}\right)^2\right] dr^2 + r^2 d\varphi^2. \tag{6}$$

Taking into account Eqs. (4) and (6) we can obtain the equation of the embedded surface. This is

$$\frac{dz}{dr} = \pm \left[\frac{r}{K(r)} - 1\right]^{-1/2}, \tag{7}$$

which diverges at the throat and tends to zero at the asymptotic limit. The throat must flare out in order to have a wormhole, which is known as the "flaring-out condition". This condition implies that the inverse of the embedded function should have an increasing derivative at the throat and close to it, i. e. $(d^2r)/(d^2z) > 0$. Therefore, taking the inverse of Eq. (7) and deriving with respect to z, it can be obtained that

$$\frac{K(r) - rK'(r)}{2(K(r))^2} > 0, \tag{8}$$

implying that $K'(r_0) < 1$.
On the other hand, considering that the energy-momentum tensor can be written in an orthonormal basis as $T_{\mu\nu} = \mathrm{diag}(\rho(r), p_r(r), p_t(r), p_t(r))$, the Einstein equations of this spacetime produce (Morris & Thorne, 1988; Visser, 1995; Lobo, n.d.)

$$\rho(r) = \frac{K'(r)}{8\pi r^2}, \tag{9}$$

$$p_r(r) = -\frac{1}{8\pi}\left[\frac{K(r)}{r^3} - 2\left(1 - \frac{K(r)}{r}\right)\frac{\Phi'(r)}{r}\right], \tag{10}$$

$$p_t(r) = \frac{1}{8\pi}\left[\Phi''(r) + (\Phi'(r))^2 + \frac{K - K'(r)r}{2r^3(1 - K(r)/r)}(r\Phi'(r) + 1) + \frac{\Phi'(r)}{r}\right]. \tag{11}$$

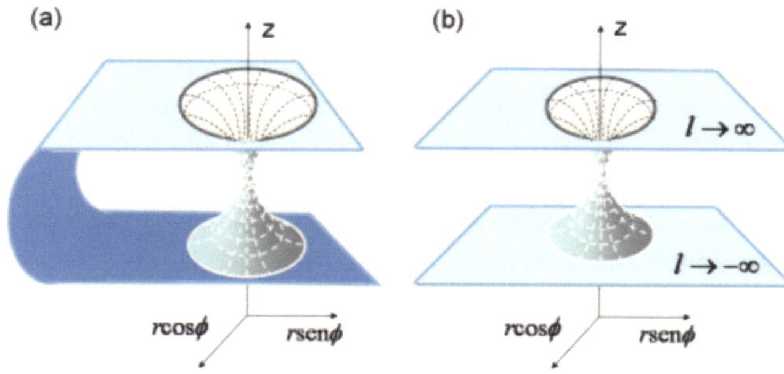

Fig. 1. The embedding diagram of an equatorial slice ($\theta = \pi/2$) of a wormhole at a given time (t =const). The wormhole can connect two regions of the same universe (a) or two different universes (b).

Evaluating the condition (8) at the throat and taking into account Eqs. (9) and (10), these equations are seen to entail $\rho(r_0) + p(r_0) < 0$. Therefore, the stuff generating this geometry violates the null energy condition at the throat and close to it. On the other hand, in order to minimize the exoticity of this stuff, it can be required that, at least, the energy density should be positive, leading to $K'(r) > 0$.

Apart from some quantum effects, as the Casimir effect, which could allow violations of the null energy condition, this violation has gained naturalness with the accelerated expansion of the universe. As we have already mentioned, some studies (Sushkov, 2005; Lobo, 2005; 2006) have extended the notion of phantom energy to inhomogeneous spherically symmetric spacetimes by regarding that the pressure related to the energy density through the equation of state parameter must be the radial pressure, calculating the transverse components by means of the Einstein equations. One can see (Lobo, 2005) that a particular specification of the redshift and shape functions in metric (2) leads to a static phantom traversable wormhole solution (where no dynamical evolution for the phantom energy is considered) which satisfies the traversability conditions (Morris & Thorne, 1988), in particular the outward flaring condition $K'(r_0) < 1$.

2.2 Trapping horizons

As a necessary tool for the development of the following sections, in the present subsection we summarize some concepts and notation of the Hayward formalism, which are based on the null dynamics and applicable to spherically symmetric spacetimes (Hayward, 1998).

First of all, it must be noticed that the metric of a spherically symmetric spacetime can always be written, at least locally, as

$$ds^2 = 2g_{+-}d\xi^+ d\xi^- + r^2 d\Omega^2, \tag{12}$$

where $r > 0$ and $g_{+-} < 0$ are functions of the null coordinates (ξ^+, ξ^-), related with the two preferred null normal directions of each symmetric sphere $\partial_\pm \equiv \partial/\partial\xi^\pm$, r is the so-called areal radius (Hayward, 1998), which is a geometrical invariant, and $d\Omega^2$ refers to the metric on the unit two-sphere. One can define the expansions in the null directions as

$$\Theta_\pm = \frac{2}{r}\partial_\pm r. \tag{13}$$

The sign of $\Theta_+\Theta_-$ is invariant, therefore it can be used to classify the spheres of symmetry. One can say that a sphere is trapped, untrapped or marginal if the product $\Theta_+\Theta_-$ is bigger, less or equal to zero, respectively. Locally fixing the orientation on an untrapped sphere such that $\Theta_+ > 0$ and $\Theta_- < 0$, ∂_+ and ∂_- will be also fixed as the outgoing and ingoing null normal vectors (or the contrary if the orientation $\Theta_+ < 0$ and $\Theta_- > 0$ is considered). A marginal sphere with $\Theta_+ = 0$ is future if $\Theta_- < 0$, past if $\Theta_- > 0$ and bifurcating[3] if $\Theta_- = 0$. This marginal sphere is outer if $\partial_- \Theta_+ < 0$, inner if $\partial_- \Theta_+ > 0$ and degenerate if $\partial_- \Theta_+ = 0$. A hypersurface foliated by marginal spheres is called a trapping horizon and has the same classification as the marginal spheres.

In spherical symmetric spacetimes a unified first law of thermodynamics can be formulated (Hayward, 1998), by using the gravitational energy in spaces with this symmetry, which is the Misner-Sharp energy (Misner & Sharp, 1964). This energy can be defined by

$$E = \frac{1}{2} r \left(1 - \nabla^a r \nabla_a r\right) = \frac{r}{2}\left(1 - 2g^{+-}\partial_+ r \partial_- r\right), \tag{14}$$

and become $E = r/2$ on a trapping horizon[4].

Two invariants are also needed in order to write the unified first law of thermodynamics. These invariants can be constructed out of the energy-momentum tensor of the background fluid which can be easily expressed in these coordinates:

$$\omega = -g_{+-} T^{+-} \tag{15}$$

and the vector

$$\psi = T^{++} \partial_+ r \partial_+ + T^{--} \partial_- r \partial_- . \tag{16}$$

The first law can be written as

$$\partial_\pm E = A \psi_\pm + \omega \partial_\pm V, \tag{17}$$

where $A = 4\pi r^2$ is the area of the spheres of symmetry and $V = 4\pi r^3/3$ is defined as the corresponding flat-space volume. The first term in the r.h.s. could be interpreted as an energy-supply term, since this term produces a change in the energy of the spacetime due to the energy flux ψ generated by the surrounding material. The second term, $\omega \partial_\pm V$, behaves like a work term, something like the work that the matter content must do to support this configuration.

The Kodama vector plays also a central role in this formalism. This vector, which was introduced by Kodama (Kodama, 1980), can be understood as a generalization from the stationary Killing vector in spherically symmetric spacetimes, reducing to it in the vacuum case. The Kodama vector can be defined as

$$k = \mathrm{rot}_2 r, \tag{18}$$

where the subscript 2 means referring to the two-dimensional space normal to the spheres of symmetry. Expressing k in terms of the null coordinates one obtains

---

[3]It must be noted that on the first part of this work we will consider future and past trapping horizons with $\Theta_+ = 0$, implying that ξ^- must be related to the ingoing or outgoing null normal direction for future ($\Theta_- < 0$) or past ($\Theta_- > 0$) trapping horizons, respectively. In Sec. 5, where we will only treat past outer trapping horizons, we will fix $\Theta_- = 0$, without lost of generality, implying that $\Theta_+ > 0$ and ξ^- related to the ingoing null direction.

[4]The reader interested in properties of E may look up Ref. (Hayward, 1994b).

$$k = -g^{+-}\left(\partial_+ r\partial_- - \partial_- r\partial_+\right), \tag{19}$$

where the orientation of k can be fixed such that in a untrapped region it is future pointing. From Eq. (14), it can be noted that the squared norm of the Kodama vector can be written as

$$||k||^2 = \frac{2E}{r} - 1. \tag{20}$$

Therefore, this vector provides the trapping horizon with the additional definition of a hypersurface where the Kodama vector is null. So, such as it happens in the case of static spacetimes, where a boundary can be generally defined as the hypersurface where the temporal Killing vector is null, in the present case we must instead use the Kodama vector. This vector has some special properties[5] similar to those of the Killing vector in static spacetimes with boundaries (Hayward, 1998), such as

$$k^a \nabla_{[a} k_{b]} = \kappa \nabla_b r, \tag{21}$$

which, evaluated on a trapping horizon, implies

$$k^a \nabla_{[a} k_{b]} = \kappa k_b \text{ on a trapping horizon}, \tag{22}$$

where the square brackets means antisymmetrization in the included scripts and

$$\kappa = \frac{1}{2}\mathrm{div}_2 \mathrm{grad}_2 r. \tag{23}$$

Due to the similarity between Eq. (22) and the corresponding one involving the Killing vector and its horizon[6], κ is known as generalized or geometric surface gravity. From the definition of this quantity (23) and the classification of the trapping horizons introduced at the beginning of this subsection, it can be seen that an outer, degenerate or inner horizon has $\kappa > 0$, $\kappa = 0$ y $\kappa < 0$, respectively. On the other hand, κ can be expressed in terms of the null coordinates as

$$\kappa = g^{+-}\partial_-\partial_+ r. \tag{24}$$

Taking into account Eq. (24), it can be seen that the projection of Eq. (17) along the vector z which generates the trapping horizon yields

$$L_z E = \frac{\kappa L_z A}{8\pi} + \omega L_z V, \tag{25}$$

where $L_z = z \cdot \nabla$ and $z = z^+\partial_+ + z^-\partial_-$. This expression allows us to relate the geometric entropy and the surface area through

$$S \propto A|_H. \tag{26}$$

Finally, the Einstein equations of interest, in terms of the null coordinates (Hayward, 1998), can be expressed using the expansions (13) as

[5]In Ref. (Hayward, 1996) other interesting properties of k are also studied.

[6]Although in the equation which relates the Killing vector with the surface gravity there is no any explicit antisymmetrization, that equation could be written in an equivalent way using an antisymmetrization. This fact is a consequence of the very definition of the Killing vector, which implies $\nabla_{(a}K_{b)} = 0$, where the brackets means symmetrization in the included scripts and K is the Killing vector.

$$\partial_\pm \Theta_\pm = -\frac{1}{2}\Theta_\pm^2 + \Theta_\pm \partial_\pm \ln(-g_{+-}) - 8\pi T_{\pm\pm}, \tag{27}$$

$$\partial_\pm \Theta_\mp = -\Theta_+\Theta_- + \frac{1}{r^2} g_{+-} + 8\pi T_{+-}. \tag{28}$$

3. 2+2-formalism applied to Morris-Thorne wormholes

The 2+2-formalism was initially introduced by Hayward for defining the properties of real black holes in terms of measurable quantities. Such a formalism can be considered as a generalization that allows the formulation of the thermodynamics of dynamical black holes by using local quantities which are physically meaningful both in static and dynamical spacetimes. In fact, this formalism consistently recovers the results obtained by global considerations using the event horizon in the vacuum static case (Hayward, 1998). Even more, as Hayward has also pointed out (Hayward, 1999), this local considerations can also be applied to dynamic wormholes spacetimes, implying that there exists a common framework for treating black holes and wormholes.

Nevertheless one of the most surprising features of the 2+2-formalism is found when applied to Morris-Thorne wormholes. Whereas in this spacetime it is not possible to obtain any property similar to those obtained in black holes physics by using global considerations, since no event horizon is present, the consideration of local quantities shows similar characteristics to those of black holes. This fact can be better understood if one notices that the Schwarzschild spacetime is the only spherically symmetric solution in vacuum and, therefore, any dynamical generalizations of black holes must be formulated in the presence of some matter content. The maximal extension of the Schwarzschild spacetime (Kruskal, 1960) can be interpreted as an Einstein-Rosen bridge (Einstein & Rosen, 1935), which corresponds to a vacuum wormhole and has associated a given thermodynamics. Nevertheless, the Einstein-Rosen bridge can not be traversed since it has an event horizon and it is unstable (Fuller & Wheeler, 1962). If we consider wormholes which can be traversed, then some matter content must be present even in the static case of Morris-Thorne. So the need of a formulation in terms of local quantities, measurable for an observer with finite life, must be related to the presence of some matter content, rather than with a dynamical evolution of the spacetime.

In this section we apply the results obtained by Hayward for spherically symmetric solutions to static wormholes, showing rigorously their consequences (Martin-Moruno & Gonzalez-Diaz, 2009b), some of which were already suggested and/or indicated by Ida and Hayward himself (Ida & Hayward, 1995).

Defining the coordinates $\xi^+ = t + r_*$ and $\xi^- = t - r_*$, with r_* such that $\mathrm{d}r/\mathrm{d}r_* = \sqrt{-g_{00}/g_{rr}} = e^{\Phi(r)}\sqrt{1-K(r)/r}$, and ξ^+ and ξ^- being related to the outgoing and ingoing direction, respectively, the metric (2) can be expressed in the form given by Eq. (12). It can be seen, by the definitions introduced in the previous section, that there is a bifurcating trapping horizon at $r = r_0$. This horizon is outer since the flaring-out condition implies $K'(r_0) < 1$.

In this spacetime the Misner-Sharp energy (14), "energy density" (15) and "energy flux" (16) can be calculated to be

$$E = \frac{K(r)}{2}, \tag{29}$$

$$\omega = \frac{\rho - p_r}{2}, \tag{30}$$

and

$$\psi = -\frac{(\rho + p_r)}{2} e^{-\Phi(r)} \sqrt{1 - K(r)/r} (-\partial_+ + \partial_-), \tag{31}$$

where we have taken into account that the components of an energy-momentum tensor which takes the form[7] $T^{(2)}_{\mu\nu} = \text{diag}(\rho, p_r)$ in an orthonormal basis, with the superscript (2) meaning the two-dimensional space normal to the spheres of symmetry, are in this basis $T_{\pm\pm} = e^{2\Phi(r)}(\rho + p_r)/4$ and $T_{+-} = T_{-+} = e^{2\Phi(r)}(\rho - p_r)/4$. The Misner-Sharp energy in this spacetime reaches its limiting value $E = r/2$ only at the wormhole throat, $r = r_0$, which corresponds to the trapping horizon, taking smaller values in the rest of the space which is untrapped. We want to emphasized that, as in the case of the studies about phantom wormholes performed by Sushkov (Sushkov, 2005) and Lobo (Lobo, 2005), any information about the transverse components of the pressure becomes unnecessary.
Deriving Eq. (29) and rising the index of Eq. (31), one can obtain

$$\partial_\pm E = \pm 2\pi r^2 \rho e^{\Phi} \sqrt{1 - K(r)/r} \tag{32}$$

and

$$\psi_\pm = \pm e^{\Phi}(r) \sqrt{1 - K(r)/r} \frac{\rho + p_r}{4}. \tag{33}$$

Therefore, we have all terms[8] of Eq. (17) for the first law particularized to the Morris-Thorne case, which vanish at the throat, what could be suspected since we are considering a wormhole without dynamic evolution. Nevertheless, the comparison of these terms in the case of Morris-Thorne wormholes with those which appear in the Schwarzschild black hole could provide us with a deeper understanding about the former spacetime, based on the exotic properties of its matter content. Of course, the Schwarzschild metric is a vacuum solution, but it could be expected that it would be a good approximation when small matter quantities are considered, which we will assume to be ordinary matter. So, in the first place, we want to point out that the variation of the gravitational energy, Eq. (32), is positive (negative) in the outgoing (ingoing) direction in both cases[9], since $\rho > 0$; therefore, this variation is positive for exotic and usual matter. In the second place, the "energy density", ω, takes positive values no matter whether the null energy condition is violated or not. Considering the "energy supply" term, in the third place, we find the key difference characterizing the wormhole spacetime. The energy flux depends on the sign of $\rho + p_r$, therefore it can be interpreted as a fluid which "gives" energy to the spacetime, in the case of usual matter, or as a fluid "receiving" or "getting" energy from the spacetime, when exotic matter is considered. This "energy removal", induced by the energy flux in the wormhole case, can never reach a value so large to change the sign of the variation of the gravitational energy.
On the other hand, the spacetime given by (2) possesses a temporal Killing vector which is non-vanishing everywhere and, therefore, there is no Killing horizon where a surface gravity can be calculated as considered by Gibbons and Hawking (Gibbons & Hawking, 1977).

---

[7] As we will comment in the next section, this energy-momentum tensor is of type I in the classification of Hawking and Ellis (Hawking & Ellis, 1973).

[8] The remaining terms can be easily obtained taking into account that $\partial_\pm r = \pm\frac{1}{2} e^{\Phi(r)} \sqrt{1 - K(r)/r}$.

[9] The factor $e^{\Phi}\sqrt{1 - K(r)/r} \equiv \alpha$, which appears by explicitly considering the Morris-Thorne solution, comes from the quantity $\alpha = \sqrt{-g_{00}/g_{rr}}$, which is a general factor at least in spherically symmetric and static cases; therefore α has the same sign both in Eq. (32) and in Eq. (33).

Nevertheless, the definition of a Kodama vector or, equivalently, of a trapping horizon implies the existence of a generalized surface gravity for both static and dynamic wormholes. In particular, in the Morris-Thorne case the components of the Kodama vector take the form

$$k^{\pm} = e^{-\Phi(r)}\sqrt{1 - K(r)/r}, \tag{34}$$

with $||k||^2 = -1 + K(r)/r = 0$ at the throat. The generalized surface gravity, (24), is

$$\kappa|_H = \frac{1 - K'(r_0)}{4r_0} > 0, \tag{35}$$

where "$|_H$" means evaluation at the throat and we have considered that the throat is an outer trapping horizon, which is equivalent to the flaring-out condition ($K'(r_0) < 1$). By using the Einstein equations (9) and (10), κ can be re-expressed as

$$\kappa|_H = -2\pi r_0 \left[\rho(r_0) + p(r_0)\right]. \tag{36}$$

with $\rho(r_0) + p(r_0) < 0$, as we have mentioned in 2.1.
It is well known that when the surface gravity is defined by using a temporal Killing vector, this quantity is understood to mean that there is a force acting on test particles in a gravitational field. The generalized surface gravity is in turn defined by the use of the Kodama vector, which can be interpreted as a preferred flow of time for observers at a constant radius (Hayward, 1996), reducing to the Killing vector in the vacuum case and recovering the surface gravity its usual meaning. Nevertheless, in the case of a spherically symmetric and static wormhole one can define both, the temporal Killing and the Kodama vector, being the Kodama vector of greater interest since it vanishes at a particular surface. Moreover, in dynamical spherically symmetric cases one can only define the Kodama vector. Therefore it could be suspected that the generalized surface gravity should originate some effect on test particles which would go beyond that corresponding to a force, and only reducing to it in the vacuum case. On the other hand, if by some kind of symmetry this effect on a test particle would vanish, then we should think that such a symmetry would also produce that the trapping horizon be degenerated.

4. Dynamical wormholes

The existence of a generalized surface gravity which appears in the first term of the r.h.s. of Eq. (25) multiplying a quantity which can be identify as something proportional to an entropy would suggest the possible formulation of a wormhole thermodynamics, as it was already commented in Ref. (Hayward, 1999). Nevertheless, a more precise definition of its trapping horizon must be done in order to settle down univocally its characteristics. With this purpose, we first have to summarize the results obtained by Hayward for the increase of the black hole area (Hayward, 2004), comparing then them with those derived from the accretion method (Babichev et al., 2004). Such comparison will shed some light for the case of wormholes.
On the one hand, the area of a surface can be expressed in terms of μ as $A = \int_S \mu$, with $\mu = r^2 \sin\theta d\theta d\varphi$ in the spherically symmetric case. Therefore, the evolution of the trapping horizon area can be studied considering

$$L_z A = \int_S \mu \left(z^+\Theta_+ + z^-\Theta_-\right), \tag{37}$$

with z the vector which generates the trapping horizon.

On the other hand, by the very definition of a trapping horizon we can fix $\Theta_+|_H = 0$, which provides us with the fundamental equation governing its evolution

$$L_z\Theta_+|_H = \left[z^+\partial_+\Theta_+ + z^-\partial_-\Theta_+\right]|_H = 0. \tag{38}$$

It must be also noticed that the evaluation of Eq. (27) at the trapping horizon implies

$$\partial_+\Theta_+|_H = -8\pi T_{++}|_H, \tag{39}$$

where $T_{++} \propto \rho + p_r$ by considering an energy-momentum tensor of type I in the classification of Hawking and Ellis[10]. (Hawking & Ellis, 1973). Therefore, if the matter content which supports the geometry is usual matter, then $\partial_+\Theta_+|_H < 0$, being $\partial_+\Theta_+|_H > 0$ if the null energy condition is violated.

Dynamic black holes are characterized by outer future trapping horizons, which implies the growth of their area when they are placed in environment which fulfill the null energy condition (Hayward, 2004). This property can be easily deduced taking into account the definition of outer trapping horizon and noticing that, when it is introduced in the condition (38), with Eq. (39) for usual matter, implies that the sign of z^+ and z^- must be different, i.e. the trapping horizon is spacelike when considering usual matter and null in the vacuum case. It follows that the evaluation of L_zA at the horizon, $\Theta_+ = 0$, taking into account that the horizon is future and that z has a positive component along the future-pointing direction of vanishing expansion, $z^+ > 0$, yields[11] $L_zA \geq 0$, where the equality is fulfilled in the vacuum case. It is worth noticing that when exotic matter is considered, then the previous reasoning would lead to a black hole area decrease.

It is well known that accretion method based on a test-fluid approach developed by Babichev et al. (Babichev et al., 2004) (and its non-static generalization (Martin-Moruno et al., 2006)) leads to the increase (decrease) of the black hole when it acreates a fluid with $p+\rho > 0$ ($p+\rho < 0$), where p could be identified in this case with p_r. These results are the same as those obtained by using the $2+2$-formalism, therefore, it seems natural to consider that both methods in fact describe the same physical process, originating from the flow of the surrounding matter into the hole.

Whereas the characterization of black holes appears in this study as a natural consideration, a reasonable doubt may still be kept about how the outer trapping horizon of wormholes may be considered. Following the same steps as in the argument relative to dynamical black holes, it can be seen that, since a traversable wormhole should necessarily be described in the presence of exotic matter, the above considerations imply that its trapping horizon should be timelike, allowing a two-way travel. However, if this horizon would be future (past) then, by Eq. (37), its area would decrease (increase) in an exotic environment, remaining constant in the static case when the horizon is bifurcating. In this sense, an ambiguity in the characterization of dynamic wormholes seems to exist.

---

[10] In general one would have $T_{++} \propto T_{00} + T_{11} - 2T_{01}$, where the components of the energy-momentum tensor on the r.h.s. are expressed in terms of an orthonormal basis. In our case, we consider an energy-momentum tensor of type I (Hawking & Ellis, 1973), not just because it represents all observer fields with non-zero rest mass and zero rest mass fields, except in special cases when it is type II, but also because if this would not be the case then either $T_{++} = 0$ (for types II and III) which at the end of the day would imply no horizon expansion, or we would be considering the case where the energy density vanishes (type IV)

[11] It must be noticed that in the white hole case, which is characterized by a past outer trapping horizon, this argument implies $L_zA \leq 0$.

Nevertheless, this ambiguity is only apparent once noticed that this method is studying the same process as the accretion method, in this case applied to wormholes (Gonzalez-Diaz & Martin-Moruno, 2008), which implies that the wormhole throat must increase (decrease) its size by accreting energy which violates (fulfills) the null energy condition. Therefore, the outer trapping horizons which characterized dynamical wormholes should be past (Martin-Moruno & Gonzalez-Diaz, 2009a;b). This univocal characterization could have been suspected from the very beginning since, if the energy which supports wormholes should violate the null energy condition, then it seems quite a reasonable implication that the wormhole throat must increase if some matter of this kind would be accreted.

In order to better understand this characterization, we could think that whereas dynamical black holes would tend to be static as one goes into the future, being their trapping horizon past, white holes, which are assumed to have born static and then allowed to evolve, are characterized by a past trapping horizon. So, in the case of dynamical wormholes one can consider a picture of them being born at some moment (at the beginning of the universe, or constructed by an advanced civilization, or any other possible scenarios) and then left to evolve to they own. Therefore, following this picture, it seems consistent to characterize wormholes by past trapping horizons.

Finally, taking into account the proportionality relation (26), we can see that the dynamical evolution of the wormhole entropy must be such that $L_z S \geq 0$, which saturates only at the static case characterized by a bifurcating trapping horizon.

5. Wormhole thermal radiation and thermodynamics

The existence of a non-vanishing surface gravity at the wormhole throat seems to imply that it can be characterized by a non-zero temperature so that one would expect that wormholes should emit some sort of thermal radiation. Although we are considering wormholes which can be traversed by any matter or radiation, passing through it from one universe to another (or from a region to another of the same single universe), what we are refereeing to now is a completely different kind of radiative phenomenon, which is not due to any matter or radiation following any classically allowed path but to thermal radiation with a quantum origin. Therefore, even in the case that no matter or radiation would travel through the wormhole classically, the existence of a trapping horizon would produce a semi-classical thermal radiation.

It has been already noticed in Ref. (Hayward et al., 2009) that the use of a Hamilton-Jacobi variant of the Parikh-Wilczek tunneling method led to a local Hawking temperature in the case of spherically symmetric black holes. Nevertheless, it was also suggested (Hayward et al., 2009) that the application of this method to past outer trapping horizon could lead to negative temperatures which, therefore, could be lacking of a well defined physical meaning. In this section we show explicitly the calculation of the temperature associated with past outer trapping horizons (Martin-Moruno & Gonzalez-Diaz, 2009a;b), which characterizes dynamical wormholes, applying the method considered in Ref. (Hayward et al., 2009). The rigorous application of this method implies a wormhole horizon with negative temperature. This result, far from being lacking in a well defined physical meaning, can be interpreted in a natural way taking into account that, as it is well known (Gonzalez-Diaz & Siguenza, 2004; Saridakis et al., 2009), phantom energy also possesses negative temperature.

We shall consider in the present study a general spherically symmetric and dynamic wormhole which, therefore, is described through metric (12) with a trapping horizon characterized by $\Theta_- = 0$ and[12] $\Theta_+ > 0$. The metric (12) can be consequently written in terms of the generalized retarded Eddington-Finkelstein coordinates, at least locally, as

$$ds^2 = -e^{2\Psi} C du^2 - 2e^{\Psi} du dr + r^2 d\Omega^2, \tag{40}$$

where $du = d\xi^-$, $d\xi^+ = \partial_u \xi^+ du + \partial_r \xi^+ dr$, and Ψ expressing the gauge freedom in the choice of the null coordinate u. Since $\partial_r \xi^+ > 0$, we have considered $e^{\Psi} = -g_{+-} \partial_r \xi^+ > 0$ and $e^{2\Psi} C = -2g_{+-} \partial_u \xi^+$. It can be seen that $C = 1 - 2E/r$, with E defined by Eq. (14). The use of retarded coordinates ensures that the marginal surfaces, characterized by $C = 0$, are past marginal surfaces.

From Eqs. (18) and (23), it can be seen that the generalized surface gravity at the horizon and the Kodama vector are

$$\kappa|_H = \frac{\partial_r C}{2} \tag{41}$$

and

$$k = e^{-\Psi} \partial_u, \tag{42}$$

respectively.

Now, similarly to as it has been done in Ref. (Hayward et al., 2009) for the dynamical black hole case, we consider a massless scalar field in the eikonal approximation, $\phi = \phi_0 \exp(iI)$, with a slowly varying amplitude and a rapidly varying action given by

$$I = \int \omega_\phi e^{\Psi} du - \int k_\phi dr, \tag{43}$$

with ω_ϕ being an energy parameter associated to the radiation. In our case, this field describes radially outgoing radiation, since ingoing radiation would require the use of advanced coordinates.

The wave equation of the field which, as we have already mentioned, fulfills the eikonal equation, implies the Hamilton-Jacobi one[13]

$$\gamma^{ab} \nabla_a I \nabla_b I = 0, \tag{44}$$

where γ^{ab} is the metric in the 2-space normal to the spheres of symmetry. Now, taking into account $\partial_u I = e^{\Psi} \omega_\phi$ and $\partial_r I = -k$, Eq. (44) yields

$$k_\phi^2 C + 2\omega_\phi k_\phi = 0. \tag{45}$$

One solution of this equation is $k_\phi = 0$, which must corresponds to the outgoing modes, since we are considering that ϕ is outgoing. On the other hand, the alternate solution, $k_\phi = -2\omega_\phi / C$, should correspond to the ingoing modes and it will produce a pole in the action integral 43, because C vanishes on the horizon. Expanding C close to the horizon, one can express the second solution in this regime as $k_\phi \approx -\omega_\phi / [\kappa|_H (r - r_0)]$. Therefore the action has an imaginary contribution which is obtained deforming the contour of integration in the lower r half-plane, which is

[12] We are now fixing, without loss of generality, the outgoing and ingoing direction as ∂_+ and ∂_-, respectively.

[13] For a deeper understanding about the commonly used approximations of this method, as the eikonal one, it can be seen, for example, Ref. (Visser, 2003).

$$\text{Im}(I)|_H = -\frac{\pi\omega_\phi}{\kappa|_H}. \tag{46}$$

This expression can be used to consider the particle production rate as given by the WKB approximation of the tunneling probability Γ along a classically forbidden trajectory

$$\Gamma \propto \exp[-2\text{Im}(I)]. \tag{47}$$

Although the wormhole throat is a classically allowed trajectory, being the wormhole a two-way traversable membrane, we can consider that the existence of a trapping horizon opens the possibility for an additional traversing phenomenon through the wormhole with a quantum origin. One could think that this additional radiation would be somehow based on some sort of quantum tunneling mechanism between the two involved universes (or the two regions of the same, single universe), a process which of course is classically forbidden. If such an interpretation is accepted, then (47) takes into account the probability of particle production rate at the trapping horizon induced by some quantum, or at least semi-classical, effect. On the other hand, considering that this probability takes a thermal form, $\Gamma \propto \exp(-\omega_\phi/T_H)$, one could compute a temperature for the thermal radiation given by

$$T = -\frac{\kappa|_H}{2\pi}, \tag{48}$$

which is negative. At first sight, one could think that we would be safe from this negative temperature because it is related to the ingoing modes. However this can no longer be the case as even if this thermal radiation is associated to the ingoing modes, they characterize the horizon temperature. Even more, the infalling radiation getting in one of the wormhole mouths would travel through that wormhole following a classical path to go out of the other mouth as an outgoing radiation in the other universe (or the other region of universe). Such a process would take place at both mouths producing, in the end of the day, outgoing radiation with negative temperature in both mouths.

Nevertheless, it is well known that phantom energy, which is no more than a particular case of exotic matter, is characterized by a negative temperature (Gonzalez-Diaz & Siguenza, 2004; Saridakis et al., 2009). Thus, this result could be taken to be a consistency proof of the used method, as a negative radiation temperature simply express the feature to be expected that wormholes should emit a thermal radiation just of the same kind as that of the stuff supporting them, such as it also occurs with dynamical black holes with respect to usual matter and positive temperature.

Now, Eq. (25) can be re-written, taking into account the temperature expressed in Eq. (48), as follows

$$L_z E = -T L_z S + \omega L_z V, \tag{49}$$

defining univocally the geometric entropy on the trapping horizon as

$$S = \frac{A|_H}{4}. \tag{50}$$

The negative sign appearing in the first term in the r.h.s. of Eq. (49) would agree with the consideration included in Sec. 3 and according to which the exotic matter supporting this spacetime "removes" energy from the spacetime itself. Following this line of thinking we can then formulate the first law of wormhole thermodynamics as:

First law: The change in the gravitational energy of a wormhole equals the sum of the energy removed from the wormhole plus the work done in the wormhole.

This first law can be interpreted by considering that the exotic matter is responsible for both the energy removal and the work done, keeping the balance always giving rise to a positive variation of the total gravitational energy.

On the other hand, as we have pointed out in Sec. 5, $L_z A \geq 0$ in an exotic environment, implying $L_z S \geq 0$ through Eq. (50), which saturates only at the static case. Thus, considering that a real, cosmological wormhole must be always in an exotic dynamical background, we can formulate the second law for wormhole thermodynamics as follows:

Second law: The entropy of a dynamical wormhole is given by its surface area which always increases, whenever the wormhole accretes exotic material.

Moreover, a wormhole is characterized by an outer trapping horizon (which must be past as has been argued in Sec. 4) which, in terms of the surface gravity, implies $\kappa > 0$. Therefore, we can formulate the third law of thermodynamic as:

Third law (first formulation): It is impossible to reach the absolute zero for surface gravity by any dynamical process.

It is worth noticing that if some dynamical process could change the outer character of a trapping horizon in such a way that it becomes an inner horizon, then the wormhole would converts itself into a different physical object. If this hypothetical process would be possible, then it would make no sense to continue referring to the laws of wormhole thermodynamics, being the thermodynamics of that new object which should instead be considered. Following this line of thinking, it must be pointed out that whenever there is a wormhole, $\kappa > 0$, its trapping horizon is characterized by a negative temperature by virtue of the arguments showed. Thus, we can re-formulate the third law of wormhole thermodynamic as:

Third law (second formulation): In a wormhole it is impossible to reach the absolute zero of temperature by any dynamical process.

It can be argued that if one could change the background energy from being exotic matter to usual one, then the causal nature of the outer trapping horizon would change[14] (Hayward, 1999). Even more, we could consider that as caused by such a process, or by a subsequent one, a past outer trapping horizon (i. e. a dynamical wormhole) should change into a future outer trapping horizon (i.e. a dynamical black hole), and vice versa. If such process would be possible, then it could be expected the temperature to change from negative (wormhole) to positive (black hole) in a way which is necessarily discontinuous due to the holding of the third law, i. e. without passing through the zero temperature, since neither of those objects is characterized by a degenerate trapping horizon.

In the hypothetical process mentioned in the previous paragraph the first law of wormholes thermodynamics would then become the first law of black holes thermodynamics, where the energy is supplied by ordinary matter rather than by the exotic one and the minus sign in Eq. (49) is replaced by a plus sign. The latter implication arises from the feature that a future outer trapping horizon should produce thermal radiation at a positive temperature. The second law would remain then unchanged since it can be noted that the variation of the horizon area, and hence of the entropy, is equivalent for a past outer trapping horizon surrounded by exotic matter and for a future outer trapping horizon surrounded by ordinary matter. And, finally, the two formulations provided for the third law would also be the same,

[14]This fact can be deduced by noticing that both, the material content and the outer property of the horizon, fix the relative sign of z^+ and z^- through Eq. (38).

but in the second formulation one would consider that the temperature takes only on positive values.

6. Conclusions and further comments

In this chapter we have first applied results related to a generalized first law of thermodynamics (Hayward, 1998) and the existence of a generalized surface gravity (Hayward, 1998; Ida & Hayward, 1995) to the case of the Morris-Thorne wormholes (Morris & Thorne, 1988), where the outer trapping horizon is bifurcating. Since these wormholes correspond to static solutions, no dynamical evolution of the throat is of course allowed, with all terms entering the first law vanishing at the throat. However, the comparison of the involved quantities (such as the variation of the gravitational energy and the energy-exchange so as work terms as well) with the case of black holes surrounded by ordinary matter actually provide us with some useful information about the nature of this spacetime (or alternatively about the exotic matter), under the assumption that in the dynamical cases these quantities keep the signs unchanged relative to those appearing outside the throat in the static cases. It follows that the variation of the gravitational energy and the "work term", which could be interpreted as the work carried out by the matter content in order to maintain the spacetime, have the same sign in spherically symmetric spacetimes supported by both ordinary and exotic matter. Notwithstanding, the "energy-exchange term" would be positive in the case of dynamical black holes surrounded by ordinary matter (i. e. it is an energy supply) and negative for dynamical wormholes surrounded by exotic matter (i. e. it corresponds to an energy removal).

That study has allowed us to show that the Kodama vector, which enables us to introduce a generalized surface gravity in dynamic spherically symmetric spacetimes (Hayward, 1998), must be taken into account not only in the case of dynamical solutions, but also in the more general case of non-vacuum solutions. In fact, whereas the Kodama vector reduces to the temporal Killing in the spherically symmetric vacuum solution (Hayward, 1998), that reduction is no longer possible for the static non-vacuum case described by the Morris-Thorne solution. That differentiation is a key ingredient in the mentioned Morris-Thorne case, where there is no Killing horizon in spite of having a temporal Killing vector and possessing a non degenerate trapping horizon. Thus, it is possible to define a generalized surface gravity based on local concepts which have therefore potentially observable consequences. When this consideration is applied to dynamical wormholes, such an identification leads to the characterization of these wormholes in terms of the past outer trapping horizons (Martin-Moruno & Gonzalez-Diaz, 2009a;b).

The univocal characterization of dynamical wormholes implies not only that the area (and hence the entropy) of a dynamical wormhole always increases if there are no changes in the exoticity of the background (second law of wormhole thermodynamics), but also that the hole appears to thermally radiate. The results of the studies about phantom thermodynamics (Gonzalez-Diaz & Siguenza, 2004; Saridakis et al., 2009) allow us to provide this possible radiation with negative temperature with a well-defined physical meaning. Therefore, wormholes would emit radiation of the same kind as the matter which supports them (Martin-Moruno & Gonzalez-Diaz, 2009a;b), such as it occurs in the case of dynamical black hole evaporation with respect to ordinary matter.

These considerations allow us to consistently re-interpret the generalized first law of thermodynamics as formulated by Hayward (Hayward, 1998) in the case of wormholes, noting that in this case the change in the gravitational energy of the wormhole throat is equal to the sum of the energy removed from the wormhole and the work done on the wormhole (first law of wormholes thermodynamics), a result which is consistent with the above mentioned results obtained by analyzing of the Morris-Thorne spacetime in the throat exterior.

At first sight, the above results might perhaps be pointing out to a way through which wormholes might be localized in our environment by simply measuring the inhomogeneities implied by phantom radiation, similarly to as initially thought for black hole Hawking radiation (Gibbons & Hawking, 1977). However, we expect that in this case the radiation would be of a so tiny intensity as the originated from black holes, being far from having hypothetical instruments sensitive and precise enough to detect any of the inhomogeneities and anisotropies which could be expected from the thermal emission from black holes and wormholes of moderate sizes.

It must be pointed out that, like in the black hole case, the radiation process would produce a decrease of the wormhole throat size, so decreasing the wormhole entropy, too. This violation of the second law is only apparent, because it is the total entropy of the universe what should be meant to increase.

It should be worth noticing that there is an ambiguity when performing the action integral in the radiation study, which depends on the r semi-plane chosen to deform the integration path. This ambiguity could be associated to the choice of the boundary conditions. Thus, had we chosen the other semi-plane, then we had obtained a positive temperature for the wormhole trapping horizon. The supposition of this second solution as physically consistent implies that the thermal radiation would be always thermodynamically forbidden in front of the accretion entropicaly favored process, since the energy filling the space has negative temperature (Gonzalez-Diaz & Siguenza, 2004; Saridakis et al., 2009) and, therefore, "hotter" than any positive temperature. Although this possibility should be mentioned, in our case we consider that the boundary conditions, in which it is natural to take into account the sign of the temperature of the surrounding material, imply that the horizon is characterized by a temperature with the same sign. However, it would be of a great interest the confirmation of this result by using an alternative method where the mentioned ambiguity would not be present.

On the other hand, we find of special interest to briefly comment some results presented during/after the publication of the works in which are based this chapter (Martin-Moruno & Gonzalez-Diaz, 2009a;b), since it could clarify some considerations adopted in our development. First of all, in a recent work by Hayward (Hayward, 2009), in which some part of the present work was also discussed following partly similar though somewhat divergent arguments, the thermodynamics of two-types of dynamic wormholes characterized by past or future outer trapping horizon was studied. Although these two types are completely consistent mathematical solutions, we have concentrated on the present work in the first one, since we consider that they are the only physical consistent wormholes solution. One of the reasons which support the previous claim has already been mentioned in this work and is based on the possible equivalence of the results coming from the 2+2 formalism and the accretion method, at least qualitatively. On the other hand, a traversable wormhole must be supported by exotic matter and it is known that it can collapse by accretion of ordinary matter. That is precisely the problem of how to traverse a traversable wormhole finding the mouth open for the back-travel, or at least avoiding a possible death by a pinched off wormhole throat during the trip. If the physical wormhole could be characterized by a future outer trapping horizon, by Eqs. 37), (38) and (39), then its size would increase (decrease) by accretion of ordinary (exotic) matter and, therefore, it would not be a problem to traverse it; even more, it would increase its size when a traveler would pass through the wormhole, contrary to what it is expected from the bases of the wormhole physics (Morris & Thorne, 1988; Visser, 1995). In the second place, Di Criscienzo, Hayward, Nadalini, Vanzo and Zerbini Ref. (Di Criscienzo et al., 2010) have shown the soundness of the method used in Ref. (Hayward et al., 2009) to study the thermal radiation of dynamical black holes, which we have considered valid, adapting it to the dynamical wormhole case; although, of course, it could be other methods which could also provide a consistent description of the process. Moreover, in this work (Di Criscienzo et al., 2010) Di criscienzo et al. have introduced a possible physical meaning for the energy parameter ω_Φ, noticing that it can be expressed in terms of the Kodama vector, which provides a preferred flow, as $\omega_\Phi = -k^\alpha \partial_\alpha I$; thus, the authors claim that ω_Φ would be the invariant energy associated with a particle. If this could be the case, then the solution presented in this chapter when considering the radiation process, $k_\Phi = -2\omega_\Phi/C$, could imply a negative invariant energy for the radiated "particles", since it seems possible to identify k_Φ with any quantity similar to the wave number, or even itself, being, therefore, a positive quantity. This fact can be understood thinking that the invariant quantity characterizing the energy of "the phantom particles" should reflect the violation of the null energy condition. Finally, we want to emphasize that the study of wormholes thermodynamics introduced in this chapter not only have the intrinsic interest of providing a better understanding of the relation between the gravitational and thermodynamic phenomena, but also it would allow us to understand in depth the evolution of spacetime structures that could be present in our Universe. We would like to once again remark that it is quite plausible that the existence of wormholes be partly based on the possible presence of phantom energy in our Universe. Of course, even though in that case the main part of the energy density of the universe would be contributed by phantom energy, a remaining 25% would still be made up of ordinary matter (dark or not). At least in principle, existing wormhole structures would be compatible with the configuration of such a universe, even though a necessarily sub-dominant proportion of ordinary matter be present, provided that the effective equation of state parameter of the universe be less than minus one.

7. References

Babichev, E., Dokuchaev, V. & Eroshenko, Y. (2004). Black hole mass decreasing due to phantom energy accretion, *Phys. Rev. Lett.* 93: 021102.

Bardeen, J. M., Carter, B. & Hawking, S. W. (1973). The four laws of black hole mechanics, *Commun. Math. Phys.* 31: 161–170.

Bronnikov, K. A. (1973). Scalar-tensor theory and scalar charge, *Acta Phys. Polon.* B4: 251–266.

Caldwell, R. R. (2002). A phantom menace?, *Phys. Lett.* B545: 23–29.

Cramer, J. G., Forward, R. L., Morris, M. S., Visser, M., Benford, G. & Landis, G. A. (1995). Natural wormholes as gravitational lenses, *Phys. Rev.* D51: 3117–3120.

Di Criscienzo, R., Hayward, S. A., Nadalini, M., Vanzo, L. & Zerbini, S. (2010). Hamilton-Jacobi tunneling method for dynamical horizons in different coordinate gauges, *Class. Quant. Grav.* 27: 015006.

Einstein, A. & Rosen, N. (1935). The particle problem in the general theory of relativity, *Phys. Rev.* 48: 73–77.

Ellis, E. G. (1973). Ether flow through a drainhole: A particle model in general relativity, *J. Math. Phys.* 14: 104–118.

Fuller, R. W. & Wheeler, J. A. (1962). Causality and multiply connected space-time, *Phys. Rev.*

128: 919–929.
Gibbons, G. W. & Hawking, S. W. (1977). Cosmological Event Horizons, Thermodynamics, And Particle Creation, *Phys. Rev.* D15: 2738–2751.
Gonzalez-Diaz, P. F. (2004). Achronal cosmic future, *Phys. Rev. Lett.* 93: 071301.
Gonzalez-Diaz, P. F. (n.d.). Is the 2008 NASA/ESA double Einstein ring actually a ringhole signature?
Gonzalez-Diaz, P. F. & Martin-Moruno, P. (2008). Wormholes in the accelerating universe, *in* H. Kleinert, R. T. Jantzen & R. Ruffini (eds), *Proceedings of the eleventh Marcel Grossmann Meeting on General Relativity*, World Scientific, pp. 2190–2192.
Gonzalez-Diaz, P. F. & Siguenza, C. L. (2004). Phantom thermodynamics, *Nucl. Phys.* B697: 363–386.
Hawking, S. W. & Ellis, G. F. R. (1973). *The large scale structure of space-time*, Cambridge University Press.
Hayward, S. A. (1994a). General laws of black hole dynamics, *Phys. Rev.* D49: 6467–6474.
Hayward, S. A. (1994b). Quasilocal Gravitational Energy, *Phys. Rev.* D49: 831–839.
Hayward, S. A. (1996). Gravitational energy in spherical symmetry, *Phys. Rev.* D53: 1938–1949.
Hayward, S. A. (1998). Unified first law of black-hole dynamics and relativistic thermodynamics, *Class. Quant. Grav.* 15: 3147–3162.
Hayward, S. A. (1999). Dynamic wormholes, *Int. J. Mod. Phys* D8: 373–382.
Hayward, S. A. (2004). Energy and entropy conservation for dynamical black holes, *Phys. Rev* D70: 104027.
Hayward, S. A. (2009). Wormhole dynamics in spherical symmetry, *Phys. Rev.* D79: 124001.
Hayward, S. A., Di Criscienzo, R., Vanzo, L., Nadalini, M. & Zerbini, S. (2009). Local Hawking temperature for dynamical black holes, *Class. Quant. Grav.* 26: 062001.
Ida, D. & Hayward, S. A. (1995). How much negative energy does a wormhole need?, *Phys. Lett.* A260: 175–181.
Kodama, H. (1980). Conserved Energy Flux For The Spherically Symmetric System And The Back Reaction Problem In The Black Hole Evaporation, *Prog. Theor. Phys.* 63: 1217.
Kodama, T. (1978). General relativistic nonlinear field: A kink solution in a generalized geometry, *Phys. Rev.* D18: 3529–3534.
Kruskal, M. D. (1960). Maximal extension of schwarzschild metric, *Phys. Rev.* 119: 1743–1745.
Lobo, F. S. N. (2005). Phantom energy traversable wormholes, *Phys. Rev.* D71: 084011.
Lobo, F. S. N. (2006). Chaplygin traversable wormholes, *Phys. Rev.* D73: 064028.
Lobo, F. S. N. (n.d.). Exotic solutions in General Relativity: Traversable wormholes and 'warp drive' spacetimes.
Martin-Moruno, P. & Gonzalez-Diaz, P. F. (2009a). Lorentzian wormholes generalizes thermodynamics still further, *Class. Quant. Grav.* 26: 215010.
Martin-Moruno, P. & Gonzalez-Diaz, P. F. (2009b). Thermal radiation from lorentzian traversable wormholes, *Phys. Rev.* D80: 024007.
Martin-Moruno, P., Jimenez Madrid, J. A. & Gonzalez-Diaz, P. F. (2006). Will black holes eventually engulf the universe?, *Phys. Lett.* B640: 117–120.
Misner, C. W. & Sharp, D. H. (1964). Relativistic equations for adiabatic, spherically symmetric gravitational collapse, *Phys. Rev.* B136: 571–576.
Misner, C. W. & Wheeler, J. A. (1957). Classical physics as geometry: Gravitation, electromagnetism, unquantized charge, and mass as properties of curved empty space, *Annals Phys.* 2: 525–603.
Morris, M. S. & Thorne, K. S. (1988). Wormholes in space-time and their use for interstellar

travel: A tool for teaching general relativity, *Am. J. Phys.* 56: 395–412.

Morris, M. S., Thorne, K. S. & Yurtsever, U. (1988). Wormholes, time machines, and the weak energy condition, *Phys. Rev. Lett.* 61: 1446–1449.

Mortlock, D. J. & Webster, R. L. (2000). The statistics of wide-separation lensed quasars, *Mon. Not. Roy. Astron. Soc.* 319: 872.

Safonova, M., Torres, D. F. & Romero, G. E. (2001). Macrolensing signatures of large-scale violations of the weak energy condition, *Mod. Phys. Lett.* A16: 153–162.

Safonova, M., Torres, D. F. & Romero, G. E. (2002). Microlensing by natural wormholes: Theory and simulations, *Phys. Rev.* D65: 023001.

Saridakis, E. N., Gonzalez-Diaz, P. F. & Siguenza, C. L. (2009). Unified dark energy thermodynamics: varying w and the -1-crossing, *Class. Quant. Grav.* 26: 165003.

Shatskiy, A. (2007). Passage of Photons Through Wormholes and the Influence of Rotation on the Amount of Phantom Matter around Them, *Astron. Rep.* 51: 81.

Shatskiy, A. (n.d.). Image of another universe being observed through a wormhole throat.

Sushkov, S. V. (2005). Wormholes supported by a phantom energy, *Phys. Rev.* D71: 043520.

Torres, D. F., Romero, G. E. & Anchordoqui, L. A. (1998a). Might some gamma ray bursts be an observable signature of natural wormholes?, *Phys. Rev.* D58: 123001.

Torres, D. F., Romero, G. E. & Anchordoqui, L. A. (1998b). Wormholes, gamma ray bursts and the amount of negative mass in the universe, *Mod. Phys. Lett.* A13: 1575–1582.

Visser, M. (1995). *Lorentzian Wormholes: From Einstein to Hawking*, American Institute of Physics Press (Woodbury, New York).

Visser, M. (2003). Essential and inessential features of hawking radiation, *Int. J. Mod. Phys.* D12: 649–661.

Wheeler, J. A. (1955). Geons, *Phys. Rev.* 97: 511–536.

8

Four Exactly Solvable Examples in Non-Equilibrium Thermodynamics of Small Systems

Viktor Holubec, Artem Ryabov, Petr Chvosta
Faculty of Mathematics and Physics, Charles University
V Holešovičkách 2, CZ-180 00 Praha
Czech Republic

1. Introduction

The diffusion dynamics in time-dependent potentials plays a central role in the phenomenon of stochastic resonance (Gammaitoni et al., 1998; Chvosta & Reineker, 2003a; Jung & Hänggi, 1990; 1991), in physics of Brownian motors (Reimann, 2002; Astumian & Hänggi, 2002; Hänggi et al., 2005; Allahverdyan et al., 2008; den Broeck et al., 2004; Sekimoto et al., 2000) and in the discussion concerning the energetics of the diffusion process (Parrondo & de Cisneros, 2002) – these papers discuss history, applications and existing literature in the domain.

Diffusion in a time-dependent potential where the dynamical system communicates with a single thermal bath can be regarded as an example of an isothermal irreversible process. Investigating the work done on the system by the external agent and the heat exchange with the heat bath (Sekimoto, 1999; Takagi & Hondou, 1999) one immediately enters the discussion of the famous Clausius inequality between the irreversible work and the free energy. If the energy considerations concern a small system, the work done on the system has been associated with individual realizations (trajectories) of the diffusive motion, i.e. the work itself is treated as a random variable whose mean value enters the thermodynamic considerations. An important achievement in the field is the discovery of new fluctuation theorems, which generalize the Clausius identity in giving the exact mean value of the exponential of the work. This Jarzynski identity (Bochkov & Kuzovlev, 1981a;b; Evans et al., 1993; Gallavotti & Cohen, 1995; Jarzynski, 1997b;a; Crooks, 1998; 1999; 2000; Maes, 2004; Hatano & Sasa, 2001; Speck & Seifert, 2004; Seifert, 2005; Schuler et al., 2005; Esposito & Mukamel, 2006; Hänggi & Thomas, 1975) enables one to specify the free energy difference between two equilibrium states. This is done by repeating real time (i.e. non-equilibrium) experiment and measuring the work done during the process. The identity has been recently experimentally tested (Mossa et al., 2009; Ritort, 2003).

In the present Chapter we discuss four illustrative, exactly solvable models in non-equilibrium thermodynamics of small systems. The examples concern: i) the *unrestricted* diffusion in the presence of the time-dependent potential (SEC. 2) (Wolf, 1988; Chvosta & Reineker, 2003b; Mazonka & Jarzynski, 1999; Baule & Cohen, 2009; Hänggi & Thomas, 1977), ii) the *restricted* diffusion of non-interacting particles in the presence of the time-dependent potential (SEC. 3) (Chvosta et al., 2005; 2007; Mayr et al., 2007), iii) the restricted diffusion of two *interacting* particles in the presence of the time-dependent potential (SEC. 4) (Rödenbeck et al., 1998; Lizana & Ambjörnsson, 2009; Kumar, 2008; Ambjörnsson et al., 2008; Ambjörnsson & Silbey, 2008; Barkai & Silbey, 2009), and iv) the two-level system with externally driven energy levels (SEC. 5) (Chvosta et al., 2010; Šubrt & Chvosta, 2007; Henrich et al., 2007; Hänggi & Thomas, 1977).

A common feature of all these examples is the following. Due to the periodic driving, the system approaches a definite steady state exhibiting cyclic energy transformations. The exact solution of underlying dynamical equations allows for the detailed discussion of the limit cycle. Specifically, in the setting i), we present the simultaneous probability density for the particle position and for the work done on the particle. In the model ii), we shall demonstrate that the cycle-averaged spatial distribution of the internal energy differs significantly from the corresponding equilibrium one. In the scenario iii), the particle interaction induces additional entropic repulsive forces and thereby influences the cycle energetics. In the two-level model iv), the system communicates with two heat baths at different temperatures. Hence it can perform a positive mean work per cycle and therefore it can be conceived as a simple microscopic motor. Having calculated the full probability density for the work, we can discuss also fluctuational properties of the motor performance.

2. Diffusion of a particle in a time-dependent parabolic potential

Consider a particle, in contact with a thermal bath at the temperature T which is dragged through the environment by a time-dependent external force. Assuming a single degree of freedom, the location of the particle at a time t is described by the time-inhomogeneous Markov process $\mathsf{X}(t)$. Let the particle moves in the time-dependent potential

$$V(x,t) = \frac{k}{2}[x - u(t)]^2 . \tag{1}$$

We can regard the particle as being attached to a spring, the other end of which moves with an instantaneous velocity $\dot{u}(t) \equiv du(t)/dt$. Furthermore, assume that the thermal forces can be modeled as the sum of the linear friction and the Langevin white-noise force. We neglect the inertial forces. Then the equation of motion for the particle position is (van Kampen, 2007):

$$\Gamma \frac{d}{dt}\mathsf{X}(t) = -\frac{\partial}{\partial x}V(x,t)\big|_{x=\mathsf{X}(t)} + \mathsf{N}(t) = -k\left[\mathsf{X}(t) - u(t)\right] + \mathsf{N}(t) , \tag{2}$$

where Γ is the particle mass times the viscous friction coefficient, and $\mathsf{N}(t)$ represents the delta-correlated white noise $\langle \mathsf{N}(t)\mathsf{N}(t')\rangle = 2D\Gamma^2\,\delta(t-t')$. Here $D = k_{\mathrm{B}}T/\Gamma$ is the diffusion constant and k_{B} is the Boltzmann constant.

We observe the motion of the particle. Assuming a specific trajectory of the particle we are interested in the total work done on the particle if it moves along the trajectory. Taking into account the whole set of all possible trajectories, the work becomes a stochastic process. We denote it as $\mathsf{W}(t)$ and it satisfies the stochastic equation (Sekimoto, 1999)

$$\frac{d}{dt}\mathsf{W}(t) = \frac{\partial}{\partial t}V(\mathsf{X}(t),t) = -k\dot{u}(t)[\mathsf{X}(t) - u(t)] \tag{3}$$

with the initial condition $\mathsf{W}(0) = 0$. Differently speaking, if the particle dwells at the position x during the time interval $[t, t+dt]$ then the work done on the particle during this time interval equals $V(x, t+dt) - V(x,t)$ (for the detailed discussion cf. also SEC. 5).

The above system of stochastic differential equations for the processes $\mathsf{X}(t)$ and $\mathsf{W}(t)$ can be translated into a single partial differential equation for the joint probability density $G(x,w,t\,|\,x_0)$. The function $G(x,w,t\,|\,x_0)$ describes the probability of achieving the position x at the time t *and* performing the work w during the time interval $[0,t]$. The partial differential equation reads (Risken, 1984; van Kampen, 2007)

$$\begin{aligned}\frac{\partial}{\partial t}G(x,w,t\,|\,x_0) = \left\{D\frac{\partial^2}{\partial x^2}+\frac{k}{\Gamma}\frac{\partial}{\partial x}[x-u(t)] \;+\; [x-u(t)]\,\dot{u}(t)\frac{\partial}{\partial w}\right\}G(x,w,t\,|\,x_0)\,,\\ G(x,w,t\,|\,x_0) \;=\; \delta(x-x_0)\delta(w)\,. \qquad (4)\end{aligned}$$

This equation can be solved by several methods. For example, one can use the Lie algebra operator methods (Wilcox, 1967; Wolf, 1988), or one can calculate the joint generating functional for the coupled process in question (Baule & Cohen, 2009). Our approach will be based on the following property of EQ. 4: if at an arbitrary fixed instant the probability density $G(x,w,t\,|\,x_0)$ is of the Gaussian form, then it will preserve this form for all subsequent times. This follows from the fact that all the coefficients on the right hand side of EQ. 4 are polynomials of the degree at most one in the independent variables x and w (van Kampen, 2007). Accordingly, the function $G(x,w,t\,|\,x_0)$ corresponds to a bivariate Gaussian distribution and it is uniquely defined by the central moments (Mazonka & Jarzynski, 1999):

$$\begin{aligned}\overline{x}(t) &= \langle \mathsf{X}(t)\rangle\,, & \overline{w}(t) &= \langle \mathsf{W}(t)\rangle\,,\\ \sigma_x^2(t) &= \left\langle[\mathsf{X}(t)]^2\right\rangle-[\overline{x}(t)]^2\,, & \sigma_w^2(t) &= \left\langle[\mathsf{W}(t)]^2\right\rangle-[\overline{w}(t)]^2\,, \qquad (5)\\ c_{xw}(t) &= \langle \mathsf{X}(t)\mathsf{W}(t)\rangle-\overline{x}(t)\overline{w}(t)\,.\end{aligned}$$

The simplest way to calculate these moments is to use EQS. (2) and (3) (Gillespie, 1992; van Kampen, 2007). The result is

$$\overline{x}(t) = u(t)-\exp\left(-\frac{k}{\Gamma}t\right)\int_0^t \mathrm{d}t'\,\dot{u}(t')\exp\left(\frac{k}{\Gamma}t'\right)+[x_0-u(0)]\exp\left(-\frac{k}{\Gamma}t\right), \qquad (6)$$

$$\sigma_x^2(t) = \frac{\Gamma D}{k}\left[1-\exp\left(-2\frac{k}{\Gamma}t\right)\right], \qquad (7)$$

$$c_{xw}(t) = -2\Gamma D\exp\left(-\frac{k}{\Gamma}t\right)\int_0^t \mathrm{d}t'\,\dot{u}(t')\sinh\left(\frac{k}{\Gamma}t'\right), \qquad (8)$$

$$\overline{w}(t) = -k\int_0^t \mathrm{d}t'\,\dot{u}(t')[\overline{x}(t')-u(t')]\,, \qquad \sigma_w^2(t) = -2k\int_0^t \mathrm{d}t'\,\dot{u}(t')c_{xw}(t')\,. \qquad (9)$$

Surprisingly, the variance $\sigma_x^2(t)$ does not depend on the function $u(t)$. Moreover, in the asymptotic regime $t \gg \Gamma/k$, the variance $\sigma_x^2(t)$ attains the saturated value $\Gamma D/k$. This means that the marginal probability density for the particle position assumes a time-independent shape.

Up to now our considerations were valid for an arbitrary form of the function $u(t)$. We now focus on the piecewise linear periodic driving. We take $u(t+\lambda)=u(t)$ and

$$u(t) = -2vt \quad \text{for} \quad t\in[0,\tau[\,, \qquad u(t) = -2v\tau+vt \quad \text{for} \quad t\in[\tau,\lambda[\,, \qquad (10)$$

where $v>0$ and $0<\tau<\lambda$. The parabola is first moving to the left with the velocity $2v$ during the time interval $[0,\tau[$. Then, at the time τ it changes abruptly its velocity and moves to the right with the velocity v during the rest of the period λ, cf. FIG. 1 d).

Due to the periodic driving the system's response (6)-(9) approaches the limit cycle. FIG. 1 illustrates the response during two such limit cycles. First, note that the mean position of the particle $\overline{x}(t)$ "lags behind" the minimum of the potential well $u(t)$ (see the panel a)). The magnitude of this phase shift is given by the second term in EQ. (6) and therefore it is proportional to the velocity v. In the adiabatic limit of the infinitely slow velocity $v \to 0$ the probability distribution for the particle position is centred at the instantaneous minimum of the parabola.

Consider now the mean work done on the system by the external agent during the time interval $[0,t[$ (panel b)). $\overline{w}(t)$ increases if either simultaneously $u(t) > \overline{x}(t)$ and $\dot{u}(t) > 0$, or if simultaneously $u(t) < \overline{x}(t)$ and $\dot{u}(t) < 0$. For instance, assume the parabola moves to the right and, at the same time, the probability packet for the particle coordinate is concentrated on the left from the instantaneous position of the parabola minimum $u(t)$. Then the dragging rises the potential energy of the particle, i.e. the work is done on it and the mean input power is positive. Similar reasoning holds if either simultaneously $u(t) > \overline{x}(t)$ and $\dot{u}(t) < 0$, or if simultaneously $u(t) < \overline{x}(t)$ and $\dot{u}(t) > 0$. Then the mean work $\overline{w}(t)$ decreases and hence the mean input power is negative. The magnitude of the instantaneous input power is proportional to the instantaneous velocity $\dot{u}(t)$. Therefore it is bigger during the first part of the period of the limit cycle in comparison with the second part of the period. Finally, let us stress that the mean work per cycle $w_{\mathrm{p}} = \overline{w}(t+\lambda) - \overline{w}(t)$ is always positive, as required by the second law of thermodynamics.

The variance of the work done on the particle by the external agent $\sigma_w^2(t)$ shows qualitatively the same behaviour as $\overline{w}(t)$.

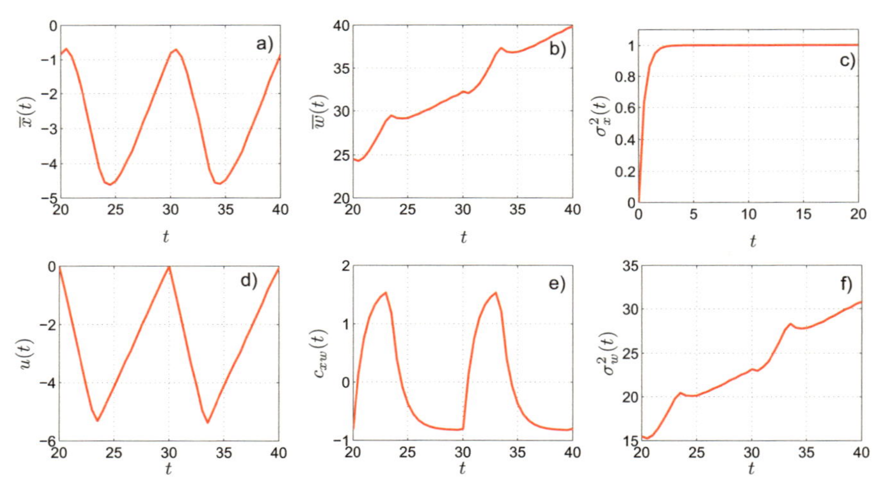

Fig. 1. The central moments (6)-(9) in the time-asymptotic regime. The driving is represented by the position of the potential minimum $u(t)$ and it is depicted in the panel d). In all panels (except of the panel c)) the curves are plotted for two periods λ of the driving. The panel a) shows the mean position of the particle, $\overline{x}(t)$, which lags behind the minimum of the potential well. The panel b) shows the mean work $\overline{w}(t)$ done on the particle by the external agent. In the panel c) we observe the saturation of the variance of the particle's position $\sigma_x^2(t)$. In the panel e) we present the correlation function $c_{xw}(t)$. The panel f) illustrates the variance $\sigma_w^2(t)$ of the work done on the particle by the external agent. The parameters used are: $k = 1\,\mathrm{kg\,s^{-2}}$, $D = 1\,\mathrm{m^2\,s^{-1}}$, $\Gamma = 1\,\mathrm{kg\,s^{-1}}$, $v = 0.825\,\mathrm{m\,s^{-1}}$, $\lambda = 10\,\mathrm{s}$, $\tau = 10/3\,\mathrm{s}$.

3. Barometric process with time-dependent force

3.1 Dynamics

In this Section we discuss a spatially restricted one-dimensional diffusion process occurring in a half-space under the influence of a harmonically oscillating and space-homogeneous driving force. We are interested in the solution of the Langevin equation

$$\Gamma \frac{d}{dt} \mathsf{X}(t) = -\frac{\partial}{\partial x} V(x,t)\big|_{x=\mathsf{X}(t)} + \mathsf{N}(t)\,, \tag{11}$$

for an overdamped Brownian particle moving in the time-dependent potential $V(x,t)$, where $V(x,t) = -xF(t)$, if $x \geq 0$, and $V(x,t) = \infty$, for $x < 0$. Here $\mathsf{N}(t)$ is the δ-correlated Langevin force, and Γ equals the particle mass times the viscous friction coefficient. Differently speaking, while localised on the positive half-line, the particle is acted upon by the Langevin force $\mathsf{N}(t)$ and by the spatially-homogeneous, time-dependent force $F(t)$. Additionally, we assume a reflecting barrier at the origin, i.e. the diffusion is restricted on the positive half-line. As an auxiliary problem, consider first the *spatially unrestricted* one-dimensional diffusion in the field of a spatially-homogeneous and time-dependent force $F(t)$. The probability density for the position of the diffusing particle reads (Hänggi & Thomas, 1975; 1977; Wolf, 1988)

$$G(x,t|x',t') = \frac{1}{\sqrt{\pi}} \frac{1}{\sqrt{4D(t-t')}} \exp\left\{ -\frac{1}{4D(t-t')} \left[x - x' - \int_{t'}^{t} v(t'')\, dt'' \right]^2 \right\}. \tag{12}$$

The Green function yields the solution of the Smoluchowski diffusion equation for the initial condition $\pi(x) = \delta(x - x')$ imposed at time t'. Qualitatively, it represents the gradually spreading Gaussian curve whose centre moves in time, the drift being controlled by the protocol (time-dependent scenario) of the external force. The momentary value of the mean particle position is given by the expression $x' + \int_{t'}^{t} ds\, v(s)$, where $v(t) = F(t)/\Gamma$ is the time-dependent drift velocity. The spreading of the Gaussian curve is controlled by the thermal-noise strength parameter $D = k_B T/\Gamma$.

We now assume that the particle is initially (i.e. at the time zero) fully localised at a fixed point $x' > 0$, and we place at the origin of the coordinate system the reflecting boundary. The Green function $U(x,t|x',0)$ which solves the problem with the reflecting boundary can be constructed in two steps (cf. the detailed derivation in REF. (Chvosta et al., 2005)). First, one has to solve the Volterra integral equation of the first kind

$$D \int_0^t G(0,t|0,t')\, U(0,t'|x',0)\, dt' = \int_{-\infty}^{0} G(x,t|x',0)\, dx\,. \tag{13}$$

Here both the kernel and the right hand side follow directly from EQ. (12). The unknown function $U(0,t|x',0)$ represents, as the designation suggests, the time evolution of the probability density for the restricted diffusion at the boundary. Secondly, the final space-resolved solution emerges after performing just one additional quadrature:

$$U(x,t|x',0) = G(x,t|x',0) - D \int_0^t \frac{\partial}{\partial x} G(x,t|0,t')\, U(0,t'|x',0)\, dt'\,. \tag{14}$$

The resulting function is properly normalized, i.e. we have $\int_0^\infty U(x,t|x',0)\, dx = 1$ for any $t \geq 0$ and for any fixed initial position $x' > 0$.

Up to now, our reasoning is valid for any form of the external driving force. A negative instantaneous force pushes the particle to the left, i.e. against the reflecting boundary at the origin. In this case, the force acts against the general spreading tendency stemming from the thermal Langevin force. A positive instantaneous force amplifies the diffusion in driving the particle to the right. We now restrict our attention to the case of a harmonically oscillating driving force $F(t) = \Gamma v(t)$ with the drift velocity $v(t) = v_0 + v_1 \sin(\omega t)$. The three parameters, v_0, v_1, and ω occurring in this formula together with the diffusion constant D yield the full description of our setting. Specifically, if $v_1 = 0$, the external force has only the static component and the explicit solution of the integral equation (13) is well known, cf. the formula (29) in (Chvosta et al., 2005). $U(0,t|x',0)$ approaches in this case either zero, if $v_0 \geq 0$, or the value $|v_0|/D$, if $v_0 < 0$. Having the oscillating force, the most interesting physics emerges if the symmetrically oscillating component superposes with a *negative* static force, i.e. if $v_1 > 0$, and $v_0 < 0$. This case is treated in the rest of the Section.

Considering the integral equation (13), the basic difficulty is related with the non-convolution structure of the integral on the left-hand side. It may appear that any attempt to perform the Laplace transformation must fail. But it has been demonstrated in REF. (Chvosta et al., 2007) that this need not be the case. The paper introduces, in full details, a special procedure which yields the exact time-asymptotic solution of EQ. (13). Here we confine ourselves to the statement of the final result and to its physical consequences.

First of all, we introduce an appropriate scaling of the time variable. Adopting any such scaling, the four model parameters will form certain dimensionless groups. However, there are just two "master" combinations of the parameters which control the substantial features of the long-time asymptotic solution. These combinations emerge after we introduce the dimensionless time $\tau = [v_0^2/(4D)]\,t$ (we assume $D > 0$). If we insert the scaled time into EQ. (12), the exponent will include solely the combinations $\kappa = |v_0|v_1/(2\omega D)$ and $\theta = 4\omega D/v_0^2$. The first of them measures the scaled amplitude of the oscillating force, the second one its scaled frequency. We now define an infinite matrix $\mathbb{R}_{-+}$ with the matrix elements

$$\langle m |\, \mathbb{R}_{-+} \,| n \rangle = \mathrm{I}_{|m-n|}(-\kappa\sqrt{1 - im\theta} + \kappa)\,. \tag{15}$$

Here m,n are integers and $\mathrm{I}_k(z)$ is the modified Bessel function of order k with argument z. We use the standard bra-ket notation. Notice that the matrix elements depend solely on the above dimensionless combinations κ and θ. As shown in REF. (Chvosta et al., 2005), the time-asymptotic dynamics can be constructed from the matrix elements of the inverse matrix $\mathbb{R}_{-+}^{-1}$. In fact, the so called *complex amplitudes*

$$f_k = \langle k |\, \mathbb{R}_{-+}^{-1} \,| 0 \rangle\,,\; k = 0, \pm 1, \pm 2, \ldots\,, \tag{16}$$

define through EQ. (17) below the full solution. The zeroth complex amplitude f_0 equals one. The amplitudes f_k and f_{-k} are complex conjugated numbers. Generally, their absolute value $|f_k|$ decreases with increasing the index k. The even (odd) amplitudes are even (odd) in the parameter κ. Summing up the whole procedure, the probability density at the origin $U(0,t|x',0)$ asymptotically approaches the function

$$U_{\mathrm{a}}(0,t) = \frac{|v_0|}{D} \sum_{k=-\infty}^{+\infty} f_k \exp(-\mathrm{i}k\theta\tau) = \frac{|v_0|}{D} \left\{ 1 + 2\sum_{k=1}^{\infty} a_k(\kappa,\theta) \cos\left[k\omega t + \phi_k(\kappa,\theta)\right] \right\}. \tag{17}$$

In the last expression, we have introduced the real amplitudes of the higher harmonics $a_k(\kappa,\theta) = |f_k|$ and the phase shifts $\phi_k(\kappa,\theta) = -\arctan(\mathrm{Im}\, f_k/\mathrm{Re}\, f_k)$. Except for the multiplicative factor $|v_0|/D$, the asymptotic form of the probability density at the boundary is

controlled solely by the parameters κ and θ. For example, changing the diffusion constant D and, at the same time, keeping a constant value of the product $D\omega$, the time-asymptotic form of the reduced function $f_a(t) = (D/|v_0|)U_a(0,t)$ will not change. Notice that, for any value of the parameters κ and θ, the time average of the probability density at the boundary equals the equilibrium value of this quantity in the problem without driving force. We have calculated the complex amplitudes (16) via a direct numerical inversion of the matrix $\mathbb{R}_{-+}$ defined in (15). Of course, the infinite-order matrix $\mathbb{R}_{-+}$ must be first reduced onto its finite-order central block. The matrix elements of the reduced matrix are again given by EQ. (15), presently, however, $m,n = 0,\pm 1,\pm 2,\ldots,\pm N$. The integer N has been taken large enough such that its further increase doesn't change the results, within a predefined precision. In this sense, the numerical results below represent the exact long-time solution of the problem in question.
Up to now, we have only discussed the time-dependence of the probability density at the boundary. As a matter of fact, the knowledge of the complex amplitudes f_k allows for a rather detailed discussion of many other features of the emerging diffusion process. First of all, we focus on the time- and space-resolved probability density for the particle coordinate. We remind that, regardless of the initial condition, the static drift towards the origin ($v_0 < 0$, $v_1 = 0$) induces the unique equilibrium density $\pi_{eq}(x) = (|v_0|/D)\exp[-x|v_0|/D]$, $x \geq 0$. Assuming the oscillating drift, we are again primarily interested in the time-asymptotic dynamics. In this regime, the probability density $U(x,t|x',0)$ does not depend on the initial condition (as represented by the variable x'), and it exhibits at any fixed point $x \geq 0$ oscillations with the fundamental frequency ω. We can write

$$U(x,t|x',0) \sim U_a(x,t) = \sum_{k=-\infty}^{+\infty} u_k(x)\exp(-ik\omega t)\,. \tag{18}$$

Presently, however, the Fourier coefficients $u_k(x)$ depend on the coordinate x. An interesting quantity will be the time-averaged value of the density in the asymptotic regime. This is simply the dc component $u_0(x)$ of the above series. We already know that the value of this function at the origin is $u_0(0) = |v_0|/D$, i.e. it equals the value of the equilibrium density in the static case at the origin, $u_0(0) = \pi_{eq}(0)$. Generically, we call the difference between the time-averaged value of a quantity in the oscillating-drift problem and the corresponding equilibrium value of this quantity in the static case as "dynamical shift". Hence we conclude that there is no dynamical shift of the density profile at the origin. But what happens for $x > 0$? Assume that the complex amplitudes f_k are known. Then we know also the time-asymptotic solution of the integral equation (13) and the subsequent asymptotic analysis can be based on the expression (14). Leaving out the details (cf. again REF. (Chvosta et al., 2007)), the x-dependent Fourier coefficients in EQ. (18) are given by the expression

$$u_k(x) = \frac{|v_0|}{D}\,\langle k\,|\,\mathbb{L}_{--}\,\mathbb{E}(x)\,\mathbb{R}_{++}\,|\,\mathbf{f}\rangle\,,\; k = 0,\pm 1,\pm 2,\ldots\,. \tag{19}$$

Here $|\,\mathbf{f}\rangle$ is the column vector of the complex amplitudes, i.e. $f_k = \langle k\,|\,\mathbf{f}\rangle$. Moreover, we have introduced the diagonal matrix $\mathbb{E}(x)$, and the two matrixes $\mathbb{L}_{--}$, $\mathbb{R}_{++}$ with the matrix elements

$$\langle m \,|\, \mathbb{E}(x) \,|\, n \rangle = \frac{\delta_{mn}}{2} \left[1 + \frac{1}{\sqrt{1 - im\theta}} \right] \exp \left[-x \frac{|v_0|}{2D} \left(\sqrt{1 - im\theta} + 1 \right) \right] , \tag{20}$$

$$\langle m \,|\, \mathbb{L}_{--} \,|\, n \rangle = \mathrm{I}_{|m-n|}(-\kappa\sqrt{1 - in\theta} - \kappa) , \tag{21}$$

$$\langle m \,|\, \mathbb{R}_{++} \,|\, n \rangle = \mathrm{I}_{|m-n|}(+\kappa\sqrt{1 - im\theta} + \kappa) , \tag{22}$$

where m and n are integers. FIG. 2 illustrates the time-asymptotic density within two periods of the external driving. Surprising features emerge provided both $\kappa \ll 1$, and $\theta \gg 1$. Under these conditions, the time-averaged probability density $u_0(x)$ exhibits in the vicinity of the boundary a strong dependence on the x-coordinate. It can even develop a well pronounced minimum close to the boundary and, simultaneously, a well pronounced maximum localized farther from the boundary. In between the two extreme values, there exists a spatial region where the time-averaged gradient of the concentration points *against* the time-averaged force. The situation is depicted in FIG. 3 where we have used the same parameters as in FIG. 2. Notice the *positive* dynamical shift $\sigma = \mu_0 - \mu_{\rm eq}$ of the mean coordinate. Here μ_0 is the

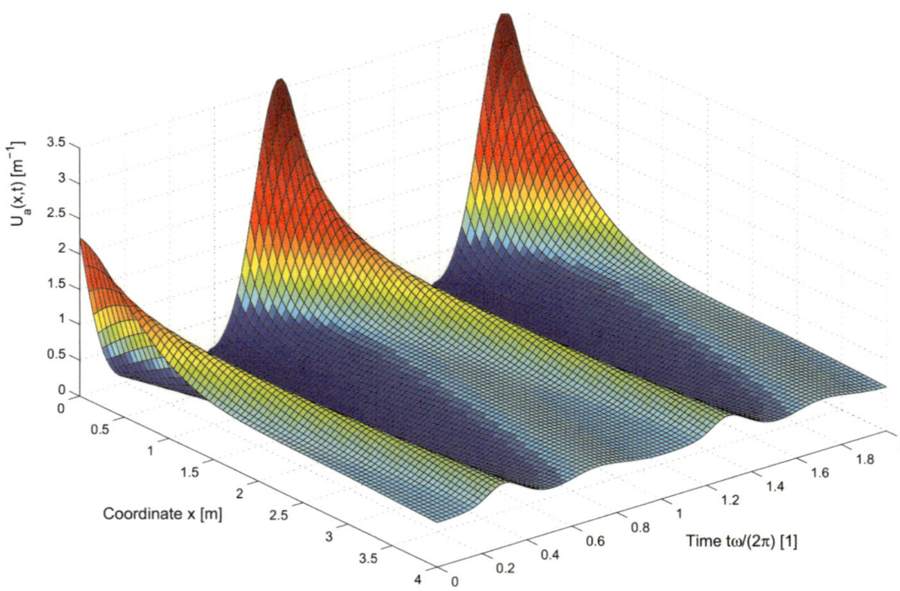

Fig. 2. Time- and space-resolved probability density in the time-asymptotic regime. For any fixed x, the function $U_{\rm a}(x,t)$ is a periodic function, the period being $2\pi/\omega$. We have plotted it for two periods. The parameters used are: $v_0 = -0.1\,{\rm m\,s^{-1}}$, $v_1 = 4.0\,{\rm m\,s^{-1}}$, $\omega = 2.0\,{\rm rad\,s^{-1}}$, and $D = 1.0\,{\rm m^2\,s^{-1}}$. These parameters yield the values $\kappa = 0.1$ and $\theta = 800$.

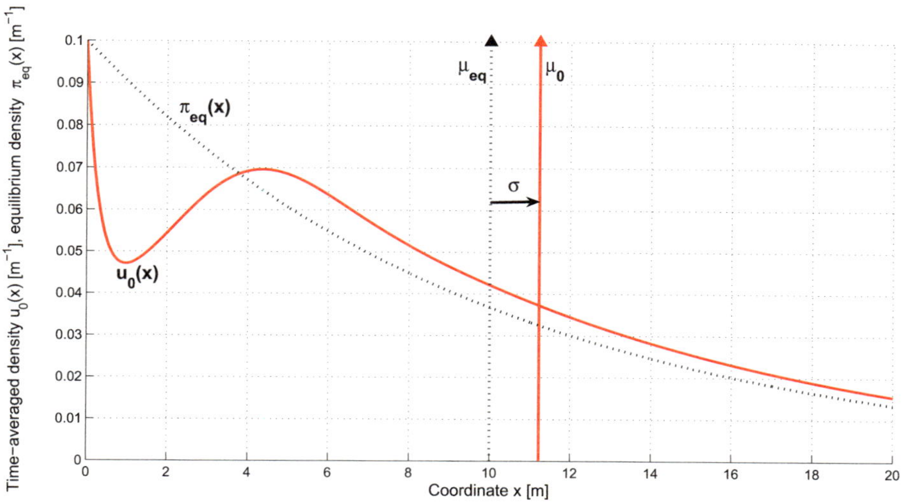

Fig. 3. Time-averaged value $u_0(x)$ of the probability density as the function of the coordinate x. We have used the same set of parameters as in FIG. 2. For comparison, we give also the equilibrium probability density $\pi_{\text{eq}}(x) = (|v_0|/D)\exp[-x|v_0|/D]$, $x \geq 0$, in the corresponding static problem. The arrows mark the time-averaged mean position μ_0 in the oscillating-force problem, and the equilibrium mean position in the static problem $\mu_{\text{eq}} = D/|v_0|$. Their difference $\sigma = \mu_0 - \mu_{\text{eq}}$ represents the dynamical shift of the mean position.

time-averaged mean coordinate

$$\mu_0 = \lim_{t\to\infty} \frac{\omega}{2\pi} \int_t^{t+2\pi/\omega} dt'\, \mu(t',x') \,, \qquad \mu(t,x') = \int_0^\infty dx\, x\, U(x,t|x',0) \,. \tag{23}$$

The exact value of the shift is determined by the complex amplitude f_1 (Chvosta et al., 2007) through $\sigma = (v_1/\omega)\text{Re}f_1$. The small-$v_1$ expansion of the dynamical shift starts with the term v_1^2, i.e. it cannot be described by a linear response theory. If we plot σ as the function of the temperature $T = D\Gamma/k_B$ (Chvosta et al., 2007), it exhibits a resonance-like maximum.

Summarizing, the approach elaborated above yields a rather complete picture of the time-asymptotic motion of the diffusing particle. Depending on its distance from the impenetrable boundary, it exhibits non-harmonic oscillations which can be represented as a linear combination of several higher harmonics. The amplitudes and the phases of the harmonics are strongly sensitive to the distance from the boundary. The calculation does not include any small-parameter expansion.

3.2 Energetics

Assuming again the time-dependent potential $V(x,t) = -x[F_0 + F_1 \sin(\omega t)]$, where $F_0 = \Gamma v_0$, $F_1 = \Gamma v_1$, the internal energy

$$E(t,x') = \int_0^\infty dx\, V(x,t)\, U(x,t|x',0) = -\Gamma\,[v_0 + v_1 \sin(\omega t)]\, \mu(t,x') \,, \tag{24}$$

asymptotically approaches a x'-independent periodic function, say $E_a(t)$. In this time-asymptotic regime, the system exhibits periodic changes of its state. The work done on the system during one such cycle equals to the heat dissipated during the period. An interesting quantity is the time-averaged internal energy

$$E_0 = \lim_{t\to\infty} \frac{\omega}{2\pi} \int_t^{t+2\pi/\omega} dt'\, E(t',x') \,. \tag{25}$$

We can show that E_0 is *always* bigger than the equilibrium internal energy $E_{\mathrm{eq}} = D\Gamma = k_B T$ in the static problem. Differently speaking, in the time-averaged sense, the external driving enforces a permanent increase of the internal energy, as compared to its equilibrium value.
Having periodic changes of the internal energy, the work done on the system during one period must equal to the heat dissipated during the period. However, their behavior during an infinitesimal time interval within the period is quite different. Generally speaking, the heat (≡ the dissipated energy) can be identified as the "work" done by the particle on the heat bath (Takagi & Hondou, 1999; Sekimoto, 1999). It arises if and only if the particle moves, i.e. it is inevitable connected with the probability density *current*. More precisely, within our setting, the heat released to the heat bath during the time interval $[0,t]$ is given as

$$Q(t,x') = \int_0^t dt' \int_0^\infty dx \left[-\frac{\partial}{\partial x} V(x,t') \right] J(x,t'|x',0) = \int_0^t dt'\, F(t')\, I(t',x') \,, \tag{26}$$

where $I(t,x') = \int_0^\infty dx\, J(x,t|x',0)$ is the integrated probability current, and $J(x,t|x',0) = \left[v(t) - D\frac{\partial}{\partial x} \right] U(x,t|x',0)$ is the local probability current. The heat released during any infinitesimal time interval is positive. Actually, at any given instant, the force $F(t)$ and the motion of the particle have the same direction. Hence the function which form the integrand in the last expression in Eq. (26) is always nonnegative.
The external agent does work on the system by increasing the potential $V(x,t)$ while the position of the particle is fixed. Thus the work done at a given instant depends on the momentary position of the particle. In the stationary regime, summing over all possible positions *and* over one time period, we get (Chvosta et al., 2005)

$$W = \int_0^{2\pi/\omega} dt' \int_0^\infty dx \left[\frac{\partial}{\partial t'} V(x,t') \right] U_{\mathrm{a}}(x,t') = -F_1 \omega \int_0^{2\pi/\omega} dt' \cos(\omega t')\, \mu_{\mathrm{a}}(t') \,. \tag{27}$$

The work done on the system per cycle equals the area enclosed by the hysteresis curve which represents the parametric plot of the oscillating force versus the mean coordinate in the stationary regime $\mu_{\mathrm{a}}(t)$. This quantity must be positive. Otherwise, the system in contact with the single heat bath would produce positive work on the environment during the cyclic process in question. On the other hand, the work done on the system during a definite time interval within the period can be both positive and negative. In order to be specific, let $W_{\mathrm{a}}^{(i)}$, $i = 1,\ldots,4$, denote the work done by the external field on the system during the ith quarter-period of the force modulation. During the first quarter-period the slope of the potential decreases and the particle does a positive work on the environment, irrespective to its momentary position. Thus we have $W_{\mathrm{a}}^{(1)} < 0$. Nevertheless, the farther is the particle from the boundary the bigger is the work done by it during the fixed time interval. Within the second and the third quarter-period, the slope of the potential increases and the positive work is done by the external agent. Hence we have $W_{\mathrm{a}}^{(2)} > 0$ and $W_{\mathrm{a}}^{(3)} > 0$. However $W_{\mathrm{a}}^{(2)}$ is bigger than $|W_{\mathrm{a}}^{(1)}|$, since during the second quarter-period, when the work $W_{\mathrm{a}}^{(2)} > 0$ is done, the mean distance of the particle from the boundary is bigger than it was during the first quarter-period. Similar reasoning holds for the comparison of the work done by the external agent during the third and the fourth quarter-period. We have $W_a^{(4)} < 0$, and $W_a^{(3)} > |W_a^{(4)}|$. On the whole, since the periodic changes of the potential are inevitably associated with the changes of the particle position, the time-averaged work done by the external agent during one fundamental period must be always positive.

4. Two interacting particles in time-dependent potential

Up to now, we have been discussing the diffusion dynamics of just one isolated Brownian particle. Let us now turn to the case of *two interacting* particles diffusing under the action of the time-dependent external force in a one-dimensional channel.

In order to incorporate the simplest inter-particle interaction, the particles can be represented as rods of the length l. *The hard-core interaction* in such system means that the space occupied by one rod is inaccessible to the neighbouring rods. Generally, the diffusion of hard rods can be mapped exactly onto the diffusion of *point particles* (particles with the linear size $l = 0$) by the simple rescaling of space variables (see e.g. (Lizana & Ambjörnsson, 2009)). Hence without loss of generality all further considerations will be done for systems of point particles.

Consider two identical hard-core interacting particles, each with the diffusion constant D, diffusing in the potential $V(x,t)$ (cf. the preceding Section). Due to the hard-core interaction, *particles cannot pass each other* and the ordering of the particles is preserved during the evolution. Starting with $y_1 < y_2$, we have

$$-\infty < \mathsf{X}_1(t) < \mathsf{X}_2(t) < +\infty \tag{28}$$

for any t. We shall call the particle with the coordinate $\mathsf{X}_1(t)$ ($\mathsf{X}_2(t)$) the left (right) one. If the instantaneous coordinates of the two particles differ ($x_1 \neq x_2$) they both diffuse as non-interacting ones. This enables to reduce the diffusion problem for two identical hard-core interacting particles onto the diffusion of one "representative" particle in the two-dimensional *half-plane* $x_1 < x_2$. Namely, it suffices to require that the probability current for this representative particle in the direction perpendicular to the line $x_1 = x_2$ vanishes at this line. Except of that, the dynamics of the representative particle inside the half-plane $x_1 < x_2$ is controlled by the Smoluchowski equation

$$\frac{\partial}{\partial t} p^{(2)}(x_1,x_2,t\,|\,y_1,y_2,t_0) = -\sum_{j=1}^{2} \frac{\partial}{\partial x_j} \left\{ v(t) - D\frac{\partial}{\partial x_j} \right\} p^{(2)}(x_1,x_2,t\,|\,y_1,y_2,t_0)\,. \tag{29}$$

Differently speaking, the hard-core interaction is implemented as the boundary condition

$$\left(\frac{\partial}{\partial x_2} - \frac{\partial}{\partial x_1} \right) p^{(2)}(x_1,x_2,t\,|\,y_1,y_2,t_0) \Big|_{x_1=x_2} = 0\,. \tag{30}$$

Returning to the original picture, *the two hard-core interacting particles in one dimension will never cross each other*.

Assuming the initial positions $y_1 < y_2$, consider the function which is defined as

$$\begin{aligned} p^{(2)}(x_1,x_2,t\,|\,y_1,y_2,t_0) &= U(x_1,t\,|\,y_1,t_0)U(x_2,t\,|\,y_2,t_0) + \\ &+ U(x_1,t\,|\,y_2,t_0)U(x_2,t\,|\,y_1,t_0)\,, \end{aligned} \tag{31}$$

within the phase space $\mathcal{R}_2 : -\infty < x_1 < x_2 < +\infty$, and which vanishes elsewhere. Here $U(x,t\,|\,y,t_0)$ is the solution of the corresponding single-particle problem. This function fulfills both EQ. (29) and EQ. (30). The proof is straightforward and it can be generalized to the N-particle diffusion problem in a general time- and space-dependent external potential.

4.1 Dynamics

Similarly as in the preceding Section, we now assume the particles are driven by the space-homogeneous and time-dependent force $F(t) = F_0 + F_1 \sin(\omega t)$. The corresponding drift velocity is $v(t) = v_0 + v_1 \sin(\omega t)$ (cf. the preceding Section). The time-independent component pushes the particles to the left against the reflecting boundary at the origin (if $F_0 < 0$), or to the right (if $F_0 > 0$). The time-dependent component $F_1 \sin(\omega t)$ harmonically oscillates with the angular frequency ω. In the rest of this Section we treat the case $F_0 < 0$. On the whole our model includes four parameters F_0, F_1, ω, and D. Notice that the hard-core interaction among particles acts as a purely geometric restriction. As such, it is not connected with any "interaction parameter".
If we integrate the joint probability density (31) over the coordinate x_1 (x_2) of the left (right) particle we obtain the marginal probability density describing the dynamics of the right (left) particle:

$$p_{\mathrm{L}}(x,t\,|\,y_1,y_2,t_0) \equiv \int_0^{+\infty} \mathrm{d}x_2\, p^{(2)}(x,x_2,t\,|\,y_1,y_2,t_0)\,, \tag{32}$$

$$p_{\mathrm{R}}(x,t\,|\,y_1,y_2,t_0) \equiv \int_0^{+\infty} \mathrm{d}x_1\, p^{(2)}(x_1,x,t\,|\,y_1,y_2,t_0)\,. \tag{33}$$

Notice that the both marginal densities depend on the initial positions of the *both* particles. Of course, this is the direct consequence of the interaction among the particles.
Let us now focus on the time-asymptotic dynamics which, as usually, includes the most important physics in the problem. If $F_0 < 0$ and $F_1 = 0$, the probability density of the single diffusing particle relaxes to the exponential function $\pi_{\mathrm{eq}}(x)$ (cf. FIG. 3). Using EQ. (31), the equilibrium two-particle joint probability density is

$$p^{(2)}_{\mathrm{eq}}(x_1,x_2) = \theta(x_2 - x_1)\left(\frac{|v_0|}{D}\right)^2 \exp\left[-(x_1 + x_2)\frac{|v_0|}{D}\right]. \tag{34}$$

Hence the equilibrium probability density of the left particle reads

$$p^{(\mathrm{eq})}_{\mathrm{L}}(x) = 2\theta(x)\frac{|v_0|}{D}\exp\left(-2x\frac{|v_0|}{D}\right). \tag{35}$$

The only difference between this density and $\pi_{\mathrm{eq}}(x)$ is the factor "2" which occurs in the above exponential and as the multiplicative prefactor. Thus $p^{(\mathrm{eq})}_{\mathrm{L}}(x)$ takes a higher value at the boundary and, as the function of the coordinate x, it decreases more rapidly than the single-particle equilibrium density $\pi_{\mathrm{eq}}(x)$. As for the right particle, its equilibrium density could not be so simply related with $\pi_{\mathrm{eq}}(x)$. It reads

$$p^{(\mathrm{eq})}_{\mathrm{R}}(x) = 2\theta(x)\frac{|v_0|}{D}\exp\left(-x\frac{|v_0|}{D}\right)\left[1 - \exp\left(-x\frac{|v_0|}{D}\right)\right]. \tag{36}$$

Notice that it vanishes at the reflecting boundary and it attains its maximum value $p^{(\mathrm{eq})}_{\mathrm{R}}(x_{\mathrm{m}}) = |v_0|/(2D)$ at the coordinate $x_{\mathrm{m}} = D\log(2)/|v_0|$.

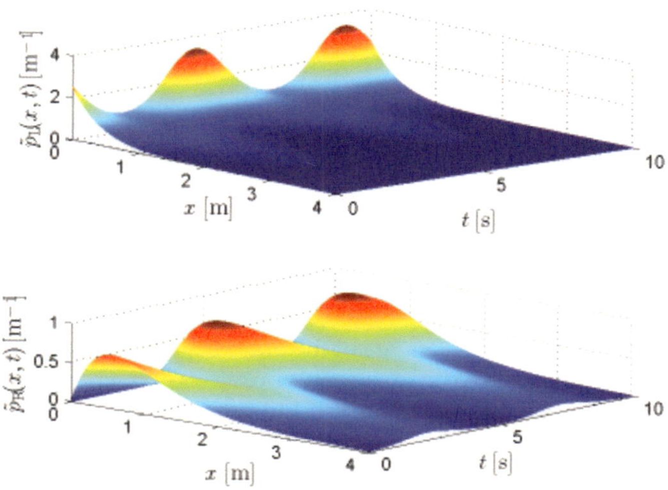

Fig. 4. Time- and space-resolved probability densities in the time-asymptotic regime. We have used the parameters $v_0 = -1.0\,\mathrm{m\,s^{-1}}$, $v_1 = 1.0\,\mathrm{m\,s^{-1}}$, $D = 1.0\,\mathrm{m^2 s^{-1}}$, $\omega = 0.4\,\pi\,\mathrm{rad\,s^{-1}}$.

Let us now take $F_0 < 0$ and $F_1 > 0$. We can simply use the expansion (18) from the preceding Section and insert it into EQ. (31). After this step, the *time-asymptotic* marginal densities (32), (33) read

$$\tilde{p}_{\rm L}(x,t) = 2 \sum_{k=-\infty}^{+\infty} \left[\sum_{n=-\infty}^{+\infty} u_{k-n}(x)\, l_n(x) \right] \exp(-ik\omega t)\,, \tag{37}$$

$$\tilde{p}_{\rm R}(x,t) = 2 \sum_{k=-\infty}^{+\infty} \left[\sum_{n=-\infty}^{+\infty} u_{k-n}(x)\, r_n(x) \right] \exp(-ik\omega t)\,, \tag{38}$$

where we have introduced the abbreviations

$$l_k(x) = \frac{|v_0|}{D} \langle k | \mathbb{L}_{--} \mathbb{E}_{\rm L}(x) \mathbb{R}_{++} | f \rangle\,, \tag{39}$$

$$r_k(x) = \frac{|v_0|}{D} \langle k | \mathbb{L}_{--} \mathbb{E}_{\rm R}(x) \mathbb{R}_{++} | f \rangle\,, \tag{40}$$

$$k = 0, \pm 1, \pm 2, \ldots\,.$$

The matrices on the right hand sides are given by EQS. (20)-(22) and by the integrals $\mathbb{E}_{\rm L}(x) \equiv \int_x^{+\infty} \mathrm{d}x' \mathbb{E}(x')$, $\mathbb{E}_{\rm R}(x) \equiv \int_0^x \mathrm{d}x' \mathbb{E}(x')$ from the matrix (20).
In order to analyse the densities $\tilde{p}_{\rm L}(x,t)$ and $\tilde{p}_{\rm R}(x,t)$ numerically, we have to curtail both the infinite vector of the complex amplitudes $|f\rangle$ and the infinite matrices $\mathbb{L}_{--}$, $\mathbb{R}_{++}$, $\mathbb{E}_{\rm L}(x)$ and $\mathbb{E}_{\rm R}(x)$. Using these controllable approximations, we obtain the full time- and space-resolved form of the functions $\tilde{p}_{\rm L}(x,t)$, $\tilde{p}_{\rm R}(x,t)$. FIG. 4 illustrates the resulting non-linear "waves".

4.2 Energetics

The equilibrium internal energy of a particle is calculated as the spatial integral from the product of the stationary potential $V(x) = -xF_0$ times the equilibrium probability density. For the single diffusing particle the result is $E^{(\rm eq)} = D\Gamma = k_{\rm B}T$. In the case of two interacting particles, the equilibrium internal energy of the left (right) particle reads $E_{\rm L}^{\rm (eq)} = k_{\rm B}T/2$ ($E_{\rm R}^{\rm (eq)} = 3k_{\rm B}T/2$). The equilibrium internal energies do not depend on the slope of the stationary potential $V(x)$ and they linearly increase with the temperature T. Notice that the effective repulsive force among the interacting particles increases (decreases) the internal energy of the right (left) particle. However, the total internal energy of the system of two interacting particles is equal to the total internal energy of the system of two non-interacting particles, i.e., $E_{\rm L}^{\rm (eq)} + E_{\rm R}^{\rm (eq)} = 2E^{\rm (eq)}$. As the hard-core interaction does not contribute to the total energy, the effective repulsive force necessarily arises from a purely entropic effect. This property stems from a zero range of the interaction and it also holds in a general (non-equilibrium) situations.

Now, consider the time-dependent potential $V(x,t) = -xF(t)$. Let the system be in the time-asymptotic regime. The internal energy of the diffusing particle at the time t is defined as the average of the potential $V(x,t)$ over all possible positions of the particle at a given instant. In the single-particle case the internal energy at the time t (say, $E(t)$) is given by EQ. (24). Similarly, in the case of two interacting particles, the internal energies of the left and the right particle are

$$E_{\rm L}(t) = -[F_0 + F_1 \sin(\omega t)]\mu_{\rm L}(t)\,, \tag{41}$$

$$E_{\rm R}(t) = -[F_0 + F_1 \sin(\omega t)]\mu_{\rm R}(t)\,, \tag{42}$$

respectively. Here, $\mu_{\rm L}(t)$ ($\mu_{\rm R}(t)$) denotes the mean position of the left (right) particle in the asymptotic regime.

Generally speaking, the internal energies $E(t)$, $E_{\rm L}(t)$, $E_{\rm R}(t)$ are periodic functions of time with the fundamental period $2\pi/\omega$. The total internal energy of two interacting particles is equal

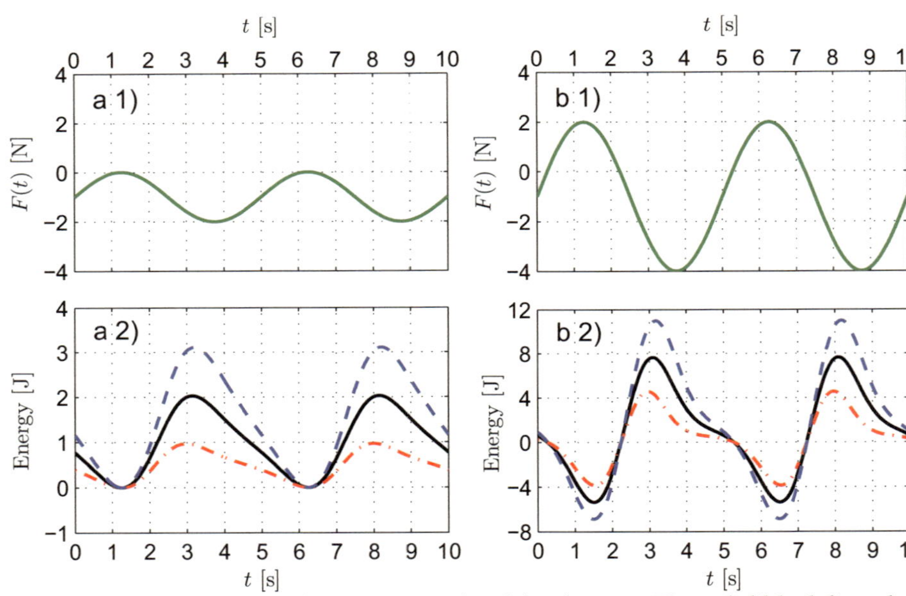

Fig. 5. The internal energies within two periods of the driving. The solid black line shows the energy $E(t)$, the dashed blue line depicts $E_{\rm R}(t)$ and the dot-dashed red line illustrates $E_{\rm L}(t)$. In the panels a1) and a2) we take $F_1 = 1.0\,$N, in the panels b1) and b2) we take $F_1 = 3.0\,$N. The static component ($F_0 = -1.0\,$N), the frequency ($\omega = 0.4\pi\,{\rm s}^{-1}$), and the diffusion constant ($D = 1.0\,{\rm m}^2{\rm s}^{-1}$) are the same in all panels.

to the total internal energy of two non-interacting particles. In symbols $E_{\mathrm{L}}(t) + E_{\mathrm{R}}(t) = 2E(t)$. FIG. 5 shows the time-dependency of the internal energies $E(t)$, $E_{\mathrm{L}}(t)$, $E_{\mathrm{R}}(t)$ within two periods of the driving force and for the different parameters F_0, F_1, ω, and D. First of all, notice the effect of the entropic repulsive force which stems from the hard-core interaction. We see that there is no *qualitative* difference between the oscillations of the function $E(t)$ and the functions $E_{\mathrm{L}}(t)$, $E_{\mathrm{R}}(t)$. Hence the hard-core interaction changes only *quantitative* features of the energetics of individual particles as compared to the diffusion without interaction. One of this quantitative changes, the most striking one at a first glance, is the change of the amplitudes of the internal energies $E_{\mathrm{L}}(t)$, $E_{\mathrm{R}}(t)$ as compared to $E(t)$.

Oscillations of the internal energies express the combine effect of both the periodically modulated *heat flow* to the bath and the periodic exchange of the *work* done on the particle by an external agent. Without loss of generality let us now analyse the energetics of the single-diffusing particle (solid black lines in FIG. 5). At the beginning of the period we choose the instant when the driving force takes the value F_0 and tends to increase. It is increasing up to the value $F_0 + F_1$. In the panel a 2) $F_0 + F_1 = 0$ N, in the panel b 2) $F_0 + F_1 = 2$ N. During this interval, the internal energy is decreasing towards its minimum due to the positive work which the system does on its surroundings. The smaller the value of the amplitude F_1 (panel a 2)) the closer is the process to the *quasi-static* one and the smaller is the work done by the system. The decreasing tendency of the internal energy is being partially compensated by the heat coming from the heat bath. On the other hand, for larger amplitudes F_1 (panel b 2)), the heat is almost entirely being released to the reservoir. Hence the greater the amplitude F_1 the lower minimum values of the internal energy are observed.

During the next part of the period, the driving force is decreasing form the value $F_0 + F_1$ to its minimum value $F_0 - F_1$. In the panel a 2) $F_0 - F_1 = -2$ N, in the panel b 2) $F_0 - F_1 = -4$ N. Within this interval the slope $-F(t)$ of the potential $V(x,t)$ is permanently increasing, hence the positive work is performed on the system. This work constitutes the most significant contribution to the changes of the internal energy. The internal energy is increasing up to its maximum and, finally, it is decreasing due to the strong heat flow from the system to the bath at the end of this time-interval.

Within the last part of the period, the driving force and the internal energy are decreasing to their initial values which they attain at the beginning of the period. Within this time interval the slope of the potential $V(x,t)$ decreases. Consequently, the positive work is performed by the system on its surrounding. Notice the sudden change of the slope of the internal energy at the beginning of this interval. This effect is more pronounced for the greater amplitudes F_1 (panel b 2)). It is connected with the fact that the system starts to exert work on its surroundings. The greater the amplitude F_1 the less significant the contribution of this work to the change of the internal energy as compared with the heat flow from the system to the reservoir (the more significant contribution of the work would cause faster decreasing of the internal energy as should be seen from the panel a2)).

Finally let us discus the internal energies averaged over the period, i.e., $\bar{E} = \frac{\omega}{2\pi}\int_0^{2\pi/\omega} \mathrm{d}t E(t)$, $\bar{E}_{\mathrm{L}} = \frac{\omega}{2\pi}\int_0^{2\pi/\omega} \mathrm{d}t E_{\mathrm{L}}(t)$, $\bar{E}_{\mathrm{R}} = \frac{\omega}{2\pi}\int_0^{2\pi/\omega} \mathrm{d}t E_{\mathrm{R}}(t)$. A remarkable fact is that differences $\bar{E} - E^{(\mathrm{eq})}$, $\bar{E}_{\mathrm{L}} - E_{\mathrm{L}}^{(\mathrm{eq})}$, and $\bar{E}_{\mathrm{R}} - E_{\mathrm{R}}^{(\mathrm{eq})}$ are always greater than zero. Differently speaking, in the time-averaged sense, the external driving induces a *permanent increase of the particle's internal energy* as compared to its equilibrium value.

5. Dynamics of a molecular motor based on the externally driven two-level system

Consider a two-level system with time-dependent energies $E_i(t)$, $i = 1,2$, in contact with a single thermal reservoir at temperature T. In general, the heat reservoir temperature T may also be time-dependent. The time evolution of the occupation probabilities $p_i(t)$, $i = 1,2$, is governed by the Master equation (Gammaitoni et al., 1998) with the time-dependent transition rates. The rates depend on the reservoir temperature but they also incorporate external parameters which control the driving protocol. To be specific the dynamics of the system is described by the time-inhomogeneous Markov process $\mathsf{D}(t)$. The state variable $\mathsf{D}(t)$ assumes the value i ($i = 1,2$) if the system resides at the time t in the ith state. Explicitly, the Master equation reads

$$\frac{\mathrm{d}}{\mathrm{d}t}\mathbb{R}(t\,|\,t') = -\begin{pmatrix} \lambda_1(t) & -\lambda_2(t) \\ -\lambda_1(t) & \lambda_2(t) \end{pmatrix} \mathbb{R}(t\,|\,t'), \qquad \mathbb{R}(t'\,|\,t') = \mathbb{I}, \tag{43}$$

where $\mathbb{I}$ is the unit matrix and $\mathbb{R}(t,t')$ is the transition matrix with the matrix elements $R_{ij}(t\,|\,t') = \langle i\,|\,\mathbb{R}(t\,|\,t')\,|\,j\rangle$. These elements are the conditional probabilities

$$R_{ij}(t\,|\,t') = \mathrm{Prob}\left\{\,\mathsf{D}(t) = i\,|\,\mathsf{D}(t') = j\,\right\}. \tag{44}$$

The occupation probabilities at the observation time t are given by the column vector $|p(t,t')\rangle = \mathbb{R}(t\,|\,t')\,|\phi(t')\rangle$. Here $\phi_i(t') = \langle i\,|\phi(t')\rangle$ denotes the occupational probabilities at the initial time t'. Due to the conservation of the total probability, the system (43) can be reduced to just one non-homogeneous linear differential equation of the first order. Therefore the Master equation (43) is exactly solvable for arbitrary functions $\lambda_1(t)$, $\lambda_2(t)$ (Šubrt & Chvosta, 2007). The rates $\lambda_1(t)$, $\lambda_2(t)$ are typically a product of an attempt frequency ν to exchange the state and an acceptance probability. We shall adopt the Glauber form

$$\lambda_1(t) = \frac{\nu}{1+\exp\{-\beta(t)\,[E_1(t)-E_2(t)]\}}\,,\ \lambda_2(t) = \lambda_1(t)\exp\{-\beta(t)\,[E_1(t)-E_2(t)]\}\,. \tag{45}$$

Here, ν^{-1} sets the elementary time scale, and $\beta(t) = 1/\left[k_B T(t)\right]$. The rates (45) satisfy the (time local) detailed balance condition (van Kampen, 2007) and they saturate at large energy differences (see (Einax et al., 2010) for a further discussion).

We now introduce the setup for the operational cycle of the engine. Within a given period, two branches with linear time-dependence of the state energies are considered with different velocities. Starting from the value h_1, the energy $E_1(t)$ linearly increases in the first branch until it attains the value $h_2 > h_1$, at the time t_+. Afterwards, in the second branch, the energy $E_1(t)$ linearly decreases and it reassumes the starting value h_1 at the time $t_- + t_+$. We always take $E_2(t) = -E_1(t)$, i.e.

$$E_1(t) = -E_2(t) = \begin{cases} h_1 + \dfrac{h_2-h_1}{t_+}\,t, & t \in [0,t_+]\,, \\ h_2 - \dfrac{h_2-h_1}{t_-}\,(t-t_+), & t \in [t_+,t_++t_-]\,. \end{cases} \tag{46}$$

This pattern will be periodically repeated, the period being $t_\mathrm{p} = t_+ + t_-$.

As the second ingredient, we need to specify the temperature schedule. The two-level system will be alternately exposed to a hot and a cold reservoir, which means that the function $\beta(t)$ in Eq. (45) will be a piecewise constant periodic function. During the first (second) branch, it assumes the value β_+ (β_-). Let us stress that the change of the heat reservoirs at the end of the individual branches is instantaneous. The switching of the reservoirs necessarily implies a finite difference between the new reservoir temperature and the actual system (effective) temperature. Even if the driving period tends to infinity (*a quasi-static limit*), we shall observe a positive entropy production originating from the relaxation processes initiated by the abrupt change of the contact temperature. Differently speaking, our engine operates in an inherently irreversible regime and there exists no reversible limit of the limit cycle.

The explicit form of the solution $\mathbb{R}(t\,|\,0)$ of the Master equation (43) with the rates (45) and the periodically modulated energies (46) can be found in (Chvosta et al., 2010). Starting from an arbitrary initial condition $|\phi(t')\rangle$ the system's response approaches a steady state. In order to specify the limit cycle we require that the system's state at the beginning of the cycle coincides with the system state at the end of the cycle. Differently speaking, we have to solve the equation $|\pi\rangle = \mathbb{R}(t_\mathrm{p}\,|\,0)\,|\pi\rangle$ for the unknown initial state $|\pi\rangle$. In the course of the limit cycle, the state of the system is described by the column vector $|p(t)\rangle = \mathbb{R}(t\,|\,0)\,|\pi\rangle$ with the elements $p_i(t) = \langle i\,|\,p(t)\rangle$, $t \in [0, t_\mathrm{p}]$.

This completes the description of the model. Any quantity describing the engine's performance can only depend on the parameters h_1, h_2, $\beta_\pm$, $t_\pm$, and ν.

5.1 Energetics of the engine

During the limit cycle, the internal energy $U(t) = \sum_{i=1}^{2} E_i(t) p_i(t)$ changes as

$$\frac{\mathrm{d}}{\mathrm{d}t}U(t) = \sum_{i=1}^{2} E_i(t)\frac{\mathrm{d}}{\mathrm{d}t}p_i(t) + \sum_{i=1}^{2} p_i(t)\frac{\mathrm{d}}{\mathrm{d}t}E_i(t) = \frac{\mathrm{d}}{\mathrm{d}t}\left[Q(t) + W(t)\right], \quad t \in [0, t_\mathrm{p}]. \tag{47}$$

Here, $Q(t)$ is the mean heat received from the reservoirs during the time interval $[0, t]$. Analogously, $W(t)$ is the mean work done on the system from the beginning of the limit cycle till the time t. If $W(t) < 0$, the positive work $-W(t)$ is done by the system on the environment. Therefore the *oriented* area enclosed by the limit cycle in FIG. 7 represents the work $W_\mathrm{out} \equiv -W(t_\mathrm{p})$ done by the engine on the environment per cycle. This area approaches its maximum absolute value in the quasi-static limit. The internal energy, being a state function, fulfils $U(t_\mathrm{p}) = U(0)$. Therefore, if the work W_out is positive, the same total amount of heat has been accepted from the two reservoirs during the limit cycle. As long as the both heat reservoirs are at the same temperature ($\beta_+ = \beta_-$), the case $W_\mathrm{out} > 0$ will never occur. That the *perpetum mobile* is actually forbidden can be traced back to the detailed balance condition in (43).

We denote the system entropy at the time t as $S_\mathrm{s}(t)$, and the reservoir entropy at the time t as $S_\mathrm{r}(t)$. They are given by

$$\frac{S_\mathrm{s}(t)}{k_\mathrm{B}} = -\left[p_1(t)\ln p_1(t) + p_2(t)\ln p_2(t)\right], \tag{48}$$

$$\frac{S_\mathrm{r}(t)}{k_\mathrm{B}} = \begin{cases} -\beta_+ \displaystyle\int_0^t \mathrm{d}t' E_1(t')\frac{d}{dt'}\left[p_1(t') - p_2(t')\right], & t \in [0, t_+], \\ S_\mathrm{r}(t_+) - \beta_- \displaystyle\int_{t_+}^t \mathrm{d}t' E_1(t')\frac{d}{dt'}\left[p_1(t') - p_2(t')\right], & t \in [t_+, t_\mathrm{p}]. \end{cases} \tag{49}$$

Upon completing the cycle, the system entropy re-assumes its value at the beginning of the cycle. On the other hand, the reservoir entropy is controlled by the heat exchange. Owing

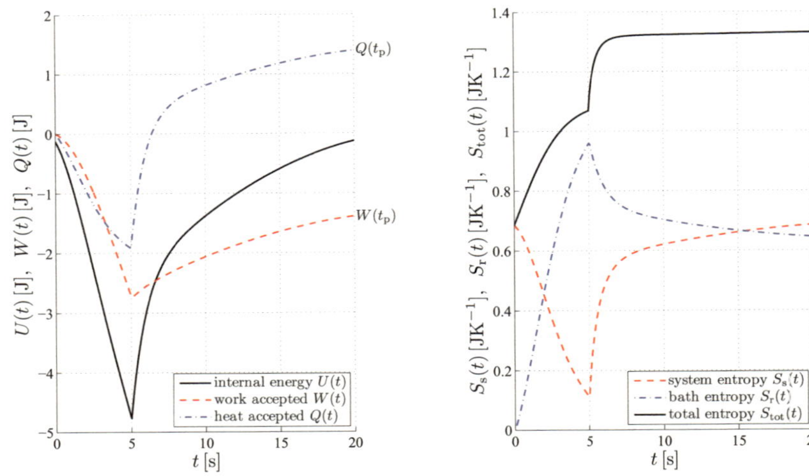

Fig. 6. Thermodynamic quantities as functions of time during the limit cycle. Left panel: internal energy, mean work done on the system, and mean heat received from both reservoirs; the final position of the mean work curve marks the work done on the system per cycle $W(t_p)$. Since $W(t_p) < 0$, the work $W_{out} = -W(t_p)$ has been done on the environment. The internal energy returns to its original value and, after completion of the cycle, the absorbed heat $Q(t_p)$ equals the negative work $-W(t_p)$. Right panel: entropy $S_s(t)$ of the system and $S_r(t)$ of the bath, and their sum $S_{tot}(t)$; after completing the cycle, the system entropy re-assumes its initial value. The difference $S_{tot}(t_p) - S_{tot}(0) > 0$ equals the entropy production per cycle. It is always positive and quantifies the degree of irreversibility of the cycle. Note that at the times t_+ and t_p, strong increase of $S_{tot}(t)$ always occurs due to the instantaneous change of the reservoirs. The parameters used are $h_1 = 1\,\mathrm{J}$, $h_2 = 5\,\mathrm{J}$, $\nu = 1\,\mathrm{s}^{-1}$, $t_+ = 5\,\mathrm{s}$, $t_- = 15\,\mathrm{s}$, $\beta_+ = 0.5\,\mathrm{J}^{-1}$ and $\beta_- = 0.1\,\mathrm{J}^{-1}$.

to the inherent irreversibility of the cycle we observe always a positive entropy production per cycle, i.e., $S_r(t_p) - S_r(0) > 0$. The total entropy $S_{tot}(t) = S_s(t) + S_r(t)$ increases for any $t \in [0, t_p]$. The rate of the increase is the greater the larger is the deviation of the representative point in the $p-E$ diagram from the corresponding equilibrium isotherm (a large deviation, e.g., can be seen in the $p-E$ diagram in FIG. 7 c). Due to the instantaneous exchanges of baths at t_+ and t_p, strong increase of $S_{tot}(t)$ always occurs after these instants. A representative example of the overall behaviour of the thermodynamic quantities (mean work and heat, and entropies) during the limit cycle is shown in FIG. 6.
Up to now, we have discussed the *averaged* thermodynamic properties of the engine. We now turn to the *fluctuations* of its performance.

5.2 Fluctuations of the engine's thermodynamic properties

By treating the state variable and work variable as the two components of a combined stochastic process, it is possible to derive a partial differential equation for the time evolution of the work probability density (or the heat probability density), see, for example, (Schuler et al., 2005; Imparato & Peliti, 2005b;c;a). For completeness, we outline the procedure in the present context.

Heuristically, the underlying time-inhomogeneous Markov process $\mathsf{D}(t)$ can be conceived as an ensemble of individual realizations (sample paths). A realization is specified by a succession of transitions between the two states. If we know the number n of the transitions during a path and the times $t_{kk=1}^{n}$ at which they occur, we can calculate the probability that this specific path will be generated. A given paths yields a unique value of the microscopic work done on the system. For example, if the system is known to remain during the time interval $[t_k, t_{k+1}]$ in the ith state, the work done on the system during this time interval is simply $E_i(t_{k+1}) - E_i(t_k)$. The probability of an arbitrary fixed path amounts, at the same time, the probability of that value of the work which is attributed to the path in question. Viewed in this way, the work itself is a stochastic process and we denote it as $\mathsf{W}(t)$. We are interested in its probability density $\rho(w,t) = \langle \delta(\mathsf{W}(t) - w) \rangle$, where $\langle \ldots \rangle$ denotes the average over all possible paths.

We now introduce the *augmented process* $\{\mathsf{W}(t), \mathsf{D}(t)\}$ which simultaneously reflects both the work variable and the state variable. The augmented process is again a time non-homogeneous Markov process. Actually, if we know at a fixed time t' both the present state variable j and the work variable w', then the subsequent probabilistic evolution of the state and the work is completely determined. The work done during the time period $[t', t]$, where $t > t'$, simply adds to the present work w' and it only depends on the succession of the states after the time t'. And this succession by itself cannot depend on the dynamics before time t'.

The one-time properties of the augmented process will be described by the functions

$$G_{ij}(w,t\,|\,w',t') = \lim_{\epsilon\to 0} \frac{\mathrm{Prob}\,\{\mathsf{W}(t) \in (w, w+\epsilon)\,\mathrm{and}\,\mathsf{D}(t) = i\,|\,\mathsf{W}(t') = w'\,\mathrm{and}\,\mathsf{D}(t') = j\}}{\epsilon}, \tag{50}$$

where $i, j = 1, 2$. We represent them as the matrix elements of a single two-by-two matrix $\mathbb{G}(w,t\,|\,w',t')$,

$$G_{ij}(w,t\,|\,w',t') = \langle i\,|\,\mathbb{G}(w,t\,|\,w',t')\,|\,j\rangle\,. \tag{51}$$

We need an equation which controls the time dependence of the propagator $\mathbb{G}(w,t\,|\,w',t')$ and which plays the same role as the Master equation (43) in the case of the simple two-state process. This equation reads (Imparato & Peliti, 2005b; Šubrt & Chvosta, 2007)

$$\frac{\partial}{\partial t}\mathbb{G}(w,t\,|\,w',t') = -\left\{ \begin{pmatrix} \frac{d\,E_1(t)}{dt} & 0 \\ 0 & \frac{d\,E_2(t)}{dt} \end{pmatrix} \frac{\partial}{\partial w} + \begin{pmatrix} \lambda_1(t) & -\lambda_2(t) \\ -\lambda_1(t) & \lambda_2(t) \end{pmatrix} \right\} \mathbb{G}(w,t\,|\,w',t'), \tag{52}$$

where the initial condition is $\mathbb{G}(w,t'\,|\,w',t') = \delta(w - w')\mathbb{I}$. The matrix equation represents a hyperbolic system of four coupled partial differential equations with the time-dependent coefficients.

Similar reasoning holds for the random variable $\mathsf{Q}(t)$ which represents the heat accepted by the system from the environment. Concretely, if the system undergoes during a time interval $[t_k, t_{k+1}]$ only one transition which brings it at an instant $\tau \in [t_k, t_{k+1}]$ from the state i to the state j, the heat accepted by the system during this time interval is $E_j(\tau) - E_i(\tau)$. The variable $\mathsf{Q}(t)$ is described by the propagator $\mathbb{K}(q,t\,|\,q',t')$ with the matrix elements

$$K_{ij}(q,t\,|\,q',t') = \lim_{\epsilon\to 0} \frac{\mathrm{Prob}\,\{\mathsf{Q}(t) \in (q, q+\epsilon) \wedge \mathsf{D}(t) = i\,|\,\mathsf{Q}(t') = q' \wedge \mathsf{D}(t') = j\}}{\epsilon}. \tag{53}$$

It turns out that there exists a simple relation between the heat propagator and the work propagator $\mathbb{G}(w,t\,|\,w',t')$. Since for each path, heat q and work w are connected by the first law of thermodynamics, we have $q = E_i(t) - E_j(t') - w$ for any path which has started at the time t' in the state j and which has been found at the time t in the state i. Accordingly,

$$\mathbb{K}(q,t\,|\,q',t') = \begin{pmatrix} g_{11}(u_{11}(t,t') - q,t\,|\,q',t') & g_{12}(u_{12}(t,t') - q,t\,|\,q',t') \\ g_{21}(u_{21}(t,t') - q,t\,|\,q',t') & g_{22}(u_{22}(t,t') - q,t\,|\,q',t') \end{pmatrix}, \tag{54}$$

where $u_{ij}(t,t') = E_i(t) - E_j(t')$.
The explicit form of the matrix $\mathbb{G}(w,t)$ which solves the dynamical equation (52) with the Glauber transition rates (45) and the periodically modulated energies (46) can be found in (Chvosta et al., 2010). Heaving the matrix $\mathbb{G}(w,t)$ for the limit cycle, the matrix $\mathbb{K}(q,t)$ is calculated using the transformation (54).
In the last step, we take into account the initial condition $|\,\pi\,\rangle$ at the beginning of the limit cycle and we sum over the final states of the process $\mathsf{D}(t)$. Then the (unconditioned) probability density for the work done on the system in the course of the limit cycle reads

$$\rho(w,t) = \sum_{i=1}^{2} \langle\, i\,|\,\mathbb{G}(w,t)\,|\,\pi\,\rangle\,. \tag{55}$$

Similarly, the probability density for the heat accepted during the time interval $[0,t]$ is

$$\chi(q,t) = \sum_{i=1}^{2} \langle\, i\,|\,\mathbb{K}(q,t)\,|\,\pi\,\rangle\,. \tag{56}$$

The form of the resulting probability densities and therefore also the overall properties of the engine critically depend on the two dimensionless parameters $a_\pm = \nu t_\pm/(2\beta_\pm|h_2 - h_1|)$. We call them *reversibility parameters* [1]. For a given branch, say the first one, the parameter a_+ represents the ratio of two characteristic time scales. The first one, $1/\nu$, describes the attempt rate of the internal transitions. The second scale is proportional to the reciprocal driving velocity. Contrary to the first scale, the second one is fully under the external control. Moreover, the reversibility parameter a_+ is proportional to the absolute temperature of the heat bath, k_B/β_+.
FIG. 7 illustrates the shape of the limit cycle together with the functions $\rho(w,t_p)$, $\chi(q,t_p)$ for various values of the reversibility parameters. Notice that the both functions $\rho(w,t_p)$ and $\chi(q,t_p)$ vanishes outside a finite support. Within their supports, they exhibit a continuous part, depicted by the full curve, and a singular part, illustrated by the full arrow. The height of the full arrow depicts the weight of the corresponding δ-function. The continuous part of the function $\rho(w,t_p)$ develops one discontinuity which is situated at the position of the full arrow. Similarly, the continuous part of the function $\chi(q,t_p)$ develops three discontinuities.
If the both reversibility parameters $a_\pm$ are small, the isothermal processes during the both branches strongly differ from the equilibrium ones. The indication of this case is a flat continuous component of the density $\rho(w,t_p)$ and a well pronounced singular part. The strongly irreversible dynamics occurs if one or more of the following conditions hold. First, if ν is small, the transitions are rare and the occupation probabilities of the individual energy

[1] The reversibility here refers to the individual branches. As pointed out above, the abrupt change in temperature, when switching between the branches, implies that there exists no reversible limit for the complete cycle.

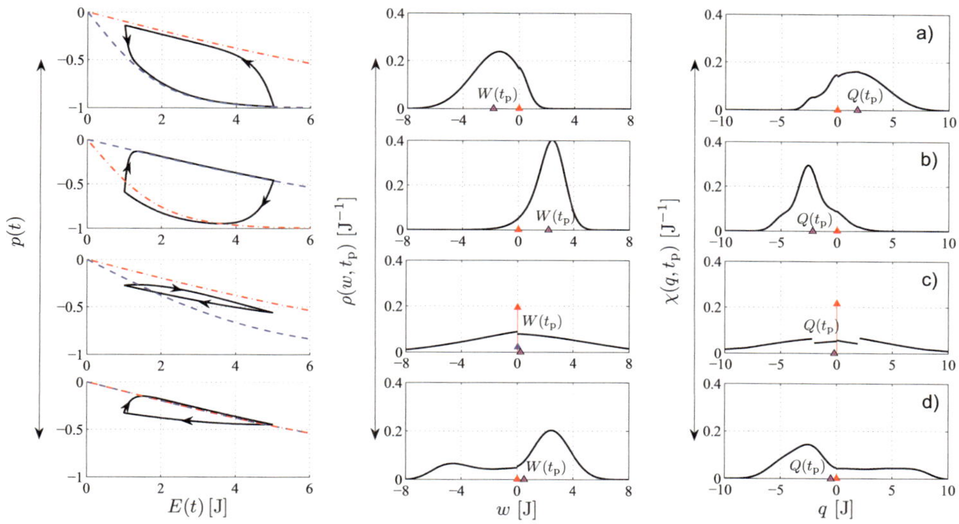

Fig. 7. Probability densities $\rho(w,t_{\rm p})$ and $\chi(q,t_{\rm p})$ for the work and the heat for four representative sets of the engine parameters (every set of parameters corresponds to one horizontal triplet of the panels). The first panel in the triplet shows the limit cycle in the $p-E$ plane ($p(t)=p_1(t)-p_2(t)$ is the occupation difference and $E(t)=E_1(t)$). In the parametric plot we have included also the equilibrium isotherm which corresponds to the first stroke (the dashed line) and to the second stroke (the dot-dashed line). In all panels we take $h_1=1\,$J, $h_2=5\,$J, and $\nu=1\,{\rm s}^{-1}$. The other parameters are the following. a in the first triplet: $t_+=50\,$s, $t_-=10\,$s, $\beta_+=0.5\,{\rm J}^{-1}$, $\beta_-=0.1\,{\rm J}^{-1}$, $a_\pm=12.5$ (the bath of the first stroke is colder than that of the second stroke). b in the second triplet: $t_+=50\,$s, $t_-=10\,$s, $\beta_+=0.1\,{\rm J}^{-1}$, $\beta_-=0.5\,{\rm J}^{-1}$, $a_+=62.5$, $a_-=2.5$ (exchange of β_+ and β_- as compared to case a, leading to a change of the traversing of the cycle from counter-clockwise to clockwise and a sign reversal of the mean values $W(t_{\rm p})\equiv\langle {\sf W}(t_{\rm p})\rangle$ and $Q(t_{\rm p})\equiv\langle {\sf Q}(t_{\rm p})\rangle$). c in the third triplet: $t_+=2\,$s, $t_-=2\,$s, $\beta_+=0.2\,{\rm J}^{-1}$, $\beta_-=0.1\,{\rm J}^{-1}$, $a_+=1.25$, $a_-=2.5$ (a strongly irreversible cycle traversed clockwise with positive work). d in the fourth triplet: $t_+=20\,$s, $t_-=1\,$s, $\beta_\pm=0.1\,{\rm J}^{-1}$, $a_+=25$, $a_-=1.25$ (no change in temperatures, but large difference in duration of the two strokes; $W(t_{\rm p})$ is necessarily positive). The height of the red arrows plotted in the panels with probability densities depicts the weight of the corresponding δ-functions.

levels are effectively frozen during long periods of time. Therefore they lag behind the Boltzmann distribution which would correspond to the instantaneous positions of the energy levels. More precisely, the population of the ascending (descending) energy level is larger (smaller) than it would be during the corresponding reversible process. As a result, the mean work done on the system is necessarily larger than the equilibrium work. Secondly, a similar situation occurs for large driving velocities $v_\pm$. Due to the rapid motion of the energy levels, the occupation probabilities again lag behind the equilibrium ones. Thirdly, the strong irreversibility occurs also in the low temperature limit. In the limit $a_\pm \to 0$, the continuous part vanishes and $\rho(w,t_{\rm p})=\delta(w)$.

In the opposite case of large reversibility parameters $a_\pm$, the both branches in the $p-E$ plane are located close to the reversible isotherms. The singular part of the density $\rho(w,t_{\rm p})$ is suppressed and the continuous part exhibits a well pronounced peak. The density $\rho(w,t_{\rm p})$ approaches the Gaussian function centered around the men work. This confirms the general considerations (Speck & Seifert, 2004). In the limit $a_{\pm} \to \infty$ the Gaussian peak collapses to the delta function located at the quasi-static work (Chvosta et al., 2010). The heat probability density $\chi(q, t_{\rm p})$ shows similar properties as $\rho(w, t_{\rm p})$.

6. Acknowledgements

Support of this work by the Ministry of Education of the Czech Republic (project No. MSM 0021620835), by the Grant Agency of the Charles University (grant No. 143610) and by the projects SVV – 2010 – 261 301, SVV – 2010 – 261 305 of the Charles University in Prague is gratefully acknowledged.

7. References

Allahverdyan, A. E., Johal, R. S. & Mahler, G. (2008). Work extremum principle: Structure and function of quantum heat engines, *Phys. Rev. E* 77(4): 041118.

Ambjörnsson, T., Lizana, L., Lomholt, M. A. & Silbey, R. J. (2008). Single-file dynamics with different diffusion constants, *J. Chem. Phys.* 129: 185106.

Ambjörnsson, T. & Silbey, R. J. (2008). Diffusion of two particles with a finite interaction potential in one dimension, *J. Chem. Phys.* 129: 165103.

Astumian, R. & Hänggi, P. (2002). Brownian motors, *Phys. Today* 55(11): 33.

Barkai, E. & Silbey, R. J. (2009). Theory of single-file diffusion in a force field, *Phys. Rev. Lett.* 102: 050602.

Baule, A. & Cohen, E. G. D. (2009). Fluctuation properties of an effective nonlinear system subject to Poisson noise, *Phys. Rev. E* 79(3): 030103.

Bochkov, G. N. & Kuzovlev, Y. E. (1981a). Nonlinear fluctuation-dissipation relations and stochastic models in nonequilibrium thermodynamics: I. Generalized fluctuation-dissipation theorem, *Physica A* 106: 443.

Bochkov, G. N. & Kuzovlev, Y. E. (1981b). Nonlinear fluctuation-dissipation relations and stochastic models in nonequilibrium thermodynamics: II. Kinetic potential and variational principles for nonlinear irreversible processes, *Physica A* 106: 480.

Chvosta, P., Einax, M., Holubec, V., Ryabov, A. & Maass, P. (2010). Energetics and performance of a microscopic heat engine based on exact calculations of work and heat distributions, *J. Stat. Mech.* p. P03002.

Chvosta, P. & Reineker, P. (2003a). Analysis of stochastic resonances, *Phys. Rev. E* 68(6): 066109.

Chvosta, P. & Reineker, P. (2003b). Diffusion in a potential with a time-dependent discontinuity, *Journal of Physics A: Mathematical and General* 36(33): 8753.

Chvosta, P., Schulz, M., Mayr, E. & Reineker, P. (2007). Sedimentation of particles acted upon by a vertical, time-oscillating force, *New. J. Phys.* 9: 2.

Chvosta, P., Schulz, M., Paule, E. & Reineker, P. (2005). Kinetics and energetics of reflected diffusion process with time-dependent and space-homogeneous force, *New. J. Phys.* 7: 190.

Crooks, G. E. (1998). Nonequilibrium measurements of free energy differences for microscopically reversible Markovian systems, *J. Stat. Phys.* 90(5–6): 1481–1487.

Crooks, G. E. (1999). Entropy production fluctuation theorem and the nonequilibrium work relation for free energy differences, *Phys. Rev. E* 60(3): 2721–2726.

Crooks, G. E. (2000). Path-ensemble averages in systems driven far from equilibrium, *Phys. Rev. E* 61(3): 2361–2366.

den Broeck, C. V., Kawai, R. & Meurs, P. (2004). Microscopic analysis of a thermal Brownian motor, *Phys. Rev. Lett.* 93(9): 090601.

Einax, M., Körner, M., Maass, P. & Nitzan, A. (2010). Nonlinear hopping transport in ring systems and open channels, *Phys. Chem. Chem. Phys.* 10: 645–654.

Esposito, M. & Mukamel, S. (2006). Fluctuation theorems for quantum master equations, *Phys. Rev. E* 73(4): 046129.

Evans, D. J., Cohen, E. G. D. & Morriss, G. P. (1993). Probability of second law violations in shearing steady states, *Phys. Rev. Lett.* 71(15): 2401–2404.

Gallavotti, G. & Cohen, E. G. D. (1995). Dynamical ensembles in nonequilibrium statistical mechanics, *Phys. Rev. Lett.* 74(14): 2694–2697.

Gammaitoni, L., Hänggi, P., Jung, P. & Marchesoni, F. (1998). Stochastic resonance, *Rev. Mod. Phys.* 70(1): 223–287.

Gillespie, D. T. (1992). *Markov Processes: an Introduction for Physical Scientists*, San Diego: Academic Press, San Diego.

Hänggi, P., Marchesoni, F. & Nori, F. (2005). Brownian motors, *Ann. Phys.* 14(11): 51–70.

Hänggi, P. & Thomas, H. (1975). Linear response and fluctuation theorems for nonstationary stochastic processes, *Z. Physik B* 22: 295–300.

Hänggi, P. & Thomas, H. (1977). Time evolution, correlations and linear response of non-Markov processes, *Z. Physik B* 26: 85–92.

Hatano, T. & Sasa, S. (2001). Steady-state thermodynamics of Langevin systems, *Phys. Rev. Lett.* 86(16): 3463–3466.

Henrich, M. J., Rempp, F. & Mahler, G. (2007). Quantum thermodynamic Otto machines: A spin-system approach, *Eur. Phys. J. Special Topics* 151: 157–165.

Imparato, A. & Peliti, L. (2005a). Work distribution and path integrals in general mean-field systems, *EPL* 70(6): 740–746.

Imparato, A. & Peliti, L. (2005b). Work probability distribution in single-molecule experiments, *EPL* 69(4): 643–649.

Imparato, A. & Peliti, L. (2005c). Work-probability distribution in systems driven out of equilibrium, *Phys. Rev. E* 72(4): 046114.

Jarzynski, C. (1997a). Equilibrium free-energy differences from nonequilibrium measurements: A master-equation approach, *Phys. Rev. E* 56(5): 5018–5035.

Jarzynski, C. (1997b). Nonequilibrium equality for free energy differences, *Phys. Rev. Lett.* 78(14): 2690–2693.

Jung, P. & Hänggi, P. (1990). Resonantly driven Brownian motion: Basic concepts and exact results, *Phys. Rev. A* 41(6): 2977–2988.

Jung, P. & Hänggi, P. (1991). Amplification of small signals via stochastic resonance, *Phys. Rev. A* 44(12): 8032–8042.

Kumar, D. (2008). Diffusion of interacting particles in one dimension, *Phys. Rev. E* 78: 021133.

Lizana, L. & Ambjörnsson, T. (2009). Diffusion of finite-sized hard-core interacting particles in a one-dimensional box: Tagged particle dynamics, *Phys. Rev. E* 80: 051103.

Maes, C. (2004). On the origin and use of fluctuation relations for the entropy, *in* J. Dalibard, B. Duplantier & V. Rivasseau (eds), *Poincaré Seminar 2003: Bose-Einstein condensation-entropy*, Birkhäuser, Basel, pp. 149–191.

Mayr, E., Schulz, M., Reineker, P., Pletl, T. & Chvosta, P. (2007). Diffusion process with two reflecting barriers in a time-dependent potential, *Phys. Rev. E* 76(1): 011125.

Mazonka, O. & Jarzynski, C. (1999). Exactly solvable model illustrating far-from-equilibrium predictions, *arXiv.org:cond-mat 9912121* .

Mossa, A., M. Manosas, N. F., Huguet, J. M. & Ritort, F. (2009). Dynamic force spectroscopy of

DNA hairpins: I. force kinetics and free energy landscapes, *J. Stat. Mech.* p. P02060.

Parrondo, J. M. R. & de Cisneros, B. J. (2002). Energetics of Brownian motors: a review, *Applied Physics A* 75: 179–191. 10.1007/s003390201332.

Reimann, P. (2002). Brownian motors: noisy transport far from equilibrium, *Phys. Rep.* 361: 57–265.

Risken, H. (1984). *The Fokker-Planck Equation. Methods of Solution and Applications*, Springer-Verlag, Berlin - Heidelberg - New York - Tokyo.

Ritort, F. (2003). Work fluctuations, transient violations of the second law, *Poincaré Seminar 2003: Bose-Einstein condensation-entropy*, Birkhäuser, Basel, pp. 195–229.

Rödenbeck, C., Kärger, J. & Hahn, K. (1998). Calculating exact propagators in single-file systems via the reflection principle, *Phys. Rev. E* 57: 4382.

Schuler, S., Speck, T., Tietz, C., Wrachtrup, J. & Seifert, U. (2005). Experimental test of the fluctuation theorem for a driven two-level system with time-dependent rates, *Phys. Rev. Lett.* 94(18): 180602.

Seifert, U. (2005). Entropy production along a stochastic trajectory and an integral fluctuation theorem, *Phys. Rev. Lett.* 95(4): 040602.

Sekimoto, K. (1999). Kinetic characterization of heat bath and the energetics of thermal ratchet models, *J. Phys. Soc. Jpn.* 66: 1234–1237.

Sekimoto, K., Takagi, F. & Hondou, T. (2000). Carnot's cycle for small systems: Irreversibility and cost of operations, *Phys. Rev. E* 62(6): 7759–7768.

Speck, T. & Seifert, U. (2004). Distribution of work in isothermal nonequilibrium processes, *Phys. Rev. E* 70(6): 066112.

Šubrt, E. & Chvosta, P. (2007). Exact analysis of work fluctuations in two-level systems, *J. Stat. Mech.* p. P09019.

Takagi, F. & Hondou, T. (1999). Thermal noise can facilitate energy conversion by a ratchet system, *Phys. Rev. E* 60(4): 4954–4957.

van Kampen, N. G. (2007). *Stochastic Process in Physics and Chemistry*, Elsevier.

Wilcox, R. M. (1967). Exponential operators and parameter differentiation in quantum physics, *J. Math. Phys.* 8: 962–982.

Wolf, F. (1988). Lie algebraic solutions of linear Fokker-Planck equations, *J. Math. Phys.* 29: 305–307.

9

Nonequilibrium Thermodynamics for Living Systems: Brownian Particle Description

Ulrich Zürcher
Physics Department, Cleveland State University
Cleveland, OH 44115
USA

1. Introduction

In most introductory physics texts, a discussion on (human) food consumption centers around the available work. For example, the altitude is calculated a person can hike after eating a snack. This connection is natural at first glance: food is burned in a *bomb calorimeter* and its energy content is measured in "Calories," which is the unit of heat. We say that "we go to the gym to burn calories." This discussion implies that the human body acts as a sort of "heat engine," with food playing the role of 'fuel.'

We give two arguments to show that this view is flawed. First, the conversion of heat into work requires a heat engine that operates between two heat baths with different temperatures T_h and $T_c < T_h$. The heat input Q_h can be converted into work W and heat output $Q_c < Q_h$ so that $Q_h = Q_c + W$ subject to the condition that entropy cannot be destroyed: $\Delta S = Q_h/T_h - Q_c/T_c > 0$. However, animals act like thermostats, with their body temperature kept at a constant value; e.g., 37°C for humans and $1-2$°C higher for domestic cats and dogs. Second, the typical diet of an adult is roughly 2,000 Calories or about 8 MJ. If we assume that 25% of caloric intake is converted into useable work, a 100-kg adult would have to climb about 2,000 m [or approximately the height of *Matterhorn* in the Swiss Alps from its base] to convert daily food intake into potential energy. While this calculation is too simplistic, it illustrates that caloric intake through food consumption is enormous, compared to mechanical work done by humans [and other animals]. In particular, the discussion ignores heat production of the skin. At rest, the rate of heat production per unit area is $\mathcal{F}/A \simeq 45\,\mathrm{W/m^2}$ (Guyton & Hall 2005). Given that the surface area of a 1.8-m tall man is about $A \simeq 2\,\mathrm{m}^2$, the rate of energy conversion at rest is approximately 90 W. Since $1\,\mathrm{d} \simeq 9 \times 10^4\,\mathrm{s}$, we find that the heat dissipated through the skin is $\mathcal{F} \simeq 8\,\mathrm{MJ/d}$, which approximately matches the daily intake of 'food calories.'

An entirely different focus of food consumption is emphasized in physiology texts. All living systems require the input of energy, whether it is in the form of food (for animals) or sun light (for plants). The chemical energy content of food is used to maintain concentration gradients of ions in the body, which is required for muscles to do useable work both inside and outside the body. Heat is the product of this energy transformation. That is, food intake is in the form of Gibbs free energy, i.e., work, and entropy is created in the form of heat and other waste products. In his classic text *What is Life?*, Schrödinger coined the expression that living systems "feed on negentropy" (Schrödinger 1967). Later, Morowitz explained that the steady state of living systems is maintained by a constant flow of energy: the input is highly organized energy [work], while the output is in the form of disorganized energy, and entropy

is produced. Indeed, energy flow has been identified as one of the principles governing all complex systems (Schneider & Sagan 2005).
As an example of the steady-state character of living systems with non-zero-gradients, we discuss the distribution of ions inside the axon and extracellular fluid. The ionic concentrations inside the axon c_i and in the extracellular fluid c_o are measured in units of millimoles per liter (Hobbie & Roth 2007):

Ion	c_i	c_o	c_o/c_i
Na^+	15	145	9.7
K^+	150	5	0.0033
Cl^-	9	125	13.9
Misc.$^-$	150	30	0.19

In thermal equilibrium, the concentration of ions across a cell membrane is determined by the Boltzmann-Nernst formula, $c_i/c_o = \exp[-ze(v_i - v_o)/k_B T]$, where $\Delta G = ze(v_i - v_o)$ is the Gibbs free energy for the potential between the inside and outside the cell, $\Delta v = v_i - v_o$. If the electrostatic potential in the extracellular fluid is chosen $v_o = 0$, the 'resting' potential inside the axon is found $v_i = -70\,\text{mV}$. For $T = 37°\text{C}$, this gives $c_i/c_o = 13.7$ and $c_i/c_o = 1/13.7 = 0.073$ for univalent positive and negative ions, respectively. That is, the sodium concentration is too low inside the axon, while there are too many potassium ions inside it. The concentration of chlorine is approximately consistent with thermal equilibrium. Non-zero gradients of concentrations and other state variables are characteristic for systems that are not in thermal equilibrium (Berry *et. al.* 2002).
A discussion of living and complex systems within the framework of physics is difficult. It must include an explanation of what is meant by the phrase "biological systems are in nonequilibrium stationary states (NESS)." This is challenging, because there is not a unique definition of 'equilibrium state;' rather entirely different definitions are used to describe closed and open systems. For a closed system, the equilibrium state can be characrterized by a (multi-dimensional) coordinate x_s, so that $x \neq x_s$ describes a nonequilibrium state. However, the notion of "state of the system" is far from obvious for open systems. For a population model in ecology, equilibrium is described by the number of animals in each species. A nonequilibrium state involves populations that are changing with time, so a 'nonequilibrium stationary state' would correspond to dynamic state with constant (positive or negative) growth rates for species. Thus, any discussion of nonequilibrium thermodynamics for biological systems must involve an explanation of 'state' for complex systems. For many-body systems, the macroscopic behavior is an "emergent behavior;" the closest analogue of 'state' in physics might be the order parameter associated with a broken symmetry near a second-order phase transition.
This chapter is not a comprehensive overview of nonequilibrium thermodynamics, or the flow of energy as a mechanism of pattern formation in complex systems. We begin by directing the reader to some of the texts and papers that were useful in the preparation of this chapter. The text by de Groot and Mazur remains an authoritative source for nonequilibrium thermodynamics (de Groot & Mazur 1962). Applications in biophysics are discussed in Ref. (Katchalsky & Curram 1965). The text by Haynie is an excellent introduction to biological thermodynamics (Haynie 2001). The texts by Kubo and coworkers are an authoritative treatment of equilibrium and nonequilibrium statistical mechanics (Toda *et al* 1983; Kubo *et al* 1983). Stochastic processes are discussed in Refs. (Wax 1954; van Kampen 1981). Sethna gives a clear explanation of complexity and entropy

(Sethna 2006). Cross and Greenside overview pattern formation in dissipative systems (Cross & Greenside 2009); a non-technical introduction to pattern formation is found in Ref. (Ball 2009). The reader is directed to Refs. (Guyton & Hall 2005) and (Nobel 1999) for background material on human and plant physiology. Some of the physics underlying human physiology is found in Refs. (Hobbie & Roth 2007; Herman 2007).

The outline of this paper is as follows. We discuss the meaning of state and equilibrium for closed systems. We then discuss open systems, and introduce the concept of order parameter as the generalization of "coordinate" for closed systems. We use the motion of a Brownian particle to illustrate the two mechanisms, namely fluctuation and dissipation, how a system interacts with a much larger heat bath. We then briefly discuss the Rayleigh-Benard convection cell to illustrate the nonequilibrium stationary states in dissipative systems. This leads to our treatment of a charged object moving inside a viscous fluid. We discuss how the flow of energy through the system determines the stability of NESS. In particular, we show how the NESS becomes unstable through a seemingly small change in the energy dissipation. We conclude with a discussion of the key points and a general overview.

2. Closed systems

The notion of 'equilibrium' is introduced for mechanical systems, such as the familiar mass-block system. The mass M slides on a horizontal surface, and is attached to a spring with constant k, cf. Fig. (1). We choose a coordinate such that $x_{eq} = 0$ when the spring force vanishes. The potential energy is then given by $U(x) = kx^2/2$, so that the spring force is given by $F_{sp}(x) = -dU/dx = -kx$. In Fig (1), the potential energy $U(x)$ is shown in black.

If the coordinate is constant, $x_{ns} = \text{const} \neq 0$, the spring-block system is in a *nonequilibrium stationary state*. Since $F_{sp} = -dU/dx|_{ns} \neq 0$, an external force must be applied to maintain the system in a steady state: $F_{net} = F_{sp} + F_{ext} = 0$. If the object with mass M also has an electric charge q, this external force can be realized by an external electric field E, $F_{ext} = qE$.

The external force can be derived from a potential energy $F_{ext} = -dU_{ext}/dx$ with $U_{ext} = -qEx$, and the spring-block system can be enlarged to include the electric field. Mathematically. this is expressed in terms of a total potential energy that incorporates the interaction with the electric field: $U \to U' = U + U_{ext}$, where

$$U'(x) = \frac{1}{2}kx^2 - F_{ext}x = \frac{1}{2}k(x - x_{ns})^2 - \frac{(qE)^2}{4k}. \tag{1}$$

The potential $U'(x)$ is shown in red. That is, the nonequilibrium state for the potential $U(x)$, x_{ns} corresponds to the equilibrium state for the potential $U'(x)$, x'_s:

$$x_{ns} = x'_s = \frac{qE}{k}. \tag{2}$$

That is, the nonequilibrium stationary state for the spring-block system is the equilibrium state for the enlarged system. We conclude that for closed systems, the notion of equilibrium and nonequilibrium is more a matter of choice than a fundamental difference between them.

For a closed system, the signature of *stability* is the oscillatory dynamics around the equilibrium state. Stability follows if the angular frequency ω is real:

$$\omega^2 = \frac{1}{M}\frac{d^2U}{dx^2} > 0 \quad \text{(stability)}, \tag{3}$$

That is, stability requires that the potential energy is a *convex* function. Since $d^2U/dx^2 = d^2U/dx^2$, the stability of the system is not affected by the inclusion of the external electric force. On the other hand, if the angular frequency is imaginary $\omega = i\tilde{\omega}$, such that

$$\frac{d^2U}{dx^2} < 0 \quad \text{(instability)}. \tag{4}$$

The corresponding potential energy is shown in Fig. (2). The solution of the equation of motion describes exponential growth. That is, a small disturbance from the stationary state is amplified by the force that drives the system towards smaller values of the potential energy for all initial deviations from the stationary state $x = 0$,

$$x(t) \longrightarrow \pm\infty, \quad t \longrightarrow \infty; \tag{5}$$

the system is dynamically unstable. We conclude that a *concave* potential energy is the condition for instability in closed systems.

3. Equilibrium thermodynamics

Open systems exchange energy (and possibly volume and particles) with a heat bath at a fixed temperature T. The minimum energy principle applies to the internal energy of the system, rather than to the potential energy. This principle states that "the equilibrium value of any constrained external parameter is such as to minimize the energy for the given value of the total energy" (Callen 1960). A thermodynamic description is based on entropy, which is a concave function of (constrained) equilibrium states. In thermal equilibrium, the extensive parameters assume value, such that the entropy of the system is *maximized*. This statement is referred to as *maximum entropy principle* [MEP]. The stability of thermodynamic equilibrium follows from the concavity of the entropy, $d^2S < 0$.
Thermodynamics describes average values, while fluctuations are described by equilibrium statistical mechanics. The distribution of the energy is given by the Boltzmann factor $p(E) = Z^{-1}\exp(-E/k_BT)$, where $Z = \int \exp(-E/k_BT)dE$ is the partition function. The equilibrium value of the energy of the system is equal to the average value, $E_{\text{eq}} = \langle E \rangle = \int dEp(E)E$. The fluctuations of the energy are $\delta E = E - \langle E \rangle$. The mean-square fluctuations can be written $\langle [\delta E]^2 \rangle = k_BT^2 \cdot d\langle E \rangle /dT$, or in terms of the inverse temperature $\beta = 1/T$, $\langle [\delta E]^2 \rangle = -k_B d\langle E \rangle /d\beta$. Thus, the variance of energy fluctuations $\langle [\delta E]^2 \rangle$ is proportional to the response of the systems $d\langle E \rangle /d\beta$. The proportionality between fluctuations and dissipation is determined by the Boltzmann constant $k_B = 1.38 \times 10^{-23}$ J/K. Einstein discussed that "the absolute constant k_B (therefore) determines the thermal stability of the system. The relationship just found is particularly interesting because it no longer contains any quantity that calls to mind the assumption underlying the theory" (Klein 1967).
In general, the state of an open system is described by an order parameter η. This concept is the generalization of coordinates used for closed systems, and was introduced by Landau to describe the properties of a system near a second-order phase transition (Landau & Lifshitz 1959a). For the Ising spin model, for example, the order parameter is the average the average magnetization (Chaikin & Lubensky 1995). In general, the choice of order parameter for a particular system is an "art" (Sethna 2006).
For simplicity, we assume a spatially homogenous system, so that $\eta(\vec{x}) = \text{const}$ and there is no term involving the gradient $\nabla\eta$. The order parameter can be chosen such that $\eta = 0$ in the symmetric phase. The thermodynamics of the system is defined by the Gibbs free energy

$G = G(\eta)$. The equation of state follows from the expression for the external field $h = \partial G/\partial\eta$. The susceptibility $\chi = \partial \langle\eta\rangle / \partial h$ characterizes the response of the system. In the absence of an external magnetic field, the appearance of a non-zero value of the order parameter is referred to as *spontaneous symmetry breaking* (Chaikin & Lubensky 1995; Forster 1975). The Gibbs free energy is written as a power series $G(\eta) = G_0 + A\eta^2 + B\eta^4$ with $B > 0$. The symmetric phase $\eta = 0$, corresponds to $A > 0$, while $A < 0$ in the asymmetric phase. The second-order coefficient vanishes at the transition point $A = 0$. We only consider the case when the system is away from the transition point, so that $A \neq 0$, and write $\bar{\eta} = 0$ and $\bar{\eta} = \pm A/2B$ for the respective minima of the Gibbs free energy, respectively. Since $\partial^2 G/\partial\eta^2\big|_{\eta=\bar{\eta}} = \chi^{-1} > 0$, the susceptibility is finite and the variance of fluctuations of the order parameter is finite as well, with $\langle[\eta - \bar{\eta}]^2\rangle \sim \chi k_B T$, which is referred to as *fluctuation-dissipation theorem* [FDT].

A Brownian particle can be used to illustrate some aspects of equilibrium statistical mechanics (Forster 1975). In a microscopic description, a heavy particle with mass M is immersed in a fluid of lighter particles of mass $m < M$. The time evolution is described by the Liouville operator for the entire system, and projection operator methods are used to eliminate the lighter particles' degrees of freedom [i.e., the heat bath]. It is shown that the interaction with a heat bath results in dissipation, described by a memory function and fluctuations characterized by stochastic forces. Because these two contributions have a common origin, it is not surprising that they are related to each other: the memory function is proportional to the autocorrelation function of the stochastic forces. The average kinetic energy of the heavy particle is given by the equipartition principle: $(M/2)\langle v^2\rangle = k_B T/2$. The memory function defines a correlation time ζ^{-1}. For times $t > \zeta^{-1}$, a Langevin equation for the velocity of the heavy particle follows (Wax 1954). In one spatial dimension,

$$\frac{\partial}{\partial t} v(t) + \zeta v(t) = \frac{1}{M}\zeta(t). \tag{6}$$

We have the averages $\langle f(t)\rangle = 0$ and $\langle f(t)v\rangle = 0$. In Eq. (6), the stochastic force ζ is Gaussian "white noise:"

$$\langle\zeta(t)\zeta(t')\rangle = 2\zeta M k_B T \delta(t - t'). \tag{7}$$

The factor $2\zeta M k_B T$ follows from the requirement that the stochastic process is "stationary." In fact, following Kubo, Eq. (7) is sometimes called the 'second fluctuation-dissipation theorem.' For long times, $t >> \zeta^{-1}$, the mean-square displacement increases *diffusively*:

$$\left\langle [x(t) - x(0)]^2 \right\rangle = \frac{2k_B T}{M\zeta} t = 2Dt. \tag{8}$$

The expression for the diffusion constant $D = k_B T/M\zeta$ is the Einstein relation for Brownian motion, and is a version of the fluctuation-dissipation theorem. The diffusion constant can be written in terms of the velocity autocorrelation, $D = \int_0^\infty dt \langle\vec{v}(t)\vec{v}(0)\rangle$.

If the Brownian particle moves in a harmonic potential well, $U(x) = M\omega_0^2 x^2/2$, the Langevin equation is written as a system of two coupled first-order differential equations: $dx/dt = v$ and $dv/dt + \zeta v + \omega_0^2 x = \zeta/M$. If the damping constant is large, the inertia of the particle can be ignored, so that the coordinate is described by the equation: $dx/dt + (\omega_0^2/\zeta)x = \zeta/M$. At zero temperature, the stochastic force vanishes, and the deterministic time evolution of the coordinate describes its relaxation towards the equilibrium $x = 0$: $dx/dt = -(\omega_0^2/\zeta)x$ so that $x(t) = x_0 \exp[-(\omega_0^2/\zeta)t]$.

In general, the relaxation of an initial nonoequilibrium state is governed by Onsager's regression hypothesis: the decay of an initial nonequilibrium state follows the same law as that of spontaneous fluctuations (Kubo *et al* 1983). The fluctuation-dissipation theorem implies that the time-dependence of equilibrium fluctuations is governed by the minimum entropy production principle. The stochastic nature of time-depedent equilibrium fluctuations is characterized by the conditional probability, or propagator $P(x,t|x_0,t_0)$. Onsager and Machlup showed that the conditional probability, or propagator, for diffusion can be written as a path integral (Hunt *et. al.* 1985):

$$P(x,t|x_0,t_0) = \int \mathcal{D}[x(t)] \exp\left[-\frac{\zeta}{2k_B T} \int_{t_0}^{t} K(s)ds \right], \tag{9}$$

where $K(t) = (M/2)(dx/dt)^2$ is the kinetic energy of the Brownian particle. The action $\int_{t_0}^{t} K(s)ds$ is minimized for the deterministic path (Feynman 1972). For fixed start (x_0,t_0) and endpoints (x,t), we find $K = (M/2)[(x-x_0)/(t-t_0)]^2$. Gaussian fluctuations around the deterministic path yields: $P(x,t|x_0,t_0) = (4\pi D[t-t_0])^{-1/2} \exp[-(x-x_0)^2/4D(t-t_0)]$, which is the Greens function for the diffusion equation in one dimension $\partial P/\partial t = D\partial^2 P/\partial x^2$ subject to the initial condtion $P(x,t_0|x_0,t_0) = \delta(x-x_0)$.

4. Systems far from equilibrium

We conclude that dissipation tends to 'dampen' the oscillatory motion around the equilibrium value $\bar{\eta}$, so that $\lim_{t\to\infty} \langle \eta(t) \rangle = \bar{\eta}$. Thus, a nonequilibrium stationary state $\eta_s \neq \bar{\eta}$ requires the input of energy through work done on the system: highly-organized energy is destroyed, and dissipated energy is associated with the production of heat.

This mechanism is often illustrated by the Rayleigh-Benard convection cell, with a fluid being placed between two horizontal plates. If the two plates are at the same temperature, there is no macroscopic fluid flow, and the system is in the symmetric phase. An energy input is used to maintain a constant temperature difference across the plates. If ΔT is large enough, the component of the velocity along the vertical is non-zero, $v_z \neq 0$. A state with $v_z \neq 0$ is the asymmetric state of the fluid. Stationary patterns such as "stripes" and "hexagons" develop inside the fluid. Thus, the temperature difference ΔT can be viewed as the "force" maintaining stationary patterns in the fluid. Swift and Hohenberg showed that a potential $V(u)$ can be defined, such that the different stationary patterns correspond to local potential minima, cf. Fig (3). The dynamic of the system is first-order in time $\partial u/\partial t = -\delta V/\delta u$, where $\delta/\delta u$ is the functional derivative. If this energy flow stops, the velocity field in the fluid dissipates and the nonequilibrium patterns disappear.

A careful study of this system provides important insights into the behavior of nonequilibrium systems. Here, we are interested in systems for which nonequilibrium states are characterized by non-zero values of dynamic variables. A particularly simple model is discussed in Ref. (Taniguchi and Cohen 2008): a Brownian particle immersed in a viscous fluid moves at a constant velocity under the the influence of an electric force. The authors refer to it as, a Brownian particle immersed in a fluid "NESS model of class A." This model was used earlier by this author to illustrate nonequilibrium stationary states (Zurcher 2008). We note, however, that this model is not appropriate to discuss important topics in nonequilibrium thermodynamics, such as pattern formation in driven-diffusive systems.

5. Nonequilibrium stationary states: brownian particle model

Our system is a particle with mass M and the "state" of the system is characterized by the velocity v. The kinetic energy plays the role of Gibbs free energy:

$$K(v) = \frac{M}{2}v^2. \tag{10}$$

The particle at rest $v = 0$ corresponds to the equilibrium state, while $v = \text{const}$ in a nonequilibrium state. We assume that the particle has an electric charge q, so that an external force is applied by an electric field $F_{\text{ext}} = qE$. Under the influence of this electric force, the (kinetic) energy of the particle would grow without bounds, $K(t) \to \infty$ for $t \to \infty$. The coupling to a 'thermostat' prevents this growth of energy. Here, we use a velocity-dependent force to describe the interaction with a thermostat. In the terminology of Ref. (Gallavotti & Cohen 2004), our model describes a *mechanical* thermostat.

Dissipation is described by velocity-dependent forces, $f = f(v)$. For a particle immersed in a fluid, the force is linear in the velocity $f_l = 6\pi a\kappa v$ for viscous flow, while turbulent flow leads to quadratic dependence $f_t = C_0\pi\rho a^2v^2/2$ for turbulent flow (Landau & Lifshitz 1959b). Here, κ is the dynamic viscosity, ρ is the density of the fluid, and a is the radius of the spherical object. These two mechanisms of dissipation are generally present at the same time; the Reynolds number determines which mechanism is dominant. It is defined as the ratio of inertial and viscous forces, i.e., $\text{Re} = f_t/f_l = \rho a v/\kappa$. Laminar flow applies to slowly moving objects, i.e., small Reynolds numbers ($\text{Re} < 1$), while turbulent flow dominates at high speeds, i.e., large Reynolds numbers ($\text{Re} > 10^5$).

In the stationary state, the velocity is constant so that the net force on the particle vanishes, $F_{\text{net}} = qE - f = 0$. We find for laminar flow,

$$v_s = \frac{qE}{6\pi\kappa a}, \quad \text{Re} < 1, \tag{11}$$

and for turbulent flow

$$v_s = \sqrt{\frac{qE}{C_0\pi\rho a^2}}, \quad Re > 10^5. \tag{12}$$

In either case, we have $F_{\text{net}} > 0$ for $v < v_s$ and $F_{\text{net}} < 0$ for $v > v_s$; we conclude that the steady state is *dynamically stable*. These are, of course, elementary results discussed in introductory texts, where the nonequilibrium stationary state v_s is referred to as "terminal speed."

In general, the "forces" acting on a complex system are not known, so that the time evolution cannot be derived from a (partial-) differential equation. We will show how a discrete version of the equation can be derived from energy fluxes. To this end, we recall that in classical mechanics, velocity-dependent forces enter via the appropriate Rayleigh's dissipation function (Goldstein 1980). We deviate from the usual definition and define $\mathcal{F}$ as the negative value of the dissipation function such that $f = \partial\mathcal{F}/\partial v$, and $\mathcal{F}$ is associated with entropy production in the fluid. If the Lagrangian is not an explicit function of time, the total energy of the system E decreases, $dE/dt = -\mathcal{F}$. For laminar flow, we have

$$\mathcal{F}_l = 3\pi a\kappa v^2 \quad \text{(laminar flow)}, \tag{13}$$

and for turbulent flow

$$\mathcal{F}_t = \frac{C_0\pi}{3}\rho a^2 v^3, \quad \text{(turbulent flow)}. \tag{14}$$

The Reynolds number can be expressed as a ratio of the dissipation functions, Re $\sim \mathcal{F}_t/\mathcal{F}_l$. Since $\mathcal{F}_l > \mathcal{F}_t$ for $v \to 0$ and $\mathcal{F}_t > \mathcal{F}_l$ for $v \to \infty$, we conclude that the dominant mechanism for dissipation in the fluid maximizes the production of entropy.
The loss of energy through dissipation must be balanced by energy input in highly organized form, i.e., work for a Brownian particle. We write $dW = qEdx$ for the work done by the electric field, if the object moves the distance dx parallel to the electric field. Since $v = dx/dt$, the energy input per unit time follows,

$$\frac{dW}{dt} = \mathcal{W} = qEv. \tag{15}$$

We thus have for the rate of change of the kinetic energy,

$$\frac{dK}{dt} = \mathcal{W} - \mathcal{F}, \tag{16}$$

cf. Ref. (Zurcher 2008). This is equivalent to Newton's second law for the object. In Fig. (5), we plot $\mathcal{F}$ (black) and $\mathcal{W}$ (in blue) as a function of the velocity v. The two curves intersect at v_s, so that $\mathcal{F} = \mathcal{W}$, and the kinetic energy of the particle is constant $dK/dt = 0$. We conclude that v_s corresponds to the nonequilibrium stationary state of the system, cf. Fig. (4).
The energy input exceeds the dissipated energy, $\mathcal{F} > \mathcal{W}$, for $0 < v_0 < v_s$ so that the excess $\mathcal{W} - \mathcal{F}$ drives the system towards the stationary state $v_0 \to v_s$. For $v_0 > v_s$, the dissipated energy is higher than the input, $\mathcal{W} > \mathcal{F}$, so that the excess damping drives the system towards v_s. That is, the nonequilibrium stationary state is stable,

$$v_0 \longrightarrow v_s. \tag{17}$$

This result is independent of detailed properties of the open system. For $\partial^2 W/\partial v^2 = 0$, the stability is a consequence of the convexity of the dissipative function

$$\frac{d^2\mathcal{F}}{dv^2} \geq 0. \tag{18}$$

While a mechanical thermostat allows for a description of the system's time evolution in terms of forces, this is not possible for an open system in contact with an arbitrary thermostat. Indeed, a discrete version of the dynamics can be found from the energy fluxes $\mathcal{W}$ and $\mathcal{F}$. We assume that the particle moves at the initial velocity $0 < v_0 < v_s$, so that $\mathcal{F}_0 > \mathcal{W}_0$. We keep the energy input fixed, and increase the velocity until the dissipated energy matches the input $\mathcal{W}_1 = \mathcal{F}_0$ at the new velocity $v_1 > v_0$. This first iteration step is indicated by a horizontal arrow in Fig (5). The energy input is now at the higher value $\mathcal{W}_1 > \mathcal{W}_0$, indicated by the vertical arrow. By construction, the inequality $\mathcal{W}_1 > \mathcal{F}_1$ holds, so that the procedure can be repeated to find the the second iteration, v_2, cf. Fig. (5). A similar scheme applies for $v_s < v_0 < \infty$. We find the sequence $\{v_i\}_i$ for $i = 0, 1, 2, ...$ with $v_{i+1} < v_i$ so that $\lim_{i\to\infty} v_i = v_s$. Thus, for both $v_0 < v_s$ and $v_0 > v_s$, the initial state converges to the stationary state,

$$v_0 \longrightarrow v_s. \tag{19}$$

We obtain a graphical representation of the dynamics by plotting the velocity v_{i+1} versus v_i. This is sometimes called a cobweb or Verhulst plot (Otto & Day 2007). The discrete version of the time evolution is indicated by the arrows, which shows that v_s is the fixed point of the time evolution, cf. Fig. (6).

We generalize this result to the case when the dissipation function is *concave*,

$$\frac{\partial^2 \mathcal{F}}{\partial v^2} < 0, \tag{20}$$

while keeping the behavior of the energy input fixed, i.e., $\partial^2 W/\partial v^2 = 0$. We assume power law behavior for the dissipative function, so that concave behavior implies

$$\mathcal{F} \sim v^{\alpha}, \quad 0 < \alpha < 1. \tag{21}$$

Since $f = \partial\mathcal{F}/\partial v \sim 1/v^{1-\alpha}$, the velocity-dependent force diverges as the system slows down, i.e., $f \to \infty$ for $v \to 0$. The concave form of the dissipation function is unphysical for $v < v_c$, where v_c is a cutoff. We ignore this cutoff in the following discussion. While this velocity-dependent function does not correspond to the behavior of any fluid, we retain the language appropriate for a Brownian particle. We now have the plot of the energy fluxes $\mathcal{F}$ and $\mathcal{W}$ as a function of velocity v, cf. Fig. (7).
The two curves intersect at the velocity v_s which characterizes the stationary state of the system. In this case, the dissipated energy exceeds the energy input for $0 < v_0 < v_s$, so that the excess dissipation drives the system towards the equilibrium state $v = 0$. For $v_0 > v_s$, the energy input is not balanced by the dissipated energy $\mathcal{W} > \mathcal{F}$. It follows that the excess input $\mathcal{W} - \mathcal{F}$ drives the state of the system away from the nonequilibrium stationary state.
We follow the same procedure as above, and assume that the initial velocity is (slightly) less than the stationary value, $0 < v_0 < v_s$ so that $\mathcal{F}_0 > \mathcal{W}_0$. We keep the energy input fixed, and decrease the velocity until $\mathcal{F}_1 = \mathcal{W}_0$ at the velocity $v_1 < v_0$. The iteration $v_0 \to v_1$ is indicated by a horizontal arrow. We now have $\mathcal{F}_1 > \mathcal{W}_1$, so that the steps can be repeated, cf. Fig. (6).
In the case $v_0 > v_s$, we have $\mathcal{W}_0 > \mathcal{F}_0$ so that the damping is not sufficient to act as a sink for the energy input into the system. Thus, the kinetic energy of the Brownian particle and the velocity increases, $v_1 > v_0$. This step is indicated by a horizontal arrow. Since $\mathcal{W}_1 > \mathcal{W}_0$, we find $\mathcal{W}_1 > \mathcal{F}_1$, and the step can be repeated to find $v_2 > v_1$. The corresponding phase portrait is shown in Fig. (7).
We conclude that the nonequilibrium stationary state v_s is *unstable* when the the dissipation function is *concave*. For $v_0 < v_s$, the initial state relaxes the towards the equilibrium state of the system,

$$v_0 \longrightarrow 0, \;\; v_0 < v_s, \tag{22}$$

while for $v_0 > v_s$, we find a runaway solution,

$$v_0 \longrightarrow \infty, \quad v_0 > v_s. \tag{23}$$

This instability is unique for nonequilibrium systems, and does not correspond to any behavior found for equilibrium systems. In fact, equilibrium thermodynamics excludes instabilities, because it is defined only for systems near local minima of the (free) energy. Exceptions are systems near a critical point, for which the free energy has a local maxima in the symmetric phase, and fluctuations diverge algebraically.

6. Discussion

A Brownian particle moving in a potential well can be used toexplain some aspects of equilibrium statistical physics. We used this model to explain certain aspects of nonequilibrium thermodynamics. A nonequilibrium stationary state corresponds to the particle moving at a constant velocity, under the influence of aexternal force. We also used this model to show how a NESS is sustained by a constant energy flow through the system. It is believedthat this is a key principle for steady states in open and complex systems (Schneider & Sagan 2005); however, the behavior of open systems cannot be explained by the second law of thermodynamics alone (Farmer 2005; Callendar 2007).

We started from a heavy sphere immersed in a viscous fluid, so that in general, both viscous and laminar forces are acting on the sphere. Laminar flow applies to slowly moving spheres, whereas turbulent flow applies when spheres are moving fast. The crossover between linear and quadratic velocity-dependent forcesis based on the Reynolds number. We showed that this criterion coincides with maximum entropy production: laminar and turbulent flows are the dominant mechanisms for entropy production at small and large flow speeds, respectively. Ifa generalized version of Onsager's regression hypothesis holds for driven diffusive systems, the analysis of competing mechanismsfor entropy production may shed insight into the origin of the MEP principle. This principle was proposed as the generalization of Onsager's regression theorem to fluctuations in nonequilibrium systems (Martyushev & Seleznev 2006; Niven 2009). MEP has beenused to explain complex behavior in ecology (Rhode 2005), earth science, and meteorology (Kleidon & Lorenz 2005).

For the Brownian particle immersed in a fluid, the dissipation function is convex, $\partial^2\mathcal{F}/\partial v^2$, and the NESS is dynamically stable. That is, $v_0 \to v_s$ for arbitrary initial velocity. We generalized our discussion to more open systems, in which the particle velocity would correspond to a growth rate. We considered thecase whenentropy production is associated with a concave dissipation function $\partial^2\mathcal{F}/\partial v^2 < 0$, and found that the NESS is dynamically unstable:The system either relaxes towards the equilibrium state$v_0 \to 0$, or approaches a runaway solution, $v_0 \to \infty$. The dynamics of an open system can be changed from stable to unstable by a variation in the dissipation function, or in the entropy production. Changes in metabolic rates have also been associated with disease (Macklem 2008): a decrease in the metabolic rate has been linked with a decrease in heart rate fluctuations in myocardial ischemia, while an increase in metabolic rate may be related to asthma.

It has also been proposed that the economy of a country or region can be considered an open system (Daley 1991), where an economy growing at a fixed rate, i.e., change of gross domestic power [GDP] per year would be in a nonequilibrium stationary state, or steady state. The population increase would correspond to an external force, while 'inefficiencies' such as wars, contribute to entropy production. If the analogue of a dissipation function for an economic system is concave, it might explain why monetary policies often fail to achieve stable growth of GDP over a sustained period. It would suggest that fluctuations of socio-economic variables are important since they can drive the system away from its steady state.

7. References

[Ball 2009] Ball, P. (2009). *Nature's Patterns:A Tapestery in Three Parts* (Oxford University Press, New York).

[Berry *et. al.* 2002] Berry, R. S.; Rice, S. A. & Ross, J. (2002). *Matter in Equilibrium:Statistical Mechanics and Thermodynamics* (Oxford University Press, New York).

[Callen & Welton 1951] Callen, H. B. & Welton, T. A. (1951). Phys. Rev. 83, 34- 40.

[Callen 1960] Callen, H. B. (1960). Thermodynamics (J. Wiley & Sons, New York).

[Callendar 2007] Callendar, C. (2007). *Not so cool.* Metascience 16, 147-151.

[Camazine *et. al.* 2001] Camazine, S.; Deneubourg, J.-L; Franks, N. R.; Sneyd, J.; Theraulaz, G. & Bonabeau, E. (2001). *Self-Organization in Biological Systems* (Princeton University

Press, Princeton, NJ).
[Chaikin & Lubensky 1995] Chaikin, P. M. & Lubensky, T. M. (1995). *Principles of Condensed Matter Physics* (Cambridge University Press, Cambridge, UK).
[Cross & Greenside 2009] Cross M. & Greenside, H. (2009). Pattern Formation and Dynamics in Nonequilibrium Systems. (Cambridge University Press, Cambridge, UK).
[Daley 1991] Dayley H. E. (1991). *Steady-State Economics: Second Edition With New Essays* (Island Press, Washington, DC).
[de Groot & Mazur 1962] deGroot, S. R. & Mazur, P. (1962). *Non-Equilibrium Thermodynamics* (North-Holland, Amsterdam).
[Farmer 2005] Farmer, J. D. (2005) *Cool is not enough*. Nature 436, 627-628.
[Feynman 1972] Feyman, R. P. (1972). *Statistical Mechanics: A Set of Lectures* (Benjamin Cummings,Reading, MA).
[Forster 1975] Forster, D. (1975). *Hydrodynamic Fluctuations, Broken Symmetry, and Correlation Functions* (Addison Wesley, Reading, MA).
[Gallavotti & Cohen 2004] Gallavotti G. & Cohen, E. G. D. (2004). *Nonequilibrium stationary states and entropy*. Phys. Rev. E 69 035104.
[Goldstein 1980] Goldstein, H. (1980). *Classical Mechanics*, 2nd Ed. (Addison-Wesley, Reading, MA).
[Guyton & Hall 2005] Guyton, A. C & Hall. J. C. (2005) *Textbook of Medical Physiology* (Elsevier Saunder, Amsterdam).
[Haynie 2001] Haynie, D. T. (2001). *Biological Thermodynamics* (Cambridge University Press, New York).
[Herman 2007] Herman, I. P. (2007). *Physics of the Human Body*. (Springer-Verlag, New York).
[Hobbie & Roth 2007] Hobbie, R. K. & Roth, B. J. (2007). *Intermediate Physics for Medicine and Biology* 4th Ed. (Springer-Verlag, New York).
[Hunt *et. al.* 1985] Hunt, K. L. C.; Hunt, P. M. & Ross, J. (1985). *Path Integral Methods in Nonequilibrium Chemical Thermodynamics:Numerical Tests of the Onsager-Machlup-Laplace Approximation and Analytic Continuation Techniques*. In *Path Integrals from meV to MeV*: Gutzwiller, M.C.; Inomata, A.; Klauder, J. R.; & Streit, L. Ed. (World Scientific, Singapore).
[Katchalsky & Curram 1965] Katchalsky, A. & Curram, P. F. (1965). *Nonequilibrium Thermodyanmics in Biology* (Harvard University Press, Cambridge, MA).
[Kleidon & Lorenz 2005] Kleidon, A. & Lorenz, R. D. (eds.) (2005). *Non-equilibrium Thermodynamics and the Production of Entropy: Life, Earth, and Beyond* (Springer-Verlag, New York).
[Klein 1967] A. Einstein as quoted in Klein, M. J. (1967). *Thermodynamics in Einstein's Thought*. Science 157, 509-516.
[Kubo *et al* 1983] R. Kubo, M. Toda and N. Hashitsume (1983). *Statistical Physics II:Nonequilibrium Statistical Mechanics* (Springer-Verlag, New York).
[Landau & Lifshitz 1959a] Landau, L. D. &Lifshitz, E. M. (1959). *Statistical Mechanics* (Pergamon Press, London).
[Landau & Lifshitz 1959b] Landau, L. D. & Lifshitz, E. M. (1959). *Hydrodynamics* (Pergamon Press, London).
[Macklem 2008] Macklem P. (2008). *Emergent Phenomena and the secrets of Life*. J. App. Physiol. 104: 1844-1846.
[Martyushev & Seleznev 2006] Martyushev, L. M. & Seleznev, V. D. (2006). *Maximum entropy production principle in physics, chemistry, and biology*. Physics Reports 426,1-45.

[Morowitz 1968] Morowitz, H. J. (1968). *Energy flow in biology* (Academic Press, New York).

[Nobel 1999] Nobel, P. S. *Physicochemical & Enviromnetal Plant Physiology* 2nd Ed. (Academic Press, San Diego, CA).

[Niven 2009] Niven, R. K. (2009). Steady state of a dissipative flow-controlledsystem andthe maximum entropy production principle.Phys. Rev. E 80, 021113.

[Onsager 1931] Onsager, L. (1931). *Reciprocal Relations in Irreversible Processes I*, Phys. Rev. 37, 405-437; and *Reciprocal Relations in Irreversible Processes II*, Phys. Rev. 38, 2265 -2279.

[Otto & Day 2007] Otto, S. P & Day, T. (2007). *A Bioloigist's Guide to Mathematical Modeling in Ecology and Evolution*.(Princeton University Press, Princeton, NJ).

[Rhode 2005] Rhode R. (2005). *Nonequilibrium Ecology* (Cambridge University Press, Cambridge, UK).

[Schneider & Sagan 2005] Schneider, E. D. & Sagan, D. (2005). *Into the Cold: Energy Flow, Thermodynamics and Life* (The University of Chicago Press, Chicago).

[Schrödinger 1967] Schrödinger, E. (1967). *What is Life* with *Mind and Matter* & Autobiographical sketches (Cambridge University Press, Cambridge UK).

[Sethna 2006] Sethna, J. P. (2006). *Entropy, Order Parameters, and Complexity* (Oxford University Press, New York).

[Taniguchi and Cohen 2008] Taniguchi, T. & Cohen, E. G. D. (2008). Nonequilibrium Steady State Thermodynamics and Fluctuationsfor Stochastic Systems. J. Stat. Phys. 130, 633-667 (2008).

[Toda *et al* 1983] Toda, M.; Kubo. R.; & Saito, N. (1983). *Statitsical Physics 1: Equilibrium Statistical Mechanics* (Springer-Verlag, New York).

[van Kampen 1981] Van Kampen, N. G. (183). *Stochastic Processes in Physics and Chemistry* (Elsevier, Amsterdam).

[Volkenstein 2009] Volkenstein, M. V. (2009). *Entropy and Information* (Birkhäuser Verlag, Basel, Switzerland).

[Wax 1954] Wax, N. Ed. (1954). *Selected Papers on Noise and Stochastic Processes* (Dover, New York).

[Zurcher 2008] Zurcher, U. (2008). *Human Food Consumption: a primer on nonequilibrium thermodynamics for college physics.* Eur. J. Phys. 29, 1183-1190.

8. Figures

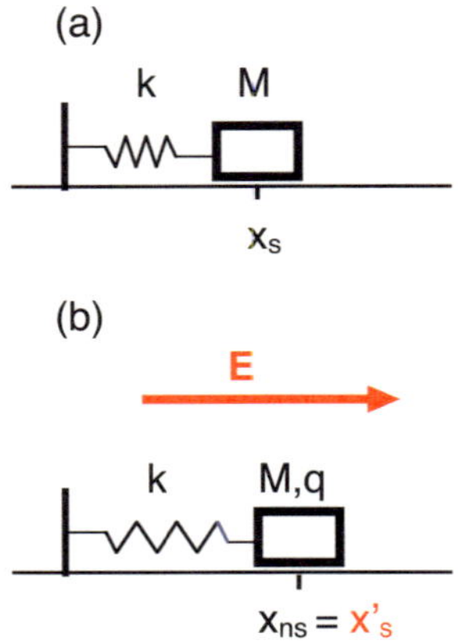

Fig. 1. The spring-block system (a) and with external force (b).

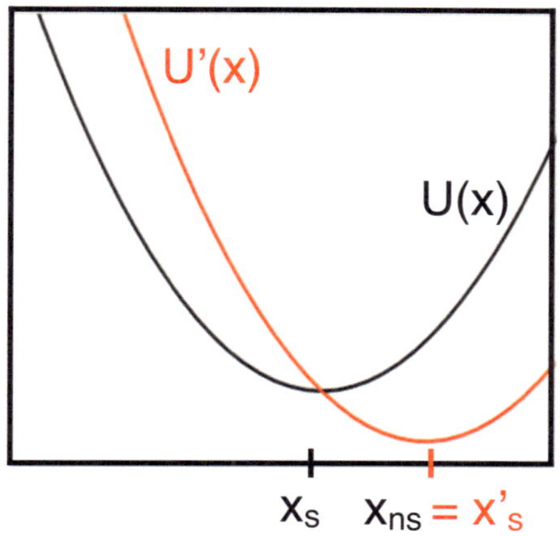

Fig. 2. The harmonic potential $U(x) = kx^2/2$ [black] and the shifted potential $U'(x) = U(x) - F_{\text{ext}}x$ [red].

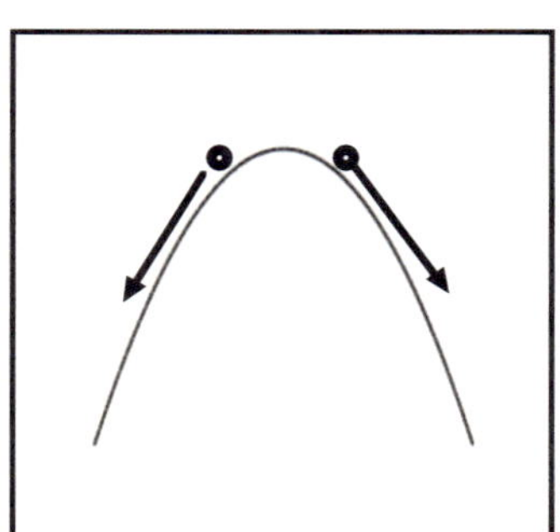

Fig. 3. The concave potential corresponding to a local maximum.

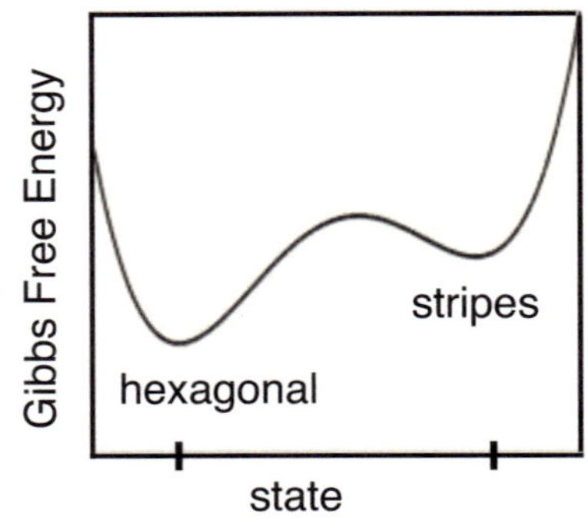

Fig. 4. The Gibbs free energy for the Rayleigh-Bernard connection cell.

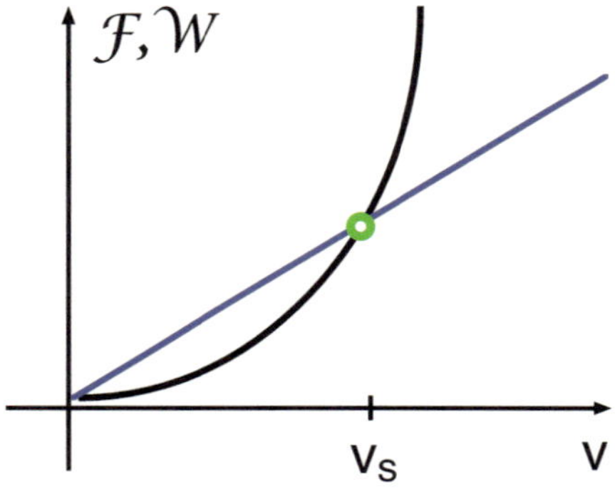

Fig. 5. The rate of energy input and energy dissipation for the Brownian particle immersed in a fluid.

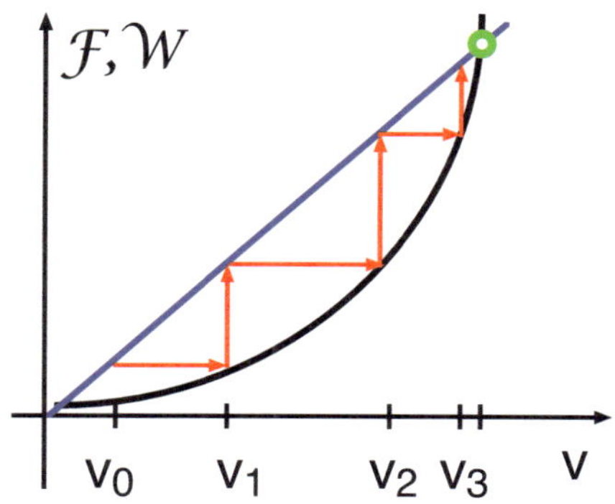

Fig. 6. Iterative time evolution for a Brownian particle with $v_0 < v_s$.

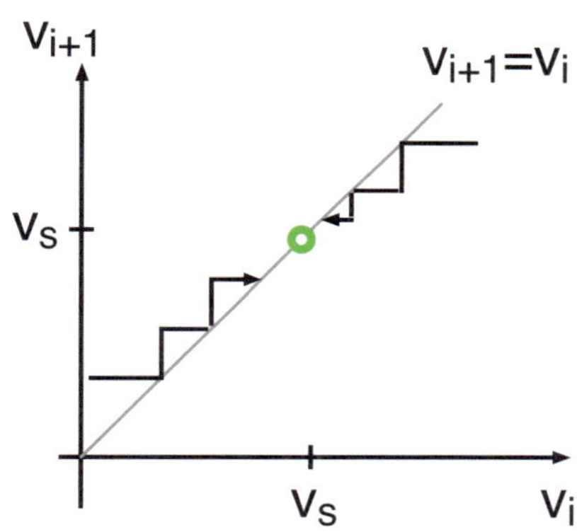

Fig. 7. The phase portrait for the discrete time evolution of the Brownian particle.

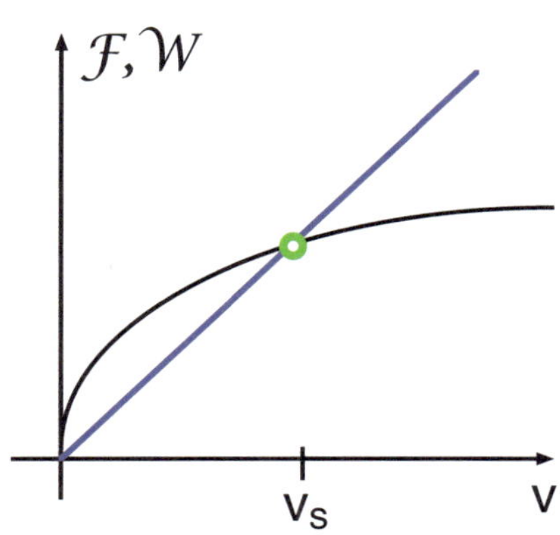

Fig. 8. The rate of energy input and energy dissipation for an open system with unstable dynamics.

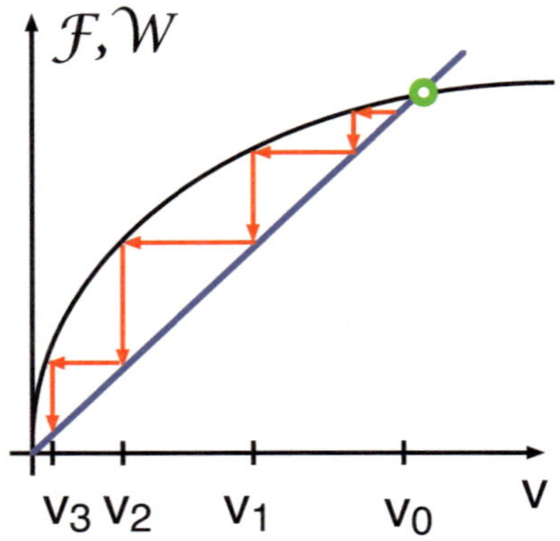

Fig. 9. Iterative time evolution for a unstable open system with $v_0 < v_s$.

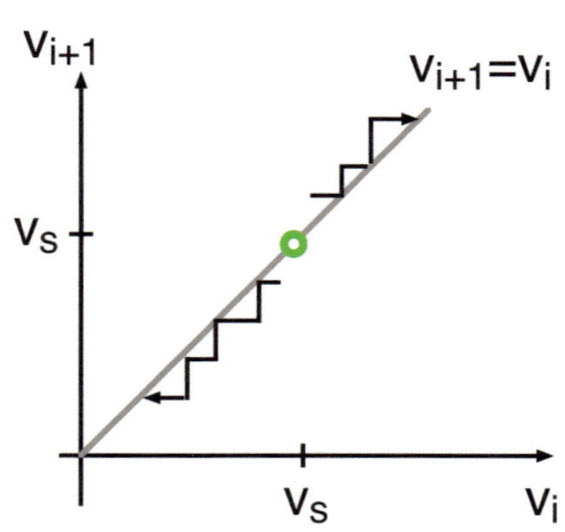

Fig. 10. The phase portrait for the discrete time evolution of an open system with unstable dynamics.

Part 2

Application of Thermodynamics to Science and Engineering

10

Mesoscopic Non-Equilibrium Thermodynamics: Application to Radiative Heat Exchange in Nanostructures

Agustín Pérez-Madrid[1], J. Miguel Rubi[1], and Luciano C. Lapas[2]
[1]*University of Barcelona,*
[2]*Federal University of Latin American Integration*
[1]*Spain*
[2]*Brazil*

1. Introduction

Systems in conditions of equilibrium strictly follow the rules of thermodynamics (Callen, 1985). In such cases, despite the intricate behaviour of large numbers of molecules, the system can be completely characterized by a few variables that describe global average properties. The extension of thermodynamics to non-equilibrium situations entails the revision of basic concepts such as entropy and its related thermodynamic potentials as well as temperature that are strictly defined in equilibrium. Non-equilibrium thermodynamics proposes such an extension (de Groot & Mazur, 1984) for systems that are in local equilibrium. Despite its generality, this theory is applicable only to situations in which the system manifests a deterministic behaviour where fluctuations play no role. Moreover, non-equilibrium thermodynamics is formulated in the linear response domain in which the fluxes of the conserved local quantities (mass, energy, momentum, etc.) are proportional to the thermodynamic forces (gradients of density, temperature, velocity, etc.). While the linear approximation is valid for many transport processes, such as heat conduction and mass diffusion, even in the presence of large gradients, it is not appropriate for activated processes such as chemical and biochemical reactions in which the system immediately enters the non-linear domain or for small systems in which fluctuations may be relevant.

To circumvent these limitations, one has to perform a probabilistic description of the system, which in turn has to be compatible with thermodynamic principles. We have recently proposed such a description aimed at obtaining a simple and comprehensive explanation of the dynamics of non-equilibrium systems at the mesoscopic scale. The theory, mesoscopic non-equilibrium thermodynamics, has provided a deeper understanding of the concept of local equilibrium and a framework, reminiscent of non-equilibrium thermodynamics, through which fluctuations in non-linear systems can be studied. The probabilistic interpretation of the density together with conservation laws in phase-space and positiveness of global entropy changes set the basis of a theory similar to non-equilibrium thermodynamics but of a much broader range of applicability. In particular, the fact of its being based on probabilities instead of densities allows for the consideration of mesoscopic systems and their fluctuations. The situations that can be studied with this formalism include, among others, slow relaxation processes, barrier crossing dynamics, chemical reactions, entropic driving, non-linear transport, and anomalous Brownian motion, processes which are generally non-linear. From the methodological point of view, given the equilibrium properties of a system, this theory provides a systematic and straightforward way to obtain stochastic non-equilibrium dynamics in terms of Fokker-Planck equations.

To set the groundwork for the development of the formalism, we discuss first the basic concepts of mesoscopic non-equilibrium thermodynamics and proceed afterwards with the application of the theory to non-equilibrium radiative transfer at the nanoscale.

2. Mesoscopic non-equilibrium thermodynamics

Mesosocopic non-equilibrium thermodynamics is based on the assumption of the validity of the second law in phase-space, which requires the appropriate definition of the non-equilibrium entropy

$$S(t) = -k_B \int \rho(\Gamma,t) \ln \frac{\rho(\Gamma,t)}{\rho_{eq.}(\Gamma)} d\Gamma + S_{eq.}, \tag{1}$$

where $\rho(\Gamma,t)$ is the probability density of the system with Γ a point of the phase space of the system, $S_{eq.}$ is the equilibrium entropy of the system plus the thermal bath and $\rho_{eq.}(\Gamma)$ is the equilibrium probability density. Note here that the non-equilibrium entropy given through Eq. (1) constitutes the expression of the Gibbs entropy postulate (de Groot & Mazur, 1984). In general, the phase-space point is a set of internal coordinates which univocally determine the state of the system. For a particle or a meso-structure, the set of internal coordinates could include the position and velocity of the particle, number of constituent atoms (as in the case of clusters), reaction coordinates, geometrical parameters, or any other mesoscopic quantity characterizing the state of the meso-structure (Pagonabarraga et al., 1997), (Rubí & Pérez-Madrid, 1999).

Changes in the entropy are related to changes in the probability density which, since the probability is conserved, are given through the continuity equation

$$\frac{\partial}{\partial t}\rho(\Gamma,t) = -\frac{\partial}{\partial \Gamma}\cdot \mathbf{J}(\Gamma,t). \tag{2}$$

The continuity equation defines the probability current $\mathbf{J} = (\Gamma,t)$ whose expression follows from the entropy production.

Assuming local equilibrium in Γ -space, variations of the entropy δS are related to changes in the probability density $\rho(\Gamma,t)$. By performing variations over our non-equilibrium entropy given through Eq. (1) and taking into account that $\delta\rho_{eq.} = 0$ and $\delta S_{eq.} = 0$, we obtain

$$\delta S = -\int \frac{\mu(\Gamma,\mathrm{t})}{T} \delta\rho(\Gamma,\mathrm{t}) d\Gamma \geq 0, \tag{3}$$

where we have introduced the non-equilibrium chemical potential

$$\mu(\Gamma,t) = k_B T \, ln \frac{\rho(\Gamma,t)}{\rho_{eq.}(\Gamma)} + \mu_{eq}. \tag{4}$$

with μ_{eq} being the equilibrium chemical potential. For the photon gas we will consider in the next section, this quantity vanishes at equilibrium due to the massless character of these quasi-particles. Since $-T\delta S = \delta F$, with F being the non-equilibrium free energy, from Eq. (3) we obtain

$$\delta F = \int \mu(\Gamma,t)\delta\rho(\Gamma,t)d\Gamma, \tag{5}$$

which shows the direct relation existing between the free energy and the non-equilibrium chemical potential. Hence, Eq. (3), which constitutes the Gibb's equation of thermodynamics formulated in the phase-space, underlines the non-equilibrium chemical potential in physical terms. From Eqs. (2) and (3) we obtain the entropy production

$$\frac{\partial}{\partial t}S = -\int \mathbf{J}(\Gamma,t)\cdot\left[\frac{\partial}{\partial\Gamma}\frac{\mu(\Gamma,t)}{T}\right]d\Gamma \geq 0 \tag{6}$$

as the product of a thermodynamic current and the conjugated thermodynamic force $\partial\left[\mu(\Gamma,t)/T\right]/\partial\Gamma$. The sign of entropy production determines the direction of evolution of the system and from this same quantity we infer linear laws relating thermodynamic currents and conjugated forces in the absence of non-local effects

$$\mathbf{J}(\Gamma,t) = -\mathbf{L}\cdot\frac{\partial}{\partial\Gamma}\frac{\mu(\Gamma,t)}{T}, \tag{7}$$

with $\mathbf{L}(\rho)$ being the matrix of Onsager coefficients which, as required for the second law, should be positive-definite. The phenomenological law, Eq. (7), together with the expression of the non-equilibrium chemical potential, Eq. (4), lead to the Fick's law of diffusion formulated in the mesoscale

$$\mathbf{J}(\Gamma,t) = -\mathbf{D}(\rho)\cdot\frac{\partial}{\partial\Gamma}\rho, \tag{8}$$

where $\mathbf{D}(\rho) = k_B\mathbf{L}/\rho$ is the matrix of diffusion coefficients. When Eq. (8) is substituted into the continuity equation (2), we obtain the diffusion equation for the probability distribution function

$$\frac{\partial}{\partial t}\rho(\Gamma,t) = \frac{\partial}{\partial\Gamma}\cdot\mathbf{D}(\rho)\cdot\frac{\partial}{\partial\Gamma}\rho. \tag{9}$$

This equation governs the evolution of the probability distribution in the space of the internal coordinates and constitutes the basis for the study of the stochastic dynamics of the non-equilibrium system.

In the case where the equilibrium probability density is a non-homogeneous quantity, *i.e.* $\rho_{eq} \sim exp\left(-\phi/k_BT\right)$, Eq. (8) becomes

$$\mathbf{J}(\Gamma,t) = -\mathbf{D}(\rho)\cdot\left(\frac{\partial}{\partial\Gamma}\rho - \frac{\rho}{k_BT}\frac{\partial}{\partial\Gamma}\phi\right) \tag{10}$$

and instead of Eq. (9) we write

$$\frac{\partial}{\partial t}\rho(\Gamma,t)=\frac{\partial}{\partial \Gamma}\cdot\left[\mathbf{D}(\rho)\cdot\left(\frac{\partial}{\partial \Gamma}\rho-\frac{\rho}{k_B T}\frac{\partial}{\partial \Gamma}\phi\right)\right], \tag{11}$$

the Fokker-Planck equation for evolution of the probability density in Γ-space which includes a drift term $\partial\phi / \partial\Gamma$ related to the potential $\phi=-k_B T \log \rho_{eq}$. In this sense, by knowing the equilibrium thermodynamic potential of a system in terms of its relevant variables it is possible to analyze its dynamics away from equilibrium. A particularly interesting circumstance is the case of a purely entropic barrier, often encountered in biophysics and soft-condensed matter.

3. Thermal radiation

Thermal radiation is a long-studied problem in the field of macroscopic physics. The analysis based on equilibrium thermodynamic grounds led to Planck's blackbody radiation law. In addition, as Planck already realized, there are some limitations to his law due to the finite character of the thermal wavelength of a photon, *i.e.* when diffraction effects are negligible (Planck, 1959). In fact, once the characteristic length scales are comparable to the wavelength of thermal radiation Planck's blackbody radiation law is no longer valid. In such a situation, the finite size of the system may give rise to non-equilibrium effects. In order to better understand these effects it becomes necessary to employ a non-equilibrium theory.

The aforementioned finite-size effects become evident in all kinds of nanostructures where radiative heat transfer occurs. Radiative heat transfer in nanostructures constitutes an issue that, owing to the rapid advancement of nanotechnology, is the object of great research activity. Understanding and predicting heat transfer at the nanoscale possesses wide implications both from the theoretical and applied points of view. There is a great variety of situations involving bodies separated by nanometric distances exchanging heat in an amount not predicted by the current macroscopic laws. We can mention the determination of the cellular temperature (Peng et al., 2010), near-field thermovoltaics (Narayanaswamy & Chen, 2003) and thermal radiation scanning tunneling microscope (De Wilde et al., 2006), just to cite some examples. In most of these cases the experimental length scales are similar to or even less than certain characteristic sizes of the system, *i.e.* the so-called near-field limit. For example, for two interacting nanoparticles (NPs) we would consider the distance between them as the experimental length scale and their diameter as the characteristic size of the system. Near-field radiative heat transfer becomes manifest through an enhancement of the power absorbed, which exceeds in several orders of magnitude the blackbody radiation limit (Rousseau et al., 2009).

The current literature on the subject of radiative energy exchange at the nanoscale is based on the validity of the fluctuation-dissipation theorem (Callen & Welton, 1951), (Landau & Lifshitz, 1980), (Joulain et al., 2005). In the dipole-dipole interaction approximation, dipole moments fluctuate since they are embedded in a heat bath. Consequently, the incident field also fluctuates as well as the energy of a pair of dipoles. Since this quantity is proportional to the dipole moment squared, its second moment is proportional to the dipole-dipole correlation function, which follows from the fluctuation-dissipation theorem. This procedure constitutes the so-called fluctuating electrodynamics (Domingues et al., 2005). Expressions for the fluctuation-dissipation theorem can also be found even when the dipolar approximation is no longer valid since due to the particular charge distribution, higher order multipoles become important (Pérez-Madrid et al., 2008). Such as in the case of two interacting NPs illustrated in Fig. 1.

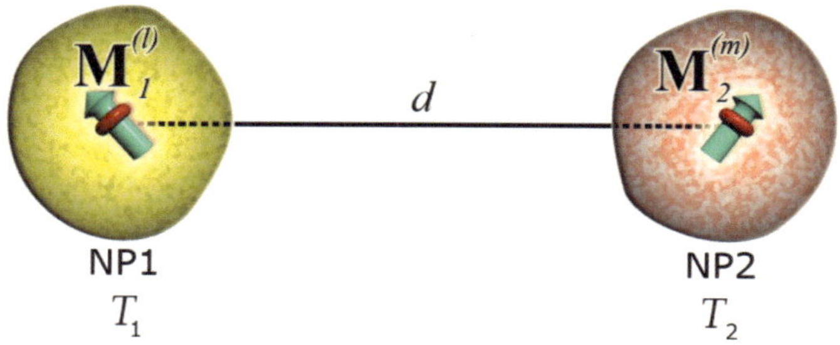

Fig. 1. Schematic illustration of the interaction between two nanoparticles (NP1 and NP2) at temperatures T_1 and T_2, respectively. Each nanoparticle is assimilated to a multipole moment (moments $\mathbf{M}_1^{(l)}$ and $\mathbf{M}_2^{(m)}$) and separated by a distance d between their centres

For extremely short length scales, since the relaxation processes involved in the absorption and emission of radiation does not follows a Debye law related to a definite relaxation time, the fluctuation-dissipation theorem ceases to be valid and a collective description becomes necessary.

In the following Sections, we will present a non-equilibrium thermo-statistical theory describing the heat exchange at the nanoscale in the framework of mesoscopic non-equilibrium thermodynamics based on the assumption of the validity of the second law and the existence of local regression laws at the mesoscale (Reguera et al., 2006).

4. Mesoscopic non-equilibrium thermodynamics of thermal radiation

In this section, we will apply the mesoscopic non-equilibrium theory developed in the previous section to study the heat exchange by thermal radiation between two parallel plates at different temperatures separated by a distance $d \gg \lambda_T$, where $\lambda_T = c\hbar / k_B T$ is the thermal wavelength of a photon (see Fig. 2). For such distances, diffraction effects can be neglected safely

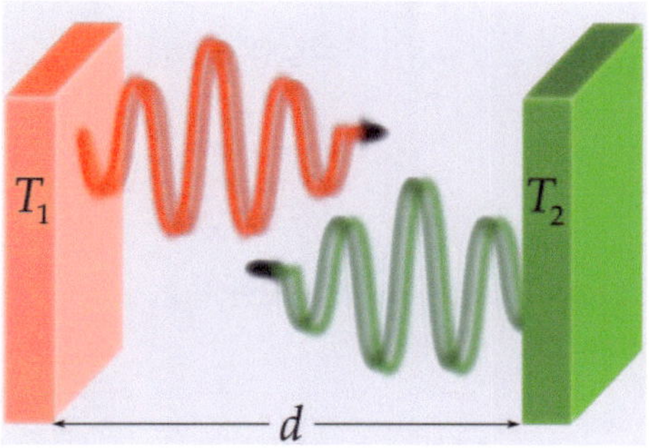

Fig. 2. Schematic illustration of the radiation exchanged between two materials maintained at different temperatures, T_1 and T_2, separated by a distance d

Let us consider the photon gas between two plates at local equilibrium in phase-space. We will assume that the photons do not interact among themselves. The gas is then homogeneous and a phase-space point is merely $\Gamma \to \mathbf{p}$ and thus, the diffusion matrix reduces to a scalar $D(\Gamma)$, the diffusion coefficient. Additionally, if there are only hot and cold photons at temperatures T_1 and T_2, respectively, then

$$J(p,t) = \hat{J}_2(t)\delta(p - p_2) + \hat{J}_1(t)\delta(p - p_1), \tag{12}$$

i.e., the system reaches a state of quasi-equilibrium. Thus, integration of Eq. (8) taking into account (12), leads to

$$\frac{\hat{J}_1(t)}{D_1} + \frac{\hat{J}_2(t)}{D_2} = \rho(p_1,t) - \rho(p_2,t), \tag{13}$$

with $\hat{J}(t) = \mathbf{u}\bullet\hat{\mathbf{J}}(t)$ and $\mathbf{u}$ being the unit vector normal to the walls. Additionally, D_1 and D_2 correspond to the diffusion coefficient of hot and could photons. From here,, by introducing the net current $J(t)$ defined through

$$\frac{J(t)}{aD_1D_2} = \frac{\hat{J}_1(t)}{D_1} + \frac{\hat{J}_2(t)}{D_2}, \tag{14}$$

where a is a dimensionality factor, or equivalently

$$J(t) = aD_1\hat{J}_2(t) + aD_2\hat{J}_1(t), \tag{15}$$

according to Eq. (13) we obtain

$$J(t) = -aD_1D_2\left[\rho(p_2,t) - \rho(p_1,t)\right]. \tag{16}$$

Term-by-term comparison of Eqs. (15) and (16) leads to the identification of the currents

$$\hat{J}_1(t) = D_1\rho(p_1,t) \tag{17}$$

and

$$\hat{J}_2(t) = -D_2\rho(p_2,t) \tag{18}$$

Therefore,

$$D_1\hat{J}_2(t) = -D_1D_2\rho(p_2,t) \tag{19}$$

represents the fraction of photons absorbed at the hot surface from the fraction $\hat{J}_2(t)$ of photons emitted at the cold surface. Likewise,

$$D_2\hat{J}_1(t) = D_1D_2\rho(p_1,t) \tag{20}$$

represents the fraction of photons absorbed at the cold surface from the fraction $\hat{J}_1(t)$ of photons emitted at the hot surface.
For a perfect absorbed, *i.e.* the ideal case, $D_1 = D_2 = 1$ and if the temperatures T_1 and T_2 remain constant, hot and cold photons will reach equilibrium with their respective baths and the probability current will attain a stationary value

$$J_{st}(\omega) = a\left[\rho_{eq.}(\omega,T_1) - \rho_{eq.}(\omega,T_2)\right] \tag{21}$$

where

$$\rho_{eq.}(\omega,T) = 2\frac{N(\omega,T)}{h^3} \tag{22}$$

with h being the Planck constant and $N(\omega,T)$ the averaged number of photons in an elementary cell of volume h^3 of the phase-space given by the Planck distribution (Planck, 1959),

$$N(\omega,T) = \frac{1}{\exp(\hbar\omega / k_B T) - 1}. \tag{23}$$

Moreover, the factor 2 in Eq. (22) comes from the polarization of photons. The stationary current (21) provides us with the flow of photons. Since each photon carries an amount of energy equal to $\hbar\omega$, the heat flow Q_{12} follows from the sum of all the contributions as

$$Q_{12} = \int \hbar\omega J_{st}(\omega) d\mathbf{p}, \tag{24}$$

where $\mathbf{p} = (\hbar\omega / c)\mathbf{\Omega}_p$, with $\mathbf{\Omega}_p$ being the unit vector in the direction of $\mathbf{p}$. Therefore it follows that by taking $a = c/4$

$$Q_{12} = \frac{c}{16\pi}\int d\omega d\mathbf{\Omega}_p \Lambda(\omega)\left[\theta(\omega,T_1) - \theta(\omega,T_2)\right], \tag{25}$$

with $\theta(\omega,T) = \hbar\omega N(\omega,T)$ being the mean energy of an oscillator and where $\Lambda(\omega) = \omega^2 / \pi^2 c^3$ plays the role of the density of states. By performing the integral over all the frequencies and orientations in Eq. (25) we finally obtain the expression of the heat interchanged

$$Q_{12} = \sigma\left(T_1^4 - T_2^4\right), \tag{26}$$

where $\sigma = \pi^2 k_B^4 / 60\hbar^3 c^2$ is the Stefan constant. At equilibrium $T_1 = T_2$, therefore $Q_{12} = 0$. This expression reveals the existence of a stationary state (Saida, 2005) of the photon gas emitted at two different temperatures. Note that for a fluid in a temperature gradient, the heat current is linear in the temperature difference whereas in our case this linearity is not observed. Despite this fact, mesoscopic non-equilibrium thermodynamics is able to derive non-linear laws for the current. In addition, if we set $T_2 = 0$ in Eq. (26), we obtain the heat radiation law of a hot plate at a temperature T_1 in vacuum (Planck, 1959)

$$Q_1 = \sigma T_1^4. \tag{27}$$

5. Near-field radiative heat exchange between two NPs

In this section, we will apply our theory to study the radiative heat exchange between two NPs in the near-field approximation, *i.e.* when the distance d between these NPs satisfies both $d < \lambda_T$ and the near-field condition $2R \lesssim d \lesssim 4R$, with R being the characteristic radius of the NPs. These NPs are thermalized at temperatures $T_1 = T_2$ (see Fig. 3). In particular we will compute the thermal conductance and compare it with molecular simulations (Domingues et al., 2005).

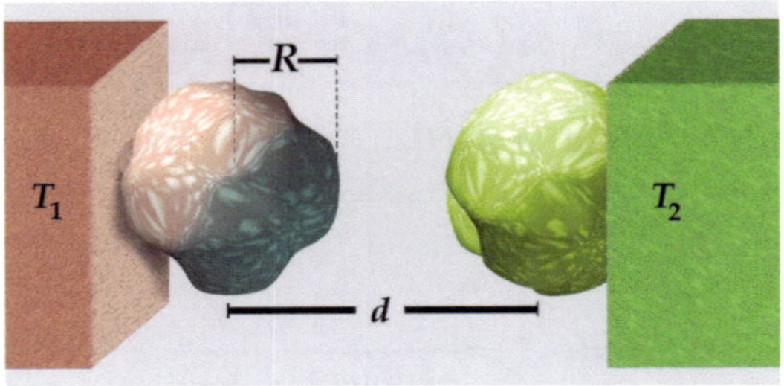

Fig. 3. Illustration of two interacting nanoparticles of characteristic radius R separated by a distance d of the order nm

Since in the present case diffraction effects cannot be ignored D_1 and D_2 must be taken as frequency dependent quantities rather than constants and hence, *Eq.* (25) also applies, now with $\Lambda(\omega) = D_1(\omega)D_2(\omega)\omega^2 / \pi^2 c^3$. This density of states differs from the Debye approximation $\omega^2 / \pi^2 c^3$ related to purely vibration modes and is a characteristic of disordered systems which dynamics is mainly due to slow relaxing modes. Analogous to similar behaviour in glassy systems, we assume here that (Pérez-Madrid et al., 2009)

$$D_1(\omega)D_2(\omega) = A\exp(B^2\omega^2)\delta(\omega - \omega_R), \tag{26}$$

where the characteristic frequency A and the characteristic time B are two fitting parameters, and $\omega_R = 2\pi c / d$ is a resonance frequency.
The heat conductance is defined as

$$G_{12}(T_0) / \pi R^2 = \lim_{T_1 \to T_2} \frac{Q_{12}(\omega)}{T_1 - T_2}, \tag{29}$$

where $T_0 = (T_1 + T_2)/2$ is the temperature corresponding to the stationary state of the system. Therefore,

$$G_{12}(T_0) = \frac{k_B \omega_R^2 R^2}{4\pi c^2} A\exp\left(B^2\omega_R^2\right)\left[\frac{\hbar\omega_R / k_B T_0}{\sinh\left(\hbar\omega_R / 2k_B T_0\right)}\right]^2. \tag{30}$$

In Fig. 4, we have represented the heat conductance as a function of the distance d between the NPs of different radii. This figure shows a significant enhancement of the heat conductance when d decreases until $2D$, which, as has been shown in a previous work by means of electromagnetic calculations and using the fluctuation-dissipation theorem (Pérez-Madrid et al., 2008), is due to multipolar interactions. In more extreme conditions when the NPs come into contact to each other, a sharp fall occurs which can be interpreted as due to an intricate conglomerate of energy barriers inherent to the amorphous character of these NPs generated by the strong interaction. In these last circumstances the multipolar expansion is no longer valid.

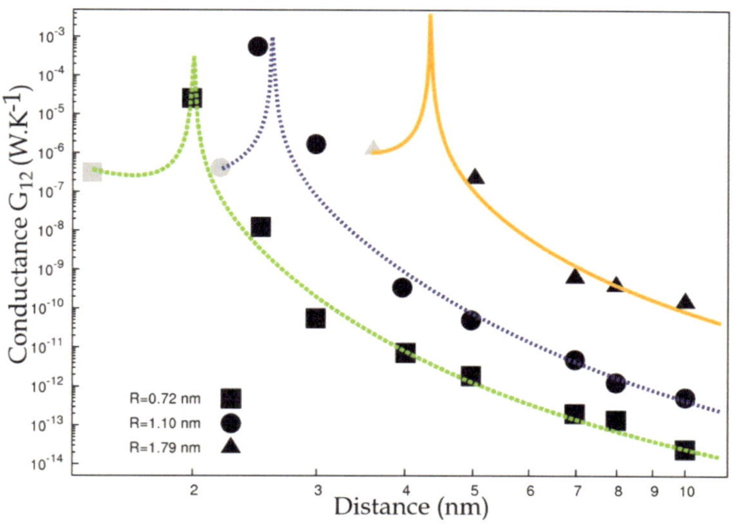

Fig. 4. Thermal conductance G_{12} vs. distance d reproducing the molecular dynamics data obtained by (Domingues et al., 2005). The grey points represent the conductance when the NPs with effective radius R = 0.72, 1.10, and 1.79 nm are in contact. The lines show the analytical result obtained from Eq. (30) by adjusting A and B to the simulation data

6. Conclusions

The classical way to study non-equilibrium mesoscopic systems is to use microscopic theories and proceed with a coarse-graining procedure to eliminate the degrees of freedom that are not relevant to the mesoscopic scale. Such microscopic theories are fundamental to understand how the macroscopic and mesoscopic behaviours of the system arise from the microscopic dynamics. However, these theories frequently involve specialized mathematical methods that prevent them from being generally applicable to complex systems; and more importantly, they use much detailed information that is lost during the coarse-graining procedure and that is actually not needed to understand the general properties of the mesoscopic dynamics.

The mesoscopic non-equilibrium thermodynamics theory we have presented here starts from mesoscopic equilibrium behaviour and adds all the dynamic details compatible with the second principle of thermodynamics and with the conservation laws and symmetries inherent to the system. Thus, given the equilibrium statistical thermodynamics of a system, it is a straightforward process to obtain Fokker-Planck equations for its dynamics. The dynamics is characterized by a few phenomenological coefficients, which can be obtained for the particular situation of interest from experiments or from microscopic theories and describes not only the deterministic properties but also their fluctuations.

Mesoscopic non-equilibrium thermodynamics has been applied to a broad variety of situations, such as activated processes in the non-linear regime, transport in the presence of entropic forces and inertial effects in diffusion. Transport at short time and length scales exhibits peculiar characteristics. One of them is the fact that transport coefficients are no longer constant but depend on the wave vector and frequency. This dependence is due to the existence of inertial effects at such scales as a consequence of microscopic conservation law. The way in which these inertial effects can be considered within a non-equilibrium thermodynamics scheme has been shown in Rubí & Pérez-Madrid, 1998.

We have presented the application of the theory to the case of radiative heat exchange, a process frequently found at the nanoscale. The obtention of the non-equilibrium Stefan-Boltzmann law for a non-equilibrium photon gas and the derivation of heat conductance between two NPs confirm the usefulness of the theory in the study of thermal effects in nanosystems.

7. References

Callen, H. (1985). Thermodynamics and an introduction to thermostatistics. New York: John Wiley and Sons.

Callen, H.B. & Welton, T.A. (1951). Irreversibility and Generalized Noise. *Phys. Rev.* , 83, 34-40.

Carminati, R. et al. (2006). Radiative and non-radiative decay of a single molecule close to a metallic NP. *Opt. Commun.* , 261, 368-375.

de Groot & Mazur (1984). Non-Equilibrium Thermodynamics. New York: Dover.

De Wilde, Y. et al. (2006). Thermal radiation scanning tunnelling microscopy. *Nature*, 444, 740-743.

Domingues G. et al. (2005). Heat Transfer between Two NPs Through Near Field Interaction. *Phys. Rev. Lett.*, 94, 085901.

Förster, T. (1948). Zwischenmolekulare Energiewanderung und Fluoreszenz. *Annalen der Physik*, 55-75.

Frauenfelder H. et al. (1991). *Science*, 254, 1598-1603.

Hess, S. & Köhler, W. (1980). Formeln zur Tensor-Rechnung. Erlangen: Palm & Enke.

Joulain, K. et al. (2005). Surface electromagnetic waves thermally excited: Radiative heat transfer, coherence properties and Casimir forces revisited in the near field. *Surf. Sci. Rep.* , 57, 59-112.

Landau, L.D. & Lifshitz, E.M. (1980). Staistical Physics (Vol. 5). Oxford: Pergamon.

Narayanaswamy, A. & Chen, G. (2003). Surface modes for near field thermophotovoltaics. *Appl. Phys. Lett.* , 82, 3544-3546.

Pagonabarraga, I. et al. (1997). Fluctuating hydrodynamics approach to chemical reactions. *Physica A* , 205-219.

Peng, H. et al. (2010). Luminiscent Europium (III) NPs for Sensing and Imaging of Temperature in the Physiological Range. *Adv. Mater.* , 22, 716-719.

Pérez-Madrid, A. et al. (2009). Heat exchange between two interacting NPs beyond the fluctuation-dissipation regime. *Phys. Rev. Lett.* , 103, 048301.

Pérez-Madrid, A. et al. (2008). Heat transfer between NPs: Thermal conductance for near-field interactions. *Phys. Rev. B* , 77, 155417.

Planck, M. (1959). The Theory of Heat Radiation. New York: Dover.

Reguera, D. et al. (2006). The Mesoscopic Dynamics of Thermodynamic Systems. *J. Phys. Chem. B*, 109, 21502-21515.

Rousseau, E. et al. (2009). Radiative heat transfer at the nanoscale. *Nature photonics* , 3, 514-517.

Rubí J.M. & Pérez-Vicente C. (1997). Complex Behaviour of Glassy Systems. Berlin: Springer.

Rubí, J.M. & Pérez-Madrid, A. (1999). Inertial effetcs in Non-equillibrium thermodynamics. *Physica A*, 492-502.

Saida, H. (2005). Two-temperature steady-state thermodynamics for a radiation field. *Physica A*, 356, 481–508.

11

Extension of Classical Thermodynamics to Nonequilibrium Polarization

Li Xiang-Yuan, Zhu Quan,
He Fu-Cheng and Fu Ke-Xiang
College of Chemical Engineering, Sichuan University
Chengdu 610065
P. R. China

1. Introduction

Thermodynamics concerns two kinds of states, the equilibrium ones (classical thermodynamics) and the nonequilibrium ones (nonequilibrium thermodynamics). The classical thermodynamics is an extremely important theory for macroscopic properties of systems in equilibrium, but it can not be fully isolated from nonequilibrium states and irreversible processes. Therefore, within the framework of classical thermodynamics, it is significant to explore a new method to solve the questions in the nonequilibrium state. Furthermore, this treatment should be helpful for getting deep comprehension and new applications of classical thermodynamics.

For an irreversible process, thermodynamics often takes the assumption of local equilibrium, which divides the whole system into a number of macroscopic subsystems. If all the subsystems stand at equilibrium or quasi-equilibrium states, the thermodynamic functions for a nonequilibrium system can be obtained by some reasonable treatments. However, the concept of local equilibrium lacks the theoretical basis and the expressions of thermodynamic functions are excessively complicated, so it is hard to be used in practice. Leontovich[1] once introduced a constrained equilibrium approach to treat nonequilibrium states within the framework of classical thermodynamics, which essentially maps a nonequilibrium state to a constrained equilibrium one by imposing an external field. In other words, the definition of thermodynamic functions in classical thermodynamics is firstly used in constrained equilibrium state, and the following step is how to extend this definition to the corresponding nonequilibrium state. This theoretical treatment is feasible in principle, but has not been paid much attention to yet. This situation is possibly resulted from the oversimplified descriptions of the Leontovich's approach in literature and the lack of practical demands. Hence on the basis of detailed analysis of additional external parameters, we derive a more general thermodynamic formula, and apply it to the case of nonequilibrium polarization. The results show that the nonequilibrium solvation energy in the continuous medium can be obtained by imposing an appropriate external electric field, which drives the nonequilibrium state to a constrained equilibrium one meanwhile keeps the charge distribution and polarization of medium fixed.

2. The equilibrium and nonequilibrium systems

2.1 Description of state

The state of a thermodynamic system can be described by its macroscopic properties under certain ambient conditions, and these macroscopic properties are called as state parameters. The state parameters should be divided into two kinds, i.e. external and internal ones. The state parameters determined by the position of the object in the ambient are the external parameters, and those parameters, which are related to the thermal motion of the particles constituting the system, are referred to the internal parameters. Consider a simple case that the system is the gas in a vessel, and the walls of the vessel are the object in the ambient. The volume of the gas is the external parameter because it concerns only the position of the vessel walls. Meanwhile, the pressure of the gas is the internal parameter since it concerns not only the position of the vessel walls but also the thermal motion of gas molecules. All objects interacting with the system should be considered as the ambient. However, we may take some objects as one part of a new system. Therefore, the distinction between external and internal parameters is not absolute, and it depends on the partition of the system and ambient. Note that whatever the division between system and ambient, the system may do work to ambient only with the change of some external parameters.

Based on the thermodynamic equilibrium theory, the thermal homogeneous system in an equilibrium state can be determined by a set of external parameters $\{a_i\}$ and an internal parameter T, where T is the temperature of the system. In an equilibrium state, there exists the caloric equation of state, $U = U(a_i, T)$, where U is the energy of the system,system capacitysystem capacity so we can choose one of T and U as the internal parameter of the system. However, for a nonequilibrium state under the same external conditions, besides a set of external parameters $\{a_i\}$ and an internal parameter U (or T), some additional internal parameters should be invoked to characterize the nonequilibrium state of an thermal homogeneous system. It should be noted that those additional internal parameters are time dependent.

2.2 Basic equations in thermodynamic equilibrium

In classical thermodynamics, the basic equation of thermodynamic functions is

$$T\mathrm{d}S = \mathrm{d}U + \sum_i A_i \mathrm{d}a_i \tag{2.1}$$

where S, U and T represent the entropy, energy and the temperature (Kelvin) of the equilibrium system respectively. $\{a_i\}$ stand for a set of external parameters, and A_i is a generalized force which conjugates with a_i. The above equation shows that the entropy of the system is a function of a set of external parameters $\{a_i\}$ and an internal parameter U, which are just the state parameters that can be used to describe a thermal homogeneous system in an equilibrium state. So the above equation can merely be integrated along a quasistatic path. Actually, $T\mathrm{d}S$ is the heat δQ_r absorbed by the system in the infinitesimal change along a quasistatic path. $\mathrm{d}U$ is the energy and $A_i \mathrm{d}a_i$ is the element work done by the system when external parameter a_i changes.

It should be noticed that the positions of any pair of A_i and a_i can interconvert through Legendre transformation. We consider a system in which the gas is enclosed in a cylinder with constant temperature, there will be only one external parameter, i.e. the gas volume V. The corresponding generalized force is the gas pressure p, so eq. (2.1) can be simplified as

$$TdS = dU + pdV \tag{2.2}$$

It shows that $S = S(U,V)$. If we define $H = U + pV$, eq. (2.2) may be rewritten as

$$TdS = d(U + pV) - Vdp = dH - Vdp \tag{2.3}$$

Thus we have $S = S(H,p)$. If we choose the gas pressure p as the external parameter, then V should be the conjugated generalized force, and the negative sign in eq. (2.3) implies that the work done by the system is positive as the pressure decreases. Furthermore, the energy U in eq. (2.2) has been changed with the relation of $U + pV = H$ in eq. (2.3), in which H stands for the enthalpy of the gas.

2.3 Nonequilibrium state and constrained equilibrium state

It is a difficult task to efficiently extend the thermodynamic functions defined in the classical thermodynamics to the nonequilibrium state. At present, one feasible way is the method proposed by Leontovich. The key of Leontovich's approach is to transform the nonequilibrium state to a constrained equilibrium one by imposing some additional external fields. Although the constrained equilibrium state is different from the nonequilibrium state, it retains the significant features of the nonequilibrium state. In other words, the constraint only freezes the time-dependent internal parameters of the nonequilibrium state, without doing any damage to the system. So the constrained equilibrium becomes the nonequilibrium state immediately after the additional external fields are removed quickly. The introduction and removal of the additional external fields should be extremely fast so that the characteristic parameters of the system have no time to vary, which provides a way to obtain the thermodynamic functions of nonequilibrium state from that of the constrained equilibrium state.

2.4 Extension of classical thermodynamics

Based on the relation between the constrained equilibrium state and the nonequilibrium one, the general idea of extending classical thermodynamics to nonequilibrium systems can be summarized as follows:

1. By imposing suitable external fields, the nonequilibrium state of a system can be transformed into a constrained equilibrium state so as to freeze the time-dependent internal parameters of the nonequilibrium state.
2. The change of a thermodynamic function between a constrained equilibrium state and another equilibrium (or constrained equilibrium) state can be calculated simply by means of classical thermodynamics.
3. The additional external fields can be suddenly removed without friction from the constrained equilibrium system so as to recover the true nonequilibrium state, which will further relax irreversibly to the eventual equilibrium state. Leontovich defined the entropy of the nonequilibrium state by the constrained equilibrium. In other words, entropy of the constrained equilibrium and that of the nonequilibrium exactly after the fast removal of the external field should be thought the same.

According to the approach mentioned above, we may perform thermodynamic calculations involving nonequilibrium states within the framework of classical thermodynamics.

3. Entropy and free energy of nonequilibrium state

3.1 Energy of nonequilibrium states

For the clarity, only thermal homogeneous systems are considered. The conclusions drawn from the thermal homogeneous systems can be extended to thermal inhomogeneous ones as long as they consist of finite isothermal parts[1]. As a thermal homogeneous system is in a constrained equilibrium state, the external parameters of the system should be divided into three kinds. The first kind includes those original external parameters $\{a_i\}$, and they have the conjugate generalized forces $\{A_i\}$. The second kind includes the additional external parameters $\{x_k\}$, which are totally different from the original ones. Correspondingly, the generalized forces $\{\xi_k\}$ conjugate with $\{x_k\}$, where ξ_k is the internal parameter originating from the nonequilibrium state. The third kind is a new set of external parameters $\{a_l'\}$, which relate to some of the original external fields and the additional external parameters, i.e.,

$$a_l' = a_l + x_l' \tag{3.1}$$

where a_l and x_l' stand for the original external parameter and the additional external parameter, respectively. Supposing a generalized force A_l' conjugates with the external parameter a_l', the basic thermodynamic equation for a constrained equilibrium state can be expressed by considering all the three kinds of external parameters, $\{a_i\}$, $\{x_i\}$, and $\{a_l'\}$, i.e.

$$TdS^* = dU^* + \sum_i A_i da_i + \sum_k \xi_k dx_k + \sum_l A_l' da_l' \tag{3.2}$$

where S^* and U^* stand for entropy and energy of the constrained nonequilibrium state, respectively, and other terms are the work done by the system due to the changes of three kinds of external parameters. Because the introduction and removal of additional external fields are so fast that the internal parameters ξ_k and A_l' may remain invariant. The transformation from the constrained equilibrium state to the nonequilibrium state can be regarded adiabatic.

Beginning with this constrained equilibrium, a fast removal of the constraining forces $\{x_k\}$ from the system then yields the true nonequilibrium state. By this very construction, the constrained equilibrium and the nonequilibrium have the same internal variables. In particular, the nonequilibrium entropy S^{non} is equal to that of the constrained equilibrium[1]

$$S^{\mathrm{non}} = S^* \tag{3.3}$$

The energy change of the system in the fast (adiabatic) process is given as follows

$$\Delta U = U^{\mathrm{non}} - U^* = -W \tag{3.4}$$

where U^{non} denotes the energies of the true nonequilibrium, and W is the work done by the system during the non-quasistatic removal of the constraining forces, i.e.,

$$U^{\text{non}} - U^* = -W = -\sum_k \xi_k \int_{x_k}^{0} \mathrm{d}x_k - \sum_l A_l' \int_{a_l + x_l'}^{a_l} \mathrm{d}a_l' = \sum_k \xi_k x_k + \sum_l A_l' x_l' \tag{3.5}$$

where $\sum_k \xi_k x_k$ and $\sum_l A_l' x_l'$ are work done by getting rid of the second and the third kinds of additional external fields quickly. Eq. (3.5) is just the relation between the energy of the nonequilibrium state and that of the constrained equilibrium state.
If $A_l' = 0$, eq. (3.5) is reduced to the Leontovich form, i.e., (Eq.3.5 of ref 1)

$$U = U^* + \sum_k \xi_k x_k \tag{3.6}$$

$A_l' = 0$ indicates that the constraining forces $\{\xi_k\}$ are new internal parameters which do not exist in the original constrained equilibrium state. This means that eq. (3.5) is an extension of Leontovich's form of eq. (3.6).
If one notes that ξ_k and A_k' remains invariant during the fast removal of their conjugate parameters, the energy change by eq. (3.5) becomes straightforward.

3.2 Free energies of the constrained equilibrium and nonequilibrium states

The free energy of the constrained equilibrium state F^* is defined as

$$F^* = U^* - TS^* \tag{3.7}$$

Differentiating on the both sides of eq. (3.7) by substituting of eq. (3.2), we have

$$\mathrm{d}F^* = -S^* \mathrm{d}T - \sum_i A_i \mathrm{d}a_i - \sum_k \xi_k \mathrm{d}x_k - \sum_l A_l' \mathrm{d}a_l' \tag{3.8}$$

The free energy of the nonequilibrium state F^{non} is defined as

$$F^{\text{non}} = U^{\text{non}} - TS^{\text{non}} \tag{3.9}$$

Subtracting eq. (3.7) from eq. (3.9), with noticing eq. (3.5), we have

$$F^{\text{non}} - F^* = \sum_k \xi_k x_k + \sum_l A_l' x_l' \tag{3.10}$$

From the above equation, F^{non} can be obtained from F^*.
A particularly noteworthy point should be that A_l' and x_l' are not a pair of conjugates, so the sum $\sum_l A_l' x_l'$ in eq. (3.10) does not satisfy the conditions of a state function. This leads to that the total differential of F^{non} does not exist.
Adding the sum $\sum_l A_l' a_l$ to both sides of eq. (3.10), the total differential can be obtained as

$$\mathrm{d}(F^{\text{non}} + \sum_l A_l' a_l) = -S^{\text{non}} \mathrm{d}T - \sum_i A_i \mathrm{d}a_i + \sum_k x_k \mathrm{d}\xi_k + \sum_l a_l' \mathrm{d}A_l' \tag{3.11}$$

If the third kind of external parameters do not exist, i.e., $a_l = 0$ and $x_l' = 0$, hence $a_l' = 0$, eq. (3.11) is identical with that given by Leontovich[1]. Eq. (3.11) shows that if there are external parameters of the third kind, the nonequilibrium free energy F^{non} which comes from the free energy F^{*} of the constrained state does not possess a total differential. This is a new conclusion. However, it will not impede that one may use eq. (3.11) to obtain F^{non}, because with this method one can transform the nonequilibrium state into a constrained equilibrium state, which can be called as state-to-state treatment. This treatment does not involve the state change with respect to time, so it can realize the extension of classical thermodynamics to nonequilibrium systems.

4. Nonequilibrium polarization and solvent reorganization energy

In the previous sections, the constrained equilibrium concept in thermodynamics, which can be adopted to account for the true nonequilibrium state, is introduced in detail. In this section, we will use this method to handle the nonequilibrium polarization in solution and consequently to achieve a new expression for the solvation free energy. In this kind of nonequilibrium states, only a portion of the solvent polarization reaches equilibrium with the solute charge distribution while the other portion can not equilibrate with the solute charge distribution. Therefore, only when the solvent polarization can be partitioned in a proper way, the constrained equilibrium state can be constructed and mapped to the true nonequilibrium state.

4.1 Inertial and dynamic polarization of solvent

Theoretical evaluations of solvent effects in continuum media have attracted great attentions in the last decades. In this context, explicit solvent methods that intend to account for the microscopic structure of solvent molecules are most advanced. However, such methods have not yet been mature for general purposes. Continuum models that can handle properly long range electrostatic interactions are thus far still playing the major role. Most continuum models are concerned with equilibrium solvation. Any process that takes place on a sufficiently long timescale may legitimately be thought of as equilibration with respect to solvation. Yet, many processes such as electron transfer and photoabsorption and emission in solution are intimately related to the so-called nonequilibrium solvation phenomena. The central question is how to apply continuum models to such ultra fast processes.

Starting from the equilibrium solvation state, the total solvent polarization is in equilibrium with the solute electric field. However, when the solute charge distribution experiences a sudden change, for example, electron transfer or light absorption/emission, the nonequilibrium polarization emerges. Furthermore, the portion of solvent polarization with fast response speed can adjust to reach the equilibrium with the new solute charge distribution, but the other slow portion still keeps the value as in the previous equilibrium state. Therefore, in order to correctly describe the nonequilibrium solvation state, it is important and necessary to divide the total solvent polarization in a proper way.

At present, there are mainly two kinds of partition method for the solvent polarization. The first one was proposed by Marcus[2] in 1956, in which the solvent polarization is divided into orientational and electronic polarization. The other one, suggested by Pekar[3], considers that the solvent polarization is composed by inertial and dynamic polarization.

The first partition method of electronic and orientational polarization is established based on the relationship between the solvent polarization and the total electric field in the solute-solvent system. We consider an electron transfer (or light absorption/emission) in solution.

Before the process, the solute-solvent system will stay in the equilibrium state "1", and then the electronic transition happens and the system will reach the nonequilibrium state "2" in a very short time, and finally the system will arrive to the final equilibrium state "2", due to the relaxation of solvent polarization. In the equilibrium states "1" and "2", the relationship between the total electric field $\mathbf{E}$ and total polarization $\mathbf{P}$ is expressed as

$$\mathbf{P}_1^{\mathrm{eq}} = \chi_{\mathrm{s}} \mathbf{E}_1^{\mathrm{eq}}, \ \mathbf{P}_2^{\mathrm{eq}} = \chi_{\mathrm{s}} \mathbf{E}_2^{\mathrm{eq}} \tag{4.1}$$

where $\chi_{\mathrm{s}} = \frac{(\varepsilon_{\mathrm{s}} - 1)}{4\pi}$ is the static susceptibility, with ε_{s} being the static dielectric constants. The superscript "eq" denotes the equilibrium state. Correspondingly, the electronic polarizations in the equilibrium states "1" and "2"are written as

$$\mathbf{P}_{1,\mathrm{op}} = \chi_{\mathrm{op}} \mathbf{E}_1^{\mathrm{eq}}, \ \mathbf{P}_{2,\mathrm{op}} = \chi_{\mathrm{op}} \mathbf{E}_2^{\mathrm{eq}} \tag{4.2}$$

where the subscript "op" represents the electronic polarization and $\chi_{\mathrm{op}} = \frac{(\varepsilon_{\mathrm{op}} - 1)}{4\pi}$ the electronic susceptibility, with $\varepsilon_{\mathrm{op}}$ being the optical dielectric constant. In solution, the electronic polarization can finish adjustment very quickly, and hence it reaches equilibrium with solute charge even if the electronic transition in the solute molecule takes place. On the other hand, it is easy to express the orientational polarization as

$$\mathbf{P}_{1,\mathrm{or}} = \mathbf{P}_1^{\mathrm{eq}} - \mathbf{P}_{1,\mathrm{op}} = \chi_{\mathrm{or}} \mathbf{E}_1^{\mathrm{eq}}, \ \mathbf{P}_{2,\mathrm{or}} = \mathbf{P}_2^{\mathrm{eq}} - \mathbf{P}_{2,\mathrm{op}} = \chi_{\mathrm{or}} \mathbf{E}_2^{\mathrm{eq}} \tag{4.3}$$

with $\chi_{\mathrm{or}} = \chi_{\mathrm{s}} - \chi_{\mathrm{op}}$. Here, χ_{or} stands for the orientational susceptibility and the subscript "or" the orientational polarization. This kind of polarization is mainly contributed from the low frequency motions of the solvents.

In the nonequilibrium state "2", we express the total electric field strength and solvent polarization as $\mathbf{E}_2^{\mathrm{non}}$ and $\mathbf{P}_2^{\mathrm{non}}$ respectively, the electronic polarization can be defined as

$$\mathbf{P}_{2,\mathrm{op}}^{\mathrm{non}} = \chi_{\mathrm{op}} \mathbf{E}_2^{\mathrm{non}} \tag{4.4}$$

At this moment, the orientational polarization keeps invariant and the value in the previous equilibrium state "1", thus the total polarization is written as

$$\mathbf{P}_2^{\mathrm{non}} = \mathbf{P}_{1,\mathrm{or}} + \mathbf{P}_{2,\mathrm{op}}^{\mathrm{non}} \tag{4.5}$$

The second partition method for the polarization is based on the equilibrium relationship between the dynamic polarization and electric field. Assuming that the solvent only has the optical dielectric constant $\varepsilon_{\mathrm{op}}$, the dynamic electric field strength and the polarization in equilibrium state "1" and "2" can be expressed as

$$\mathbf{P}_{1,\mathrm{dy}} = \chi_{\mathrm{op}} \mathbf{E}_{1,\mathrm{dy}}, \tag{4.6}$$

$$\mathbf{P}_{2,\mathrm{dy}} = \chi_{\mathrm{op}} \mathbf{E}_{2,\mathrm{dy}} \tag{4.7}$$

Then the inertial polarization in an equilibrium state is defined as

$$\mathbf{P}_{1,\text{in}} = \mathbf{P}_1^{\text{eq}} - \mathbf{P}_{1,\text{dy}}\,,\ \mathbf{P}_{2,\text{in}} = \mathbf{P}_2^{\text{eq}} - \mathbf{P}_{2,\text{dy}} \tag{4.8}$$

where the subscripts "dy" and "in" stand for the quantities due to the dynamic and inertial polarizations. In a nonequilibrium state, the inertial polarization will be regarded invariant, and hence the total polarization is decomposed to

$$\mathbf{P}_2^{\text{non}} = \mathbf{P}_{1,\text{in}} + \mathbf{P}_{2,\text{dy}} \tag{4.9}$$

With the dynamic-inertial partition, the nonequilibrium polarization is of the following form, i.e.,

$$\mathbf{P}_2^{\text{non}} = \mathbf{P}_{1,\text{in}} + \mathbf{P}_{2,\text{dy}} = \mathbf{P}_1^{\text{eq}} - \mathbf{P}_{1,\text{dy}} + \mathbf{P}_{2,\text{dy}} = \chi_s \mathbf{E}_1^{\text{eq}} - \chi_{\text{op}} \mathbf{E}_{1,\text{dy}} + \chi_{\text{op}} \mathbf{E}_{2,\text{dy}} \tag{4.10}$$

According to the inertial-dynamic partition, the picture of the nonequilibrium state "2" is very clear that the invariant part from equilibrium to nonequilibrium is the inertial polarization and the dynamic polarization responds to the solute charge change without time lag in nonequilibrium state, being equal to the dynamic polarization in equilibrium state "2".

4.2 Constrained equilibrium by external field and solvation energy in nonequilibrium state

Based on the inertial-dynamic polarization partition, the thermodynamics method introduced in the previous sections can be adopted to obtain the solvation energy in nonequilibrium state, which is a critical problem to illustrate the ultra-fast dynamical process in the solvent.

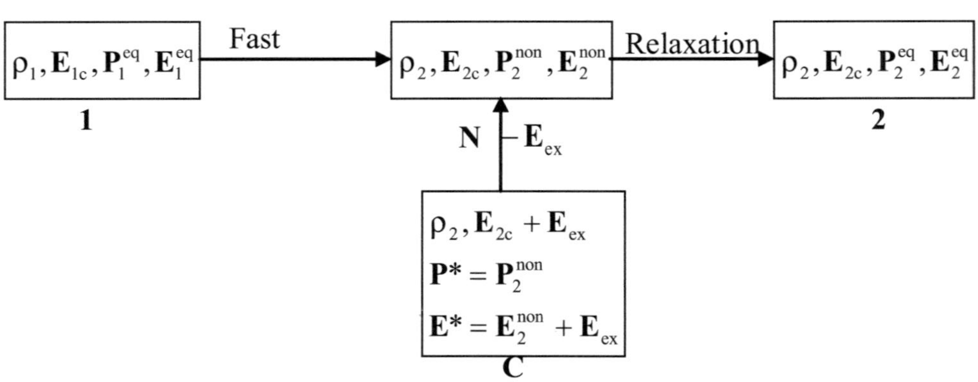

Scheme 1.

In the real solvent surroundings, the solvation energy is composed of three contributions: the cavitation energy, the dispersion-repulsion energy and electrostatic solvation energy. The cavitation energy, needed to form the solute cavity, will not change from the equilibrium "1" to the nonequilibrium state "2" due to the fixed solute structure. At the same time, the dispersion-repulsion energy is supposed invariant here. Therefore, the most important contribution to the solvation energy change from equilibrium to nonequilibrium is the electrostatic part, and the electrostatic solvation energy, which measures the free energy change of the medium, simplified as solvation energy in the following paragraphs, is the research focus for the ultrafast process in the medium.

As shown in Scheme 1, we adopt the letter "N" to denote the nonequilibrium state, which has the same solute electric field $\mathbf{E}_{2c}$ as equilibrium state "2". The differences of polarization strength and polarization field strength between states "N" and "2" in scheme 1 can be expressed as

$$\mathbf{P}' = \mathbf{P}_2^{non} - \mathbf{P}_2^{eq} = \Delta\mathbf{P}_{dy} - \Delta\mathbf{P}_{eq} = -\Delta\mathbf{P}_{in} \tag{4.11}$$

$$\mathbf{E}'_p = \mathbf{E}_2^{non} - \mathbf{E}_2^{eq} = \Delta\mathbf{E}_{dy} - \Delta\mathbf{E}_{eq} = -\Delta\mathbf{E}_{in} \tag{4.12}$$

with

$$\Delta\mathbf{M}_k = \mathbf{M}_{2,k} - \mathbf{M}_{1,k} \qquad (\text{k= "dy", "in" or "eq"})$$

where $\mathbf{M}$ can be electric filed $\mathbf{E}$ or polarization $\mathbf{P}$. In eqs. (4.11) and (4.12), $\mathbf{P}'$ is hereafter called the residual polarization which will disappear when the polarization relaxation from state "N" to the final equilibrium state "2" has finished after enough long time. $\mathbf{E}'_p$ is actually a polarization field resulted from $\mathbf{P}'$.

In order to obtain the solvation energy for the nonequilibrium state "N", we can construct a constrained equilibrium state, denoted as state "C" in scheme 1, by imposing an external field $\mathbf{E}_{ex}$ from the ambient on the equilibrium state "2", which produces the residual polarization $\mathbf{P}'$ and the corresponding polarization field $\mathbf{E}'_p$. It is clear that $\mathbf{P}' = \chi_s(\mathbf{E}_{ex} + \mathbf{E}'_p)$ in the medium with the dielectric constant ε_s. Thus the total electric field $\mathbf{E}'$ due to the external field in the medium with the dielectric constant ε_s can be expressed as

$$\mathbf{E}' = \mathbf{E}_{ex} + \mathbf{E}'_p = \frac{\mathbf{P}'}{\chi_s} = \frac{\Delta\mathbf{P}_{dy} - \Delta\mathbf{P}_{eq}}{\chi_s} = \frac{\chi_{op}}{\chi_s}\Delta\mathbf{E}_{dy} - \Delta\mathbf{E}_{eq} \tag{4.13}$$

Combining eqs. (4.12) and (4.13), the external field strength can be defined as[4-6]

$$\mathbf{E}_{ex} = \mathbf{E}' - \mathbf{E}'_p = \frac{\chi_{op} - \chi_s}{\chi_s}\Delta\mathbf{E}_{dy} = \frac{\varepsilon_{op} - \varepsilon_s}{\varepsilon_s - 1}\Delta\mathbf{E}_{dy} \tag{4.14}$$

Through the introduction of the external field, the constrained equilibrium state has been constructed as

$$\begin{aligned} \mathbf{E}_c^* &= \mathbf{E}_{2c} + \mathbf{E}_{ex} \\ \mathbf{E}^* &= \mathbf{E}_2^{non} + \mathbf{E}_{ex} \\ \mathbf{P}^* &= \mathbf{P}_2^{non} = \mathbf{P}_2^{eq} + \mathbf{P}' \end{aligned} \tag{4.15}$$

where $\mathbf{E}_c^*$ is the solute electric field in vacuum. In constrained equilibrium state, the polarization, entropy and solute charge distribution are the same as the nonequilibrium state "N". It is shown in eq. (4.15) that nonequilibrium polarization $\mathbf{P}_2^{\text{non}}$ equilibrates with solute and external electric field $\mathbf{E}_{2c}+\mathbf{E}_{\text{ex}}$ in the medium with static dielectric constant. Therefore, the only difference between the nonequilibrium state and constrained equilibrium state is the external field $\mathbf{E}_{\text{ex}}$.

Now we can analyze the equilibrium and constrained equilibrium states from the view of thermodynamics. For clarity, we take the medium (or solvent) as the "system" but both the solute (free) charge and the source of $\mathbf{E}_{\text{ex}}$ as the "ambient". This means that the thermodynamic system is defined to only contain the medium, while the free charges and the constraining field act as the external field. The exclusion of the free charges from the "thermodynamic system" guarantees coherent thermodynamic treatment.

Given the above definition on the "system", we now turn to present the free energy F_{sol} of the medium. Here we use the subscript "sol" to indicate the quantities of the medium, or solvent. Let us calculate the change in F_{sol} resulting from an infinitesimal change in the field which occurs at constant temperature and does not destroy the thermodynamic equilibrium of the medium. The free energy change of the medium for an equilibrium polarization is equal to the total free energy change of the solute-solvent system minus the self-energy change of the solute charge, i.e.,

$$\delta F_{\text{sol}}=\frac{1}{4\pi}\int \mathbf{E}\cdot\delta\mathbf{D}\mathrm{d}V-\frac{1}{4\pi}\int \mathbf{E}_{\text{c}}\cdot\delta\mathbf{E}_{\text{c}}\mathrm{d}V \tag{4.16}$$

where $\mathbf{E}$ is the total electric field while $\mathbf{E}_{\text{c}}$ is the external field by the solute charge in the vacuum. $\mathbf{D}$ is the electric displacement with the definition of $\mathbf{D}=\mathbf{E}+4\pi\mathbf{P}=\varepsilon\mathbf{E}$. Eq.(4.16) gives the free energy of the medium for an equilibrium polarization as

$$\begin{aligned}F_{\text{sol}}&=\frac{1}{8\pi}\int(\mathbf{D}\cdot\mathbf{E}-\mathbf{E}_{\text{c}}\cdot\mathbf{E}_{\text{c}})\mathrm{d}V\\&=\frac{1}{8\pi}\int(\mathbf{E}\cdot\mathbf{E}_{\text{c}}-\mathbf{D}\cdot\mathbf{E}_{\text{c}})\mathrm{d}V+\frac{1}{8\pi}\int(\mathbf{E}+\mathbf{E}_{\text{c}})\cdot(\mathbf{D}-\mathbf{E}_{\text{c}})\mathrm{d}V\end{aligned} \tag{4.17}$$

We note that

$$\mathbf{E}=-\nabla\Phi\,,\ \mathbf{E}_{\text{c}}=-\nabla\psi_{\text{c}} \tag{4.18}$$

where Φ is the total electric potential produced and ψ_{c} is the electric potential by the solute (free) charge in vacuum. With eq.(4.18), the last term in the second equality of eq.(4.17) becomes

$$-\frac{1}{8\pi}\int\nabla(\Phi+\psi_{\text{c}})\cdot(\mathbf{D}-\mathbf{E}_{\text{c}})\mathrm{d}V \tag{4.19}$$

The volume integral (4.19) can be change to the following form by integration by parts:

$$\frac{1}{8\pi}\int(\Phi+\psi_{\text{c}})\nabla\cdot(\mathbf{D}-\mathbf{E}_{\text{c}})\mathrm{d}V=0 \tag{4.20}$$

Thus eq.(4.17) can be rewritten as[7,8]

$$F_{\mathrm{sol}} = -\frac{1}{2}\int \mathbf{P}\cdot\mathbf{E}_{\mathrm{c}}\mathrm{d}V \tag{4.21}$$

We consider our nonequilibrium polarization case. For the solvent system in the constrained equilibrium state "C", the external field strength $\mathbf{E}_{2\mathrm{c}}$ takes the role of the external parameter a, $\mathbf{E}_{\mathrm{ex}}$ takes the role of χ', and solvent polarization $\mathbf{P}^* = \mathbf{P}_2^{\mathrm{non}} = \mathbf{P}_2^{\mathrm{eq}} + \mathbf{P}'$ takes the role of A'. The total external (vacuum) electric field in this state is $\mathbf{E}_{\mathrm{c}}^* = \mathbf{E}_{2\mathrm{c}} + \mathbf{E}_{\mathrm{ex}}$. A constrained equilibrium can be reached through a quasistatic path, so the electrostatic free energy by an external field is of the form like eq.(4.21),

$$F_{\mathrm{sol}}^* = -\frac{1}{2}\int \mathbf{P}^*\cdot\mathbf{E}_{\mathrm{c}}^*\mathrm{d}V = -\frac{1}{2}\int (\mathbf{P}_2^{\mathrm{eq}} + \mathbf{P}')\cdot(\mathbf{E}_{\mathrm{ex}} + \mathbf{E}_{2\mathrm{c}})\mathrm{d}V \tag{4.22}$$

Similarly, the electrostatic free energy of the final equilibrium state "2" is given by

$$F_{2,\mathrm{sol}}^{\mathrm{eq}} = -\frac{1}{2}\int \mathbf{P}_2^{\mathrm{eq}}\cdot\mathbf{E}_{2\mathrm{c}}\mathrm{d}V \tag{4.23}$$

Starting form the constrained equilibrium "C", we prepare the nonequilibrium state "N" by removing the external $\mathbf{E}_{\mathrm{ex}}$ suddenly without friction. In this case, the constrained equilibrium will return to the nonequilbirium state. According to eq. (3.10), the nonequilibrium solvation energy is readily established as

$$F_{2.\mathrm{sol}}^{\mathrm{non}} = F_{\mathrm{sol}}^* + \int (\mathbf{P}_2^{\mathrm{eq}} + \mathbf{P}')\cdot\mathbf{E}_{\mathrm{ex}}\mathrm{d}V \tag{4.24}$$

Substituting eq. (4.22) into eq. (4.24), the electrostatic solvation energy (it is just the electrostatic free energy of the medium) for the nonequilibrium state "N" is given by

$$\begin{aligned} F_{2,\mathrm{sol}}^{\mathrm{non}} &= -\frac{1}{2}\int (\mathbf{E}_{2\mathrm{c}} + \mathbf{E}_{\mathrm{ex}})\cdot(\mathbf{P}_2^{\mathrm{eq}} + \mathbf{P}')\mathrm{d}V + \int \mathbf{E}_{\mathrm{ex}}\cdot(\mathbf{P}_2^{\mathrm{eq}} + \mathbf{P}')\mathrm{d}V \\ &= \frac{1}{2}\int (\mathbf{E}_{\mathrm{ex}} - \mathbf{E}_{2\mathrm{c}})\cdot(\mathbf{P}_2^{\mathrm{eq}} + \mathbf{P}')\mathrm{d}V \end{aligned} \tag{4.25}$$

Eq. (4.25) can be further simplified as

$$F_{2,\mathrm{sol}}^{\mathrm{non}} = -\frac{1}{2}\int \mathbf{E}_{2\mathrm{c}}\cdot\mathbf{P}_2^{\mathrm{eq}}\mathrm{d}V + \frac{1}{2}\int \mathbf{E}_{\mathrm{ex}}\cdot\mathbf{P}'\mathrm{d}V \tag{4.26}$$

with the relationship of $\int \mathbf{E}_{2\mathrm{c}}\cdot\mathbf{P}'\mathrm{d}V = \int \mathbf{E}_{\mathrm{ex}}\cdot\mathbf{P}_2^{\mathrm{eq}}\mathrm{d}V$, which is proved in Appendix A. Here, the first term on the right hand side of eq. (4.26) stands for the solvation energy of equilibrium state "2", and the second term is just the solvent reorganization energy, i.e.,

$$\lambda_{\mathrm{s}} = \frac{1}{2}\int \mathbf{E}_{\mathrm{ex}}\cdot\mathbf{P}'\mathrm{d}V \tag{4.27}$$

Therefore, it can be seen from eqs. (4.26) and (4.27) that the solvent reorganization energy is the energy stored in the medium from equilibrium state "2" to nonequilbirium state "2", that is, the energy change of the medium resulted from the addition of $\mathbf{P}'$ in the equilibrium state "2" by imposing the external field $\mathbf{E}_{ex}$.

Combining eqs. (4.11), (4.14) and (4.27), we obtain the final form for the solvent reorganization energy as

$$\lambda_s = \frac{1}{2}\frac{\varepsilon_s - \varepsilon_{op}}{\varepsilon_s - 1}\int_V \Delta\mathbf{E}_{dy} \cdot (\Delta\mathbf{P}_{eq} - \Delta\mathbf{P}_{dy})\mathrm{d}V \tag{4.28}$$

4.3 Solvent reorganization energy and its application

4.3.1 Solvent reorganization energy and spectral shift

Electron transfer reactions play an important role in chemistry and biochemistry, such as the break and repair of DNA, the function of enzyme and the breath of the life body. In Marcus' electron transfer theory, the total reorganization energy is composed of two contributions: the internal reorganization λ_{in} due to the change of the reactant structure and the solvent reorganization energy λ_s due to the change of the solvent structure, i.e.

$$\lambda = \lambda_{in} + \lambda_s \tag{4.29}$$

Marcus defined the solvent reorganization energy between the difference of the electrostatic solvation free energy between the nonequilibrium "2" and equilibrium "2" state, i.e.[9]

$$\lambda_s = F_{2,sol}^{non} - F_{2,sol}^{eq} \tag{4.30}$$

In the above derivation, we have obtained the solvent reorganization energy in electric field-polarization representation as shown in eq.(4.27) and we also can derive another form of charge-potential representation as

$$\lambda_s = \frac{1}{2}\frac{\varepsilon_s - \varepsilon_{op}}{\varepsilon_s - 1}\oint_S \Delta\Phi_{dy}(\Delta\sigma_{dy} - \Delta\sigma_{eq})\mathrm{d}S \tag{4.31}$$

The detailed derivation can be found in Appendix B.

For the different solute size, shape and charge distribution, we simplify the solute charge distribution as the multipole expansion located as the center of a spherical cavity. In the case of the solute monopole, we can obtain the concise form as

$$\lambda_s = \frac{1}{2}q_{ex}(\Delta\varphi_{eq} - \Delta\varphi_{dy}) \tag{4.32}$$

where q_{ex} is the external charge located at the center of the cavity to produce $\mathbf{P}'$. For the point charge q_D and q_A locating at the centers of the electron donor and accepter spherical cavities, the form of solvent reorganization energy in two-sphere model is given as

$$\lambda_s = \frac{1}{2}q_{D,ex}(\Delta\varphi_{D,eq} - \Delta\varphi_{D,dy}) + \frac{1}{2}q_{A,ex}(\Delta\varphi_{A,eq} - \Delta\varphi_{A,dy}) \tag{4.33}$$

where $q_{D,ex}$ and $q_{A,ex}$ are the imposed external charge at the center of donor's and acceptor's spheres.

In the case of solute charge being point dipole moment at the center of a sphere, the solvent reorganization energy can be derived to

$$\lambda_s = \frac{1}{2}\boldsymbol{\mu}_{ex} \cdot (\Delta \mathbf{E}_{p,dy} - \Delta \mathbf{E}_{p,eq}) \tag{4.34}$$

where $\boldsymbol{\mu}_{ex}$ is the external dipole at the sphere center and the subscript "p" denote the field strength produced by the polarization. The derivation for eqs. (4.32)-(4.33) is detailed in appendix C.

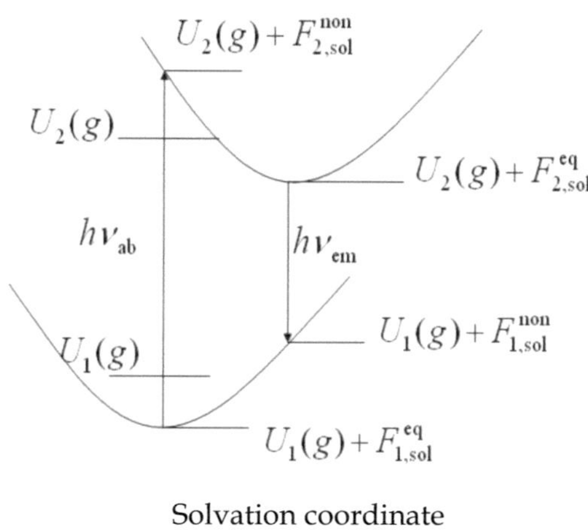

Fig. 1. Spectral shift for the absorption and emission spectrum

Similar to the definition for the solvent reorganization energy, the spectral shifts for light absorption and emission also can be defined as shown in Figure 1. Due to the Franck-Condon transition of the solute in medium, the solute-solvent system will experience the following change: starting from the equilibrium ground state, then reaching the nonequilibrium excited state, and then relaxing to the equilibrium excited state, following by the nonequilibrium ground state, and finally reaching the starting equilibrium ground state. Here we use subscripts "1" and "2" to denote the different charge distributions in ground and excited state respectively. In Figure 1, $U_i(\mathrm{g})\,(i=1,2)$ stands for the internal energies of the solute in ground state "1" and excited state "2" in vacuum. $h\nu_{ab}$ and $h\nu_{em}$ are the absorption and emission energy in medium respectively.

According to the traditional nonequilibrium solvation theory[2,9], the absorption spectral shift is defined as the free energy difference between nonequilibrium excited state "2" and equilibrium ground state "1". Ignoring the self-consistence between the solute and solvent charge, the spectral shift for the absorption spectrum can be defined as the solvation energy difference between nonequilibrium excited state "2" and equilibrium ground state "1", i.e.

$$\Delta h\nu_{ab} = F_{2,sol}^{non} - F_{1,sol}^{eq} \tag{4.35}$$

Correspondingly, for the inversed process, namely, emission (or fluorescence) spectrum, the spectral shift can be expressed as

$$\Delta h\nu_{\text{em}} = F_{1,\text{sol}}^{\text{non}} - F_{2,\text{sol}}^{\text{eq}} \tag{4.36}$$

According to the definitions given in eqs. (4.35) and (4.36), the positive value of $\Delta h\nu_{\text{ab}}$ is blue shift, while the positive value of $\Delta h\nu_{\text{em}}$ is red shift. The solvation energies for the equilibrium ground and excited states in the charge-potential presentation can be given as

$$F_{1,\text{sol}}^{\text{eq}} = \frac{1}{2}\int_V \rho_1 \varphi_1^{\text{eq}} \mathrm{d}V \tag{4.37}$$

$$F_{2,\text{sol}}^{\text{eq}} = \frac{1}{2}\int_V \rho_2 \varphi_2^{\text{eq}} \mathrm{d}V \tag{4.38}$$

where φ is the polarization potential and ρ the charge density of the solute. According to eq. (4.26), the nonequilibrium solvation energy can be expressed in charge-potential form as

$$F_2^{\text{non}} = \frac{1}{2}\int_V \rho_2 \varphi_2^{\text{eq}} \mathrm{d}V + \lambda_{\text{s}} \tag{4.39}$$

$$F_1^{\text{non}} = \frac{1}{2}\int_V \rho_1 \varphi_1^{\text{eq}} \mathrm{d}V + \lambda_{\text{s}} \tag{4.40}$$

Together with eqs. (4.35)-(4.40), the general forms for the absorption and emission spectral shift can be obtained as

$$\Delta h\nu_{\text{ab}} = \Delta F_2^{\text{non}} - \Delta F_1^{\text{eq}} = \lambda_{\text{s}} + \frac{1}{2}\int_V (\rho_2 \varphi_2^{\text{eq}} - \rho_1 \varphi_1^{\text{eq}}) \mathrm{d}V \tag{4.41}$$

$$\Delta h\nu_{\text{em}} = \Delta F_1^{\text{non}} - \Delta F_2^{\text{eq}} = \lambda_{\text{s}} - \frac{1}{2}\int_V (\rho_2 \varphi_2^{\text{eq}} - \rho_1 \varphi_1^{\text{eq}}) \mathrm{d}V \tag{4.42}$$

4.3.2 The two-sphere model for the solvent reorganization energy

For the electron transfer reaction between the electron donor D with charges of q_{D} and electron acceptor A with charge of q_{A}, the reaction process of transferring the charge of Δq can be described by the following equation

$$\mathrm{D}^{q_{\text{D}}} + \mathrm{B} + \mathrm{A}^{q_{\text{A}}} \rightarrow \mathrm{D}^{q_{\text{D}}+\Delta q} + \mathrm{B} + \mathrm{A}^{q_{\text{A}}-\Delta q} \tag{4.43}$$

where B is bridge between the donor and acceptor, $q_{\text{D}} + \Delta q$ and $q_{\text{A}} - \Delta q$ are the charge brought by the donor and acceptor after the electron transfer reaction. Here, we assume that all the point charges q_{D}, q_{A}, $q_{\text{D}} + \Delta q$ and $q_{\text{A}} - \Delta q$ locate at the centers of the two spheres shown in Figure 2. r_{D} and r_{A} are the radii for donor and acceptor spheres respectively. The two spheres are surrounded by the solvent with ε_{s}, and the distance between the two spherical centers is *d*, which is assumed much larger than the radius of r_{D} and r_{A}.

Similar to the treatment by Marcus [2], ignoring the image charge effect due to the surface polarization charge, the polarization potential due to charge variation $\Delta q_{\mathrm{D}} = \Delta q$ on the surface of sphere D can be expressed as

$$Q_{\mathrm{D,s}} = \Delta q(\frac{1}{\varepsilon_{\mathrm{s}}} - 1) \text{, } Q_{\mathrm{D,dy}} = \Delta q(\frac{1}{\varepsilon_{\mathrm{op}}} - 1) \tag{4.44}$$

in the medium of ε_{s} and $\varepsilon_{\mathrm{op}}$. Correspondingly the charge variation $\Delta q_{\mathrm{A}} = -\Delta q$ in sphere A will induce the polarized charge on the surface of sphere A as

$$Q_{\mathrm{A,s}} = -\Delta q(\frac{1}{\varepsilon_{\mathrm{s}}} - 1) , Q_{\mathrm{A,dy}} = -\Delta q(\frac{1}{\varepsilon_{\mathrm{op}}} - 1) \tag{4.45}$$

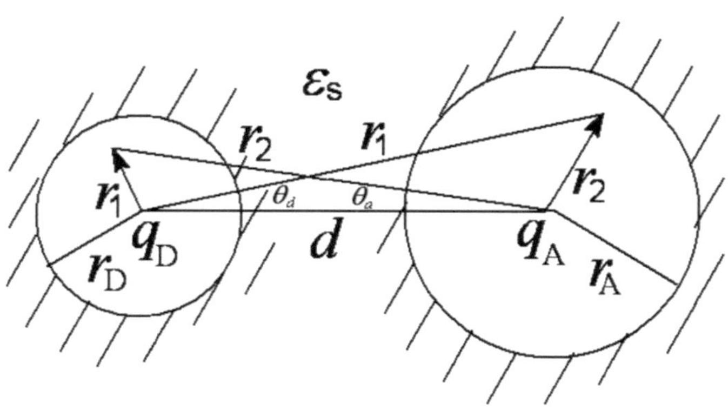

Fig. 2. Two-sphere model

Thus, in the medium of $\varepsilon_{\mathrm{op}}$, the polarization charge $Q_{D,\mathrm{dy}}$ due to Δq_{D} will generate the polarization potential $\frac{\Delta q}{r_D}(\frac{1}{\varepsilon_{\mathrm{op}}} - 1)$ at the center of sphere D, and $Q_{\mathrm{A,dy}}$ due to Δq_{A} will generate the polarization potential $-\frac{\Delta q}{d}(\frac{1}{\varepsilon_{\mathrm{op}}} - 1)$ at the center of sphere D. Based on the principle of potential superposition, the total polarization potential at the center of sphere D can be expressed as

$$\Delta\varphi_{\mathrm{D,dy}} = (\frac{\Delta q}{r_D} - \frac{\Delta q}{d})(\frac{1}{\varepsilon_{\mathrm{op}}} - 1) \tag{4.46}$$

With the similar treatment, the total polarization potential at the center of sphere A is

$$\Delta\varphi_{\mathrm{A,dy}} = -(\frac{\Delta q}{r_A} - \frac{\Delta q}{d})(\frac{1}{\varepsilon_{\mathrm{op}}} - 1) \tag{4.47}$$

For the solvent with dielectric constant ε_s, we have

$$\Delta\varphi_{D,eq} = (\frac{\Delta q}{r_D} - \frac{\Delta q}{d})(\frac{1}{\varepsilon_s} - 1) \quad (4.48)$$

$$\Delta\varphi_{A,eq} = -(\frac{\Delta q}{r_A} - \frac{\Delta q}{d})(\frac{1}{\varepsilon_s} - 1) \quad (4.49)$$

With the zeroth approximation of multipole expansion for the solute charge distribution, the external charges at the position of donor and acceptor can be derived from eq. (C2) as

$$q_{D,ex} = \frac{\varepsilon_{op} - \varepsilon_s}{(\varepsilon_s - 1)\varepsilon_{op}}\Delta q, \; q_{A,ex} = -\frac{\varepsilon_{op} - \varepsilon_s}{(\varepsilon_s - 1)\varepsilon_{op}}\Delta q \quad (4.50)$$

Substituting eqs.(4.46)-(4.50) into eq.(4.33), the solvent reorganization energy in the two-sphere and point charge model can be obtained as

$$\lambda_s = \frac{(\Delta q)^2}{2}\frac{(\varepsilon_s - \varepsilon_{op})^2}{(\varepsilon_s - 1)\varepsilon_s\varepsilon_{op}^2}(\frac{1}{r_D} + \frac{1}{r_A} - \frac{2}{d}) \quad (4.51)$$

It is different from the traditional Marcus result [2,9]

$$\lambda_M = \frac{(\Delta q)^2}{2}\frac{(\varepsilon_s - \varepsilon_{op})}{\varepsilon_s\varepsilon_{op}}(\frac{1}{r_D} + \frac{1}{r_A} - \frac{2}{d}) \quad (4.52)$$

The two sphere model is widely used to investigate the electron transfer reactions in solvent for its brief and simple expression. It is clear that the present two-sphere model will predict the solvent reorganization energy to be smaller than that by Marcus formula by a factor of $\frac{\varepsilon_s - \varepsilon_{op}}{\varepsilon_{op}(\varepsilon_s - 1)}$.

4.3.3 The spectral shift of photo-induced ionization energy in a single sphere

Now we consider the simplest case for the nonequilibrium state: the solute charge distribution is point charge located at the center of the sphere with radius *a*, surrounded by the solvent with dielectric constant ε_s. This model can be adopted to treat the spectral shift of the vertical ionization energy. The atomic (or ionic) photo-induced ionization process in the medium with dielectric constants ε_s or ε_{op} can be represented as

$$B^{Q_1} \xrightarrow{-e^-} B^{Q_2} \quad (4.53)$$

where Q_2 and Q_1 are the solute charges before and after the ionization respectively. Induced by the charge change $\Delta Q = Q_2 - Q_1$, the polarization charge on the sphere surface can be obtained as

$$\Delta Q_{eq}^{surf} = \Delta Q(\frac{1 - \varepsilon_s}{\varepsilon_s}) \quad (4.54)$$

$$\Delta Q_{dy}^{surf} = \Delta Q(\frac{1-\varepsilon_{op}}{\varepsilon_{op}}) \tag{4.55}$$

in the medium of ε_s and ε_{op}, and it will generate the polarization potential at the sphere center in these two cases as

$$\Delta\varphi_{eq} = \frac{Q_2 - Q_1}{a}(\frac{1}{\varepsilon_s} - 1) \tag{4.56}$$

$$\Delta\varphi_{dy} = \frac{Q_2 - Q_1}{a}(\frac{1}{\varepsilon_{op}} - 1) \tag{4.57}$$

Recalling eq. (C2), the external charge condensed at the center can be achieved as

$$q_{ex} = \frac{\varepsilon_{op} - \varepsilon_s}{(\varepsilon_s - 1)\varepsilon_{op}}(Q_2 - Q_1) \tag{4.58}$$

Thus eqs. (4.39) and (4.41) can be simplified as

$$\lambda_s = \frac{1}{2} q_{ex}(\Delta\varphi_{eq} - \Delta\varphi_{dy}) \tag{4.59}$$

$$\begin{aligned} \Delta h\nu_{ab} &= \lambda_s + \frac{1}{2}(Q_2\varphi_2^{eq} - Q_1\varphi_1^{eq}) \\ &= \frac{1}{2}\frac{\varepsilon_{op} - \varepsilon_s}{(\varepsilon_s - 1)\varepsilon_{op}}(Q_2 - Q_1)\frac{Q_2 - Q_1}{a}(\frac{1-\varepsilon_s}{\varepsilon_s} - \frac{1-\varepsilon_{op}}{\varepsilon_{op}}) + \frac{1}{2}[Q_2\frac{Q_2}{a}(\frac{1}{\varepsilon_s} - 1) - Q_1\frac{Q_1}{a}(\frac{1}{\varepsilon_s} - 1)] \end{aligned} \tag{4.60}$$

Further we have the form of the spectral shift in the vertical ionization of the charged particle,

$$\Delta h\nu_{ab} = \frac{1}{2a}(Q_1 - Q_2)^2 \frac{(\varepsilon_s - \varepsilon_{op})^2}{(\varepsilon_s - 1)\varepsilon_{op}^2\varepsilon_s} + \frac{1}{2a}(Q_1^2 - Q_2^2)(1 - \frac{1}{\varepsilon_s}) \tag{4.61}$$

4.3.4 Spectral shift of point dipole in a sphere cavity

Here we will adopt Onsager model of sphere cavity and point dipole moment to treat the nonequilibrium polarization in spectrum. The solute charge distribution is considered as the point dipole, locating at the center of single vacuum sphere with the radius a, as shown in Figure 3. The solute cavity is surrounded by the solvent with dielectric constant ε_s. The solute dipole will change from $\boldsymbol{\mu}_1$ to $\boldsymbol{\mu}_2$ due to the Franck-Condon transition in the light absorption process, and the light emission will lead to the inversed change of the solute dipole.

First, the reaction field in the sphere cavity will be derived. In Figure 2, the total electric potential Φ_1^{eq} in equilibrium ground state satisfies the following differential equations and boundary conditions:

$$\begin{cases} \nabla^2 \varphi_{1,\text{in}}^{\text{eq}} = 0 \quad (r < a); \qquad \nabla^2 \Phi_{1,\text{out}}^{\text{eq}} = 0 \quad (r > a) \\ \Phi_{1,\text{in}}^{\text{eq}} = \varphi_{1,\text{in}}^{\text{eq}} + \dfrac{\mu_1}{r^2}\cos\theta \end{cases} \tag{4.62}$$

$$\begin{cases} \varphi_{1,\text{in}}^{\text{eq}} |_{r\to 0} \ \text{ is finite}, \quad \Phi_{1,\text{out}}^{\text{eq}} |_{r\to\infty} = 0 \\ \Phi_{1,\text{ln}}^{\text{eq}} = \Phi_{1,\text{out}}^{\text{eq}}, \quad \dfrac{\partial \Phi_{1,\text{in}}^{\text{eq}}}{\partial r} = \varepsilon_s \dfrac{\partial \Phi_{1,\text{out}}^{\text{eq}}}{\partial r} \qquad (r = a) \end{cases} \tag{4.63}$$

where the subscripts "in" and "out" stand for inside and outside the sphere cavity and "θ" is the angle between the vectors of solute dipole and **r**. We assume that Φ_1^{eq} has the following form:

$$\begin{cases} \Phi_{1,\text{in}}^{\text{eq}} = -E_{1\text{p}} r \cos\theta + \dfrac{\mu_1}{r^2}\cos\theta \\ \Phi_{1,\text{out}}^{\text{eq}} = \dfrac{\mu_1 + \mu_{1\text{p}}}{r^2}\cos\theta \end{cases} \tag{4.64}$$

where the unknown $E_{1\text{p}}$ and $\mu_{1\text{p}}$ are polarization field strength and equivalent dipole for the solvent polarization. The above eq. (4.64) can satisfy the differential equation (4.62). So, substituting eq. (4.64) into (4.63) leads to

$$\begin{cases} -E_{1\text{p}} a + \dfrac{\mu_1}{a^2} = \dfrac{\mu_1 + \mu_{1\text{p}}}{a^2} \\ -E_{1\text{p}} - \dfrac{2\mu_1}{a^3} = -\varepsilon_s \dfrac{2(\mu_1 + \mu_{1\text{p}})}{a^3} \end{cases} \tag{4.65}$$

then we can obtain

$$E_{1\text{p}} = \frac{2(\varepsilon_s - 1)}{(2\varepsilon_s + 1)} \frac{\mu_1}{a^3}, \quad \mu_{1\text{p}} = -\frac{2(\varepsilon_s - 1)}{2\varepsilon_s + 1}\mu_1 \tag{4.66}$$

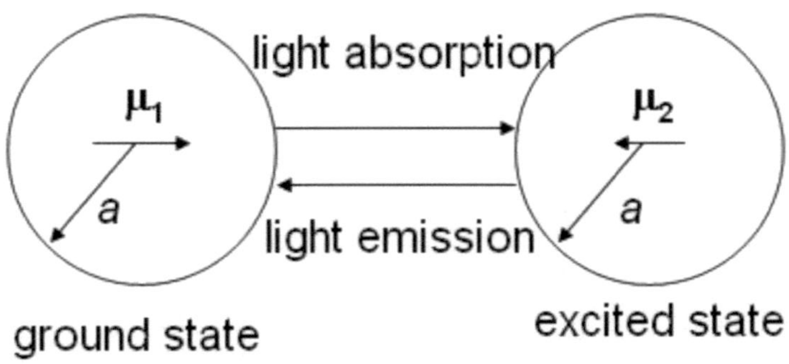

Fig. 3. Solvation model for single sphere

By substituting eq. (4.66) into eq. (4.64), the potentials inside and outside the sphere are

$$\begin{cases} \Phi_{1,\text{in}}^{\text{eq}} = -\dfrac{2(\varepsilon_s - 1)}{(2\varepsilon_s + 1)} \dfrac{\mu_1 r}{a^3} \cos\theta + \dfrac{\mu_1}{r^2} \cos\theta \\ \Phi_{1,\text{out}}^{\text{eq}} = \dfrac{3}{(2\varepsilon_s + 1)} \dfrac{\mu_1}{r^2} \cos\theta \end{cases} \tag{4.67}$$

Therefore, the polarization potential inside the sphere cavity is

$$\varphi_1 = -\frac{2(\varepsilon_s - 1)}{(2\varepsilon_s + 1)} \frac{\mu_1}{a^3} r \cos\theta \tag{4.68}$$

Correspondingly, we can obtain the polarization field strength $\mathbf{E}_{1\text{p}}$ inside the sphere and the total field outside the sphere for equilibrium ground state as

$$\mathbf{E}_{1\text{p}} = \frac{2(\varepsilon_s - 1)}{(2\varepsilon_s + 1)} \frac{\boldsymbol{\mu}_1}{a^3} \tag{4.69}$$

$$\mathbf{E}_{1,\text{out}} = \frac{3}{(2\varepsilon_s + 1)} \frac{\boldsymbol{\mu}_1 \cdot \mathbf{r}}{r^3} \tag{4.70}$$

Thus, the polarization field due to the dipole change $\Delta\boldsymbol{\mu} = \boldsymbol{\mu}_2 - \boldsymbol{\mu}_1$ in medium with ε_s and ε_{op} can be achieved as

$$\Delta\mathbf{E}_{\text{p,eq}} = \frac{2(\varepsilon_s - 1)}{(2\varepsilon_s + 1)} \frac{\Delta\boldsymbol{\mu}}{a^3}, \quad \Delta\mathbf{E}_{\text{p,dy}} = \frac{2(\varepsilon_{\text{op}} - 1)}{(2\varepsilon_{\text{op}} + 1)} \frac{\Delta\boldsymbol{\mu}}{a^3} \tag{4.71}$$

Recalling eq. (C5), the introduced external dipole moment is

$$\boldsymbol{\mu}_{\text{ex}} = \frac{\varepsilon_{\text{op}} - \varepsilon_s}{\varepsilon_s - 1} \frac{3}{2\varepsilon_{\text{op}} + 1} \Delta\boldsymbol{\mu} \tag{4.72}$$

According to the definition of the solvent reorganization energy in eq. (4.34), we can obtain

$$\begin{aligned} \lambda_s &= \frac{1}{2} \frac{\varepsilon_{\text{op}} - \varepsilon_s}{\varepsilon_s - 1} \frac{3}{2\varepsilon_{\text{op}} + 1} \Delta\boldsymbol{\mu} \cdot \left[\frac{2(\varepsilon_{\text{op}} - 1)}{(2\varepsilon_{\text{op}} + 1)} \frac{\Delta\boldsymbol{\mu}}{a^3} - \frac{2(\varepsilon_s - 1)}{(2\varepsilon_s + 1)} \frac{\Delta\boldsymbol{\mu}}{a^3}\right] \\ &= \frac{(\Delta\mu)^2}{a^3} \frac{9(\varepsilon_s - \varepsilon_{\text{op}})^2}{(\varepsilon_s - 1)(2\varepsilon_s + 1)(2\varepsilon_{\text{op}} + 1)^2} \end{aligned} \tag{4.73}$$

According to the definition in eq. (4.41), we obtain the final form for the absorption spectral shift with single sphere and point dipole approximation as

$$\Delta h\nu_{ab} = \lambda_s + \frac{1}{2}(\boldsymbol{\mu}_1 \cdot \mathbf{E}_{1p}^{eq} - \boldsymbol{\mu}_2 \cdot \mathbf{E}_{2p}^{eq}) = \frac{(\Delta\mu)^2}{a^3}\frac{9(\varepsilon_s - \varepsilon_{op})^2}{(\varepsilon_s - 1)(2\varepsilon_s + 1)(2\varepsilon_{op} + 1)^2} + \frac{(\varepsilon_s - 1)}{(2\varepsilon_s + 1)}\frac{\mu_1^2 - \mu_2^2}{a^3} \tag{4.74}$$

The similar treatment can lead to the emission spectral shift as

$$\Delta h\nu_{em} = \lambda_s - \frac{1}{2}(\boldsymbol{\mu}_1 \cdot \mathbf{E}_{1}^{eq} - \boldsymbol{\mu}_2 \cdot \mathbf{E}_{2}^{eq}) = \frac{(\Delta\mu)^2}{a^3}\frac{9(\varepsilon_s - \varepsilon_{op})^2}{(\varepsilon_s - 1)(2\varepsilon_s + 1)(2\varepsilon_{op} + 1)^2} - \frac{(\varepsilon_s - 1)}{(2\varepsilon_s + 1)}\frac{\mu_1^2 - \mu_2^2}{a^3} \tag{4.75}$$

4.4 Comments on traditional nonequilibrium solvation theory

Nowadays, accompanying the development of computational methods and the progresses in computer science, solvent effect calculations at different levels have attracted much attention. Because most of chemistry and biochemistry reactions occur in solution, incorporation of the solvent effects into chemical models has been of great interest for several decades. Owing to the competitive advantages, continuum models are still playing a key role so far, although more and more explicit solvent methods, which take the microscopic structures of the solvent molecules into account, have been explored. There are two principal advantages of the continuum models. The first one is the reduction of the system's numbers of freedom degrees. If we take explicitly a few of solvent layers which involve hundreds of solvent molecules, a huge number of degrees of freedom will be added. The first thing we must face with is a large number of conformations. In addition, the observable structural and dynamical properties of some specific solute we most concern will be averaged. In fact, if one realizes that the complementary methods based on some explicit solvent methods are also not perfectly accurate, one will find the continuum model accounts for the dominant parts of solvent effects. So the second advantage of the macroscopic continuum models provides rather good ways to treat the strong and long-range electrostatic forces that dominate many solvation phenomena.

There are many circumstances in molecular modeling studies where a simplified description of solvent effects has advantages over the explicit modeling of each solvent molecule. The solute charge distribution and its response to the reaction field of the solvent dielectric, can be modeled either by quantum mechanics or by partial atomic charges in a molecular mechanics description. In spite of the severity of approximation of continuum models, it often gives a good account of equilibrium solvation energy, and hence widely used to estimate pKs, redox potential, and the electrostatic contributions to molecular solvation energy.

Up to now, several models for the equilibrium solvation based on the continuous medium theory were developed. The simplest one is the Onsager model with a point-dipole of solute in a spherical cavity. One of the most remarkable successes of the calculation of equilibrium solvation for arbitrary solute cavity is the establishment of polarizable continuum model (PCM) by Tomasi. Thereafter, different procedures for solute-solvent system have been developed. Introducing the numerical solution of the appropriate electrostatic potential into the popular quantum chemical packages yields different equilibrium solvation models. At present, a feature common to all the continuum solvation approximations is that the solute-solvent interactions are described in terms of the solute-reaction field interactions. The reaction field is due to the solvent polarization perturbed by the presence of solute, and the reaction field in turn perturbs the solute, until self-consistence is achieved. The reaction field is usually computed by solving the suitable Poisson equations.

So far, most of continuum models are properly referred to as equilibrium solvation models. Besides the structures and properties of a thermodynamically equilibrated solute-solvent system, the processes that take place on longer timescales may thus be legitimately thought of as equilibrium processes with respect to solvation. However, the question arises how to apply continuum models to the very fast processes. For instance, the transition state structures in principle live for only a single vibrational period. In such cases, the solvent response may not have time to equilibrate with the electronic state change at the position of transition state. Hence, a continuum model developed based on the fully equilibrated solvation would overestimate the solvation free energy by the assumed equilibration. In fact, many cases concern the nonequilibrium solvation problems in solution. The typical examples are: condensed-phase electron transfer, spectral shifts of photon absorption and emission in solution, and vibrational spectrum in solution. Among them, the solvent reorganization energy of the electron transfer and the spectral shifts attracted the most attention. So, in the present comments, we confine ourselves to these two kinds of nonequilibrium solvent effects, although the nonequilibrium solvation problem exists in some other processes such as proton transfer.

Let us date back to the beginning of the establishment of the nonequilibrium solvation theories. A brief overview on this topic will be helpful for us to clarify what fundamental defects exist in the present theories and application models. The concept of nonequilibrium solvation led to great progresses for people to understand the physics of fast processes in solution. Based on the separation of the two kinds of polarizations, orientational and electronic, Marcus applied the reversible work method to the establishment of the electrostatic free energy expression of the nonequilibrium solvation state. In Marcus' original treatments, the electrostatic free energy of nonequilibrium state of solution was defined as the sum of reversible works done during charging process involving two steps as follows[2]

$$[\rho=0,\Phi=0]\xrightarrow{A1,\varepsilon_S}[\rho_1,\Phi_1^{eq}]\xrightarrow{A2,\varepsilon_{op}}[\rho_2,\Phi_2^{non}] \tag{4.76}$$

In eq.(4.76), Φ denotes the total electric potential, including both the potential ψ by the solute charge in vacuum and potential φ due to the polarization of the medium. We confine our discussions only to the solute charge and the bound charge at present. The solute charge, which refers to the "free charge" from the viewpoint of electrodynamics, in principle represents the charge that can move about through the material. In practice what this ordinarily means is that free charges are not associated with any particular nucleus, but roam around at will. By contrast, the bound charges in dielectrics are attached to specific atoms or molecules. They are on a tight leash, and what they can do is to move a bit within the atom or molecule. Such microscopic displacements are not as dramatic as the wholesale rearrangement of solute charge, but their cumulative effects account for the characteristic behaviors of dielectric materials. For convenience, we call hereafter ψ the vacuum potential and φ the polarization potential. We do not distinguish "free charge" and "solute charge". Two terms, bound charge and polarized charge, are undistinguished in our previous works, but we use "bound charge" here.

In the establishment of the nonequilibrium state, the first step, A1, charges the solute to ρ_1, and Φ reaches an equilibrium in solvent of a static dielectric constant ε_s. In step A2, the solute is charged from ρ_1 to ρ_2 but only the electronic component of the solvent polarization, which corresponds to the optical dielectric constant ε_{op} of the solvent, responds. The system arrives at a new state in which the electronic polarization of solvent reaches equilibrium with ρ_2 but the orientaional polarization does not. This state, we denote it by $[\rho_2,\Phi_2^{non}]$, is referred to as the "nonequilibrium" state. If we note that the potential change in step A2 is caused by the change of solute charge, but only the electronic polarization responds, we can take the nonequilibrium as a superposition of two "equilibrium" states, $[\rho_1,\Phi_1^{eq}]$ and $[\Delta\rho,\Delta\Phi_{op}]$. The former is a state in which ρ_1 equilibrates with the medium of dielectric constant of ε_s, but the latter is such that the solute charge difference $\Delta\rho$ equilibrates in the hypothetical medium of a dielectric constant ε_{op}. Here we define the solute charge change and the potential change as

$$\begin{aligned}\Delta\rho &= \rho_2 - \rho_1 \\ \Delta\Phi_{op} &= \Phi_2^{non} - \Phi_1^{eq}\end{aligned} \tag{4.77}$$

As mentioned above, we divide the total potential Φ into two constitutive parts: ψ due to the solute charge in vacuum and φ due to the bound charge. We need to distinguish φ^{eq} of equilibrium from φ^{non} of nonequilibrium for the polarization potential but this is unnecessary for ψ. Therefore we have

$$\Phi_i^{eq} = \psi_i + \varphi_i^{eq}, \quad \Phi_i^{non} = \psi_i + \varphi_i^{non} \qquad (i=1,2) \tag{4.78}$$

If we consider the inverse process of eq.(4.76) (denoted as as process B), we can write the analogue as

$$[\rho=0,\Phi=0] \xrightarrow{\text{B1},\varepsilon_S} [\rho_2,\Phi_2^{eq}] \xrightarrow{\text{B2},\varepsilon_{op}} [\rho_1,\Phi_1^{non}] \tag{4.79}$$

As mentioned above, we ignore the influence of solvent polarization upon the solute free charge, hence the charge distributions ρ_1 and ρ_2 in eq.(4.79) are supposed to be exactly the same as given in eq.(4.76).

If the properties of the dielectric do not vary during the process, it is very common to integrate the work done in the charging process by the following equation,

$$\delta W = \int_V \Phi \delta\rho \mathrm{d}V \tag{4.80}$$

The integration is over the whole space. Throughout this review, we use W to denote the work done and G the total free energy. But if we ignore the penetration of ρ into the medium region, the integration will be in fact only carried out within the cavity occupied by the solute. Introducing a charging fraction α during step A1 of eq.(4.76), the electrostatic free energy of equilibrium state $[\rho_1, \Phi_1^{\mathrm{eq}}]$ was expressed in the well-known form, i.e.,

$$G_1^{\mathrm{eq}} = W_{\mathrm{A1}} = (1/2)\int_V \rho_1 \Phi_1^{\mathrm{eq}} \mathrm{d}V \tag{4.81}$$

On the basis of step A1, step A2 introduces the further charge distribution change $\Delta\rho$, and the potential accordingly responds, so the charge distribution ρ^{α} and the total electric potential Φ^{α} during step A2 were expressed by Marcus as[2]

$$\rho^{\alpha} = \rho_1 + \alpha(\rho_2 - \rho_1) \quad \text{and} \quad \Phi^{\alpha} = \Phi_1^{\mathrm{eq}} + \alpha(\Phi_2^{\mathrm{non}} - \Phi_1^{\mathrm{eq}}) \quad (\alpha = 0 \sim 1) \tag{4.82}$$

Therefore, the electrostatic free energy of nonequilibrium state was expressed by Marcus as the sum of work done in steps A1 and A2 [eq.(17) of ref.2], i.e.

$$G_2^{\mathrm{non}}(\mathrm{A}) = (1/2)\int_V (\rho_2 \Phi_2^{\mathrm{non}} + \rho_2 \Phi_1^{\mathrm{eq}} - \rho_1 \Phi_2^{\mathrm{non}}) \mathrm{d}V \tag{4.83}$$

Our following arguments will make it clear that eq.(4.83) is incorrect owing to the different response properties of the medium in equilibrium and nonequilibrium cases.

In the work of Marcus, the solvent reorganization energy is defined as the difference of electrostatic free energies between the nonequilibrium state and the equilibrium state subject to the same solute charge distribution, i.e.,

$$\lambda_{\mathrm{o}} = G_2^{\mathrm{non}} - G_2^{\mathrm{eq}} \tag{4.84}$$

Introducing the two-sphere approximation, the famous two-sphere model (as given by eq.(4.52) of estimating the reorganization energy was consequently developed and widely applied for decades. However, the Marcus two-sphere model often overestimates the solvent reorganization energy was, by a factor of about two for many electron transfer reactions[10,11]. For example, Basilevsky[12] developed a numerical method to evaluate the reorganization energy and applied it to the well-known Closs-Miller ET systems by using the conventional Marcus theory. However, the calculated values for the biphenyl-bridge-naphthalene system were exaggerated by a factor of about 2 than those fitted from the experimental rate constants.

The classical issue on the electrostatic free energy of nonequilibrium solvation in a continuous medium is revisited. The central idea, which has never been considered before, is to introduce a constrained equilibrium that is required to have the same charge distribution, polarization and entropy as the true nonequilibrium state (see Sections 1~3). Such a reference is certainly realizable via a quasistatic procedure. The location of the source for the tuning electric field $\mathbf{E}_{\mathrm{ex}}$ is yet completely irrelevant. From this reference, the electrostatic free energy of nonequilibrium solvation can directly be obtained in strict accordance with the principle of thermodynamics. It is also shown that the long lasting

problem that the solvent reorganization energy is always overestimated by the previous continuum models is solved in a natural manner. It is believed that the present paradigm is completely general and can be used to derive other thermodynamic quantities of the isothermal nonequilibrium system as well.
The freezing of the state variables here is quite different from the treatment by Marcus. In fact, to freeze the variables of any nonequilibrium state is not only an abstract idea, but also a proper arrangement which can be used to realize the freezing. The fundamental difference between the Marcus approach and the present strategy is obvious. The freezing of the inertial polarization in Marcus work is just an idea without any measure, while our work realizes the freezing by introducing an external field. In our work, the whole polarization is kept frozen, not only the inertial part.
We mention that here the problem in the traditional nonequilibrium solvation theory arises from the simple reversible work integration, without consideration of any variable that describes the nonequilibrium state. A reversible work method applying to a non-quasistatic process is obviously arbitrary and lack of thermodynamic support. On the contrary, in our treatment, we rigorously obey the thermodynamics and a crucial external variable $\mathbf{E}_{\mathrm{ex}}$, which is used to constrain the nonequilibrium state to an "equilibrium" one, enters the expression of solvent reorganization energy. More details can be found in the references 4~6.

5. Appendix

5.1 Appendix A : Proof of $\int \mathbf{E}_{2\mathrm{c}} \cdot \mathbf{P}'\mathrm{d}V = \int \mathbf{E}_{\mathrm{ex}} \cdot \mathbf{P}_2^{\mathrm{eq}}\mathrm{d}V$

In the constrained equilibrium state, there is the relations of $\mathbf{P}' = \chi_s \mathbf{E}'$ and $\mathbf{P}_2^{\mathrm{eq}} = \chi_s \mathbf{E}_2^{\mathrm{eq}}$, thus we have

$$\int_V \mathbf{E}_2^{\mathrm{eq}} \cdot \mathbf{P}'\mathrm{d}V = \int_V \chi_s \mathbf{E}_2^{\mathrm{eq}} \cdot \mathbf{E}'\mathrm{d}V = \int_V \mathbf{E}' \cdot \mathbf{P}_2^{\mathrm{eq}}\mathrm{d}V \tag{A1}$$

$$\int_V \mathbf{E}_{2\mathrm{c}} \cdot \mathbf{P}'\mathrm{d}V + \int_V \mathbf{E}_{2\mathrm{p}} \cdot \mathbf{P}'\mathrm{d}V = \int_V \mathbf{E}_{\mathrm{ex}} \cdot \mathbf{P}_2^{\mathrm{eq}}\mathrm{d}V + \int_V \mathbf{E}_{\mathrm{p}'} \cdot \mathbf{P}_2^{\mathrm{eq}}\mathrm{d}V \tag{A2}$$

Applying the formulas of $\nabla \cdot \mathbf{P} = 0$, and $\mathbf{n} \cdot \mathbf{P} = \sigma$ with σ being the surface polarized charge density, the second term on the left hand side of the above equation can be rewritten as

$$\begin{aligned}\int_V \mathbf{E}_{2\mathrm{p}} \cdot \mathbf{P}'\mathrm{d}V &= -\int_V \nabla \varphi_{2\mathrm{p}} \cdot \mathbf{P}'\mathrm{d}V = -\int_V \nabla \cdot (\varphi_{2\mathrm{p}} \mathbf{P}')\mathrm{d}V \\ &= -\int_S \varphi_{2\mathrm{p}} \mathbf{n} \cdot \mathbf{P}'\mathrm{d}S = -\int_S \varphi_{2\mathrm{p}}(\mathbf{r})\sigma'(\mathbf{r})\mathrm{d}S = -\int_S \int_{S'} \sigma'(\mathbf{r}) \frac{\sigma_{2\mathrm{p}}(\mathbf{r}')}{|\mathbf{r} - \mathbf{r}'|}\mathrm{d}S\mathrm{d}S'\end{aligned} \tag{A3}$$

where $\sigma_{2\mathrm{p}}$ and σ' represent the polarization surface charges corresponding to $\mathbf{E}_{2\mathrm{p}}$ and $\mathbf{P}'$ respectively, while $\varphi_{2\mathrm{p}}$ denotes the equilibrium polarization potential, $\mathbf{E}_{2\mathrm{p}} = -\nabla \varphi_{2\mathrm{p}}$. In the same way, the second term on the right hand side of eq. (A2) can be changed to the following form,

$$\int_V \mathbf{E}'_{\mathrm{p}} \cdot \mathbf{P}_2^{\mathrm{eq}} \mathrm{d}V = -\int_V \nabla \varphi' \cdot \mathbf{P}_2^{\mathrm{eq}} \mathrm{d}V = -\int_S \int_{S'} \sigma'(\mathbf{r}) \frac{\sigma_{2\mathrm{p}}(\mathbf{r}')}{|\mathbf{r}-\mathbf{r}'|} \mathrm{d}S \mathrm{d}S' \tag{A4}$$

Substituting eqs. (A3) and (A4) into eq. (A2), we obtain the desired equality

$$\int \mathbf{E}_{2\mathrm{c}} \cdot \mathbf{P}' \mathrm{d}V = \int \mathbf{E}_{\mathrm{ex}} \cdot \mathbf{P}_2^{\mathrm{eq}} \mathrm{d}V \tag{A5}$$

5.2 Appendix B: The proof for the solvent reorganization energy in charge-potential form

In the equilibrium medium, the divergency of the solvent polarization and surface polarized charge σ can be expressed as

$$\nabla \cdot \mathbf{P} = 0 \text{ , } \mathbf{n} \cdot \mathbf{P} = \sigma \tag{B1}$$

Further by using $\Delta \mathbf{E}_{\mathrm{dy}} = -\nabla(\Delta \Phi_{\mathrm{dy}})$, we have

$$\begin{aligned} \lambda_{\mathrm{s}} &= \frac{1}{2} \frac{\varepsilon_{\mathrm{s}} - \varepsilon_{\mathrm{op}}}{\varepsilon_{\mathrm{s}} - 1} \int_V \nabla(\Delta \Phi_{\mathrm{dy}}) \cdot (\Delta \mathbf{P}_{\mathrm{dy}} - \Delta \mathbf{P}_{\mathrm{eq}})] \mathrm{d}V \\ &= \frac{1}{2} \frac{\varepsilon_{\mathrm{s}} - \varepsilon_{\mathrm{op}}}{\varepsilon_{\mathrm{s}} - 1} \int_V \nabla \cdot [\Delta \Phi_{\mathrm{dy}} (\Delta \mathbf{P}_{\mathrm{dy}} - \Delta \mathbf{P}_{\mathrm{eq}})] \mathrm{d}V \\ &= \frac{1}{2} \frac{\varepsilon_{\mathrm{s}} - \varepsilon_{\mathrm{op}}}{\varepsilon_{\mathrm{s}} - 1} \oint_S \Delta \Phi_{\mathrm{dy}} (\Delta \mathbf{P}_{\mathrm{dy}} - \Delta \mathbf{P}_{\mathrm{eq}}) \cdot \mathbf{n} \mathrm{d}S \end{aligned} \tag{B2}$$

that is,

$$\lambda_{\mathrm{s}} = \frac{1}{2} \frac{\varepsilon_{\mathrm{s}} - \varepsilon_{\mathrm{op}}}{\varepsilon_{\mathrm{s}} - 1} \oint_S \Delta \Phi_{\mathrm{dy}} (\Delta \sigma_{\mathrm{dy}} - \Delta \sigma_{\mathrm{eq}}) \mathrm{d}S \tag{B3}$$

which is applicable to solute cavities of general shapes and sizes.

5.3 Appendix C：The brief expression for the solvent reorganization energy in sphere cavity model

If the point charge q of the solute locate at the center of the sphere cavity with the radius of r, it will produce the electric field strength in vacuum as

$$\mathbf{E}_{\mathrm{c}} = \frac{q\mathbf{r}}{r^2} \tag{C1}$$

Then we can set a point charge q_{ex} at the center of the solute sphere, defined as

$$q_{\text{ex}} = \frac{\varepsilon_{\text{op}} - \varepsilon_{\text{s}}}{\varepsilon_{\text{s}} - 1} \frac{1}{\varepsilon_{\text{op}}} \Delta q \tag{C2}$$

with $\Delta q = q_2 - q_1$, it can generate the needed external field strength in vacuum as

$$\mathbf{E}_{\text{ex}} = \frac{\varepsilon_{\text{op}} - \varepsilon_{\text{s}}}{\varepsilon_{\text{s}} - 1} \frac{\Delta q \mathbf{r}}{\varepsilon_{\text{op}} r^2} = (\frac{\chi_{\text{op}}}{\chi_{\text{s}}} - 1) \Delta \mathbf{E}_{\text{dy}} \tag{C3}$$

If the solute charge can be regarded as the point dipole $\boldsymbol{\mu}$ at the sphere center, the field strength produce by it in vacuum is

$$\mathbf{E}_{\text{c}} = \frac{\boldsymbol{\mu} \cdot \mathbf{r}}{r^3} \tag{C4}$$

If we can place another point dipole $\boldsymbol{\mu}_{\text{ex}}$ at the center, defined by

$$\boldsymbol{\mu}_{\text{ex}} = \frac{\varepsilon_{\text{op}} - \varepsilon_{\text{s}}}{(\varepsilon_{\text{s}} - 1)} \frac{3}{2\varepsilon_{\text{op}} + 1} \Delta \boldsymbol{\mu} \tag{C5}$$

then there will be the needed external field strength $\mathbf{E}_{\text{ex}}$ as

$$\mathbf{E}_{\text{ex}} = \frac{\varepsilon_{\text{op}} - \varepsilon_{\text{s}}}{(\varepsilon_{\text{s}} - 1)} \frac{3}{2\varepsilon_{\text{op}} + 1} \frac{\Delta \boldsymbol{\mu} \cdot \mathbf{r}}{r^3} = (\frac{\chi_{\text{op}}}{\chi_{\text{s}}} - 1) \Delta \mathbf{E}_{\text{dy}} \tag{C6}$$

by using the relation of $\Delta \mathbf{E}_{\text{dy}} = \frac{3}{2\varepsilon_{\text{op}} + 1} \frac{\Delta \boldsymbol{\mu} \cdot \mathbf{r}}{r^3}$. It should be noticed that $\mathbf{E}_{\text{ex}}$ is the vacuum field strength due to external charge and it will generate the additional polarization $\mathbf{P}'$, polarization field $\mathbf{E}_{\text{p}}'$ and polarization potential φ' as

$$\mathbf{P}' = \Delta \mathbf{P}_{\text{dy}} - \Delta \mathbf{P}_{\text{eq}}, \ \mathbf{E}_{\text{p}}' = \Delta \mathbf{E}_{\text{dy}} - \Delta \mathbf{E}_{\text{eq}}, \ \varphi' = \Delta \varphi_{\text{dy}} - \Delta \varphi_{\text{eq}} \tag{C7}$$

in the medium with dielectric constant ε_{s}. By using $\mathbf{E}_{\text{ex}} = -\nabla \psi_{\text{ex}}$ with ψ_{ex} being the vacuum potential due to $\mathbf{P}'$, we can obtain

$$\begin{aligned} \lambda_{\text{s}} &= \frac{1}{2} \int \mathbf{E}_{\text{ex}} \cdot \mathbf{P}' \mathrm{d}V = -\frac{1}{2} \int \mathbf{P}' \cdot \nabla \psi_{\text{ex}} \mathrm{d}V = \frac{1}{2} \int \psi_{\text{ex}} \nabla \cdot \mathbf{P}' \mathrm{d}V = -\frac{1}{2} \int \psi_{\text{ex}}(\mathbf{r}) \rho'(\mathbf{r}) \mathrm{d}V \\ &= -\frac{1}{2} \int_V \int_{V'} \frac{\rho_{\text{ex}}(\mathbf{r}') dV'}{|\mathbf{r} - \mathbf{r}'|} \rho'(\mathbf{r}) \mathrm{d}V = -\frac{1}{2} \int_V \int_{V'} \frac{\rho'(\mathbf{r}) dV}{|\mathbf{r} - \mathbf{r}'|} \rho_{\text{ex}}(\mathbf{r}') \mathrm{d}V' = -\frac{1}{2} \int \rho_{\text{ex}} \varphi' \mathrm{d}V \end{aligned} \tag{C8}$$

where ρ' is the polarized charge due to $\mathbf{E}_{\text{ex}}$ in the medium. Substituting eq. (C7) into the above equation, it can be obtained that

$$\lambda_{\text{s}} = \frac{1}{2}\int \rho_{\text{ex}}(\Delta\varphi_{\text{eq}} - \Delta\varphi_{\text{dy}})\text{d}V \tag{C9}$$

This equation is the brief expression for the solvent reorganization energy with sphere cavity approximation.
In the case of solute charge being point charge, eq. (C9) can be simplified as

$$\lambda_{\text{s}} = \frac{1}{2}q_{\text{ex}}(\Delta\varphi_{\text{eq}} - \Delta\varphi_{\text{dy}}) \tag{C10}$$

In another case with point charges q_{D} and q_{A} locating at the centers of electron donor's and acceptor's spheres, eq. (C9) can be rewritten as

$$\lambda_{\text{s}} = \frac{1}{2}q_{\text{D,ex}}(\Delta\varphi_{\text{D,eq}} - \Delta\varphi_{\text{D,dy}}) + \frac{1}{2}q_{\text{A,ex}}(\Delta\varphi_{\text{A,eq}} - \Delta\varphi_{\text{A,dy}}) \tag{C11}$$

In the case of solute point dipole, the dipole can be expressed as the product of the charge q and distance $d\mathbf{l}$, i.e., $\boldsymbol{\mu} = qd\mathbf{l}$, thus we have

$$q\varphi_{+} - q\varphi_{-} = q\text{d}\varphi = qd\mathbf{l}\cdot\nabla\varphi = -\boldsymbol{\mu}\cdot\mathbf{E} \tag{C12}$$

According to eqs. (C9) and (C12), the solvent reorganization energy with point dipole and sphere cavity approximation can be expressed as

$$\lambda_{\text{s}} = \frac{1}{2}\boldsymbol{\mu}_{\text{ex}}\cdot(\Delta\mathbf{E}_{\text{p,dy}} - \Delta\mathbf{E}_{\text{p,eq}}) \tag{C13}$$

6. References

[1] (a) Leontovich M. A. *An Introduction to Thermodynamics*, 2nd ed, Gittl Publ, Moscow, 1950 (in Russian). (b) Leontovich M. A. *Introduction to Thermodynamics, Statistical Physics 2nd*; Nauka: Moscow, 1983(in Russian).

[2] Marcus R. A. *J. Chem. Phys.* 1956, 24: 979.

[3] Pekar S. I. *Introduction into Electronic Theory of Crystals*, Technical Literature Publishers, Moscow, 1951.

[4] Li X.-Y., He F.-C., Fu K.-X., Liu W. *J. Theor. Comput. Chem.* 2010, 9(supp.1): 23.

[5] Wang X.-J., Zhu Q., Li Y.-K., Cheng X.-M., Fu K.-X., Li X.-Y. *J. Phys. Chem. B.* 2010, 114: 2189.

[6] Li X.-Y., Wang Q.-D., Wang J.-B., Ma J.-Y., Fu K.-X., He F.-C. *Phys. Chem. Chem. Phys.* 2010, 12: 1341.

[7] Jackson J. D. *Classical Electrodynamics,* Third Edition, John Wiley & Sons, Inc. New York, 1999: 165-168.

[8] Landau L. D., Lifshitz E. M., Pitaevskii L. P. *Eletronynamics of Continous Media*, 2nd ed. Butterworth-Heinemman, Ltd, 1984.

[9] Marcus R. A. *J. Phys. Chem.* 1994, 98: 7170.

[10] Johnson M. D., Miller J. R., Green N. S., Closs G. L. *J. Phys. Chem.* 1989, 93: 1173.

[11] Formasinho S. J., Arnaut L. G., Fausto R. *Prog. Reaction. Kinetics.* 1998, 23: 1.

[12] Basilevsky M. V., Chudinov G. E., Rostov I. V., Liu Y., Newton M. D. *J. Mol. Struct. Theochem.* 1996, 371: 191.

12

Hydrodynamical Models of Superfluid Turbulence

D. Jou[1], M.S. Mongiovì[2], M. Sciacca[2], L. Ardizzone[2] and G. Gaeta[2]
[1]*Departament de Física, Universitat Autònoma de Barcelona, Bellaterra, Catalonia*
[2]*Dipartimento di Metodi e Modelli Matematici, Università di Palermo, Palermo*
[1]*Spain*
[2]*Italy*

1. Introduction

Turbulence is almost the rule in the flow of classical fluids. It is a complex nonlinear phenomenon for which the development of a satisfactory theoretical framework is still incomplete. Turbulence is often found in the flow of quantum fluids, especially superfluid Helium 4, known as liquid helium II (Donnelly, 1991), (Nemirovskii & Fiszdon, 1995), (Barenghi et al., 2001), (Vinen & Niemela, 2002).
In recent years there has been growing interest in superfluid turbulence, because of its unique quantum peculiarities and of its similarity with classical turbulence to which it provides a wide range of new experimental possibilities at very high Reynolds numbers (Vinen, 2000), (Barenghi, 1999), and because of their influence in some practical applications, as in refrigeration by means of superfluid helium. We will consider here the turbulence in superfluid ^{4}He, for which many detailed experimental techniques have been developed.
The behavior of liquid helium, below the *lambda point* ($T_c \simeq 2.17$ K), is very different from that of ordinary fluids. One example of non-classical behavior is the possibility to propagate the second sound, a wave motion in which temperature and entropy oscillate. A second example of non-classical behavior is heat transfer in counterflow experiments. Using an ordinary fluid (such as helium I), a temperature gradient can be measured along the channel, which indicates the existence of a finite thermal conductivity. If helium II is used, and the heat flux inside the channel is not too high, the temperature gradient is so small that it cannot be measured, so indicating that the liquid has an extremely high thermal conductivity (three million times larger than that of helium I). This is confirmed by the fact that helium II is unable to boil. This effect explains the remarkable ability of helium II to remove heat and makes it important in engineering applications.
The most known phenomenological model, accounting for many of the properties of He II, given by Tisza and Landau (Tisza, 1938), (Landau, 1941) is called the two-fluid model. The basic assumption is that the liquid behaves as a mixture of two fluids: the normal component with density ρ_n and velocity $\mathbf{v}_n$, and the superfluid component with density ρ_s and velocity $\mathbf{v}_s$, with total mass density ρ and barycentric velocity $\mathbf{v}$ defined by $\rho = \rho_s + \rho_n$ and $\rho\mathbf{v} = \rho_s\mathbf{v}_s + \rho_n\mathbf{v}_n$. The second component is related to the quantum coherent ground state and it is an ideal fluid, which does not experience dissipation neither carries entropy. The superfluid component, which is absent above the lambda transition temperature, was originally considered to be composed by particles in the Bose-Einstein state and is an ideal

fluid, and the normal component by particles in the excited state (phonons and rotons) and is a classical Navier-Stokes viscous fluid.

The two-fluid model explains the experiment described above in the following way: in the absence of mass flux ($\rho_n \mathbf{v}_n + \rho_s \mathbf{v}_s = 0$ and $\mathbf{v}_n$ and $\mathbf{v}_s$ averaged on a small mesoscopic volume Λ), in helium II the heat is carried toward the bath by the normal fluid only, and $\mathbf{q} = \rho s T \mathbf{v}_n$ where s is the entropy per unit mass and T the temperature. Being the net mass flux zero, there is superfluid motion toward the heater ($\mathbf{v}_s = -\rho_n \mathbf{v}_n / \rho_s$), hence there is a net internal counterflow $\mathbf{V}_{ns} = \mathbf{v}_n - \mathbf{v}_s = \mathbf{q}/(\rho_s s T)$ which is proportional to the applied heat flux $\mathbf{q}$.

An alternative model of superfluid helium is the one-fluid model (Lebon & Jou, 1979), (Mongiovì, 1993), (Mongiovì, 2001) based on extended thermodynamics (Müller & Ruggeri, 1998), (Jou et al., 2001), (Lebon et al., 2008). Extended Thermodynamics (E.T.) is a thermodynamic formalism proposed in the last decades, which offers a natural framework for the macroscopic description of liquid helium II. The basic idea underlying E.T. is to consider the physical fluxes as independent variables. In previous papers, the E.T. has been applied to formulate a non-standard one-fluid model of liquid helium II, for laminar flows. This model is recalled in Section 2, in the absence of vortices (laminar flow) and in Section 3 both in rotating containers and in counterflow situations.

Quantum turbulence is described as a chaotic tangle of quantized vortices of equal circulation

$$\kappa = \oint \mathbf{u}_s \cdot dl \tag{1}$$

($\mathbf{u}_s$ microscopic velocity of the superfluid component) called *quantum of vorticity* and results $\kappa = h/m_4$, with h the Planck constant, and m_4 the mass of ^{4}He atom: $\kappa \simeq 9.97\ 10^{-4}\mathrm{cm}^2/\mathrm{s}$. Since the vorticity is quantized, the increase of turbulence is manifested as an increase of the total length of the vortex lines, rather than with a faster spinning of the vortices. Thus, the dynamics of the vortex length is a central aspect of quantum turbulence.

A preliminary study of these interesting phenomena was made in (Jou et al., 2002), where the presence of vortices was modeled through a pressure tensor $\mathbf{P}_\omega$ for which a constitutive relation was written. In homogeneous situations, the vortex tangle is described by introducing a scalar quantity L, the average vortex line length per unit volume (briefly called *vortex line density*). The evolution equation for L in counterflow superfluid turbulence has been formulated by Vinen (Vinen, 1958), (Donnelly, 1991), (Barenghi et al., 2001)

$$\frac{dL}{dt} = \alpha_v V_{ns} L^{3/2} - \beta_v \kappa L^2, \tag{2}$$

with V_{ns} the modulus of the counterflow velocity $\mathbf{V}_{ns} = \mathbf{v}_n - \mathbf{v}_s$, which is proportional to the heat flux $\mathbf{q}$, and α_v and β_v dimensionless parameters. This equation assumes homogeneous turbulence, i.e. that the value of L is the same everywhere in the system. In fact, homogeneity may be expected if the average distance between the vortex filaments, of the order of $L^{-1/2}$, is much smaller than the size of the system.

Recent experiments show the formation of a new type of superfluid turbulence, which has some analogies with classical one, as for instance using towed or oscillating grids, or stirring liquid helium by means of propellers. In this situation, which has been called *co-flow*, both components, normal and superfluid, flow along the same direction. To describe these experiments it is necessary to build up a hydrodynamic model of quantum turbulence, in which the interactions between both fields can be studied and the role of inhomogeneities is explicitly taken into account.

Our aim in this review is to show hydrodynamical models for turbulent superfluids, both in linear and in non linear regimes. To this purpose, in Section 4 we will choose as fundamental fields the density ρ, the velocity $\mathbf{v}$, the internal energy density E, in addition to the heat flux $\mathbf{q}$, and the averaged vortex line density L (Mongiovì & Jou, 2007), (Ardizzone & Gaeta, 2009). We will write general balance equations for the basic variables and we will determine the constitutive equations for the fluxes; the nonlinear relations which constrain the constitutive quantities will be deduced from the second law of thermodynamics, using the Liu method of Lagrange multipliers (Liu, 1972). The physical meaning of the Lagrange multipliers both near and far from equilibrium will be also investigated. Under the hypothesis of homogeneity in the vortex tangle, the propagation of second sound in counterflow is studied, with the aim to determine the influence of the vortex tangle on the velocity and attenuation of this wave.
In this model the diffusion flux of vortices $\mathbf{J}^L$ is considered as a dependent variable, collinear with the heat flux $\mathbf{q}$. But, in general, this feature is not strictly verified because the vortices move with a velocity $\mathbf{v}_L$, which is not collinear with the counterflow velocity. For this reason, a more detailed model of superfluid turbulence would be necessary, by choosing as fundamental fields, in addition to the fields previously used, also the velocity of the vortex line $\mathbf{v}_L$. In Section 5 we aim to study the interaction between second sound and vortex density wave, a model which choose as field variables, the internal energy density E, the line density L, and the vortex line velocity $\mathbf{v}_L$ (Sciacca et al, 2008).
The paper is the first general review of the hydrodynamical models of superfluid turbulence inferred using the procedures of E.T. Furthermore, the text is not exclusively a review of already published results, but it contains some new interpretations and proposals which are formulated in it for the first time.

2. The one-fluid model of liquid helium II derived by extended thermodynamics

Extended Thermodynamics (E.T.) is a macroscopic theory of non-equilibrium processes, which has been formulated in various ways in the last decades (Müller & Ruggeri, 1998), (Jou et al., 2001), (Lebon et al., 2008). The main difference between the ordinary thermodynamics and the E.T. is that the latter uses dissipative fluxes, besides the traditional variables, as independent fields. As a consequence, the assumption of local equilibrium is abandoned in such a theory. In the study of non equilibrium thermodynamic processes, an extended approach is required when one is interested in sufficiently rapid phenomena, or else when the relaxation times of the fluxes are long; in such cases, a constitutive description of these fluxes in terms of the traditional field variables is impossible, so that they must be treated as independent fields of the thermodynamic process.
From a macroscopic point of view, an extended approach to thermodynamics is required in helium II because the relaxation time of heat flux is comparable with the evolution times of the other variables; this is confirmed by the fact that the thermal conductivity of helium II cannot be measured. As a consequence, this field cannot be expressed by means of a constitutive equation as a dependent variable, but an evolution equation for it must be formulated.
From a microscopic point of view, E.T. offers a natural framework for the (macroscopic) description of liquid helium II: indeed, as in low temperature crystals, using E.T., the dynamics of the relative motion of the excitations is well described by the dynamics of the heat flux.
The conceptual advantage of the one-fluid model is that, in fact, from the purely macroscopic point of view one sees only a single fluid, rather than two physically different fluids. Indeed the variables $\mathbf{v}$ and $\mathbf{q}$ used in E.T. are directly measurable, whereas the variables $\mathbf{v}_n$ and $\mathbf{v}_s$,

are only indirectly measured, usually from the measurements of $\mathbf{q}$ and $\mathbf{v}$. The internal degree of freedom arising from the relative motion of the two fluids is here taken into account by the heat flux, whose relaxation time is very long. However, the two-fluid model provides a very appealing image of the *microscopic* helium behavior, and therefore is the most widely known.

2.1 Laminar flows

A non standard one-fluid model of liquid helium II deduced by E.T. was formulated in (Mongiovì, 1991). The model chooses as fundamental fields the mass density ρ, the velocity $\mathbf{v}$, the absolute temperature T and the heat flux density $\mathbf{q}$. Neglecting, at moment, dissipative phenomena (mechanical and thermal), the linearized evolution equations for these fields are:

$$\begin{cases} \dot{\rho} + \rho \nabla \cdot \mathbf{v} = 0, \\ \rho \dot{\mathbf{v}} + \nabla p^* = 0, \\ \rho \dot{\epsilon} + \nabla \cdot \mathbf{q} + p \nabla \cdot \mathbf{v} = 0, \\ \dot{\mathbf{q}} + \zeta \nabla T = 0. \end{cases} \tag{3}$$

In these equations, the quantity ϵ is the specific internal energy per unit mass, p the thermostatic pressure, and $\zeta = \lambda_1/\tau$, being τ the relaxation time of the heat flux and λ_1 the thermal conductivity. As it will be shown, coefficient ζ characterizes the second sound velocity, and therefore it is a measurable quantity. Upper dot denotes the material time derivative.

Equations (3) describe the propagation in liquid helium II of two waves, whose speeds w are the solutions of the following characteristic equation:

$$\left(w^2 - V_1^2\right)\left(w^2 - V_2^2\right) - W_1 W_2 u^2 = 0, \tag{4}$$

where

$$V_1^2 = p_\rho, \qquad V_2^2 = \frac{\zeta}{\rho c_V}, \qquad W_1 = \frac{p_T}{\rho}, \qquad W_2 = \frac{T p_T}{\rho c_V}, \tag{5}$$

and with $c_V = \partial\epsilon/\partial T$ the constant volume specific heat and $p_T = \partial p/\partial T$ and $p_\rho = \partial p/\partial \rho$. Neglecting thermal expansion ($W_1 = 0$, $W_2 = 0$) equation (4) admits the solutions $w_{1,2} = \pm V_1$ and $w_{3,4} = \pm V_2$, corresponding to the two sounds typical of helium II: $w = \pm V_1$ implies vibration of only density and velocity; while $w = \pm V_2$ implies vibration of only temperature and heat flux. This agrees with the experimental observations. The coefficient ζ can be determined by the second equation in 5, once the expression of the second sound velocity is known.

Finally, we observe that the Gibbs equation for helium II can be written as

$$T ds = d\epsilon - \frac{p}{\rho^2} d\rho - \frac{1}{\rho \zeta T} \mathbf{q} \cdot d\mathbf{q}, \tag{6}$$

where s is the specific entropy.

2.2 The viscous pressure tensor

It is experimentally known that dissipative effects both of mechanical and thermal origin are present in the propagation of the two sounds in liquid helium II, also in the absence of vortices. To take into account of these effects, a symmetric dissipative pressure tensor $\mathbf{P}_K$ must be introduced:

$$[\mathbf{P}_K]_{ik} = p_{<ik>} + p_V \delta_{ik}. \tag{7}$$

In (Mongiovì, 1993) for the two fields $p_{<ij>}$ and p_V, respectively deviator and trace of the stress tensor, the following constitutive relations were determined:

$$p_V = -\lambda_0 \frac{\partial v_j}{\partial x_j} + \beta' T \lambda_0 \frac{\partial q_j}{\partial x_j}, \tag{8}$$

$$p_{<ik>} = -2\lambda_2 \frac{\partial v_{<i}}{\partial x_{k>}} + 2\beta T \lambda_2 \frac{\partial q_{<i}}{\partial x_{k>}}. \tag{9}$$

In these equations λ_0 and λ_2 are the bulk and the shear viscosity, while β and β' are coefficients appearing in the general expression of the entropy flux in E.T. and take into account of the dissipation of thermal origin.
Equations (8)–(9) contain, in addition to terms proportional to the gradient of velocity (the classical viscous terms), terms depending on the gradient of the heat flux (which take into account of the dissipation of thermal origin). The first terms in (8)–(9) allow us to explain the attenuation of the first sound, the latter the attenuation of the second sound.
In the presence of dissipative phenomena, the field equations (3) are modified in:

$$\begin{cases} \dot{\rho} + \rho \nabla \cdot \mathbf{v} = 0, \\ \dot{\mathbf{v}} + \frac{1}{\rho}\nabla p + \frac{1}{\rho}\nabla p_V + \frac{1}{\rho}\nabla p_{<ji>} = 0, \\ \dot{T} + \frac{T p_T}{\rho c_V}\nabla \cdot \mathbf{v} + \frac{1}{\rho c_V}\nabla \cdot \mathbf{q} = 0, \\ \dot{\mathbf{q}} + \zeta \nabla T - \beta' T^2 \zeta \nabla p_V + \beta T^2 \zeta \nabla p_{<ji>} = 0. \end{cases} \tag{10}$$

The propagation of small amplitude waves was studied in (Mongiovì, 1993). Supposing zero thermal expansion under the hypothesis of small dissipative losses (viscous and thermal) approximation, one sees that in helium II two waves propagate (the first and the second sound), whose velocities are identical to that found in the absence of dissipation, and the attenuation coefficients are found to be:

$$k_s^{(1)} = \frac{\omega^2}{2\rho w_1^3}\left(\lambda_0 + \frac{4}{3}\lambda_2\right), \qquad k_s^{(2)} = \frac{\omega^2 T^3 \zeta}{2 w_2^3}\left(\lambda_0 \beta'^2 + \frac{4}{3}\lambda_2 \beta^2\right). \tag{11}$$

2.3 Comparison with the two-fluid model

Comparing these results with the results of the two-fluid model (Mongiovì, 1993), we observe that the expression of the attenuation coefficient $k_s^{(1)}$ of the first sound is identical to the one inferred by Landau and Khalatnikov, using the two-fluid model (Khalatnikov, 1965). The attenuation coefficient of the second sound appears different from the one obtained in (Khalatnikov, 1965). However, it contains a term proportional to the square of the frequency ω, in agreement with the experimental results.
The main difference between the results of the one-fluid theory and the two-fluid model is that, while in the latter the thermal dissipation (needed to explain the attenuation of the

second sound) is due to a dissipative term of a Fourier type, in the extended model it is a consequence of terms dependent on the gradient of the heat flux q_i (which are present in the expressions of the trace and the deviator of non equilibrium stress, besides the traditional viscous terms).

3. Vortices in liquid helium II

From the historical and conceptual perspectives, the first observations of the peculiar aspects of rotation in superfluids arose in the late 1950's, when it was realized that vorticity may appear inside superfluids and that it is quantized, its quantum κ being $\kappa = h/m_4$, with h the Planck constant and m_4 the mass of the particles. According to the two-fluid model of Tisza and Landau (Tisza, 1938), (Landau, 1941), the superfluid component cannot participate to a rigid rotation, owing to its irrotationality. Consequently, owing to the temperature dependence of the normal component fraction, different forms of the liquid free surface should be observed at different temperatures. In order to check this prediction, Osborne (Osborne, 1950) put in rotation a cylindrical vessel containing helium II, but no dependence of the form of the free surface of temperature was observed. Feynman (Feynman, 1955) gave an explanation of the rigid rotation of helium II without renouncing to the hypothesis of the irrotationality of the velocity of the superfluid. Following the suggestion of the quantization of circulation by Onsager (Onsager, 1949), he supposed that the superfluid component, although irrotational at the microscopic level, creates quantized vortices at an intermediate level; these vortices yield a non-zero value for the curl of the macroscopic velocity of the superfluid component.

Another interesting experiment was performed by Hall and Vinen (Hall & Vinen,, 1956), (Hall & Vinen,, 1956) about propagation of second sound in rotating systems. A resonant cavity is placed inside a vessel containing He II, and the whole setting rotates at constant angular velocity Ω. When the second sound propagates at right angles with respect to the rotation axis, it suffers an extra attenuation compared to a non-rotating vessel of an amount proportional to the angular velocity. On the other hand, a negligible attenuation of the second sound is found when the direction of propagation is parallel to the axis of rotation. The large increase of the attenuation observed by Hall and Vinen when the liquid is rotated can be explained by the *mutual friction*, which finds its origin in the interaction between the flow of excitations (phonons and rotons) and the array of straight quantized vortex filaments in helium II. Indeed, such vortices have been directly observed and quantitatively studied.

In fact, vortices are always characterized by the same quantum of vorticity, in such a way that for higher rotation rates the total length of the vortices increases. The vortices are seen to form a regular array of almost parallel lines. This has strong similarities with electrical current vortex lines appearing in superconductors submitted to a high enough external magnetic field. In fact, this analogy has fostered the interest in vortices in superfluids, which allow one to get a better understanding of the practically relevant vortices in superconductors (Fazio & van der Zant, 2001).

The situation we have just mentioned would scarcely be recognized as "turbulence", because its highly ordered character seems very far from the geometrical complexities of usual turbulence. In fact, it only shares with it the relevance of vorticity, but it is useful to refer to it, as it provides a specially clear understanding of the quantization of vorticity.

The interest in truly turbulent situations was aroused in the 1960's in counterflow experiments (Vinen, 1957), (Vinen, 1958). In these experiments a random array of vortex filaments appears, which produces a damping force: the mutual friction force. The measurements of vortex

lines are described as giving a macroscopic average of the vortex line density L. There are essentially two methods to measure L in superfluid ^{4}He: observations of temperature gradients in the channel and of changes in the attenuation of the second-sound waves (Donnelly, 1991), (Barenghi et al., 2001).
In the present section, our attention is focused on the study of the action of vortices on second sound propagation in liquid helium II. This will be achieved by using the one-fluid model of liquid helium II derived in the framework of E.T., modified in order to take into account of the presence of vortices.

3.1 The vorticity tensor

To take into account the dissipation due to vortices, a dissipative pressure tensor $\mathbf{P}_\omega$ can be introduced in equations (3) (Jou et al., 2002)

$$\mathbf{P} = \mathbf{P}_K + \mathbf{P}_\omega, \tag{12}$$

where $\mathbf{P}_K$ designates the kinetic pressure tensor introduced in the previous section (equation (7)). In contrast with $\mathbf{P}_K$ (a symmetric tensor), $\mathbf{P}_\omega$ is in general nonsymmetric. The decomposition (12) is analogous to the one performed in real gases and in polymer solutions, where particle interaction or conformational contributions are respectively included as additional terms in the pressure tensor (Jou et al., 2001).
As in the description of the one-fluid model of liquid helium II made in Section 2 (see also (Mongiovì, 1991), (Mongiovì, 1993)), the relative motion of the excitations may still be described by the dynamics of the heat flux, but now the presence of the vortices modifies the evolution equation for heat flux. For the moment, we will restrict our attention to stationary situations, in which the vortex filaments are supposed fixed, and we focus our attention on their action on the second sound propagation. In other terms, in this section, we do not assume that $\mathbf{P}_\omega$ is itself governed by an evolution equation, but that it is given by a constitutive relation. Furthermore, we neglect $\mathbf{P}_K$ as compared to $\mathbf{P}_\omega$, because the mutual friction effects are much greater than bulk and shear forces acting inside the superfluid.
Let us now reformulate the evolution equation for the heat flux $\mathbf{q}$. The experimental data show that the extra attenuation due to the vortices is independent of the frequency. Therefore, a rather natural generalization of the last equation in system (3) for the time evolution of the heat flux $\mathbf{q}$ is the following:

$$\dot{\mathbf{q}} + 2\Omega \times \mathbf{q} + \zeta \nabla T = -\mathbf{P}_\omega \cdot \mathbf{q}. \tag{13}$$

This relation is written in a noninertial system, rotating at uniform velocity $\boldsymbol{\Omega}$; the influence of the vortices on the dynamics of the heat flux is modeled by the last term in the r.h.s. of (13). In this equation all the non linear terms have been neglected, with the exception of the production term $\vec{\sigma}_q = -\mathbf{P}_\omega \cdot \mathbf{q}$, which takes into account the interaction between vortex lines and heat flux.
To close the set of equations, we need a constitutive relation for the tensor $\mathbf{P}_\omega$. The presence of quantized vortices leads to a interaction force with the excitations in the superfluid known as mutual friction. From a microscopic point of view, the major source of mutual friction results from the collision of rotons with the cores of vortex lines: the quasiparticles scatter off the vortex filaments and transfer momentum to them. The collision cross-section is clearly a strong function of the direction of the roton drift velocity relative to the vortex line: it is a maximum when the roton is travelling perpendicular to this line and a minimum (in fact zero)

when the roton moves parallel to the line. The microscopic mechanism is the same in rotating helium II and in superfluid turbulence.
We are therefore led to take:

$$\mathbf{P}_\omega = \lambda < \omega >< \mathbf{U} - \mathbf{s}' \otimes \mathbf{s}' > + \lambda' < \omega >< \mathbf{W} \cdot \mathbf{s}' >, \tag{14}$$

where brackets denote (spatial and temporal) macroscopic averages. The unspecified quantities introduced in (14) are the following: $\vec{\omega}$ is the microscopic vorticity vector, $\omega = |\vec{\omega}|$; $\lambda = \lambda(\rho, T)$ and $\lambda' = \lambda'(\rho, T)$ are coefficients relating the internal energy of the liquid to the microscopic vorticity (Khalatnikov, 1965), $\mathbf{s}'$ is a unit vector tangent to the vortices, $\mathbf{U}$ the unit second order tensor and $\mathbf{W}$ the Ricci tensor, an antisymmetric third order tensor such that $\mathbf{W} \cdot \mathbf{s}' \cdot \mathbf{q} = -\mathbf{s}' \times \mathbf{q}$. Finally, the quantity $< \omega >$ depends on the average vortex line length per unit volume L. Neglecting the bulk and shear viscosity and under the hypothesis of small thermal dilation (which in helium II are very small), the linearized system of field equations for liquid helium II, in a non inertial frame and in absence of external force, is (Jou et al., 2002):

$$\begin{cases} \frac{\partial \rho}{\partial t} + \rho \frac{\partial v_j}{\partial x_j} = 0, \\ \rho \frac{\partial v_i}{\partial t} + \frac{\partial p}{\partial x_i} + \mathbf{i}_i^0 + 2\rho \left(\mathbf{\Omega} \wedge \mathbf{v} \right)_i = 0, \\ \frac{\partial T}{\partial t} + \frac{1}{\rho c_V} \frac{\partial q_j}{\partial x_j} = 0, \\ \frac{\partial q_i}{\partial t} + \zeta \frac{\partial T}{\partial x_i} + 2 \left(\mathbf{\Omega} \wedge \mathbf{q} \right)_i = \left(\vec{\sigma}_q \right)_i = - \left(\mathbf{P}_\omega \cdot \mathbf{q} \right)_i, \end{cases} \tag{15}$$

where $\mathbf{i}^0 + 2\rho \left(\mathbf{\Omega} \wedge \mathbf{v} \right)_i$ stands for the inertial force.
In this section we consider the three most characteristic situations: the wave propagation in a rotating frame, the wave propagation in a cylindrical tube in presence of stationary thermal counterflow (no mass flux), and the wave propagation in the combined situation of rotation and thermal counterflow.

3.2 Rotating frame

Rotating helium II is characterized by straight vortex filaments, parallel to the rotation axis, when the angular velocity exceeds a critical value. The amount of these vortices is proportional to the absolute value of the angular velocity Ω of the cylinder by the Feynman's rule: $L_R = 2|\Omega|/\kappa$. Therefore

$$< \omega > = \kappa L = 2|\Omega|. \tag{16}$$

In this situation the averaged unit vector tangent to the vortices is $< \mathbf{s}' > = \mathbf{\Omega}/\Omega$.
But, the state with all the vortex lines parallel to the rotation axis will not be reached, because the vortex lines will always exhibit minuscule deviations with respect to the straight line, and such deviations produce a mutual friction force parallel to the rotation axis. Indeed, in an another experiment (Snyder & Putney, 1966) the component of the mutual friction along the rotational axis was studied, and their result shows that this component is very small compared with the orthogonal components but not exactly zero. In this subsection, in order to include the axial component of the mutual friction force, the following more general expression for vorticity tensor $\mathbf{P}_\omega$ is used:

$$\mathbf{P}^R_\omega = \frac{1}{2}\kappa L_R \left[(B - B'')\left(\mathbf{U} - \hat{\mathbf{\Omega}} \otimes \hat{\mathbf{\Omega}}\right) + B'\mathbf{W}\cdot\hat{\mathbf{\Omega}} + 2B''\hat{\mathbf{\Omega}} \otimes \hat{\mathbf{\Omega}}\right], \tag{17}$$

where B and B' are the Hall-Vinen coefficients (Hall & Vinen,, 1956) describing the orthogonal dissipative and non dissipative contributions while B'' is the friction coefficient along the rotational axis. The production term in (15d) can be expressed as (Donnelly, 1991), (Jou & Mongiovì, 2005), (Jou & Mongiovì, 2006):

$$\vec{\sigma}^R_q = \frac{1}{2}\kappa L_R \left[(B - B'')\hat{\mathbf{\Omega}} \wedge \left(\hat{\mathbf{\Omega}} \wedge \mathbf{q}\right) + B'\hat{\mathbf{\Omega}} \wedge \mathbf{q} - 2B''\hat{\mathbf{\Omega}} \otimes \hat{\mathbf{\Omega}} \cdot \mathbf{q}\right]. \tag{18}$$

Assuming the rotation axis as first axis, the vorticity tensor (17) can be written as:

$$\mathbf{P}_\omega = \frac{1}{2}B\kappa L \left\{ \begin{pmatrix} 2b & 0 & 0 \\ 0 & 1-b & 0 \\ 0 & 0 & 1-b \end{pmatrix} + \begin{pmatrix} 0 & 0 & 0 \\ 0 & 0 & c \\ 0 & -c & 0 \end{pmatrix} \right\}. \tag{19}$$

where we have put $b = B''/B$ and $c = B'/B$. Comparing (19) with (14): if $B'' = 0$ then $B = 2\lambda$, $B' = 2\lambda'$, $< (s'_{x_1})^2 >= 1$ and $< (s'_{x_2})^2 >=< (s'_{x_3})^2 >= 0$; if $B'' \neq 0$ then the previous identification is not possible but it results $< (s'_{x_1})^2 >= 1 - 2B''/B$ and $< (s'_{x_2})^2 >=< (s'_{x_3})^2 >= 2B''/B$.

3.2.1 Wave propagation in a rotating frame

In the following we assume that Ω is small, so that the term $\mathbf{i}_0$ in (15b) can be neglected. Substituting the expression (18) into the system (15) and choosing $\mathbf{\Omega} = (\Omega, 0, 0)$, the system assumes the following form:

$$\begin{cases} \frac{\partial \rho}{\partial t} + \rho \frac{\partial v_j}{\partial x_j} = 0, \\ \rho \frac{\partial v_i}{\partial t} + \frac{\partial p}{\partial x_i} + 2\rho\Omega v_j W_{1ji} = 0, \\ \frac{\partial T}{\partial t} + \frac{1}{\rho c_V}\frac{\partial q_j}{\partial x_j} = 0, \\ \frac{\partial q_i}{\partial t} + \zeta \frac{\partial T}{\partial x_i} + \left(2\Omega - \frac{1}{2}B'\kappa L_R\right) q_j W_{1ji} = \frac{1}{2}\kappa L_R[(B - B'')(-q_i + q_1\delta_{i1}) - 2B'' q_1 \delta_{i1}], \end{cases} \tag{20}$$

where δ_{ij} is the unit tensor and W_{kji} the Ricci tensor.
It is easily observed that a stationary solution of this system is:

$$\rho = \rho_0, \quad \mathbf{v} = \mathbf{0}, \quad T = T_0, \quad \mathbf{q} = \mathbf{0}. \tag{21}$$

In order to study the propagation of plane harmonic waves of small amplitude (Whitham, 1974), we linearize system (20) in terms of the fields $\Gamma = (\rho, v_i, T, q_i)$, and we look for solutions of the form:

$$\Gamma = \Gamma_0 + \tilde{\Gamma} e^{i(Kn_j x_j - \omega t)}, \tag{22}$$

where $\Gamma_0 = (\rho_0, 0, T_0, 0)$ denotes the unperturbed state, $\tilde{\Gamma} = (\tilde{\rho}, \tilde{v}_i, \tilde{T}, \tilde{q}_i)$ are small amplitudes whose products can be neglected, $K = k_r + ik_s$ is the wavenumber, $\omega = \omega_r + i\omega_s$ the frequency and $\mathbf{n} = (n_i)$ the unit vector orthogonal to the wave front. For the sake of simplicity, the subscript 0, which denotes quantities referring to the unperturbed state Γ_0, will be dropped out.

First case: **n** *parallel to* $\mathbf{\Omega}$.

Assuming that the unit vector **n** orthogonal to the wave front is parallel to the rotating axis (x_1−axis), it follows that longitudinal and transversal modes evolve independently. The study of the longitudinal modes ($\tilde{\rho}$, $\tilde{v}_1$, $\tilde{T}$ and $\tilde{q}_1$) furnishes the existence of two waves: the first sound (or pressure wave) in which density and velocity vibrate with velocity $V_1 := \frac{\omega_{1,2}}{k_r} = \sqrt{p_\rho}$ (ω real), and the second sound (or temperature wave) in which temperature and heat flux vibrate with velocity

$$w^2 = \left(\frac{\omega}{k_r}\right)^2 = V_2^2 - \frac{B''^2 \kappa L_R^2}{4V_2^2 k_r^2 + B''^2 \kappa L_R^2} \quad \text{and} \quad k_s = \frac{w B'' \kappa L_R}{2V_2^2}, \tag{23}$$

where $V_2^2 = \frac{\zeta}{\rho c_V}$ is the velocity of the second sound in the absence of vortices and k_s is the attenuation. The longitudinal modes are

$\omega_{1,2} = \pm k_r V_1$	$\omega_{3,4} = \pm\sqrt{\frac{4V_2^4 k_r^4}{4V_2^2 k_r^2 + B''^2 \kappa L_R^2}}$
$\tilde{\rho} = \psi$ $\tilde{v}_1 = \pm\frac{V_1}{\rho}\psi$ $\tilde{T}_0 = 0$ $\tilde{q}_1 = 0$	$\tilde{\rho} = 0$ $\tilde{v}_1 = 0$ $\tilde{T} = T_0 \psi$ $\tilde{q}_1 = \pm\rho c_V T_0 \sqrt{\frac{4V_2^4 k_r^4}{4V_2^2 k_r^2 + B''^2 \kappa L_R^2}}\, \psi$

Therefore, as observed in (Snyder & Putney, 1966), when the wave is propagated parallel to the rotation axis, the longitudinal modes are influenced by the rotation only through the axial component of the mutual friction (B'' coefficient).

On the contrary, the transversal modes ($\tilde{v}_2$, $\tilde{v}_3$, $\tilde{q}_2$ and $\tilde{q}_3$) are influenced by the rotation. In fact, the ones of velocity **v** admit nontrivial solutions if and only if $\omega_{5,6} = \pm 2|\Omega|$, while the ones related to **q** require the following dispersion relation:

$$\omega_{7,8} = \pm(2\Omega - \frac{1}{2}\kappa L_R B') - \frac{i}{2}\kappa L_R (B - B''). \tag{24}$$

These transversal modes are influenced from both dissipative and nondissipative contributions B, B' and B'' in the interaction between quasi-particles and vortex lines (Peruzza & Sciacca, 2007).

Second case: **n** *orthogonal to* $\mathbf{\Omega}$.

In the case in which the direction of propagation of the waves (for instance along x_2) is orthogonal to the rotation axis (along x_1), the longitudinal and transversal modes do not evolve independently. The first sound is coupled with one of the two transversal modes in which velocity vibrates, whereas fields v_1, T and q do not vibrate.

$\omega_1 = 0$	$\omega_{2,3} \simeq \pm K V_1 + O(\Omega^2)$
$\tilde{\rho} = \psi$ $\tilde{v}_2 = 0$ $\tilde{v}_3 = i \frac{K V_1^2}{2\Omega\rho} \psi$	$\tilde{\rho} = \psi$ $\tilde{v}_2 = \frac{\pm V_1}{\rho} \psi$ $\tilde{v}_3 = -\frac{2i\Omega}{\rho K} \psi$

Second sound is coupled with a transversal mode in which T, q_2 and q_3 vibrate. Neglecting the second-order terms in Ω, the dispersion relation becomes:

$$\left(-\omega - \frac{i}{2}\kappa L_R (B - B'')\right) \left[-\omega \left(-\omega - \frac{i}{2}\kappa L_R (B - B'')\right) - K^2 V_2^2\right] = 0. \tag{25}$$

For $\omega \in \Re$ and $K = k_r + i k_s$ complex, one gets the solution $\omega_4 = 0$, which represents a stationary mode; and two solutions which furnish the following phase velocity and attenuation coefficient of the temperature wave (approximated with respect to $(B - B'')\kappa L_R / \omega$):

$$w \simeq \pm V_2 \left(1 - \frac{(B - B'')^2 \kappa^2 L_R^2}{32\omega^2}\right) + O\left(\frac{(B - B'')^4 \kappa^4 L_R^4}{\omega^4}\right), \tag{26}$$

$$k_s \simeq \frac{(B - B'')\kappa L_R}{4 V_2} + O\left(\frac{(B - B'')^3 \kappa^3 L_R^3}{\omega^2}\right). \tag{27}$$

The corresponding modes are $\tilde{\rho} = \tilde{q}_1 = \tilde{v}_1 = \tilde{v}_2 = \tilde{v}_3 = 0$ and

$\omega_4 = 0$	$\omega_{5,6} \simeq \pm k_r V_2 \left(1 - \frac{(B-B'')^2\kappa^2 L_R^2}{32\omega^2}\right) + O\left(\frac{(B-B'')^3\kappa^3 L_R^3}{\omega^2}\right)$
$\tilde{T} = -\frac{i(2\Omega - \frac{1}{2}\kappa L_R B')}{\zeta K} \psi$ $\tilde{q}_2 = 0$ $\tilde{q}_3 = \psi$	$\tilde{T} = T_0 \psi$ $\tilde{q}_2 = \frac{T_0 \zeta}{V_2} \left(1 - \frac{(B-B'')^2\kappa^2 L_R^2}{32\omega^2}\right) \psi$ $\tilde{q}_3 = \frac{i(2\Omega - \frac{1}{2}\kappa L_R B') T_0 \zeta \left(1 - \frac{(B-B'')^2\kappa^2 L_R^2}{32\omega^2}\right)}{V_2 \left[\pm k_r V_2 \left(1 - \frac{(B-B'')^2\kappa^2 L_R^2}{32\omega^2}\right) - \frac{i}{2}(B-B'')\kappa L_R\right]} \psi$

We note that in the mode of frequency $\omega_4 = 0$, only the transversal component of the heat flux is involved.
For $\omega = \omega_r + i\omega_s$ complex and $K \in \Re$, the first solution of the dispersion relation (25) becomes $\omega_4 = -\frac{i}{2}(B - B'')\kappa L_R$. This first mode corresponds to an extremely slow relaxation phenomenon involving the temperature and the transversal component of the heat flux

$\omega_4 = -\frac{i}{2}(B - B'')\kappa L_R$
$\tilde{\rho} = \tilde{v}_1 = \tilde{v}_2 = \tilde{v}_2 = \tilde{v}_3 = 0$ $\tilde{T} = -\frac{i(2\Omega - \frac{1}{2}\kappa L_R B')}{\zeta K} \psi$ $\tilde{q}_2 = 0$ $\tilde{q}_3 = \psi$

which, when $\Omega \to 0$, converges to a stationary mode.

3.3 Counterflow in a cylindrical tube

Here we apply the model proposed in Section 2 to study the superfluid turbulence, in a cylindrical channel filled with helium II and submitted to a longitudinal stationary heat flux; for simplicity we suppose that the vortex distribution is described as an isotropic tangle. This allows us to suppose that the *microscopic vorticity* $\vec{\omega}$ (hence the unit vector $\mathbf{s}'$) is isotropically distributed, so that

$$< \mathbf{U} - \mathbf{s}' \otimes \mathbf{s}' > = \frac{2}{3}\mathbf{U}. \tag{28}$$

while $< \omega >$ depends on the average vortex line length L per unit volume, through the simple proportionality law $< \omega > = \kappa L$ and $\lambda = B/2$, $\lambda' = 0$. As a consequence, the pressure tensor (14) takes the simplified form

$$\mathbf{P}_\omega = \lambda\, \kappa L \frac{2}{3}\mathbf{U} \quad \Rightarrow \quad \vec{\sigma}_q^H = -K_1 L \mathbf{q}, \tag{29}$$

where $K_1 = \frac{1}{3}\kappa B$.

3.3.1 Wave propagation in presence of thermal counterflow

Consider a cylindrical channel filled with helium II, submitted to a longitudinal heat flux $\mathbf{q}_0$, exceeding the critical value $\mathbf{q}_c$. We refer now to the experimental device (Donnelly & Swanson, 1986), (Donnelly, 1991) in which second sound is excited transversally with respect to the channel. In this case, the heat flux $\mathbf{q}$ can be written as $\mathbf{q} = \mathbf{q}_0 + \mathbf{q}'$, with $\mathbf{q}'$ the contribution to the heat flux, orthogonal to $\mathbf{q}_0$, due to the temperature wave. Suppose that the longitudinal heat flux $\mathbf{q}_0$ down the channel is much greater than the perturbation $\mathbf{q}'$. Under these hypotheses, neglecting second order terms in $\mathbf{q}'$, the production term is linear in the perturbation $\mathbf{q}'$.

To study the second sound attenuation in the experiment described above, we use simplified field equations, where all the nonlinear contributions are neglected. Under the above hypotheses, omitting also the thermal dilation, the linearized set of field equations read as

$$\begin{cases} \frac{\partial \rho}{\partial t} + \rho \frac{\partial v_j}{\partial x_j} = 0, \\ \rho \frac{\partial v_i}{\partial t} + \frac{\partial p}{\partial x_i} = 0, \\ \frac{\partial T}{\partial t} + \frac{1}{\rho c_V} \frac{\partial q_j}{\partial x_j} = 0, \\ \frac{\partial q_i}{\partial t} + \zeta \frac{\partial T}{\partial x_i} = -\frac{1}{3}\kappa B L q_i. \end{cases} \tag{30}$$

A stationary solution of the system (30) is (Jou et al., 2002):

$$\rho = \rho_0, \quad \dot{\mathbf{v}} = \mathbf{0}, \quad T = T(x_1) = T_0 - \frac{\kappa B L}{3\zeta} q_0 x_1, \quad \mathbf{q} = \mathbf{q}_0, \tag{31}$$

where x_1 is the direction of the heat flux $\mathbf{q} = \mathbf{q}_0$. In order to study the propagation of harmonic plane waves in the channel, we look for solutions of the system (30) of the form (22) with $\Gamma_0 = (\rho_0, \mathbf{0}, T(x_1), \mathbf{q}_0)$. The longitudinal modes are obtained projecting the vectorial equations for the small amplitudes of velocity and heat flux on the direction orthogonal to the wave front. It is observed that the first sound is not influenced by the thermal counterflow, while the velocity and the attenuation of the second sound are influenced by the presence of the vortex tangle. The results are (Peruzza & Sciacca, 2007):

$$w_{1,2} = \pm\sqrt{p_\rho},$$

with p_ρ standing for $\partial p/\partial \rho$ and:

$$w_{3,4} = \pm V_2 \sqrt{\left(1 + \frac{k_s^2 V_2^2}{\omega^2}\right)^{-1}} \simeq \pm V_2 \left(1 - k_s^2 \frac{V_2^2}{2\omega^2}\right), \qquad k_s = \frac{1}{6}\kappa B L w. \tag{32}$$

The transversal modes are obtained projecting the vectorial equations for the small amplitudes of velocity and heat flux on the wave front. The solutions of this equation are: $\omega_5 = 0$ and $\omega_6 = \frac{i}{3}\kappa BL$. The mode $\omega_5 = 0$ is a stationary mode.

3.4 Combined situation of rotating counterflow

The combined situation of rotation and heat flux, is a relatively new area of research (Jou & Mongiovì, 2004), (Mongiovì & Jou, 2005), (Tsubota et al., 2004). The first motivation of this interest is that from the experimental observations one deduces that the two effects are not merely additive; in particular, for $\mathbf{q}$ or $\mathbf{\Omega}$ high, the measured values of L are always less than $L_H + L_R$ (Swanson et al., 1983).

Under the simultaneous influence of heat flux $\mathbf{q}$ and rotation speed Ω, rotation produces an ordered array of vortex lines parallel to rotation axis, whereas counterflow velocity causes a disordered tangle. In this way the total vortex line is given by the superposition of both contributions so that the vortex tangle is anisotropic. Therefore, assuming that the rotation is along the x_1 direction $\mathbf{\Omega} = (\Omega,0,0)$ and isotropy in the transversal $(x_2 - x_3)$ plane, for the vorticity tensor $\mathbf{P}_\omega$, in combined situation of counterflow and rotation, the following explicit expression is taken

$$\mathbf{P}_\omega = \frac{B}{2}\kappa L \left\{ \frac{2}{3}(1-D)\mathbf{U} + D\left[\left(1 - \frac{B''}{B}\right)\left(\mathbf{U} - \hat{\mathbf{\Omega}} \otimes \hat{\mathbf{\Omega}}\right) + \frac{B'}{B}\mathbf{W}\cdot\hat{\mathbf{\Omega}} + 2\frac{B''}{B}\hat{\mathbf{\Omega}} \otimes \hat{\mathbf{\Omega}}\right]\right\}, \tag{33}$$

where D is a parameter between 0 and 1 related to the anisotropy of vortex lines, describing the relative weight of the array of vortex lines parallel to $\mathbf{\Omega}$ and the disordered tangle of counterflow (when $D = 0$ we recover an isotropic tangle – right hand side of Eq. (30d) –, whereas when $D = 1$ the ordered array – Eq. (17)). Assuming $b = \frac{1}{3}(1-D) + \frac{DB''}{B}$ and $c = \frac{B'D}{B}$, the vorticity tensor (33) can be written as:

$$\mathbf{P}_\omega = \frac{B}{2}\kappa L \left\{ \begin{pmatrix} 2b & 0 & 0 \\ 0 & 1-b & 0 \\ 0 & 0 & 1-b \end{pmatrix} + \begin{pmatrix} 0 & 0 & 0 \\ 0 & 0 & c \\ 0 & -c & 0 \end{pmatrix} \right\}. \tag{34}$$

Note that the isotropy in the $x_2 - x_3$ plane may only be assumed when both $\mathbf{\Omega}$ and $\mathbf{V}_{ns}$ are directed along the x_1 axis. A more general situations was studied in (Jou & Mongiovì, 2006).

3.4.1 Wave propagation with simultaneous rotation and counterflow

Substituting the expression (34) into the linearized set of field equations (15), it becomes

$$\begin{cases} \frac{\partial \rho}{\partial t} + \rho \frac{\partial v_j}{\partial x_j} = 0, \\ \rho \frac{\partial v_i}{\partial t} + \frac{\partial p}{\partial x_i} + 2\rho\Omega v_j W_{1ji} = 0, \\ \frac{\partial T}{\partial t} + \frac{1}{\rho c_V}\frac{\partial q_j}{\partial x_j} = 0, \\ \frac{\partial q_i}{\partial t} + \zeta \frac{\partial T}{\partial x_i} + 2\Omega q_j W_{1ji} = -\frac{B}{2}\kappa L \left\{ 2bq_1\delta_{1i} + \left[(1-b)q_2 + cq_3\right]\delta_{2i} + \left[(1-b)q_3 - cq_2\right]\delta_{3i} \right\}, \end{cases} \tag{35}$$

A stationary solution of this system is:

$$\rho = \rho_0, \quad \dot{\mathbf{v}} = \mathbf{0}, \quad \mathbf{q} = \mathbf{q}_0 \equiv (q_0, 0, 0), \quad T = T(x_i) = T_0 - 2\frac{B\kappa L}{2\zeta} b q_0 \delta_{1i} x_i.$$

In order to study the propagation of harmonic plane waves, we look for solutions of (35) of the form (22), with $\Gamma_0 = (\rho_0,\ 0,\ T(x_i),\ \mathbf{q}_0)$.
Now, we investigate two different cases: $\mathbf{n}$ parallel to $\boldsymbol{\Omega}$ and $\mathbf{n}$ orthogonal to $\boldsymbol{\Omega}$; the latter is the only case for which experimental data exist (Swanson et al., 1983).
First case: $\mathbf{n}$ *parallel to* $\boldsymbol{\Omega}$.
Let x_1 be the direction of the rotation axis and of the unit vector $\mathbf{n}$ orthogonal to the wave front. In this case the longitudinal and transversal modes evolve independently. In particular, we can observe that the first sound is not influenced by the presence of the vortex tangle $k_s^{(1)} = 0$ and $\tilde{T} = 0$, $\mathbf{q} = 0$

$$\boxed{\omega_{1,2} = \pm k_r V_1}$$

$$\boxed{\begin{array}{l} \tilde{\rho} = \psi \\ \tilde{v}_1 = \frac{V_1}{\rho}\psi \end{array}}$$

whereas the second sound suffers an extra attenuation due to the vortex tangle. This is confirmed by the approximate solutions of the dispersion relation

$$w_{3,4} \simeq \pm V_2 \left(1 - \frac{B^2\kappa^2 L^2 b^2}{8\omega^2}\right) + O\left(\frac{B^4\kappa^4 L^4 b^4}{16\omega^4}\right), \tag{36}$$

$$k_s^{(2)} \simeq \frac{B\kappa L b}{2V_2} + O\left(\frac{B^3\kappa^3 L^3 b^3}{8\omega^2}\right). \tag{37}$$

where ω is assumed real and $K = k_r + ik_s$ complex. When $\Omega = 0$ and $b = 1/3$ the results of the Section 3.3 are obtained again.
Now, we study the transversal modes, corresponding to $\omega_{5,6} = \pm 2|\Omega|$; in this case $\tilde{\rho} = \tilde{T} = \tilde{q}_1 = \tilde{q}_2 = \tilde{q}_3 = \tilde{v}_1 = 0$ and

$\omega_{5,6} = \pm 2\lvert\Omega\rvert$
$\tilde{v}_3 = \psi$ $\tilde{v}_2 = \pm i\psi$

They correspond to extremely slow phenomena, which, when $\Omega \to 0$, tend to stationary modes. Finally, the dispersion relation

$$\omega_{7,8} = \pm\left(2\Omega - \frac{B}{2}\kappa Lc\right) - i\frac{B}{2}\kappa L\,(1-b) \tag{38}$$

corresponds to the vibration of only these fields

$\omega_{7,8} = \pm\left(2\Omega - \frac{B}{2}\kappa Lc\right) - i\frac{B}{2}\kappa L\,(1-b)$
$\tilde{q}_3 = \psi$ $\tilde{q}_2 = \pm i\psi$

From (36), (37) and (38) one may obtain the following quantities L, b and c:

$$L = \frac{-\omega_s w + V_2^2 k_s}{\kappa w B/2}, \quad b = \frac{V_2^2 k_s}{-\omega_s w + V_2^2 k_s}, \quad c = \frac{-\omega_r w + 2\Omega w}{-\omega_s w + V_2^2 k_s}, \tag{39}$$

where we have put $\omega_7 = \omega_r + i\omega_s$.
The results of this section imply that measurement in a single direction are enough to give information on all the variables describing the vortex tangle.
Second case: **n** *orthogonal to* $\boldsymbol{\Omega}$.
Now we assume that the direction of propagation of the waves is orthogonal to the rotation axis (axis x_1), i.e. for example, $\mathbf{n} = (0,1,0)$. In this case the longitudinal and the transversal modes do not evolve independently. In particular, the first sound is coupled with one of the two transversal modes in which velocity vibrates, while the second sound is coupled with a transversal mode in which heat flux vibrates.
Fields $\tilde{\rho}$, $\tilde{v}_2$, $\tilde{v}_3$ have the same solutions and the same dispersion relation to the case of pure rotation

$$-\,\omega\left[\omega^2 - 4\Omega^2 - K^2 p_\rho\right] = 0. \tag{40}$$

The dispersion relation of fields $\tilde{T}$, $\tilde{q}_2$, $\tilde{q}_3$ is instead:

$$\left(-\omega - i\gamma\frac{B}{2}\kappa L(1-b)\right)\left[\omega\left(-\omega - i\frac{B}{2}\kappa L(1-b)\right) + K^2 V_2^2\right] + \omega\left(2i\Omega - i\frac{B}{2}\kappa Lc\right)^2 = 0. \tag{41}$$

Assuming $\omega \in \Re$ and $K = k_r + ik_s$ and in the hypothesis of small dissipation ($k_r^2 \gg k_s^2$), one obtains:

$$k_s = \frac{B}{2}\kappa L(1-b)\left(\frac{2w^2 - V_2^2}{2wV_2^2}\right), \tag{42}$$

$$\left(\frac{\omega}{k_r}\right)^2 = 0, \quad \text{and}$$
$$\left(\frac{\omega}{k_r}\right)^2 = w^2 = V_2^2 \frac{(\omega^2 - \tilde{B})}{(\omega^2 + \tilde{A})} = V_2^2 \frac{1}{1 - \frac{\left(2\Omega - \frac{B}{2}\kappa Lc\right)^2}{\omega^2 + (B/2)^2 \kappa^2 L^2 (1-b)^2}}. \tag{43}$$

where $\tilde{A} = -\left[\left(2\Omega - \frac{B}{2}\kappa Lc\right)^2 - \frac{B^2}{4}\kappa^2 L^2 (1-b)^2\right]$ and $\tilde{B} = -\frac{B^2}{4}\kappa^2 L^2 (1-b)^2$.
We can remark that the coefficients $\tilde{A}$ and $\tilde{B}$ are negative and that $w^2 \geq V_2^2$ because $\omega^2 + \tilde{A} \leq \omega^2 - \tilde{B}$ and, in particular, $w^2 = V_2^2$ for $\Omega = \frac{B\kappa Lc}{4}$.
Now, studying the transversal modes, i.e. that ones corresponding to non zero $\tilde{v}_1$ and $\tilde{q}_1$, we obtain $\omega_7 = 0$, which corresponds to a stationary mode, and

$$\omega_8 = -iB\kappa Lb. \tag{44}$$

Summarizing, also in this case measurements in a single direction are enough to given information on all the variables describing the vortex tangle, namely L, b and c, from equations (42), (43) and (44)

$$L = \frac{4k_s w V_2^2 - \omega_s \left(2w - V_2^2\right)}{\left(2w^2 - V_2^2\right) B\kappa}, \qquad b = -\frac{\omega_s \left(2w^2 - V_2^2\right)}{4k_s w V_2^2 - \omega_s \left(2w - V_2^2\right)},$$
$$c = \frac{4\Omega(2w^2 - V_2^2) - \sqrt{(1 - V_2^2)(4k_r^2(2w^2 - V_2^2)^2 + 16k_s^2 V_2^4)}}{4k_s w V_2^2 - \omega_s (2w^2 - V_2^2)}, \tag{45}$$

where we have put $\omega_8 = i\omega_s$ and $\omega_s = -\kappa LbB$.
In this subsection we have analyzed wave propagation in the combined situation of rotation and counterflow with the direction n orthogonal to Ω. In (Swanson et al., 1983) authors experimented the same situation, but they didn't represent the attenuation neither the speed of the second sound but only the vortex line density L as function of Ω and V_{ns}. Therefore, it is unknown how they plotted these graphics, which hypothesis they made and what the anisotropy considered. Instead, the results of these two subsections allow to know the spatial distribution of the vortex tangle simply by performing experiments on waves propagating orthogonally to Ω (equations (39)) or parallelly to Ω (equations (45)).
From the physical point of view it is interesting to note that our detailed analysis in this subsection shows that, in contrast to which one could intuitively expect, measurements in a single direction are enough to give information on all the variables describing the vortex tangle, namely L, b and c, for instance, from one of (36)-(37) and (38) or of (42)-(43) and (44). This is not an immediate intuitive result.

3.5 Comparison with the two-fluid model

To compare the one-fluid model of liquid helium II in a non-inertial frame with the two-fluid one, we recall that in (Mongiovì, 1991), (Mongiovì, 1993) it is shown that the linearized field equations (3) can be identified with those of the two-fluid non dissipative model if we define

$$\zeta = \rho \frac{\rho_s}{\rho_n} T s^2, \tag{46}$$

and we make the following change of variables:

$$\mathbf{q} = \rho_s Ts \mathbf{V}_{ns}, \tag{47}$$

$$\mathbf{v} = \frac{\rho_s}{\rho}\mathbf{v}_s + \frac{\rho_n}{\rho}\mathbf{v}_n, \tag{48}$$

where, we recall, $\mathbf{v}_n$ and $\mathbf{v}_s$ are the mesoscopic velocities of the normal and superfluid components and $\mathbf{V}_{ns} = \mathbf{v}_n - \mathbf{v}_s$ is the counterflow velocity.
If we perform in the field equations (15) the change of variables (47–48), we check immediately that the first three equations are identical to the ones of the two-fluid model for helium II, even in non-inertial frame (Peruzza & Sciacca, 2007). We concentrate therefore on the field equation for the heat flux. To the first order approximation with respect to the relative velocity $\mathbf{V}_{ns}$ and the derivatives of the field variables, we obtain:

$$\frac{\partial \mathbf{V}_{ns}}{\partial t} + \frac{\zeta}{\rho_s Ts}\nabla T + 2\mathbf{\Omega} \times \mathbf{V}_{ns} = \frac{1}{\rho_s Ts}\vec{\sigma}_q, \tag{49}$$

where $\vec{\sigma}_q$ stands for $\vec{\sigma}_q^R$ in rotation case, $\vec{\sigma}_q^H$ in counterflow case and $\vec{\sigma}_q^{HR}$ in rotating counterflow. We multiply equation (49) by ρ_n/ρ and add it to the balance equation (15 b). Making use of the result $\mathbf{v}_s = \mathbf{v} - (\rho_n/\rho)\mathbf{V}_{ns}$, we find

$$\frac{\partial \mathbf{v}_s}{\partial t} - s\nabla T + \frac{1}{\rho}\nabla p + 2\mathbf{\Omega} \times \mathbf{v}_s + \frac{\rho_n}{\rho}\frac{1}{\rho_s Ts}\vec{\sigma}_q = 0. \tag{50}$$

In virtue of equation $d\mu = (1/\rho)dp - sdT$, which relates the chemical potential $\mu = \epsilon - Ts + (p/\rho)$ to the equilibrium variables, the field equation for the superfluid velocity takes the form

$$\rho_s \frac{\partial \mathbf{v}_s}{\partial t} + \rho_s \nabla \mu + 2\rho_s \mathbf{\Omega} \times \mathbf{v}_s + \frac{\rho_n}{\rho}\frac{1}{Ts}\vec{\sigma}_q = 0. \tag{51}$$

Expression (51) is identical to the corresponding field equation for $\mathbf{v}_s$, obtained in the two-fluid model. Of course in the pure counterflow case $\mathbf{\Omega}$ has to be set zero in (51). This result is a confirmation of the results derived in the framework of the one-fluid model based on E.T..
In counterflow experiments, equation (51) can be written as:

$$\rho_s \frac{\partial \mathbf{v}_s}{\partial t} + \rho_s \nabla \mu = \mathbf{F}_{ns}^{(E)}, \qquad \text{where} \qquad \mathbf{F}_{ns}^{(E)} = \frac{1}{3}\frac{\rho_s \rho_n}{\rho}\kappa B \gamma^2 V_{ns}^2 \mathbf{V}_{ns} \tag{52}$$

and relation $L = \gamma^2 V_{ns}^2$ has been used.
To interpret the experimental results on stationary helium flow through channels using the two-fluid model, Gorter and Mellink (Gorter & Mellink, 1949) and Vinen (Vinen, 1957) postulate the existence, in the field equation for the superfluid component, of a dissipative term proportional to the cube of the relative velocity $\mathbf{V}_{ns}$:

$$\mathbf{F}_{ns}^{(GM)} = \rho_s \rho_n \bar{A} V_{ns}^2 \mathbf{V}_{ns}, \tag{53}$$

$\bar{A}$ being a temperature dependent coefficient. It is interesting to note that, setting $\bar{A} = \kappa B \gamma^2/(3\rho)$ in (52b), and using (47), the results of the present work are in full agreement with those of Gorter and Mellink.

4. Hydrodynamical model of inhomogeneous superfluid turbulence

In Section 3 a first model of superfluid turbulence was presented, where the vortices were modeled through the pressure tensor $\mathbf{P}_\omega$ for which a constitutive relation was written.
Experiments (Vinen, 2000), (Vinen & Niemela, 2002), show the formation of a new type of superfluid turbulence, which has some analogies with classical one, as for instance using towed or oscillating grids, or stirring liquid helium by means of propellers. In this situation (named co-flow) both components, normal and superfluid, flow along the same direction. To describe these experiments it is necessary to build up a hydrodynamic model of quantum turbulence, in which the interactions between both fields can be studied and the role of inhomogeneities is explicitly taken into account (Mongiovì & Jou, 2007), (Ardizzone & Gaeta, 2009).
In a more complete hydrodynamic model of superfluid turbulence the line density L acquires field properties: it depends on the coordinates, it has a drift velocity $\mathbf{v}_L$, and it has associated a diffusion flux. These features are becoming increasingly relevant, as the local vortex density may be measured with higher precision, and the relative motion of vortices is observed and simulated. Thus it is important to describe situations going beyond the usual description of the vortex line density averaged over the volume. Our aim, in this Section, is to formulate a hydrodynamical framework sufficiently general to encompass vortex diffusion and to describe the interactions between the second sound waves and the vortices, instead of considering the latter as a rigid framework where such waves are simply dissipated. This is important because second sound provides the standard method of measuring the vortex line density L, and the mentioned dynamical mutual interplay between second sound and vortex lines may modify the standard results.

4.1 The line density and Vinen's equation

The most well known equation in the field of superfluid turbulence is Vinen's equation (Vinen, 1958), which describes the evolution of L, the total length of vortex lines per unit volume, in counterflow situations characterized by a heat flux q. Vinen suggested that in homogeneous counterflow turbulence there is a balance between generation and decay processes, which leads to a steady state of quantum turbulence in the form of a self-maintained vortex tangle. The Vinen's equation (2), written in terms of the variable q, is:

$$\frac{dL}{dt} = \alpha_q |q| L^{3/2} - \beta_q L^2, \tag{54}$$

with $\alpha_q = \alpha_v \rho_s s T$ and $\beta_q = \kappa \beta_v$.
Vinen considered homogeneous superfluid turbulence and assumed that the time derivative dL/dt is composed of two terms:

$$\frac{dL}{dt} = \left[\frac{dL}{dt}\right]_f - \left[\frac{dL}{dt}\right]_d, \tag{55}$$

the first is responsible for the growth of L, the second for its decay. Vinen assumes that the production term $[dL/dt]_f$ depends linearly on the instantaneous value of L and the force $\mathbf{f}$ between the vortex line and the normal component, which is linked to the modulus $|q|$ of the heat flux, and he obtained:

$$\left[\frac{dL}{dt}\right]_f = \alpha_v \kappa V_{ns} L^{3/2} = \alpha_q |q| L^{3/2}. \tag{56}$$

The form of the term responsible for the vortex decay was determined assuming Feynman's model of vortex breakup, analogous to Kolmogorov's cascade in classical turbulence

$$\left[\frac{dL}{dt}\right]_d = -\beta_v \kappa L^2 = -\beta_q L^2, \tag{57}$$

thus obtaining equation (54). A microscopic derivation of this equation was made by Schwarz (Schwarz, 1988).
The stationary solutions of this equation are $L = 0$ and $L^{1/2} = (\alpha_q/\beta_q)|q|$. The non-zero solution is proportional to the square of the heat flux and describes well the full developed turbulence.

4.2 Derivation of the hydrodynamical model

The starting point here is to formulate a theory for a turbulent superfluid, which uses the averaged vortex line density L in addition to the fields ρ, $\mathbf{v}$, E and $\mathbf{q}$, used in Sections 2 and 3. Because we want to formulate a general nonlinear theory, we will suppose that the dynamics of the excitations is described by a vector field m_i, which must be considered as an internal variable, linked to the heat flux q_i through a constitutive relation, but not identical to it.
We consider for the fields ρ, $\mathbf{v}$, E and $\mathbf{m}$ and L the following balance equations written in terms of the non-convective terms (Ardizzone & Gaeta, 2009):

$$\begin{cases} \dot{\rho} + \rho \frac{\partial v_k}{\partial x_k} = 0, \\ \rho \dot{v}_i + \frac{\partial J^v_{ik}}{\partial x_k} = 0, \\ \dot{E} + E\frac{\partial v_k}{\partial x_k} + \frac{\partial q_k}{\partial x_k} + J^v_{ik}\frac{\partial v_i}{\partial x_k} = 0, \\ \dot{m}_i + m_i \frac{\partial v_k}{\partial x_k} + \frac{\partial J^m{}_{ik}}{\partial x_k} = \sigma^m_i, \\ \dot{L} + L\frac{\partial v_k}{\partial x_k} + \frac{\partial J^L_k}{\partial x_k} = \sigma^L. \end{cases} \tag{58}$$

where J^v_{ij} is the stress tensor, J^m_{ij} the flux of the field m_i, and J^L_i the flux of vortex lines; σ^m_i and σ^L are terms describing the net production of the field m_i characterizing the dynamics of the excitations and the production of vortices. Dot denotes the material time derivative.
Since in the system (58) there are more unknowns than equations, it is necessary to complete it by adding constitutive equations, relating the variables m_i, J^v_{ik}, J^m_{ik} and J^L_i to the independent fields ρ, E, q_i and L. As a consequence of the material objectivity principle, the constitutive equations can be expressed in the form:

$$\begin{aligned} m_i &= \alpha(\rho, E, q^2, L) q_i, \\ J^v_{ik} &= p(\rho, E, q^2, L)\delta_{ik} + a(\rho, E, q^2, L) q_{<i}q_{k>}, \\ J^m_{ik} &= \beta(\rho, E, q^2, L)\delta_{ik} + \gamma(\rho, E, q^2, L) q_{<i}q_{k>}, \\ J^L_i &= \nu(\rho, E, q^2, L) q_i. \end{aligned} \tag{59}$$

where α, β, a, p, γ, ν are scalar functions, δ_{ik} is the Kronecker symbol and $q_{<i}q_{k>} = q_i q_k - \frac{1}{3}q^2\delta_{ik}$ is the deviatoric part of the diadic product $q_i q_j$.

4.2.1 Restrictions imposed by the entropy principle

Further restrictions on these constitutive relations are deduced from the second law of thermodynamics. Accordingly, there exists a convex function $S = S(\rho, E, q^2, L)$, the entropy per unit volume, and a vector function $J_k^S = \phi(\rho, E, q^2, L)q_k$, the entropy flux density, such that the rate of production of entropy σ^S is non-negative

$$\sigma^S = \dot{S} + S\frac{\partial v_k}{\partial x_k} + \frac{\partial J_k^S}{\partial x_k} \geq 0. \tag{60}$$

Note that this inequality does not hold for any value of the fundamental variables, but only for the thermodynamic processes, i.e. only for those values which are solution of the system (58). This means that we can consider the equations (58) as constraints for the entropy inequality to hold. A way to take these constraints into account was proposed by Liu (Liu, 1972): he showed that the entropy inequality becomes totally arbitrary provided that we complement it by the evolution equations for the fields ρ, v_i, E, m_i and L affected by Lagrange multipliers: $\Lambda_\rho = \Lambda_\rho(\rho, E, q^2, L)$, $\Lambda_i^v = \Lambda_v(\rho, E, q^2, L)q_i$, $\Lambda_E = \Lambda_E(\rho, E, q^2, L)$, $\Lambda_i^m = \lambda(\rho, E, q^2, L)q_i$, $\Lambda_L = \Lambda_L(\rho, E, q^2, L)$. One obtains the following inequality, which is satisfied for arbitrary values of the field variables:

$$\begin{aligned}
\dot{S} + S\frac{\partial v_k}{\partial x_k} + \frac{\partial J_k^S}{\partial x_k} \quad &- \quad \Lambda_\rho\left[\dot{\rho} + \rho\frac{\partial v_k}{\partial x_k}\right] - \Lambda_i^v\left[\dot{v}_i + \frac{1}{\rho}\frac{\partial J_{ik}^v}{\partial x_k}\right] \\
&- \quad \Lambda_E\left[\dot{E} + E\frac{\partial v_k}{\partial x_k} + \frac{\partial q_k}{\partial x_k} + J_{ik}^v\frac{\partial v_i}{\partial x_k}\right] \\
&- \quad \Lambda_i^m\left[\dot{m}_i + m_i\frac{\partial v_k}{\partial x_k} + \frac{\partial J_{ik}^m}{\partial x_k} - \sigma_i^m\right] \\
&- \quad \Lambda_L\left[\dot{L} + L\frac{\partial v_k}{\partial x_k} + \frac{\partial J_k^L}{\partial x_k} - \sigma^L\right] \geq 0.
\end{aligned} \tag{61}$$

Imposing that the coefficients of the time derivatives of ρ, v_i, E, q_i and L vanish, one gets: $\Lambda_v = 0$ and

$$dS = \Lambda_\rho d\rho + \Lambda_E dE + \Lambda_L dL + \Lambda_i^m \, dm_i, \tag{62}$$

Imposing that the coefficients of space derivatives of ρ, E, q_i and L vanish, one finds:

$$dJ_k^S = \Lambda_i^m dJ_{ik}^m + \Lambda_L dJ_k^L + \Lambda_E dq_k. \tag{63}$$

From these relations in (Ardizzone et al., 2009) we have found:

$$\Lambda_v = 0, \qquad\qquad a = 0, \tag{64}$$

$$dS = \Lambda_\rho d\rho + \Lambda_E dE + \lambda q_i d(\alpha q_i) + \Lambda_L dL, \tag{65}$$

$$S - \rho\Lambda_\rho - \Lambda_E(E + p) - \lambda\alpha q^2 - \Lambda_L L = 0, \tag{66}$$

$$d\phi = \lambda\left(d\beta + \frac{1}{6}\gamma dq^2 + \frac{2}{3}q^2 d\gamma\right) + \Lambda_L dv, \tag{67}$$

$$\phi = \Lambda_E + \lambda\gamma q^2 + \Lambda_L \nu. \tag{68}$$

We note that all the relations (65)-(68) are exact, because no approximation has been used for their determination and maintain their validity also far from equilibrium.
It remains the following residual inequality for the entropy production:

$$\sigma^S = \Lambda_i^m \sigma_i^m + \Lambda^L \sigma^L \geq 0. \tag{69}$$

Introducing the specific internal energy $\epsilon = E/\rho$, substituting the constitutive equations (59) in system (58) and the restriction $a = 0$, the following system of field equations is obtained:

$$\begin{cases} \dot{\rho} + \rho\frac{\partial v_k}{\partial x_k} = 0, \\ \rho\dot{v}_i + \frac{\partial p}{\partial x_i} = 0, \\ \rho\dot{\epsilon} + \frac{\partial q_k}{\partial x_k} + p\frac{\partial v_k}{\partial x_k} = 0, \\ \dot{\widehat{\alpha q_i}} + \alpha q_i\frac{\partial v_j}{\partial x_j} + \frac{\partial[\beta\delta_{ik} + \gamma q_{<i}q_{k>}]}{\partial x_k} = \sigma_i^m, \\ \dot{L} + L\frac{\partial v_k}{\partial x_k} + \frac{\partial(\nu q_k)}{\partial x_k} = \sigma^L. \end{cases} \tag{70}$$

Observe that in these equations there are the unknown quantities α, p, ϵ, β, γ and ν, which are not independent, because they must satisfy relations (65)-(68), and the productions σ_i^m and σ^L which must satisfy inequality (69).
In (Ardizzone et al., 2009) it is shown that, using a Legendre transformation, the constitutive theory is determined by the choice of only two scalar functions S' and ϕ' of the intrinsic Lagrange multipliers, defined as:

$$S' = -S + \Lambda_\rho \rho + \Lambda_E E + \Lambda_L L + \Lambda_i^m m_i, \tag{71}$$

$$\Phi_k' = \phi' \Lambda_k^m = -J_k^S + \Lambda_E q_k + \Lambda_L J_k^L + \Lambda_i^m J_{ik}^m, \tag{72}$$

Furthermore, if one chooses as state variables the fields

$$\tilde{\Lambda}_\rho = \Lambda_\rho + \frac{1}{2}\Lambda_E v^2, \; \tilde{\Lambda}_{v_i} = -\Lambda_E v_i, \; \tilde{\Lambda}_{m_i} = \Lambda_{m_i}, \; \tilde{\Lambda}_E = \Lambda_E, \; \tilde{\Lambda}_L = \Lambda_L, \tag{73}$$

the system of field equation (58) assumes the form of a symmetric hyperbolic system and, therefore, for it the Cauchy problem is well posed, i.e. the existence, uniqueness and continuous dependence of its solutions by the initial data is assured.

4.2.2 Physical interpretation of the constitutive quantities and of the Lagrange multipliers

As shown, the use of the Lagrange multipliers as independent variables results very useful from a mathematical point of view. In order to single out the physical meaning of the constitutive quantities and of the Lagrange multipliers, we analyze now in detail the relations obtained in the previous section. First we will determine the equilibrium values for these multipliers. Denoting with Y any of the scalar quantities α, h, ϕ, p, β, γ, ν, Λ_ρ, Λ_E, λ, Λ_L and making the position

$$Y_0(\rho, E, q^2, L) = Y_0(\rho, E, L) + O(q^2), \tag{74}$$

the following relations are obtained:

$$\begin{aligned} dS_0 &= \Lambda_0^\rho d\rho + \Lambda_0^E dE + \Lambda_0^L dL, \\ S_0 - \rho\Lambda_0^\rho - \Lambda_0^E(E + p_0) - \Lambda_0^L L &= 0, \\ d\phi_0 &= \lambda_0 d\beta_0 + \Lambda_0^L d\nu_0, \\ \phi_0 &= \Lambda_0^E + \Lambda_0^L \nu_0. \end{aligned} \tag{75}$$

Introduce now a "generalized temperature" as the reciprocal of the first-order part of the Lagrange multiplier of the energy

$$\Lambda_0^E = \left[\frac{\partial S_0}{\partial E}\right]_{\rho,L} = \frac{1}{T} \tag{76}$$

and observe that, in the laminar regime (when $L = 0$), Λ_0^E reduces to the absolute temperature of thermostatics. In the presence of a vortex tangle the quantity (76) depends also on the line density L. Writing equations (75a) and (75b) as

$$dE = TdS_0 - T\Lambda_0^\rho d\rho - T\Lambda_0^L dL, \tag{77}$$

$$-T\Lambda_0^\rho = \frac{E}{\rho} - T\frac{S_0}{\rho} + \frac{p_0 + LT\Lambda_0^L}{\rho}, \tag{78}$$

and defining the quantity $-\Lambda_0^\rho/\Lambda_0^E = -T\Lambda_0^\rho$ as the "mass chemical potential" in turbulent superfluid

$$-T\Lambda_0^\rho = -T\left[\frac{\partial S_0}{\partial \rho}\right]_{E,L} = \mu_0^\rho, \tag{79}$$

and the quantity $-\Lambda_0^L/\Lambda_0^E = -T\Lambda_0^L$ as the "chemical potential of vortex lines", which is denoted with μ_0^L,

$$-T\Lambda_0^L = -T\left[\frac{\partial S_0}{\partial L}\right]_{\rho,L} = \mu_0^L, \tag{80}$$

one can write equations (77) and (78) in the following form:

$$dS_0 = \frac{1}{T}dE - \frac{1}{T}\mu_0^\rho d\rho - \frac{1}{T}\mu_0^L dL, \tag{81}$$

$$\rho\mu_0^\rho + L\mu_0^L = E - Th_0 + p_0. \tag{82}$$

Indeed, in absence of vortices ($L = 0$) equation (77) is just Gibbs equation of thermostatics and the quantity (79) is the equilibrium chemical potential. The presence of vortices modifies the energy density E, and introduce a new chemical potential.

Consider now the consequences of equations (75c) and (75d) which concern the expressions of the fluxes. Using definitions (76) and (80), we get:

$$\lambda_0 d\beta_0 = d\left(\frac{1}{T}\right) - \nu_0 d\left(\frac{\mu_0^L}{T}\right). \tag{83}$$

From this equation, recalling that in (Mongiovì & Jou, 2007) it has shown that μ_0^L depends only on T and L, one obtains $\partial\beta_0/\partial\rho = 0$ and

$$\frac{\partial\beta_0}{\partial T} = \zeta_0 = -\frac{1}{T^2\lambda_0}\left[1 + \nu_0 T^2 \frac{\partial}{\partial T}\left(\frac{\mu_0^L}{T}\right)\right], \qquad \frac{\partial\beta_0}{\partial L} = \chi_0 = -\frac{\nu_0}{T\lambda_0}\frac{\partial\mu_0^L}{\partial L}. \tag{84}$$

In (Mongiovì & Jou, 2007) it was shown also that it results $\lambda_0 < 0$, $\zeta_0 \geq 0$, $\nu_0 \leq 0$ and $\chi_0 \leq 0$.

4.2.3 The constitutive relations far from equilibrium

Finally, we analyze the complete mathematical expressions far from equilibrium of the constitutive functions and of the Lagrange multipliers.

Non-equilibrium temperature. First, we introduce the following quantity:

$$\theta = \frac{1}{\Lambda_E(\rho, E, L, q^2)}, \tag{85}$$

which, near equilibrium ($L = 0$, $q_i = 0$) can be identified with the local equilibrium absolute temperature. In agreement with (Jou et al., 2001), we will call θ "non-equilibrium temperature", a topic which is receiving much attention in current non-equilibrium Thermodynamics (Casas-Vázquez & Jou, 2003).

Using this quantity, the scalar potential S' is expressed as:

$$S' = -\frac{p}{\theta}. \tag{86}$$

Non-equilibrium Chemical Potentials. As we have seen, at equilibrium the quantities $-\Lambda_\rho/\Lambda_E$ and $-\Lambda_L/\Lambda_E$ can be interpreted as the equilibrium mass chemical potential and the equilibrium vortex line density chemical potential. Therefore, we define as non-equilibrium chemical potentials the quantities:

$$\mu_\rho = -\frac{\Lambda_\rho}{\Lambda_E}, \qquad \text{and} \qquad \mu_L = -\frac{\Lambda_L}{\Lambda_E}. \tag{87}$$

Generalized Gibbs equation. Using equations (65) and (66) and defining $s = S/\rho$ the non-equilibrium specific entropy, one obtains

$$\theta d(\rho s) = dE - \mu_\rho d\rho - \mu_L dL + \theta\lambda q_i d(\alpha q_i), \tag{88}$$

$$\mu_\rho + \frac{L}{\rho}\mu_L = \epsilon - \theta s + \frac{p}{\rho} + \frac{\theta}{\rho}\alpha\lambda q^2. \tag{89}$$

One gets also:

$$dp = \rho d\mu_\rho + L d\mu_L + \rho s d\theta - \alpha q_i d(\theta\lambda q_i). \tag{90}$$

For the interested reader, in (Ardizzone & Gaeta, 2009), the complete constitutive theory can be found.

Non-equilibrium Entropy Flux. The theory developed here furnishes also the complete non-equilibrium expression of the entropy flux J_k^S. Remembering relation (68), we can write:

$$J_i^S = \left(\frac{1}{\theta} + \nu\Lambda_L + \gamma\lambda q^2\right) q_i = \frac{1}{\theta}\left(q_i - \mu_L J_i^L + \theta\gamma\lambda q^2 q_i\right). \tag{91}$$

This equation shows that, in a nonlinear theory of Superfluid Turbulence, the entropy flux is different from the product of the reciprocal non-equilibrium temperature and the heat flux, but it contains additional terms depending on the flux of heat flux and on the flux of line density.

4.3 Linearized field equations

Now we will apply the general set of equations derived to the analysis of two specific situations: vortex diffusion and wave propagation. First of all, we note that, substituting in (70) the constitutive expressions obtained in Subsection 4.2.2, and neglecting nonlinear terms in the fluxes, the following system is obtained (Mongiovì & Jou, 2007):

$$\begin{cases} \dot{\rho} + \rho\nabla\cdot\mathbf{v} = 0, \\ \rho\dot{\mathbf{v}} + \nabla p_0 = 0, \\ \rho\dot{\epsilon} + \nabla\cdot\mathbf{q} + p_0\nabla\cdot\mathbf{v} = 0, \\ \dot{\mathbf{q}} + \zeta_0\nabla T + \chi_0\nabla L = \sigma^{\mathbf{q}}, \\ \dot{L} + L\nabla\cdot\mathbf{v} + \nabla\cdot(\nu_0\mathbf{q}) = \sigma^L, \end{cases} \tag{92}$$

with ζ_0 and χ_0 defined by (84) and satisfying $(\nu_0/\chi_0) = -(T\lambda_0^q L/\epsilon_V)$.
The total pressure of the turbulent superfluid has the form (Mongiovì & Jou, 2007):

$$p_0 = p^* + \epsilon_V L, \tag{93}$$

p^* being the pressure of the bulk superfluid and $\epsilon_V L$ the contribution of the tangle, with ϵ_V the energy per unit length of the vortices (Donnelly, 1991).
For the production terms $\sigma^{\mathbf{q}}$ and σ^L, we will take

$$\sigma^{\mathbf{q}} = -K_1 L\mathbf{q} \qquad \sigma^L = -\beta_q L^2 + \alpha_q|q|L^{3/2}, \tag{94}$$

where $K_1 = \frac{1}{3}\kappa B$. In this approximation, the unknown coefficients, which must be determined from experimental data, are the specific energy ϵ, the pressure p_0, and the three coefficients ζ_0, χ_0 and ν_0, which are functions only of T and L. Here, we will focus a special attention on the coefficients χ_0 and ν_0, which are the new ones appearing in the present formulation, as compared with the formulation presented in (58).

4.3.1 The drift velocity of the tangle

As observed, in a hydrodynamical model of turbulent superfluids, the line density L acquires field properties and its rate of change must obey a balance equation of the general form:

$$\frac{\partial L}{\partial t} + \nabla\cdot(L\mathbf{v}^L) = \sigma^L, \tag{95}$$

with $\mathbf{v}^L$ the drift velocity of the tangle. If we now observe that the last equation of system (92) can be written:

$$\frac{\partial L}{\partial t} + \nabla \cdot (L\mathbf{v} + \nu_0 \mathbf{q}) = \sigma^L, \tag{96}$$

we conclude that the drift velocity of the tangle, with respect to the container, is given by

$$\mathbf{v}^L = \mathbf{v} + \frac{\nu_0}{L}\mathbf{q}. \tag{97}$$

Note that the velocity $\mathbf{v}^L$ does not coincide with the microscopic velocity of the vortex line element, but represents an averaged macroscopic velocity of this quantity. It is to make attention to the fact that often in the literature the microscopic velocity $\dot{\mathbf{s}}$ is denoted with $\mathbf{v}_L$.

Observing that in counterflow experiments ($\mathbf{v} = 0$) results $\mathbf{v}^L = \nu_0 \mathbf{q}/L$, and recalling that measurements (in developed superfluid turbulence) show that the vortex tangle drifts as a whole toward the heater, we conclude that $\nu_0 \leq 0$. The measurement of the drift velocity $\mathbf{v}^L$ of the vortex tangle, together with the measurement of $\mathbf{q}$ and L, would allow one to obtain quantitative values for the coefficient ν_0. In the following section we will propose a way to measure the coefficient χ_0 too.

Another possibility is to interpret $\nu_0 \mathbf{q} = \mathbf{J}^L$ as the diffusion flux of vortices, which since $\nu_0 \leq 0$, would be opposite to the direction of $\mathbf{q}$. Note that, in this model, if $\mathbf{q} = 0$, $\mathbf{J}^L$ is also zero.

4.3.2 Vortex diffusion

An interesting physical consequence from the generalized equations (92) is the description of vortex diffusion. A diffusion equation for the vortex line density was proposed for the first time by van Beelen et al. (van Beelen et al., 1988), in an analysis of vorticity in capillary flow of superfluid helium, in situations with a step change in L arising when the tube is divided in a region with laminar flow and another one with turbulent flow. Assume, for the sake of simplicity, that $T = constant$ and that $\mathbf{q}$ varies very slowly, in such a way that $\dot{\mathbf{q}}$ may be neglected. We find from (92d) and (94a) that

$$\mathbf{q} = -\frac{\chi_0}{K_1 L}\nabla L. \tag{98}$$

Introducing this expression in equation (92e), we find:

$$\frac{dL}{dt} + L\nabla \cdot \mathbf{v} - \frac{\nu_0 \chi_0}{K_1}\nabla \cdot \left(\frac{\nabla L}{L}\right) = \sigma^L = -\beta_q L^2 + \alpha_q q L^{3/2}, \tag{99}$$

where q denotes the modulus of (98). Equation (99) can be written (if $\nabla L \neq 0$)

$$\frac{dL}{dt} + L\nabla \cdot \mathbf{v} - \frac{\nu_0 \chi_0}{K_1 L}\Delta L + \left(\frac{\nu_0 \chi_0}{K_1 L^2}\right)(\nabla L)^2 = \sigma^L. \tag{100}$$

Then, we have for L a reaction-diffusion equation, which generalizes the usual Vinen's equation (54) to inhomogeneous situations. The diffusivity coefficient is (Mongiovì & Jou, 2007)

$$D_{diff} = \frac{\nu_0 \chi_0}{K_1 L}. \tag{101}$$

Since $K_1 > 0$, it turns out that $D_{diff} > 0$, as it is expected. Thus, the vortices will diffuse from regions of higher L to those of lower L. Note that D_{diff} must have dimensions $(length)^2/time$, the same dimensions as κ. Then, a dimensional ansatz could be $D_{diff} \propto \kappa$. Indeed, Tsubota et al. (Tsubota et al., 2003b) have studied numerically the spatial vortex diffusion in a localized initial tangle allowed to diffuse freely, and they found for D_{diff} at very low temperatures (when there is practically no normal fluid), a value $D_{diff} \approx (0.1 \pm 0.05)\kappa$.

If $\mathbf{v}$ vanishes, or if its divergence vanishes, equation (99), neglecting also the term in $(\nabla L)^2$, yields

$$\dot{L} = -\beta_q L^2 + \alpha_q q L^{3/2} + D_{diff}\Delta L. \tag{102}$$

Equation (102) indicates two temporal scales for the evolution of L: one of them is due to the production-destruction term (τ_{decay}) and another one to the diffusion

$$\tau_{decay} \approx [\beta_q L - \alpha_q q L^{1/2}]^{-1}, \qquad \tau_{diff} \approx \frac{X^2}{D_{diff}}, \tag{103}$$

where X is the size of the system. For large values of L, τ_{decay} will be much shorter and the production-destruction dynamics will dominate over diffusion; for small L, instead, diffusion processes may be dominant. This may be also understood from a microscopic perspective because the mean free path of vortex motion is of the order of intervortex spacing, of the order of $L^{-1/2}$, and therefore it increases for low values of L.

A more general situation for the vortex diffusion flux is to keep the temperature gradient in (92d). In this more general case, $\mathbf{q}$ is not more parallel to ∇L but results

$$\mathbf{q} = -\frac{\chi_0}{K_1 L}\nabla L - \frac{\zeta_0}{K_1 L}\nabla T, \tag{104}$$

in which case, it is:

$$\mathbf{J}^L = \nu_0 \mathbf{q} = -D_{diff}\nabla L - D_{diff}\frac{\zeta_0}{\chi_0}\nabla T. \tag{105}$$

Thus, if $\nabla L = 0$, (105) will yield

$$\mathbf{q} = -\lambda_{eff}\nabla T, \tag{106}$$

with an effective thermal conductivity $\lambda_{eff} = \frac{D_{diff}\zeta_0}{\chi_0 \nu_0} > 0$. As in the case of the diffusion coefficient D_{diff}, λ_{eff} is expected to be positive. Note that $\lambda_{eff} > 0$ implies again $\chi_0 \nu_0 > 0$.

The second term in (105) plays a role analogous to thermal diffusion — or Soret effect — in usual diffusion of particles. In this case, equation (99) modifies as

$$\frac{dL}{dt} + L\nabla \cdot \mathbf{v} - \frac{\nu_0 \chi_0}{K_1 L}\Delta L - \frac{\nu_0 \zeta_0}{K_1 L}\Delta T = \sigma^L. \tag{107}$$

This kind of situations have not been studied enough in the context of vortex tangles, but they would arise in a natural way when trying to understand the behavior of quantum turbulence in the presence of a temperature gradient.

Expression (105) yields a coupling between the heat flux and an inhomogeneity in L; in other terms, it means that a heat flux may influence the vortex line density. In view of (104), we have

$$-\chi_0 \nabla L = K_1 L \mathbf{q} + \zeta_0 \nabla T. \tag{108}$$

From here, it follows that there should be a slight inhomogeneity in L in such a way that ∇L points in the same direction as $K_1 L\mathbf{q} + \zeta_0 \nabla T$. It follows, in contrast with the standard assumption that the vortex line density is longitudinally homogeneous in counterflow experiments, that the vortex tangle would be slightly inhomogeneous. Thus, an experiment suggested by our formalism would be to carefully measure the longitudinal profile of L along the heat flux, to check whether there is a slight increase in L. Furthermore, equation (108) would allow one to measure the coefficient χ_0, in the linear approximation.
Since below (97) we have mentioned a way to measure ν_0, it turns out that the new coefficients ν_0 and χ_0 could be measured independently of each other.

4.3.3 Propagation of plane harmonic waves

Here, we will study wave propagation in counterflow vortex tangles, assuming that only fields T, $\mathbf{q}$ and L are involved. The equations for these fields, expressing the energy in terms of T and L, are simply:

$$\begin{cases} \rho c_V \dot{T} + \rho \epsilon_L \dot{L} + \nabla \cdot \mathbf{q} = 0, \\ \dot{\mathbf{q}} + \zeta_0 \nabla T + \chi_0 \nabla L = -K_1 L \mathbf{q}, \\ \dot{L} + \nabla \cdot (\nu_0 \mathbf{q}) = -\beta_q L^2 + \alpha_q q L^{3/2}. \end{cases} \tag{109}$$

where $c_V = \partial \epsilon / \partial T$ is the specific heat at constant volume, $\epsilon_L = \partial \epsilon / \partial L \simeq \epsilon_V$.
These equations are enough for the discussion of the physical effects of the coupling of second-sound and the distortion of the vortex tangle (represented by the inhomogeneities in L), which must be taken into account in an analysis of the vortex tangle by means of second sound. In fact, some of the previous hydrodynamical analyses of turbulent superfluids had this problem as one of their main motivations (Tsubota et al., 2003a), (Tsubota et al., 2003b).
As we can easily see, a stationary solution of system (109) is:

$$\mathbf{q} = \mathbf{q}_0 = (q_{10}, 0, 0), \quad L = L_0 = \frac{\alpha_q^2}{\beta_q^2}[q_{10}]^2, \quad T = T_0(\mathbf{x}) = T^* - \frac{K_1 L_0 q_{10}}{\zeta_0} x_1, \tag{110}$$

with $q_{10} > 0$.
To study the wave propagation in a neighborhood of this solution, we substitute $\sigma^{\mathbf{q}}$ and σ^L with

$$\sigma^{\mathbf{q}} \simeq -K_1 \left[L_0 \mathbf{q} + \mathbf{q}_0 (L - L_0) \right], \tag{111}$$

$$\sigma^L \simeq -\left[2\beta_q L_0 - \frac{3}{2} \alpha_q q_{10} L^{1/2} \right] (L - L_0) + \alpha_q L_0^{3/2} \hat{\mathbf{q}}_0 \cdot (\mathbf{q} - \mathbf{q}_0), \tag{112}$$

obtaining:

$$\begin{cases} \rho c_V \frac{\partial T}{\partial t} + \rho \epsilon_L \frac{\partial L}{\partial t} + \nabla \cdot \mathbf{q} = 0, \\ \frac{\partial \mathbf{q}}{\partial t} + \zeta_0 \nabla T + \chi_0 \nabla L = -K_1 \left[L_0 \mathbf{q} + \mathbf{q}_0 (L - L_0) \right], \\ \frac{\partial L}{\partial t} + \nu_0 \nabla \cdot \mathbf{q} = -\left[2\beta_q L_0 - \frac{3}{2} \alpha_q q_{10} L^{1/2} \right] (L - L_0) + \alpha_q L_0^{3/2} \hat{\mathbf{q}}_0 \cdot (\mathbf{q} - \mathbf{q}_0). \end{cases} \tag{113}$$

Consider the propagation of harmonic plane waves, seeking for solutions of equations (109) of the form

$$T = T_0(\mathbf{x}) + \tilde{T}e^{i(k\mathbf{n}\cdot\mathbf{x}-\omega t)}, \quad \mathbf{q} = \mathbf{q}_0 + \tilde{\mathbf{q}}e^{i(k\mathbf{n}\cdot\mathbf{x}-\omega t)}, \quad L = L_0 + \tilde{L}e^{i(k\mathbf{n}\cdot\mathbf{x}-\omega t)}, \tag{114}$$

where $K = k_r + ik_s$ is the complex wave number, ω the real frequency and $\mathbf{n}$ the unit vector in the direction of the wave propagation. Furthermore, we suppose that the oversigned quantities denote small amplitudes, whose products can be neglected. Inserting (114) in the linearized field equations (109), and making the positions:

$$N_1 = K_1 L_0, \quad N_2 = 2\beta_q L_0 - \frac{3}{2}\alpha_q L_0^{1/2} q_{10}, \quad N_3 = K q_{10}, \quad N_4 = \alpha_q q_{10} L_0^{3/2}, \tag{115}$$

we obtained in (Mongiovì & Jou, 2007), when the wave is collinear with the heat flux $\mathbf{q}$, the following dispersion relation:

$$\omega^2 = K^2\left[V_2^2(1-\rho\epsilon_L\nu_0) + \nu_0\chi_0\right] + \\ -i\omega(N_1+N_2) + i\frac{K^2}{\omega}V_2^2 N_2 - iK\left[(\chi_0 + V_2^2\rho\epsilon_L)N_4 - \nu_0 N_3\right], \tag{116}$$

while, when the wave is orthogonal with the heat flux $\mathbf{q}$, we obtained

$$\omega^2 = K^2\left[V_2^2(1-\rho\epsilon_L\nu_0) + \nu_0\chi_0\right] + N_1N_2 - i\omega(N_1+N_2) + i\frac{K^2}{\omega}V_2^2\left(N_2 - \frac{N_3N_4}{\omega + iN_1}\right). \tag{117}$$

We compare these results with the result obtained in Section 3 (see also (Jou et al., 2002)), where we supposed L a fixed quantity, and the term ν_0 was assumed to vanish, eliminating in this way the effects of the oscillations of $\mathbf{q}$ on the vortex line density L of the tangle. Assuming $\nu_0 = N_2 = N_4 = 0$ in (116) and (117), the same dispersion relation for the second sound is obtained

$$\omega^2 = V_2^2 K^2 - i\omega K_1 L_0. \tag{118}$$

Comparison of (116) and (117) with (118) shows that the distortion of the vortex tangle under the action of the heat wave, and its corresponding back reaction on the latter, implies remarkable changes in the velocity and the attenuation of the second sound, the latter effect depending on the relative direction between $\mathbf{q}_0$ and $\mathbf{n}$.

4.3.4 High-frequency waves

In the hypothesis of high-frequency waves, which means $\omega \gg N_1$, $\omega \gg N_2$ and $|\bar{K}| \gg max\left(\frac{K_1 q_{10}}{\chi_0}, \frac{\alpha_q q_{10} L_0^{3/2}}{\nu_0}\right)$, the dispersion relation ($\mathbf{n}$ parallel or orthogonal to the initial heat flux) is

$$w^2 = V_2^2(1 - \nu_0\rho\epsilon_L) + v_\infty^2, \tag{119}$$

where $w = \omega/k_r$ is the speed of the wave, $V_2^2 = \zeta_0/\rho c_V$ is the second sound speed in the absence of vortices and

$$v_\infty^2 = \chi_0\nu_0 \tag{120}$$

is the speed of the vortex density wave, which was found in (Jou et al., 2007).
If we try to read the relation (119) in terms of the second sound, we note that the vortex vibrations modify this second sound speed through the two contributions $-V_2^2 \nu_0 \rho \epsilon_L$ and v_∞^2, the latter due to the presence of the vortex density waves and the former due to the reciprocal existence of two waves. The correction for the speed of the second sound is not important, because V_2 is of the order of 20 m/s near 1,7 K, whereas, for $L_0 = 10^6 \mathrm{cm}^{-2}$, the speed of vortex density waves would be of the order 0,25 cm/s, much lower than V_2 (Jou et al., 2007).
To obtain approximate solutions of equations (116) and (117), we assume that the quantities N_1, N_2, N_3 and N_4 are coefficients small enough to assume them as perturbations of the physical system. This is reasonable at high-frequencies, since we have assumed that $\omega \gg N_1$, $\omega \gg N_2$ and $|\bar{K}| \gg max\left(\frac{Kq_{10}}{\chi_0}, \frac{Aq_{10}L_0^{3/2}}{\nu_0}\right)$.
Therefore, let assume that the speed of the wave has the following expression

$$\tilde{w} = \frac{\omega}{k_r} = w + \delta, \tag{121}$$

for which substituting it in the equations (116) and (117), we obtain at the lowest order the relation (119) for the speed w. From the next order follows that $\delta = 0$, that is the perturbations due to the coefficients N_1, N_2, N_3 and N_4 do not modify the speed of the wave while they modify the coefficients k_s related to the attenuation in the form, which in the parallel case is

$$k_s^{\parallel} = \frac{N_2\left(w^2 - V_2^2\right) + w\left(\rho\epsilon_L N_4 V_2^2 + N_1 w + N_3 \nu_0 - N_4 \chi_0\right)}{2w^3}. \tag{122}$$

and in the orthogonal case is

$$k_s^{\perp} = \frac{N_2\left(w^2 - V_2^2\right) + N_1 w^2}{2w^3}, \tag{123}$$

From a comparison between the two relations of k_s, (122) and (123), one may note that

$$k_s^{\parallel} = k_s^{\perp} + \frac{\left(\rho\epsilon_L V_2^2 - \chi_0\right) N_4 + N_3 \chi_0}{2w^2}. \tag{124}$$

Anyway, the modification of the attenuation coefficients, due to N_i, will be small because $w^2 - V_2^2$ is small and the coefficients N_i are also small in the considered situation. This is in contrast with what happens at low frequency, or when the vortex tangle is assumed as perfectly rigid, not affected by the second sound, in which case the relative motion of the normal fluid with respect to the vortex lines yields an attenuation which allows to determine the vortex line density L of the tangle. However, the wave character of vortex density perturbations at high frequency makes that vortex lines and the second sound become two simultaneous waves with a low joint dissipation, in the first-order approach. Thus, from the practical point of view, it seems that, at high frequency, second sound will not provide much information on the vortex tangle because the influence of the average vortex line density L is small both in the speed as in the attenuation. In the next section we will propose an extended hydrodynamical model which includes flux of vortices as independent variable in order to study vortex density waves.

5. The flux of line density L as new independent variable, vortex density waves

The vortex lines and their evolution are investigated by second sound waves, so that it is necessary to analyze in depth their mutual interactions. In particular, high-frequency second

sound may be of special interest to probe small length scales in the tangle, which is necessary in order to explore, for instance, the statistical properties of the vortex loops of several sizes. In fact, the reduction of the size of space averaging is one of the active frontiers in second sound techniques applied to turbulence, but at high-frequencies, the response of the tangle to the second sound is expected to be qualitatively different than at low frequencies, as its perturbations may change from diffusive to propagative behavior (Mongiovì & Jou, 2007), (Nemirovskii & Lebedev, 1983), (Yamada et al, 1989), (Jou et al., 2007).
In Section 4 and in the paper (Mongiovì & Jou, 2007) a thermodynamical model of inhomogeneous superfluid turbulence was built up with the fundamental fields: density ρ, velocity $\mathbf{v}$, internal energy density E, heat flux $\mathbf{q}$ and average vortex line length per unit volume L. In (Jou et al., 2007), starting from this model, a semiquantitative expression for the vortex diffusion coefficient was obtained and the interaction between second sound and the tangle in the high-frequency regime was studied. In both these works the diffusion flux of vortices $\mathbf{J}$ was considered as a dependent variable, collinear with the heat flux $\mathbf{q}$, which is proportional to the counterflow velocity $\mathbf{V}_{ns}$. But, in general, this feature is not strictly verified because the vortices move with a velocity $\mathbf{v}_L$, which is not collinear with the counterflow velocity (for more detail see paper (Sciacca et al, 2008)).

5.1 Balance equations and constitutive theory

In this section we build up a thermodynamical model of inhomogeneous counterflow superfluid turbulence, which chooses as fundamental fields the energy density E, the heat flux $\mathbf{q}$, the averaged vortex line length per unit volume L, and the vortex diffusion flux $\mathbf{J}$. Because experiments in counterflow superfluid turbulence in the linear regime are characterized by a zero value of the barycentric velocity $\mathbf{v}$, in this paper one does not consider $\mathbf{v}$ as independent variable. In a more complete model $\mathbf{v}$ and ρ will be also fundamental fields.
Consider the following balance equations

$$\begin{cases} \frac{\partial E}{\partial t} + \frac{\partial q_k}{\partial x_k} = 0 \\ \frac{\partial q_i}{\partial t} + \frac{\partial J^q_{ik}}{\partial x_k} = \sigma^q_i \\ \frac{\partial L}{\partial t} + \frac{\partial J_k}{\partial x_k} = \sigma^L \\ \frac{\partial J_i}{\partial t} + \frac{\partial F_{ik}}{\partial x_k} = \sigma^{\mathbf{J}}_i \end{cases} \tag{125}$$

where E is the specific energy per unit volume of the superfluid component plus the normal component plus the vortex lines, J^q_{ij} the flux of the heat flux, J_i the flux of vortex lines (which was denoted with J^L_i in the previous Section), and F_{ij} the flux of the flux of vortex lines; σ^q_i, σ_L and $\sigma^{\mathbf{J}}_i$ are the respective production terms. Since here one is interested to study the linear propagation of the second sound and vortex density waves, the convective terms have been neglected.
If one supposes that the fluid is isotropic, the constitutive equations for the fluxes J^q_{ij} and F_{ij}, to the first order in q_i and J_i, can be expressed in the form

$$J^q_{ik} = \beta_1(E,L)\delta_{ik}, \qquad F_{ik} = \psi_1(E,L)\delta_{ik}. \tag{126}$$

Restrictions on these relations are obtained, as in Section 4, imposing the validity of the second law of thermodynamics, applying Liu's procedure.
In order to make the theory internally consistent, one must consider for entropy density S and entropy flux density J_k^S approximate constitutive relations to second order in q_i and J_i

$$S = S_0(E,L) + S_1(E,L)q^2 + S_2(E,L)J^2 + S_3(E,L)q_iJ_i, \qquad J_k^s = \phi^q(E,L)q_k + \phi^J(E,L)J_k. \quad (127)$$

The quantities Λ^E, Λ_i^q, Λ^L and Λ_i^J are Lagrange multipliers, which are also objective functions of E, q_i, L and J_i; in particular, one sets

$$\Lambda^E = \Lambda^E(E,L,q_i,J_i) = \Lambda_0^E(E,L) + \Lambda_1^E(E,L)q^2 + \Lambda_2^E(E,L)J^2 + \Lambda_3^E(E,L)q_iJ_i,$$
$$\Lambda^L = \Lambda^L(E,L,q_i,J_i) = \Lambda_0^L(E,L) + \Lambda_1^L(E,L)q^2 + \Lambda_2^L(E,L)J^2 + \Lambda_3^L(E,L)q_iJ_i,$$

$$\Lambda_i^q = \lambda_{11}q_i + \lambda_{12}J_i \quad \text{and} \quad \Lambda_i^J = \lambda_{21}q_i + \lambda_{22}J_i, \quad (128)$$

with $\lambda_{mn} = \lambda_{mn}(E,L)$. Imposing that the coefficients of the time derivatives are zero, one obtains

$$dS = \Lambda^E dE + \Lambda_i^q\, dq_i + \Lambda^L dL + \Lambda_i^J dJ_i. \quad (129)$$

In the same way, imposing that the coefficients of space derivatives vanish, one finds

$$dJ_k^S = \Lambda^E dq_k + \Lambda_i^q dJ_{ik}^q + \Lambda^L dJ_k + \Lambda_i^J dF_{ik}. \quad (130)$$

Substituting now (126), (127) and (128) in (129-130), one gets

$$S_1 = \frac{1}{2}\lambda_{11}, \qquad S_2 = \frac{1}{2}\lambda_{22}, \qquad S_3 = \lambda_{12} = \lambda_{21}, \quad (131)$$

$$\phi^q = \Lambda_0^E, \qquad \phi^J = \Lambda_0^L, \quad (132)$$

$$dS_0 = \Lambda_0^E dE + \Lambda_0^L dL, \qquad dS_1 = \Lambda_1^E dE + \Lambda_1^L dL, \quad (133)$$

$$dS_2 = \Lambda_2^E dE + \Lambda_2^L dL, \qquad dS_3 = \Lambda_3^E dE + \Lambda_3^L dL, \quad (134)$$

$$d\phi^q = \lambda_{11} d\beta_1 + \lambda_{21} d\psi_1, \qquad d\phi^J = \lambda_{12} d\beta_1 + \lambda_{22} d\psi_1. \quad (135)$$

In particular, one obtains to the second order in $\mathbf{q}$ and $\mathbf{J}$ the following expressions for the entropy and for the entropy flux

$$S = S_0 + \frac{1}{2}\lambda_{11}q^2 + \frac{1}{2}\lambda_{22}J^2 + \lambda_{12}q_iJ_i, \qquad J_k^s = \Lambda_0^E q_k + \Lambda_0^L J_k. \quad (136)$$

It remains the following residual inequality for the entropy production

$$\sigma^S = \Lambda_i^q \sigma_i^q + \Lambda^L \sigma_L + \Lambda_i^J \sigma_i^J \geq 0. \quad (137)$$

Now, the obtained relations are analyzed in detail. As in Section 4 we first introduce a generalized temperature as the reciprocal of the first-order part of the Lagrange multiplier of the energy:

$$\Lambda_0^E = \left[\frac{\partial S_0}{\partial E}\right]_L = \frac{1}{T}. \tag{138}$$

and the chemical potential of vortex lines (near equilibrium):

$$-T\Lambda_0^L = \mu^L. \tag{139}$$

Neglecting in (129) second order terms in **q** and **J**, and using relations (131), (138) and (139), the following expression for the entropy density S is obtained

$$dS = \frac{1}{T}dE - \frac{\mu^L}{T}dL + \lambda_{11}q_i dq_i + \lambda_{22}J_i dJ_i + \lambda_{12}(J_i dq_i + q_i dJ_i). \tag{140}$$

Consider now equations (135), which one rewrites using (132) and (139) as

$$d\left(\frac{1}{T}\right) = \lambda_{11}d\beta_1 + \lambda_{21}d\psi_1, \qquad d\left(-\frac{\mu^L}{T}\right) = \lambda_{12}d\beta_1 + \lambda_{22}d\psi_1. \tag{141}$$

After some calculations (Sciacca et al, 2008), we find:

$$\zeta_1 = \frac{\partial \beta_1}{\partial T} = \frac{1}{N}\left[-\frac{1}{T^2}\lambda_{22} + \lambda_{12}\frac{\partial}{\partial T}\left(\frac{\mu_0^L}{T}\right)\right], \qquad \chi_1 = \frac{\partial \beta_1}{\partial L} = \frac{1}{T}\frac{\lambda_{12}}{N}\frac{\partial \mu_0^L}{\partial L}, \tag{142}$$

$$\eta_1 = \frac{\partial \psi_1}{\partial T} = \frac{1}{N}\left[\frac{1}{T^2}\lambda_{12} - \lambda_{11}\frac{\partial}{\partial T}\left(\frac{\mu_0^L}{T}\right)\right], \qquad \nu_1 = \frac{\partial \psi_1}{\partial L} = -\frac{1}{T}\frac{\lambda_{11}}{N}\frac{\partial \mu_0^L}{\partial L}, \tag{143}$$

where $N = \lambda_{11}\lambda_{22} - {\lambda_{12}}^2$. A physical meaning for the coefficient λ_{22} was furnished in (Sciacca et al, 2008). Finally, one obtains for the entropy flux

$$J_k^s = \frac{1}{T}q_k - \frac{\mu_0^L}{T}J_k, \tag{144}$$

which is analogous to the usual expression of the entropy flux in the presence of a mass flux and heat flux, but with the second term related to vortex transport rather than to mass transport.
Substituting the constitutive equations (126) in system (125), using the relations (142-143), and expressing the energy E in terms of T and L, the following system of field equations is obtained

$$\begin{cases} \rho c_V \frac{\partial T}{\partial t} + \rho \epsilon_L \frac{\partial L}{\partial t} + \frac{\partial q_j}{\partial x_j} = 0 \\ \frac{\partial q_i}{\partial t} + \zeta_1 \frac{\partial T}{\partial x_i} + \chi_1 \frac{\partial L}{\partial x_i} = \sigma_i^q \\ \frac{\partial L}{\partial t} + \frac{\partial J_j}{\partial x_j} = \sigma^L \\ \frac{\partial J_i}{\partial t} + \eta_1 \frac{\partial T}{\partial x_i} + \nu_1 \frac{\partial L}{\partial x_i} = \sigma^{J_i} \end{cases} \tag{145}$$

where c_V is the specific heat at constant volume and $\epsilon_L = \partial E/\partial L$. The coefficients χ_1 and η_1 describe cross effects linking the dynamics of **q** and **J** with L and T, respectively. Thus, they are expected to settle an interaction between heat waves and vortex density waves. These

equations are analogous to those proposed in Section 4 (Mongiovì & Jou, 2007) except for the choice of J_i: in fact here J_i is assumed to be an independent field whereas in Section 4 (Mongiovì & Jou, 2007) J_i was assumed as dependent on q_i. However, at high frequency, J_i will become dominant and will play a relevant role, as shown in the following.
The production terms σ must also be specified. Regarding σ_i^q and σ^L, since only counterflow situation is considering, as in Section 4, we assume

$$\sigma^{\mathbf{q}} = -K_1 L \mathbf{q}, \qquad \sigma^L = -\beta_q L^2 + \alpha_q |q| L^{3/2}, \tag{146}$$

where $K_1 = \frac{1}{3}\kappa B$. For the production term of vortex line diffusion, one assumes the following relaxational expression:

$$\vec{\sigma}^J = -\frac{\mathbf{J}}{\tau_J} = -\gamma_1 \kappa L \mathbf{J}, \tag{147}$$

where γ_1 is a positive coefficient which can depend on the temperature T (Sciacca et al, 2008). Note that in (146) one has assumed that the production terms of $\mathbf{q}$ and $\mathbf{J}$ depend on $\mathbf{q}$ and $\mathbf{J}$, respectively, but not on both variables. In more general terms, one could assume that both production terms depend on the two fields $\mathbf{q}$ and $\mathbf{J}$ simultaneously.
In order to determine the physical meaning of the coefficients appearing in equations (145)–(147), concentrate first the attention on the equations for L and $\mathbf{J}$. Supposing that $\mathbf{J}$ varies very slowly, one obtains (Sciacca et al, 2008)

$$\frac{\partial}{\partial t} L = \frac{\eta_1}{\gamma_1 \kappa L} \nabla^2 T + \frac{\nu_1}{\gamma_1 \kappa L} \nabla^2 L + \sigma^L. \tag{148}$$

It is then seen that the coefficient $\frac{\nu_1}{\gamma_1 \kappa L} \equiv \overline{D}_1$ represents the diffusion coefficient of vortices. Coefficient $\frac{\eta_1}{\gamma_1 \kappa L} \equiv \overline{D}_2$ may be interpreted as a thermodiffusion coefficient of vortices because it links the temperature gradient to vortex diffusion. In other terms, this implies a drift of the vortex tangle. Detailed measurements have indeed shown [(Donnelly, 1991), pag.216] a slow drift of the tangle towards the heater; this indicates that $\eta_1 < 0$ and small. The hypothesis $\eta_1 = 0$ corresponds to $\overline{D}_2 = 0$, i.e. the vortices do not diffuse in response to a temperature gradient.

5.2 Interaction of second sound and vortex density waves

In this Section wave propagation in counterflow vortex tangles is studied, with the aim to discuss the physical effects of the interaction between high-frequency second sound and vortex density waves. A stationary solution of the system (145), with the expressions of the production terms (146–147), is

$$\mathbf{q} = \mathbf{q_0} = (q_{01}, 0, 0), \quad L = L_0 = \frac{\alpha_q^2}{\beta_q^2} q_{01}^2, \tag{149}$$

$$T = T_0(\mathbf{x}) = T^* - \frac{1}{3}\frac{\kappa B}{\zeta_1} L_0 q_{01} x_1, \quad \mathbf{J}_0 = \left(\frac{1}{3}\frac{\kappa B}{\zeta_1 \gamma_1 \kappa} q_{01}, 0, 0\right), \tag{150}$$

with $q_{01} > 0$.
Consider the propagation of harmonic plane waves of the four fields of the equation (145) in the following form

$$\begin{cases} T = T_0(\mathbf{x}) + \tilde{T}e^{i(K\mathbf{n}\cdot\mathbf{x}-\omega t)} \\ \mathbf{q} = \mathbf{q}_0 + \tilde{\mathbf{q}}e^{i(K\mathbf{n}\cdot\mathbf{x}-\omega t)} \\ L = L_0 + \tilde{L}e^{i(K\mathbf{n}\cdot\mathbf{x}-\omega t)} \\ \mathbf{J} = \mathbf{J}_0 + \tilde{\mathbf{J}}e^{i(K\mathbf{n}\cdot\mathbf{x}-\omega t)} \end{cases} \tag{151}$$

where $K = k_r + ik_s$ is the wave number, ω the real frequency, $\mathbf{n}$ the unit vector along the direction of the wave propagation.
Substituting (151) in the system (145), and linearizing the quantities (146), and (147) around the stationary solutions, the following equations for the small amplitudes are obtained

$$\begin{cases} -\omega[\rho c_V]_0\tilde{T} - \omega[\rho\epsilon_L]_0\tilde{L} + K\tilde{\mathbf{q}}\cdot\mathbf{n} = 0 \\ \left[-\omega - \frac{i}{3}\kappa B L_0\right]\tilde{\mathbf{q}} + \zeta_{10}K\tilde{T}\mathbf{n} - \left(-\chi_{10}K\mathbf{n} + \frac{i}{3}\kappa B\mathbf{q}_0\right)\tilde{L} = 0 \\ \left[-\omega - i\left(2\beta_q L_0 - \frac{3}{2}\alpha_q L_0^{1/2}q_{01}\right)\right]\tilde{L} + K\tilde{\mathbf{J}}\cdot\mathbf{n} + i\alpha_q L_0^{3/2}\tilde{q}_1 = 0 \\ (-\omega - i\gamma_1\kappa L_0)\tilde{\mathbf{J}} + \eta_{10}K\mathbf{n}\tilde{T} + (\nu_{10}K\mathbf{n} - i\gamma_1\kappa\mathbf{J}_0)\tilde{L} = 0 \end{cases} \tag{152}$$

Note that the subscript 0 refers to the unperturbed state; in what follows, this subscript will be dropped out to simplify the notation.
First case: $\mathbf{n}$ *parallel to* $\mathbf{q_0}$.
Now, impose the condition that the direction of the wave propagation $\mathbf{n}$ is parallel to the heat flux $\mathbf{q_0}$, namely $\mathbf{n} = (1,0,0)$. Through these conditions the system (152) becomes

$$\begin{cases} -\omega\rho c_V\tilde{T} + K\tilde{q}_1 - \omega\rho\epsilon_L\tilde{L} = 0 \\ \zeta_1 K\tilde{T} - \left(\omega + \frac{i}{3}\kappa B L\right)\tilde{q}_1 - \left(-\chi_1 K + i\kappa\frac{B}{3}q_1\right)\tilde{L} = 0 \\ i\alpha_q L^{3/2}\tilde{q}_1 - \left(\omega + i\tau_L^{-1}\right)\tilde{L} + K\tilde{J}_1 = 0 \\ \eta_1 K\tilde{T} + (\nu_1 K - i\gamma_1\kappa J_1)\tilde{L} + (-\omega - i\gamma_1\kappa L)\tilde{J}_1 = 0 \\ \\ \left(-\omega - \frac{i}{3}\kappa B L\right)\tilde{q}_2 = 0 \\ \left(-\omega - \frac{i}{3}\kappa B L\right)\tilde{q}_3 = 0 \\ (-\omega - i\gamma_1\kappa L)\tilde{J}_2 = 0 \\ (-\omega - i\gamma_1\kappa L)\tilde{J}_3 = 0 \end{cases} \tag{153}$$

where $\tau_L^{-1} = \left(2\beta_q L - \frac{3}{2}\alpha_q L^{1/2}q_1\right)$.
Note that the transversal modes, those corresponding to the four latter equations, evolve independently with respect to the longitudinal ones, corresponding to the four former equations.
One will limit the study to the case in which ω and the modulus of the wave number K assume values high enough to make considerable simplification in the system. Indeed, it is for high values of the frequency that the wave behavior of the vortex tangle can be evidenced because the first term in (145c) will become relevant. Note that the assumption $|K| = |k_r + ik_s|$ large

$w_{1,2} = \pm V_2$	$w_{3,4} = \pm\sqrt{\nu_1}$
$\tilde{T} = \psi$	$\tilde{T} = -\frac{1}{\rho c_V}\left(\frac{\chi_1 - \nu_1\rho\epsilon_L}{V_2^2 - \nu_1}\right)\psi$
$\tilde{q}_1 = \pm V_2 \rho c_V \psi$	$\tilde{q}_1 = \pm\frac{\sqrt{\nu_1}(\rho\epsilon_L V_2^2 - \chi_1)}{V_2^2 - \nu_1}\psi$
$\tilde{L} = 0$	$\tilde{L} = \psi$
$\tilde{J}_1 = 0$	$\tilde{J}_1 = \pm\sqrt{\nu_1}\psi$

Table 1. Modes corresponding to second sound velocity and vortex density waves, respectively.

refers to a large value of its real part k_r, which is related to the speed of the vortex density wave, whereas the imaginary part k_s, corresponding to the attenuation factor of the wave, will be assumed small.
This problem is studied into two steps: first assuming $|K|$ and ω extremely high to neglect all terms which do not depend on them. Then, the solution so obtained is perturbed in order to evaluate the influence of the neglected terms on the velocity and the attenuation of high-frequency waves.
Denoting with $w = \omega/k_r$ the speed of the wave, and assuming $|K|$ and ω large, the following dispersion relation is obtained:

$$w^4 - \left[V_2^2 + \nu_1 - \frac{\eta_1}{\rho c_V}\left(\rho\epsilon_L - \frac{\chi_1}{\nu_1}\right)\right]w^2 + V_2^2\nu_1 = 0, \tag{154}$$

where $V_2 = \left(-\lambda_{22}T^2\rho c_V\right)^{-1/2}$ is the second sound speed in the absence of vortex tangle (see previous sections) and from (142b) it is related to the coefficient ζ_1 by the relation $\zeta_1 = V_2^2\rho c_V - \lambda_{12}\eta_1/\lambda_{11}$. Further, if one assumes that the coefficient η_1 is zero

$$\eta_1 = 0 \qquad \Rightarrow \qquad \frac{\lambda_{12}}{\lambda_{11}} = T^2\frac{\partial}{\partial T}\left(\frac{\mu^L}{T}\right) = \frac{2S_3}{S_2} = -\frac{\chi_1}{\nu_1}, \tag{155}$$

then the dispersion relation (154) has the solutions

$$w_{1,2} = \pm V_2, \qquad w_{3,4} = \pm\sqrt{\nu_1}, \tag{156}$$

to which correspond the propagation modes shown in Table 1.
As one sees from the first column of Table 1, under the hypothesis (155) the high-frequency wave of velocity $w_{1,2} = \pm V_2$ is a temperature wave (i.e. the second sound) in which the two quantities $\tilde{L}$ and $\tilde{J}_1$ are zero, whereas in the second column the high-frequency wave of velocity $w_{3,4} = \pm\sqrt{\nu_1}$ is a wave in which all fields vibrate. The latter result is logic because when the vortex density wave is propagated in the superfluid helium, temperature T and heat flux q_1 cannot remain constant. This behavior is different from that obtained in Section 4, because using that model in the second sound also the line density L vibrates. In fact, there the flux of vortices $\mathbf{J}$ was chosen proportional to $\mathbf{q}$, so that vibrations in the heat flux (second sound) produce vibrations in the vortex tangle. Experiments on high-frequency second sound are needed to confirm this new result.
Now we consider all the neglected terms of the system (153) and the coefficient η_1 as small perturbations of the velocity w of the wave and of the attenuation term k_s of the wave number K. Substituting the following assumptions

$$\bar{w} = \frac{\omega}{k_r} = w + \delta \quad \text{and} \quad K = k_r + ik_s$$

in the system (153), one find the expression (156), at the zeroth order in δ and k_s, whereas at the first order in δ and k_s, one obtains

$$\bar{w}^{\|}_{1,2} = \left(1 - \frac{\eta_1}{2\rho c_V\left(w^2_{1,2} - w^2_{3,4}\right)}\left(\rho\epsilon_L - \frac{\chi_1}{w^2_{3,4}}\right)\right) w_{1,2}, \tag{157}$$

$$\bar{w}^{\|}_{3,4} = \left(1 + \frac{\eta_1}{2\rho c_V\left(w^2_{1,2} - w^2_{3,4}\right)}\left(\rho\epsilon_L - \frac{\chi_1}{w^2_{3,4}}\right)\right) w_{3,4}, \tag{158}$$

and

$$k_s^{(1,2)} = \frac{\kappa LB}{6w_{1,2}} + \frac{\alpha_q L^{3/2}\left(w^2_{1,2}\rho\epsilon_L - \chi_1\right)}{2\left(w^2_{1,2} - w^2_{3,4}\right)}, \tag{159}$$

$$k_s^{(3,4)} = \frac{\kappa L\gamma_1 + \tau_L^{-1}}{2w_{3,4}} - \frac{\alpha_q L^{3/2}\left(w^2_{1,2}\rho\epsilon_L - \chi_1\right)}{2\left(w^2_{1,2} - w^2_{3,4}\right)} + \frac{J_1\kappa\gamma_1}{2w^2_{3,4}}. \tag{160}$$

Observe that in this approximation all thermodynamical fields vibrate simultaneously and the attenuation coefficients k_s are influenced by the choice of **J** as independent variable, as one easily sees by comparing expressions (159–160) with those obtained in Section 4 (Jou et al., 2007). Looking at these results, in particular the two speeds (157–158), one sees that these velocities are not modified when one makes the simplified hypothesis that the coefficient η_1 is equal to zero. In Section 4 (Jou et al., 2007) it was observed that the second sound velocity is much higher than that of the vortex density waves, so that the small quantity η_1 should influence the two velocities (157-158) in a different way: negligible for the second sound velocity but relevant for the vortex density waves. Regarding the attenuation coefficients (159–160), one sees that the first term in (159) is identical to that obtained in (Jou et al., 2002), when the vortices are considered fixed. The new term, proportional to α_q, comes from the interaction between second sound and vortex density waves.

Note that the second term of the dissipative coefficient $k_s^{(1,2)}$ is the same as the third term of $k_s^{(3,4)}$, but with an opposite sign. This means that this term contributes to the attenuation of the two waves in opposite ways; and its contribution depends also on whether the propagation of forward waves or of backward waves is considered. The first term of $k_s^{(3,4)}$ produces always an attenuation of the wave, while the behavior of the third term is analogous to the first one.

Second case: **n** *orthogonal to* $\mathbf{q_0}$

In order to make a more detailed comparison with the model studied in Section 4 (Mongiovì & Jou, 2007), (Jou et al., 2007), one proceeds to analyze another situation, in which the direction of the wave propagation is perpendicular to the heat flux, that is, for example, assuming $\mathbf{n} = (0,0,1)$. This choice simplifies the system (152) in the following form

$$\left\{\begin{array}{l}-\omega\rho c_V\tilde{T}+K\tilde{q}_3-\omega\rho\epsilon_L\tilde{L}=0\\ \left(-\omega-\frac{i}{3}\kappa BL\right)\tilde{q}_1-\frac{i}{3}\kappa Bq_1\tilde{L}=0\\ \zeta_1K\tilde{T}-\left(\omega+\frac{i}{3}\kappa BL\right)\tilde{q}_3+\chi_1K\tilde{L}=0\\ i\alpha_qL^{3/2}\tilde{q}_1-\left(\omega+i\tau_L^{-1}\right)\tilde{L}+K\tilde{J}_3=0\\ \eta_1K\tilde{T}+\nu_1K\tilde{L}+(-\omega-i\gamma_1\kappa L)\tilde{J}_3=0\\ \\ \left(-\omega-\frac{i}{3}\kappa BL\right)\tilde{q}_2=0\\ -i\gamma_1\kappa J_1\tilde{L}+(-\omega-i\gamma_1\kappa L)\tilde{J}_1=0\\ (-\omega-i\gamma_1\kappa L)\tilde{J}_2=0\end{array}\right. \tag{161}$$

Note that, in contrast with what was seen before, but in agreement with the corresponding situation of the model described in Section 4, here the transversal and the longitudinal modes in general do not evolve independently, as shown from the first five equations. However, one will see that this is the case if high-frequency waves are considered.
As in the previous situation, we assume that the values of the frequencies ω and of the real part of the wave number, k_r, are high enough, such that the system (161) may be easily solved. The other terms will be considered as perturbations to $w=\omega/k_r$ and k_s. Note that in this special case, as in the previous case and in (Jou et al., 2007), only the longitudinal modes are present, so that the dispersion relation assumes the form

$$w\left(w^4-\left[V_2^2+\nu_1-\frac{\eta_1}{\rho c_V}\left(\rho\epsilon_L+\frac{\lambda_{12}}{\lambda_{11}}\right)\right]w^2+V_2^2\nu_1\right)=0, \tag{162}$$

which is similar to equation (154).
Assuming the same hypothesis (155), the dispersion relation (162) takes the form

$$w(w^2-\nu_1)(w^2-V_2^2)=0, \tag{163}$$

where V_2 is the second sound velocity and $\sqrt{\nu_1}$ is the velocity of the vortex density waves in helium II. The conclusions which one achieves here are the same to those of the previous situation. Indeed, $\omega_0=0$ corresponds to $\tilde{q}_1=\psi$ and $\tilde{T}=\tilde{q}_3=\tilde{L}=\tilde{J}_3=0$; while $w_{1,2}=\pm V_2$ and $w_{3,4}=\pm\sqrt{\nu_1}$ correspond to those in Table 1.
Now, as in the previous case, we assume that all the neglected terms in (161) modify w and K by small quantities δ and k_s, that is

$$\bar{w}=\frac{\omega}{k_r}=w+\delta \quad \text{and} \quad K=k_r+ik_s.$$

Substituting them in the dispersion relation of the system (161), one finds the relation (163), at the zeroth order in δ and k_s, and the following two expressions at the first order in δ and k_s

$$\bar{w}_{1,2}^{\perp}=\bar{w}_{1,2}^{\parallel}, \tag{164}$$

$$\bar{w}_{3,4}^{\perp}=\bar{w}_{3,4}^{\parallel}, \tag{165}$$

and

$$k_s^{(1,2)}=\frac{\kappa LB}{6w_{1,2}}, \tag{166}$$

$$k_s^{(3,4)} = \frac{\tau_L^{-1} + \kappa L \gamma_1}{2w_{3,4}}. \tag{167}$$

As regards the expression (166) for the dissipative term $k_s^{(1,2)}$, note that it is the same as the expression obtained when the vortices are assumed fixed (Jou & Mongiovì, 2006), (Peruzza & Sciacca, 2007), whereas the attenuation term $k_s^{(3,4)}$ is the same as the second term of $k_s^{(3,4)}$ of the first case ($\mathbf{n}$ parallel to $\mathbf{q}_0$). As in (Mongiovì & Jou, 2007), (Jou et al., 2007), in this case one has the propagation of two kinds of waves, namely heat waves and vortex density waves, which cannot be considered as propagating independently from each other. In fact, the uncoupled situation (equation (156)), in which the propagation of the second sound is not influenced by the fluctuations of the vortices, is no more the case when the quantities $\overline{N}_1 = \frac{1}{3}\kappa BL$, $\overline{N}_2 = \frac{1}{3}\kappa Bq_1$, $\overline{N}_3 = AL^{3/2}$, $\overline{N}_4 = \gamma_1 \kappa J_1$, $\overline{N}_5 = \gamma_1 \kappa L$, τ_L^{-1} and η_1, are considered. Indeed, from (157–158) and from the results of (Jou et al., 2007) one makes in evidence that heat and vortex density waves cannot be considered separately, that is as two different waves, but as two different features of the same phenomena. Of course, the results obtained here are more exhaustive than those of Section 4: in fact, comparing the velocities at the first order of approximation in both models, one deduces that the expressions (157–158) depend not only on the velocities of heat waves and vortex density waves, as in (Mongiovì & Jou, 2007), (Jou et al., 2007), but also on the coefficient η_1, which comes from the equation (145d) of the vortex flux $\mathbf{J}$, and whose physical meaning is a thermodiffusion coefficient of vortices. The fourth equation of the system (153) shows that the vortex flux $\tilde{J}_1$ is not proportional to the heat flux, as it was assumed in Sections 2 and 4, but it satisfies an equation in which also the fields $\tilde{L}$ and $\tilde{T}$, through η_1, are present.

It is to note that the attenuation of the second sound depends on the relative direction of the wave with respect to the heat flux: in some experiments this dependence was shown for parallel and orthogonal directions (Awschalom et al, 1984). These results were explained assuming an anisotropy of the tangle of vortices. But, looking at the expressions (159) and (166) of the attenuation of the second sound in the high-frequency regime, one notes that these expressions are not equal. In particular, the term

$$\frac{\alpha_q L^{3/2}\left(w_{1,2}^2 \rho \epsilon_L - \chi_1\right)}{2\left(w_{1,2}^2 - w_{3,4}^2\right)} \tag{168}$$

in (159) causes a dependence of the attenuation depending on whether the wave direction agrees with the direction of the heat flux $\mathbf{q}$ or not. This term is absent if the wave propagates orthogonal to the heat flux. In (Sciacca et al, 2008) vortex tangle was assumed to be anysotropic. The result was that $\bar{w}_{1,2,3,4}^{\perp} = \bar{w}_{1,2,3,4}^{\parallel}$ and that the behavior of speed of propagation is isotropic and does not depend on the isotropy or anisotropy of the tangle.

In conclusion, it could be that an anisotropy of the behavior of high-frequency second sound does not necessarily imply an actual anisotropy of the tangle in pure counterflow regime, but only a different behavior of the second sound due to the interaction with the vortex density waves. This may be of interest if one wants to explore the degree of isotropy at small spatial scales. Of course, some more experiments are needed in order to establish the presence and the sign of these additional terms.

6. Conclusions and perspectives

Helium behaves in a very strange way when temperature is dropped down below the lambda line, different to any classical fluid. This review is a first attempt to put together some of our results concerning the application of the Extended Thermodynamics to superfluid helium, both in laminar and turbulent flows.

In Section 2 a one-fluid model for superfluid helium in absence of vortices is shown, which chooses heat flux as an independent variable, and a comparison between this non-standard model and the more well-known two-fluid model is faced. The main part of the review is devoted to the macroscopic description of the interesting behaviour of this liquid in the presence of quantized vortex lines. They are very thin dynamical defects of superfluid helium, which are usually sketched by geometrical lines, representing the quantized vorticity of the superfluid motion. The amount of quantized vortices is high enough in turbulent superfluid helium, so they are usually expressed by means of the line length per unit volume L. Different hydrodynamical models of superfluids in the presence of vortices are dealt with, that have more detailed successive descriptions. First, in Section 3 the one-fluid model for laminar flow (no presence of vortices) is extended introducing a vorticity tensor (in the heat flux equation), which takes into account the presence of vortices as a fixed structure. The influence of vortices to the main fields is studied, mainly in the three experimental situations: rotating helium (vortices are basically straight lines parallel to the rotating axis), pure counterflow (an enough high heat flux, without mass flux, which causes an almost isotropic vortex tangle), and then the combined situation of rotating counterflow turbulence.

Since vortex lines density may experimentally be detected by means of the second sound (temperature waves), the propagation of harmonic waves is investigated in all the situations above mentioned. Section 4 is devoted to build up a new model in which the line density L acquires field properties: it depends on the coordinates, it has a drift velocity, and it has associated a diffusion flux. These features are becoming increasingly relevant today, as the local vortex density may be measured with higher precision, and the relative motion of vortices is observed and simulated. The hydrodynamical model built in this section is sufficiently general to encompass vortex diffusion and to describe the interactions between the usual waves and the vortices, which in Section 3 were simply considered as a rigid framework where second sound waves are dissipated. A hint about vortex density waves is also shown, which is then better considered in Section 5. In this section we further generalize the model, in order to include the velocity of the vortex tangle as a new independent variable. This is motivated by the fact that this velocity (or the flux of the vortex line density) is not always properly parallel to the heat flux, so it needs an own evolution equation. Also this model is formulated using Extended Thermodynamics, determining the restrictions imposed by the second law of Thermodynamics by means of the Liu's procedure. One of the results of this section is that when the high frequency harmonic plane waves are considered, vortex density waves are found out. The interesting thing is that heat waves and vortex density waves cannot be considered separately, that is as two different waves, but as two different features of the same phenomena. Another interesting result is that attenuation of the second sound depends on the relative direction of the wave with respect to the heat flux: it seems that the anisotropy in the behavior of high-frequency second sound does not mean anisotropy in the tangle in counterflow regime.

These results are important because second sound provides the standard methods of measuring the vortex line density L, and we have shown that the dynamical mutual interplay between second sound and vortex lines modifies the standard results. In the case when there is a net motion of the mass, the model is useful to study Couette and Poiseuille flow, where the bulk motion of the system contribute to the production of new vortex lines (Jou et al., 2008). The renewed interest in superfluid turbulence lies on the fact that at some length scales it appears similar to classical hydrodynamic turbulence, and therefore a better understanding of it can throw new light on problems in classical turbulence. Our results are relevant also to modelize the influence of the bulk motion on the vortex production in Couette and Poiseuille flows, and in towed or oscillating grids, including the important application of superfluid helium as a coolant for superconducting devices.

7. Acknowledgments

The authors acknowledge the support of the Università di Palermo (grant 2007.ORPA07LXEZ and Progetto CoRI 2007, Azione D, cap. B.U. 9.3.0002.0001.0001) and the collaboration agreement between Università di Palermo and Universitàt Autònoma de Barcelona. MS acknowledges the "Assegno di ricerca" of the University of Palermo. DJ acknowledges the financial support from the Dirección General de Investigación of the Spanish Ministry of Education under grant FIS 2009-13370-C02-01 and of the Direcció General de Recerca of the Generalitat of Catalonia, under grant 2009 SGR-00164.

8. References

Ardizzone, L. & Gaeta, G. (2009). Far from equilibrium constitutive relations in a nonlinear model of Superfluid turbulence, In: *Boll. di Mat. Pura ed Appl. Vol. II*, Brugarino T., Mongiovì M.S., (Ed.), 59–67, ARACNE, Roma.

Ardizzone, L., Gaeta, G. & Mongiovì M.S. (2009). A Continuum Theory of Superfluid Turbulence based on Extended Thermodynamics. *J. Non-Equilib. Thermodyn.*, Vol. 34, 277–297.

Awschalom, D.D., Milliken, F.P. & Schwarz, K.W. (1984). Properties of superfluid turbulence in a large channel. *Phys. Rev. Lett.*, Vol. 53, 1372–1375.

Barenghi, C.F. (1999). Classical aspects of quantum turbulence. *J. Phys. Cond. Matter*, Vol. 11, 7751–7759.

Barenghi, C.F., Donnelly, R.J. & Vinen, W.F. (2001). *Quantized Vortex Dynamics and Superfluid Turbulence*, Springer, Berlin.

Casas-Vázquez, J.& Jou, D. (2003). Temperature in non-equilibrium states: a review of open problems and current proposals. *Rep. Prog. Phys.*, Vol. 66, 1937–2023.

Donnelly, R. J. & Swanson, C. E. (1986). Quantum turbulence. *J. Fluid Mech.*, Vol. 173, 387–429.

Donnelly, R.J. (1991). *Quantized Vortices in Helium II*, Cambridge University Press, Cambridge.

Fazio, R. & van der Zant, H. (2001). Quantum phase transition and vortex dynamics in superconducting networks. *Physics Reports*, Vol. 355, 235–334.

Feynman, R.P. (1955). Application of quantum mechanics to liquid helium II, In: *Progress in Low Temperature Physics Vol. I*, Gorter, C.J., (Ed.), 17–53, North-Holland, Amsterdam.

Gorter, C. J. & Mellink, J. H. (1949). On the irreversible processes in liquid helium II. *Physica*, Vol. 15, 285–304.

Hall, H.E. & Vinen, W.F (1956). The rotation of liquid helium II, I. Experiments the propagation of second sound in uniformly rotating helium II. *Proc. Roy. Soc.*, Vol. A238, 215–234.

Hall, H.E. & Vinen, W.F (1956). The rotation of liquid helium II, II. The theory of mutual friction in uniformly rotating helium II. *Proc. Roy. Soc.*, Vol. A238, 204–214.

Jou D., Casas-Vázquez, J. & Lebon, G. (2001). *Extended Irreversible Thermodynamics*,

Springer-Verlag, Berlin.
Jou, D., Lebon, G. & Mongiovì, M.S. (2002). Second sound, superfluid turbulence and intermittent effects in liquid helium II. *Phys. Rev. B*, Vol. 66, 224509 (9 pages).
Jou, D. & Mongiovì, M.S. (2004). Phenomenological description of counterflow superfluid turbulence in rotating containers. *Phys. Rev. B*, Vol. 69, 094513 (7 pages).
Jou, D. & Mongiovì, M.S. (2005). Non-Equilibrium Thermodynamics in Counterflow and Rotating Situations. *Phys. Rev. B*, Vol. 72, 144517 (8 pages).
Jou, D. & Mongiovì, M.S. (2005). Description and evolution of anisotropy in superfluid vortex tangles with counterflow and rotation. *Phys. Rev. B*, Vol. 74, 054509 (11 pages).
Jou, D., Mongiovì, M.S. & Sciacca, M. (2007). Vortex density waves and high-frequency second sound in superfluid turbulence hydrodynamics. *Phys. Lett. A*, Vol. 368, 7–12.
Jou, D., Sciacca, M. & Mongiovì, M.S. (2008). Vortex dynamics in rotating counterflow and plane Couette and Poiseuille turbulence in superfluid helium, *Phys. Rev. B*, Vol. 78, 024524 (12 pages).
Khalatnikov, I.M.(1965). *An Introduction to the Theory of Superfluidity*, Benjamin, New York.
Landau, L.D. (1941). The theory of superfluidity of He II. *J. Phys.*, Vol. 5, 71–90.
Lebon, G., Jou, D. (1979). A continuum theory of liquid helium II based on the classical theory of irreversible processes. *J. Non-Equilib.Thermodyn.*, Vol. 4, 259–276.
Lebon, G., Jou D. & Casas-Vázquez, J. (2008). *Understanding non-equilibrium thermodynamics*, Springer-Verlag, Berlin.
Liu,I. (1972). Method of Lagrange multipliers for exploitation of entropy principle. *Arch. Rat. Mech. Anal.*, Vol. 46, 131–148.
Mongiovì, M.S. (1991). Superfluity and entropy conservation in extended thermodynamics. *J. Non-Equilib. Thermod.*, Vol. 16, 225–239.
Mongiovì, M.S. (1993). Extended Irreversible Thermodynamics of Liquid Helium II. *Phys. Rev. B*, Vol. 48, 6276–6283.
Mongiovì, M.S. (2001). Extended irreversible thermodynamics of liquid helium II: boundary condition and propagation of fourth sound. *Physica A*, Vol. 291, 518–537.
Mongiovì, M.S. & Jou, D. (2005). Superfluid turbulence in rotating containers: phenomenological description of the influence of the wall. *Phys. Rev. B*, Vol. 72, 104515 (8 pages).
Mongiovì, M.S. & Jou, D. (2007). Thermodynamical derivation of a hydrodynamical model of inhomogeneous superfluid turbulence. *Phys. Rev. B*, Vol. 75, 024507 (14 pages).
Müller, I. & Ruggeri, T. (1998). *Rational Extended Thermodynamics*, Springer-Verlag, New York.
Nemirovskii, S.K. & Lebedev, V.V.(1983). The hydrodynamics of superfluid turbulence. *Sov. Phys. JETP*, Vol. 57, 1009–1016.
Nemirovskii, S.K. & Fiszdon, W.(1995). Chaotic quantized vortices and hydrodynamic processes in superfluid helium. *Reviews of Modern Physics*, Vol. 67, 37–84.
Onsager, L.(1949). Statistical hydrodynamics. *Nuovo Cimento*, Vol. 6, 249–250.
Osborne, D.V.(1950). The Rotation of Liquid Helium II. *Proc. Phys. Soc. A*, Vol. 63, 909–912.
Peruzza & Sciacca, (2007). Waves propagation in turbulent superfluid helium in presence of combined rotation and counterflow. *Physica B*, Vol. 398, 8–17.
Schwarz, K.W.(1982). Generating superfluid turbulence from simple dynamical rules. *Phys. Rev. Lett.*, Vol. 49, 283–285.
Schwarz, K.W.(1985). Three-dimensional vortex dynamics in superfluid ^{4}He, I. Line-line and line boundary interactions. *Phys. Rev. B*, Vol. 31, 5782–5804.
Schwarz, K.W.(1988). Three-dimensional vortex dynamics in superfluid ^{4}He. *Phys. Rev. B*, Vol.

38, 2398–2417.

Sciacca, M., Mongiovì, M.S. & Jou, D. (2008). A mathematical model of counterflow superfluid turbulence describing heat waves and vortex-density waves. *Math. Comp. Mod.*, Vol. 48, 206–221.

Snyder, H.A., Putney, Z. (1966). Angular dependence of mutual friction in rotating He II. *Phys. Rev.*, Vol. 150, 110–117.

Sonin, E.B. (1987). Vortex oscillations and hydrodynamics of rotating superfluids. *Rev. Mod. Phys.*, Vol. 59, 87–155.

Swanson, C.E., Barenghi, C.F.& Donnelly, R.J.(1983). Rotation of a tangle of quantized vortex lines in He II. *Phys. Rev. Lett.*, Vol. 50, 190–193.

Tisza, L. (1938). Transport phenomena in He II. *Nature*, Vol. 141, 913.

Tough, J.T., (1982). Superfluid turbulence, In: *Progress of Low Temperature Physics Vol. III*, Brewer, D.F., (Ed.), 133, North-Holland, Amsterdam.

Tsubota, M., Araki, T. & Barenghi, C.F.(2003). Rotating superfluid turbulence. *Phys. Rev. Lett.*, Vol. 90, 205301 (4 pages).

Tsubota, M., Araki, T. & Vinen, W.F.(2003). Diffusion of an inhomogeneous vortex tangle. *Physica B*, Vol. 329–333, 224–225.

Tsubota, M., Araki, T. & Barenghi, C.F.(2004). Vortex Tangle Polarized by Rotation. *Jour. Low Temp. Phys.*, Vol. 134, 471–476.

Tsubota, M., Barenghi, C.F., Araki, T. & Mitani, A. (2004). Instability of vortex array and transitions to turbulence in rotating helium II. *Phys. Rev. B*, Vol. 69, 134515 (12 pages).

van Beelen, H., van Joolingen, W. & Yamada, K. (1988). On a balance equation for superfluid vorticity in capillary flow of helium II. *Physica B*, Vol. 153, 248–253.

Vinen, W.F. (1957). Mutual friction in a heat current in liquid helium II. I. Experiments on steady heat current. *Proc. Roy. Soc.*, Vol. A240, 114–127.

Vinen, W.F. (1957). Mutual friction in a heat current in liquid helium II. II. Experiments on transient effects. *Proc. Roy. Soc.*, Vol. A242, 128–143.

Vinen, W.F. (1957). Mutual friction in a heat current in liquid helium II. III. Theory of the mutual friction. *Proc. Roy. Soc.*, Vol. A240, 493–515.

Vinen, W.F. (1958). Mutual friction in a heat current in liquid helium II. IV. Critical heat current in wide channels. *Proc. Roy. Soc.*, Vol. A243, 400–413.

Vinen, W.F. (2000). Classical character of turbulence in quantum fluid. *Phys. Rev. B*, Vol. 61, 1410–1420.

Vinen & Niemela, (2002). Quantum Turbulence. *Jour. Low Temp. Phys.*, Vol. 128, 167–231.

Whitham, J. (1974). *Linear and Nonlinear Waves*, Wiley, New York.

Yamada, K., Kashiwamura, S.& Miyake, K. (1989). Stochastic theory of vortex tangle in superfluid turbulence. *Physica B*, Vol. 154, 318–326.

13

Thermodynamics of Thermoelectricity

Christophe Goupil
ENSICAEN, UMS-CNRT, CRISMAT
France

1. Introduction

Discovered at the beginning of the nineteen century, the thermoelectric effects nicely reflect the richness of the out of equilibrium thermodynamics of coupled phenomena. In 1821 by John Seebeck (Seebeck 1821), (Seebeck 1823), (Seebeck 1826) firstly observed the coupling of two potentials, the electrochemical potential and the temperature. In 1834 Jean Peltier (Peltier 1834) demonstrated that heat flux and electrical current could be coupled. Then in 1855, W. Thompson, future Lord Kelvin, using thermodynamic arguments, discovered that the Seebeck and Peltier effects were in fact not independent (Thompson 1848), (Thompson 1849), (Thompson 1852), (Thompson 1854), (Thompson 1856) giving decisive arguments in favor of a complete and compact description of all these phenomena. Only latter, in 1931, these coupled thermodynamic forces and fluxes where described in a very general form when Lars Onsager proposed a theoretical description of linear out of equilibrium thermodynamic processes. In two major articles the fundamentals of thermodynamics of dissipative transport were developed in a consistent way (Onsager 1931a), (Onsager 1931b). Next, In 1948, Callen developed the Onsager theory in the case of thermal and electrical coupled fluxes, leading to a coherent thermodynamical description of the thermoelectric processes(Callen 1948), (Domenicali 1954). Then, in the middle of the last century Abraham Ioffe, considering both thermodynamics and solid state approaches, extended the previous developments to the microscopic area, opening the door for material engineering and practical applications (Ioffe 1960). He introduced the so-called "figure of merit" ZT, which, as a material parameter, gather the different transport coefficients, leading to an efficient classification of the various thermoelectric materials. The contains of this chapter is divided in six sections. In a first section we remind the basic thermodynamics of thermoelectricity from classical thermodynamic cycle. The second section is devoted to the Onsager description of out equilibrium thermodynamics of coupled transport processes. In a third section the consequence of the Onsager theory are derived leading to the expressions of heat and entropy production. The fourth section is devoted to the presentation of the general conductance matrix. Using these latter the concept of relative current and thermoelectric potential are exposed in the fifth section.Then in the final section the traditional expressions of the efficiencies and Coefficients Of Performance (COP) are revisited using the thermoelectric potential approach.

2. The thermoelectric engine

In a first approach we propose here to consider the analogies between a classical steam engine, and a thermoelectric material (Vinning 1997). The principle analogy is the fact that, in both systems, the entropy is transported by a fluid, which, in the present case, is a gas of electron, also called "Fermi gas" since electrons are Fermions. At first this Fermi gas can be considered to be a perfect gas. Then the equivalent partial pressure of the fluid in the system is the electrochemical potential μ_e.

$$\mu_e = \mu_c + eV$$

where μ_c is the chemical potential, e the electron charge, and V the electrical potential. Then the "gas" equivalences for the steam and thermoelectric engines are,

Molecular gas	P	T
Fermi gas	$\mu_e(T,r)$	T

Then we can draw a picture of a schematic thermoelectric cell:

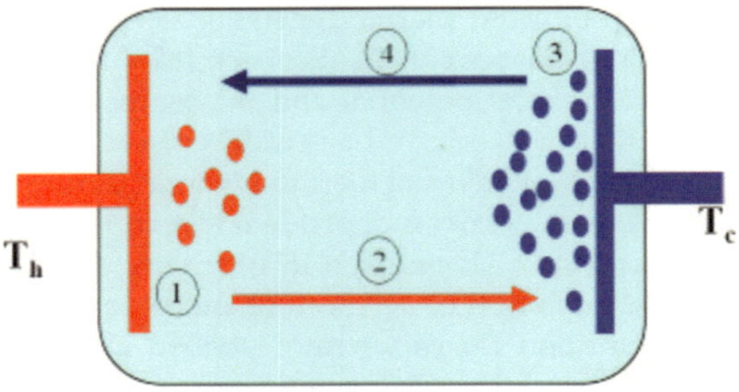

Fig. 1. Thermoelectric "Engine".

One end of the cell is maintained at the temperature $T_h$and the other at temperature $T_c$with $T_h > T_c$. From basic gas kinetic theory we know that we get large velocities and small gas density at the hot end and reduced velocities and large gas density at the cold end. It should be noticed here that, since heat flows from hot to cold end, the system cannot be considered under equilibrium condition meanwhile the average carrier flux is zero. We see that the gradient of carrier density is directly driven by the temperature difference. Since the carriers are charged particles, it produces an electrochemical difference, commonly called voltage difference, which is induced by the application of a temperature difference. This illustrates the coupling of the electrochemical potential gradient and the temperature gradient. Next, since the average carrier flux is zero, meanwhile heat is transported, we get the same values of local fluxes of carriers from hot to cold and cold to hot. From this observation we can conclude that heat and carrier fluxes are coupled. While very simple, this description contains the principal contributions to the thermoelectric processes[1] . The thermodynamical cycle is a Carnot cycle with two "'isothermal branches"' (1 and 3) and to "'adiabatic branches"' (2 and 4). We can then estimate the principle sources of entropy of the working system that are the non-isothermal heat transfers and the non-adiabatic travel of the carriers from cold to hot sources and hot to cold. The principle sources of entropy in the adiabatic branches are the collisions between carriers and the interactions of the carriers with the cristal lattice of the material. It can be noticed that the entropy is transported by the carriers of the Fermi gas. We

[1]In the case of uncharged particles we would also observed a similar process, called "Soret effect" with the diffusion of particles but, of course, no electrical drift contribution.

will see latter that we can defined an "entropy per carrier" which is a fundamental parameter of the thermoelectric process. The principle weakness of this schematic presentation is the complete absence of a correct out of equilibrium thermodynamic in the description. Then we now need to define properly the conditions of this working system. This is the subject of the next section.

3. The Onsager-Callen theory

The thermoelectric processes have been nicely derived by Onsager and Callen, on the framework of the Thermodynamics of dissipative system by the link between out of equilibrium coupled processes (Onsager 1931a), (Onsager 1931b) and (Callen 1948). This description is based on the assumption that the system evolution is driven by a minimal production of entropy where each fluctuation of any thermodynamical potential undergoes a restoring force to equilibrium (Rocard 1967). This leads to a stationary picture where all the thermodynamical potential can be defined, though the system itself produces dissipation. It should be point out here that this description is also a definition of a quasi-static process because the different intensive parameters of the system are defined at each time and location. This implies that the intrinsic local time constants are much shorter than the time evolution of the macroscopic system itself. As a consequence, the classical quasi-static relation $dS = \frac{dQ_{qs}}{T}$ between the heat and the entropy variation can be extended in a flux form expression,

$$\vec{J}_S = \frac{\vec{J}_Q}{T} \tag{1}$$

The quasi-static conditions, previously defined, leads to a continuous thermodynamical description where the thermodynamical equilibrium, with all average fluxes equal zero, just becomes one possible thermodynamical state for the system.

3.1 Forces and fluxes

Let us consider the complete energy flux. From first principle of thermodynamics the energy flux $\vec{J}_E$, heat flux $\vec{J}_Q$ and the particles flux $\vec{J}_N$ are link through the conservation of the energy expression,

$$\vec{J}_E = \vec{J}_Q + \mu_e \vec{J}_N \tag{2}$$

Moreover each of these fluxes is conjugated to its thermodynamics potential gradients, acting as a generalized force. According to energy and particle conservation, in the case of a Fermion gas the correct potentials for energy and particles are $\frac{1}{T}$ and $\frac{\mu_e}{T}$.Then the corresponding forces are:

$$\vec{F}_N = \vec{\nabla}(-\frac{\mu_e}{T}) \tag{3}$$

$$\vec{F}_E = \vec{\nabla}(\frac{1}{T}) \tag{4}$$

If we now write down the linear coupling of forces and fluxes we get then the transport of energy and particles is given by a linear set of coupled equation (Pottier 2007),

$$\begin{bmatrix} \vec{J}_N \\ \vec{J}_E \end{bmatrix} = \begin{bmatrix} L_{NN} & L_{NE} \\ L_{EN} & L_{EE} \end{bmatrix} \begin{bmatrix} \vec{\nabla}(-\frac{\mu_e}{T}) \\ \vec{\nabla}(\frac{1}{T}) \end{bmatrix}$$

with,

$$L_{NE} = L_{EN} \tag{5}$$

This expression of symmetry inside the out of equilibrium process is fundamental in the Onsager description since it is equivalent to the assumption of a minimal entropy production of the system under out of equilibrium conditions. It should be noticed that the minimal entropy production is not a general property of out of equilibrium processes at all and this imposes a precise delimitation of the validity of the model. The Onsager description gives very strong arguments in favor of this assumption. First, from the linear response theory it is known that linear response and fluctuations inside a dissipative system are closely linked. In the present case this implies that every fluctuating potential undergoes a restoring force from the other in a symmetric form. From a purely thermodynamic point of view this coincides with the Lechatelier-Braun principle. Then the $L_{NE} = L_{EN}$ equality traduces the intrinsic symmetry of the coupled fluctuations process. From a microscopic point of view this equality also implies the time reversal symmetry of the processes at micro scale, which should be "micro-reversible".

3.2 Energy flux, heat flux

For sake of simplicity it is more convenient to use the heat fluw $\overrightarrow{J}_Q$ instead of the energy flux $\overrightarrow{J}_E$. Then, using [2] we get,

$$\begin{bmatrix} \overrightarrow{J}_N \\ \overrightarrow{J}_Q \end{bmatrix} = \begin{bmatrix} L_{11} & L_{12} \\ L_{21} & L_{22} \end{bmatrix} \begin{bmatrix} -\frac{1}{T}\overrightarrow{\nabla}(\mu_e) \\ \overrightarrow{\nabla}(\frac{1}{T}) \end{bmatrix}$$

$$L_{12} = L_{21} \tag{6}$$

with the kinetic coefficients equivalence.

$$L_{11} = L_{NN} \tag{7}$$

$$L_{12} = L_{NE} - \mu_e L_{NN} \tag{8}$$

$$L_{22} = L_{EE} - \mu_e(L_{EN} + L_{NE}) + \mu_e L_{NN} \tag{9}$$

3.3 Basic transport results

From the two expressions of particle and heat flux we can now derive the expressions of uncoupled and coupled transport processes, depending on the working conditions, isothermal, adiabatic, open or close systems.

3.3.1 Ohm's law

If considering charged particles we then get the expression of the current density,

$$\overrightarrow{J} = -eL_{11}\frac{1}{T}\overrightarrow{\nabla}(\mu_e) \tag{10}$$

Under isothermal conditions we can directly consider the particle flux in a simplified form,

$$\overrightarrow{J} = e\overrightarrow{J}_N \tag{11}$$

Since the electric field derives from the electrochemical potential we obtain,

$$\vec{E} = -\frac{\vec{\nabla}(\mu_e)}{e} \tag{12}$$

which finally gives the expression of the isothermal electrical conductivity,

$$\sigma_T = \frac{e^2}{T} L_{11} = \frac{e^2}{T} L_{NN} \tag{13}$$

then,

$$L_{11} = \frac{T}{e^2} \sigma_T \tag{14}$$

3.3.2 Fourier law

In order to estimate the thermal conductivity we now consider the heat flux density in the absence of any particle transport, or under zero electrical current in the case of charged particles.

$$\vec{J_N} = \vec{0} = -L_{11}\left(\frac{1}{T}\vec{\nabla}(\mu_e)\right) + L_{12}\vec{\nabla}(\frac{1}{T})$$

then,

$$\left(\frac{1}{T}\vec{\nabla}(\mu_e)\right) = \frac{L_{12}}{L_{11}}\vec{\nabla}(\frac{1}{T}) \tag{15}$$

The heat flux density becomes,

$$\vec{J}_{Q_J} = \frac{1}{T^2}\left[\frac{L_{21}L_{12} - L_{11}L_{22}}{L_{11}}\right]\vec{\nabla}(T) \tag{16}$$

and finally the thermal conductivity under zero electrical current is,

$$\kappa_J = \frac{1}{T^2}\left[\frac{L_{11}L_{22} - L_{21}L_{12}}{L_{11}}\right] \tag{17}$$

We can also consider the thermal conductivity under zero electrochemical gradient, κ_E, and after calculation we get,

$$\vec{J}_{Q_E} = \frac{L_{22}}{T^2}\vec{\nabla}(T) = \kappa_E \vec{\nabla}(T) \tag{18}$$

then

$$\kappa_E = \frac{L_{22}}{T^2} \tag{19}$$

3.3.3 Seebeck coefficient

We now consider the interaction between thermal and electrochemical processes, in the absence of particle transport. The basic expression is already known since it is given by equation [15]. We define the Seebeck coefficient as the ratio between the two forces, electrochemical gradient and temperature gradient, then the Seebeck coefficient expression is given by,

$$\alpha \equiv -\frac{1}{e}\frac{\vec{\nabla}(\mu_e)}{\vec{\nabla}(T)} \tag{20}$$

$$\alpha = \frac{1}{eT}\frac{L_{12}}{L_{11}} \tag{21}$$

3.3.4 Peltier coefficient

If we consider an isothermal configuration we can derive the expression of the coupling term between current density and heat flux, usually called the Peltier coefficient. Since

$$\vec{J} = eL_{11}\left(-\frac{1}{T}\vec{\nabla}(\mu_e)\right) \tag{22}$$

$$\vec{J}_Q = L_{21}\left(-\frac{1}{T}\vec{\nabla}(\mu_e)\right) \tag{23}$$

then,

$$\vec{J}_Q = \frac{1}{e}\frac{L_{12}}{L_{11}}\vec{J} \tag{24}$$

The Peltier coefficient Π is then given by,

$$\vec{J}_Q = \Pi\vec{J} \tag{25}$$

$$\Pi = \frac{1}{e}\frac{L_{12}}{L_{11}} \tag{26}$$

As one can see we have the equality,

$$\Pi = T\alpha \tag{27}$$

This latter expression reveal the close connexion between Peltier and Seebeck effects. We will see latter that another effect, called Thomson effect is also strongly linked with theses two effect. The reason is that all of these effect depends on a quantity called the "'entropy per carrier"' which is now presented.

3.3.5 Entropy per carrier

If we get back to the first description of a thermoelectric cell as a Carnot like process, it is clear that the carriers inside the thermodynamic cycle transport a certain amount of entropy. Let us consider this by writing the entropy flux density,

$$\vec{J}_S = \frac{\vec{J_Q}}{T} = \frac{1}{T}\left[L_{21}\left(-\frac{1}{T}\vec{\nabla}(\mu_e)\right) + L_{22}\vec{\nabla}\left(\frac{1}{T}\right)\right] \tag{28}$$

This expression can be simplified considering the Ohm law expression [10]

$$\vec{\nabla}(\mu_e) = -\vec{J}\frac{T}{eL_{11}} \tag{29}$$

Then,

$$\vec{J}_S = \frac{L_{21}}{TeL_{11}}\vec{J} + \frac{1}{T}L_{22}\nabla\left(\frac{1}{T}\right) \tag{30}$$

As we can observe the entropy flux is driven by two terms coming from thermal and electrochemical contributions. The latter term shows that a fraction of the entropy is transported by the flux of carriers. We then get the expression of the "'entropy per carrier"',

$$S_J = \frac{L_{21}}{TL_{11}} \tag{31}$$

As one can see the Seebeck coefficient is directly proportional to S_J since we have,

$$S_J = e\alpha \tag{32}$$

The entropy per particle is a fundamental parameter since all the thermoelectric effects derive from its value, Seebeck effect, or its variations Peltier and Thomson effects.

3.3.6 Kinetic coefficients and transport parameters

From the previous results we have now the complete expressions of the kinetic coefficients,

$$\begin{aligned} L_{11} &= \tfrac{T}{e^2}\sigma_T \\ L_{12} &= \tfrac{T^2}{e^2}\sigma_T S_J \\ L_{22} &= \tfrac{T^3}{e^2}\sigma_T S_J^2 + T^2\kappa_J \end{aligned} \tag{33}$$

The thermal conductivities under zero electrochemical gradient or zero particle transport can then be expressed:

$$\kappa_E = T\alpha^2\sigma_T + \kappa_J \tag{34}$$

We see that κ_J is a purely conductive contribution meanwhile κ_E contains both conductive and convective terms.

4. Heat and entropy

4.1 Heat flux

Let us consider again the coupled Onsager expressions:

$$\vec{J} = -\sigma_T\left(\frac{\vec{\nabla}(\mu_e)}{e}\right) + \frac{\sigma_T S_J T^2}{e^2}\left(\vec{\nabla}(\frac{1}{T})\right) \tag{35}$$

$$\vec{J}_Q = -T\sigma_T S_J\left(\frac{\vec{\nabla}(\mu_e)}{e}\right) + \left[\frac{T^3}{e^2}\sigma_T S_J^2 + T^2\kappa_J\right]\left(\vec{\nabla}(\frac{1}{T})\right) \tag{36}$$

Since the heat flux $\vec{J}_Q$ contains both conduction and convection contribution to heat flux. We can rewrite these equations in the form,

$$\vec{J}_Q = TS_J\vec{J} + T^2\kappa_J\left(\vec{\nabla}(\frac{1}{T})\right)$$

where we identify a conductive and a convective contribution,

$$\vec{J}_{Q_{cond}} = T^2\kappa_J\left(\vec{\nabla}(\frac{1}{T})\right) \tag{37}$$

$$\vec{J}_{Q_{conv}} = TS_J\vec{J} \tag{38}$$

4.2 Volumic heat production

The volumic heat production can be estimated from the total energy flux,

$$\vec{J}_E = \vec{J}_Q + \mu_e \vec{J}_N \tag{39}$$

According to energy and particle conservation we have

$$\vec{\nabla}.\vec{J}_E = 0 \tag{40}$$

$$\vec{\nabla}.\vec{J}_N = 0 \tag{41}$$

Then,

$$\vec{\nabla}.\vec{J}_Q = -\vec{\nabla}\mu_e.\vec{J}_N$$

or in an equivalent form,

$$\vec{\nabla}.\vec{J}_Q = \vec{E}.\vec{J} \tag{42}$$

This summarizes the possible transformation of the energy since it shows that heat can be produced by the degradation of the electrochemical potential μ_e, and electrical power can be extracted from heat.

4.3 Entropy production density

If we consider the entropy flux density we can calculate the entropy production from,

$$\vec{\nabla}.\vec{J}_S = \overset{\circ}{S} = \vec{\nabla}\left(\frac{\vec{J}_Q}{T}\right) = \vec{\nabla}\left(\frac{1}{T}\right).\vec{J}_Q + \frac{1}{T}\vec{\nabla}.\vec{J}_Q \tag{43}$$

then,

$$\overset{\circ}{S} = \vec{\nabla}\frac{1}{T}.\vec{J}_Q - \frac{\vec{\nabla}\tilde{\mu}_e}{T}.\vec{J}_N \tag{44}$$

A shown above, the entropy production is due to non isothermal heat transfers and electrical Joule production. This latter corresponds to a non adiabatic transfer of the entropy by the particles. The previous expression can be rewritten in the form,

$$\overset{\circ}{S} = \vec{\nabla}(\frac{1}{T}).\vec{J}_E + \vec{\nabla}(-\frac{\mu_e}{T}).\vec{J}_N \tag{45}$$

In this form, we get the illustration of one major result of the Onsager description:
The total entropy production is given by the summation of the force-flux products,

$$\vec{\nabla}.\vec{J}_S = \overset{\circ}{S} = \Sigma\overrightarrow{force}.\overrightarrow{flux} \tag{46}$$

This general result can also be obtained from minimal entropy production assumptions.

5. General conductivity matrix

5.1 Presentation

As we have seen from Ohm law and Fourier law, the kinetic coefficients can be written in the form of conductances and it is tempting to express the force-flux expressions in the form of a general conductance matrix. It should be notice that such description has been derived in very general way by Callen and Greene in 1952 (Callen 1952), (Greene 1952). The complete derivation is out of the scope of this presentation and we only consider here the basic derivation of a general conductivity. Starting from equations [6] and [33] we have,

$$\begin{bmatrix} \vec{J}_N \\ \vec{J}_Q \end{bmatrix} = \begin{bmatrix} \frac{T}{e^2}\sigma_T & \frac{T^2}{e^2}\sigma_T S_J \\ \frac{T^2}{e^2}\sigma_T S_J & T^2\kappa_E \end{bmatrix} \begin{bmatrix} -\frac{1}{T}\vec{\nabla}(\mu_e) \\ \vec{\nabla}(\frac{1}{T}) \end{bmatrix} \tag{47}$$

Using

$$\vec{E} = -\frac{\vec{\nabla}(\mu_e)}{e}$$

and considering that at local state $\vec{\nabla}(\frac{1}{T}) = -\frac{1}{T^2}\vec{\nabla}(T)$ then we get ,

$$\begin{bmatrix} \vec{J} \\ \vec{J}_Q \end{bmatrix} = \begin{bmatrix} \sigma_T & \alpha\sigma_T \\ T\alpha\sigma_T & \kappa_E \end{bmatrix} \begin{bmatrix} \vec{E} \\ -\vec{\nabla}(T) \end{bmatrix}$$

Then the electrical and heat fluxes are totally described through the general conductivity matrix,

$$[\sigma] = \begin{bmatrix} \sigma_T & \alpha\sigma_T \\ T\alpha\sigma_T & \kappa_E \end{bmatrix} \tag{48}$$

One can notice that the Seebeck coefficient appears clearly as the coupling term between the electrical and thermal processes. In the case where $\alpha = 0$, the conductivity matrix reduces to a diagonal form $\begin{bmatrix} \sigma_T & 0 \\ 0 & \kappa_E \end{bmatrix}$ where Ohm law and Fourier law are then totally decoupled. In other words the Seebeck coefficient, or more precisely the entropy per carrier is the "tunning parameter" of the coupling between electrical and thermal fluxes.

5.2 Heat transformation

In the previous paragraphs we have consider the volumic heat transformation from the calculation of the divergence of the heat flux $\vec{\nabla}.\vec{J}_Q$. We propose now to get more deeply into this expression by considering its different terms. First, by elimination of the electrical field $\vec{E}$ from the previous set of equations we get,

$$\vec{J}_Q = \alpha T\vec{J} - \kappa_J\vec{\nabla}T \tag{49}$$

Then, calculating the heat flux divergence,

$$\vec{\nabla}.\vec{J}_Q = \vec{\nabla}.\left[\alpha T\vec{J} - \kappa_J\vec{\nabla}T\right] = \left[T\vec{J}.\vec{\nabla}\alpha + \alpha\vec{\nabla}T.\vec{J} + \alpha T\vec{\nabla}.\vec{J} + \vec{\nabla}.\left[\kappa_J\left(-\vec{\nabla}T\right)\right]\right] \tag{50}$$

we find four terms which can be identified:

- $\alpha T\vec{\nabla}.\vec{J}$: equal zero due to particle conservation.

- $T\vec{J}.\vec{\nabla}\alpha$: "Peltier-Thomson" term.
- $\vec{J}.\alpha\vec{\nabla}T := \vec{J}.\left[\vec{E} - \frac{\vec{J}}{\sigma_T}\right] = \vec{J}.\vec{E} - \frac{J^2}{\sigma_T}$: Electrical work production and dissipation.
- $\vec{\nabla}.\left[\kappa_J\left(-\vec{\nabla}T\right)\right]$: heat production from non isothermal transfer.

then,

$$\vec{\nabla}.\vec{J}_Q = T\vec{J}.\vec{\nabla}\alpha + \vec{J}.\vec{E} - \frac{J^2}{\sigma_T} - \vec{\nabla}.\left[\kappa_J\vec{\nabla}T\right] \tag{51}$$

Most of these terms are common, but less intuitive is the Peltier-Thomson term which is now considered.

5.3 The Peltier-Thomson term

As we will show, the $T\vec{J}.\vec{\nabla}\alpha$ term contains both the Thomson contribution (local, temperature gradient effect), and the Peltier contribution (isothermal, spatial gradient effect). Using the equivalence $\Pi = \alpha T$ we have,

$$T\vec{J}.\vec{\nabla}\alpha = T\vec{J}.\vec{\nabla}\frac{\Pi}{T} = T\vec{J}.\left[\frac{1}{T}\vec{\nabla}\Pi - \frac{1}{T^2}\Pi\vec{\nabla}T\right] = \vec{J}.\left[\vec{\nabla}\Pi - \alpha\vec{\nabla}T\right] \tag{52}$$

Then the traditional separation of the Peltier and Thomson contribution is artificial since they both refer to the same physic of the gradient of the entropy per particle, temperature driven gradient or spatially driven gradient.The isothermal configuration leads to Peltier expression meanwhile a spatial gradient gives the Thomson result.

- Pure Peltier ($\nabla T = 0$):

$$\vec{J}.\left[\vec{\nabla}\Pi - \alpha\vec{\nabla}T\right] = \vec{J}.\left[\vec{\nabla}\Pi\right]$$

- Pure Thomson ($\alpha = f(T)$) :

$$= \vec{J}.\left[\vec{\nabla}\Pi - \alpha\vec{\nabla}T\right] = \vec{J}.\left[\frac{d\Pi}{dT} - \alpha\right]\vec{\nabla}T = \tau\vec{J}.\vec{\nabla}T \tag{53}$$

with,

$$\vec{\nabla}\Pi = \frac{d\Pi}{dT}\vec{\nabla}T \tag{54}$$

$$\tau = \frac{d\Pi}{dT} - \alpha \tag{55}$$

then the heat flux divergence takes the form,

$$\vec{\nabla}.\vec{J}_Q = \tau\vec{J}.\vec{\nabla}T + \vec{J}.\vec{E} - \frac{J^2}{\sigma_T} - \vec{\nabla}.\left[\kappa_J\vec{\nabla}T\right] \tag{56}$$

If we consider a configuration $\kappa_J \neq f(T)$ then it reduces to,

$$\vec{\nabla}.\vec{J}_Q = \tau\vec{J}.\vec{\nabla}T + \vec{J}.\vec{E} - \frac{J^2}{\sigma_T} - \kappa_J\nabla^2 T \tag{57}$$

5.4 Energy balance

Using the expression [42] $\vec{\nabla}.\vec{J}_Q = \vec{E}.\vec{J}$ the local energy balance can be expressed from expression [57] (Landau 1984):

$$\vec{\nabla}.\vec{J}_Q - \vec{E}.\vec{J} = \kappa_T \nabla^2 T + \frac{J^2}{\sigma_T} - \tau \vec{J}.\vec{\nabla} T = 0 \tag{58}$$

It should be notice that this derivation does not need any assumption concerning the behavior of the particle, in equilibrium or not. In the case of transient configuration the energy balance equation should be corrected using ρC_p where C_p is the heat capacitance and ρ the volumic mass.

$$\kappa_J \nabla^2 T + \frac{J^2}{\sigma_T} - \tau \vec{J}.\vec{\nabla} T = \rho C_p \frac{\partial T}{\partial t} \tag{59}$$

5.5 Entropy flux, volumic entropy production

Using the expression [49] for the heat flux $\vec{J}_Q = \alpha T \vec{J} - \kappa_J \vec{\nabla} T$ we can derive the expression of the entropy flux,

$$\vec{J}_S = \frac{\vec{J}_Q}{T} = \alpha \vec{J} - \frac{\kappa_J}{T} \vec{\nabla} T \tag{60}$$

The calculation of the volumic entropy production is then straightforward,

$$\vec{\nabla}.\vec{J}_S = \vec{\nabla}\left(\frac{1}{T}\right).\vec{J}_Q + \frac{1}{T}\vec{\nabla}. \tag{61}$$

$$\vec{J}_Q = \vec{\nabla}\left(\frac{1}{T}\right).\left[\alpha T \vec{J} - \kappa_J \vec{\nabla} T\right] + \frac{1}{T}\vec{E}.\vec{J} \tag{62}$$

but since $\vec{E} = \frac{\vec{J}}{\sigma_T} + \alpha \vec{\nabla} T$ then,

$$\vec{\nabla}\vec{J}_S = -\kappa_J \vec{\nabla}(T).\vec{\nabla}\left(\frac{1}{T}\right) + \frac{1}{T}\frac{J^2}{\sigma_T} \tag{63}$$

And we verify that $\vec{\nabla}.\vec{J}_S$ is in agreement with the force-flux description since,

$$\vec{\nabla}.\vec{J}_S = -\kappa_J \vec{\nabla}(T).\vec{\nabla}\left(\frac{1}{T}\right) + \frac{1}{T}\frac{J^2}{\sigma_T} \tag{64}$$

$$= \left[-\kappa_J \vec{\nabla}(T) + \alpha \vec{J} T\right].\vec{\nabla}\left(\frac{1}{T}\right) + \vec{J}.\left(\frac{\vec{E}}{T}\right) \tag{65}$$

$$= \vec{J}_Q.\vec{\nabla}\left(\frac{1}{T}\right) - \vec{J}.\frac{\vec{\nabla}\tilde{\mu}_e}{eT} = \sum flux.force \tag{66}$$

5.6 Basic thermogenerator

Let us consider a bar of thermoelectric material inserted between a hot and a cold sources respectively at temperatures T_h and T_c, giving a very schematic picture of thermogenerator. As we know, electrical power can be extracted from the heat going through the bar. In such a device only a fraction of of the heat going trough is converted into electricital energy, and the maximum fraction is given by the Carnot efficiency

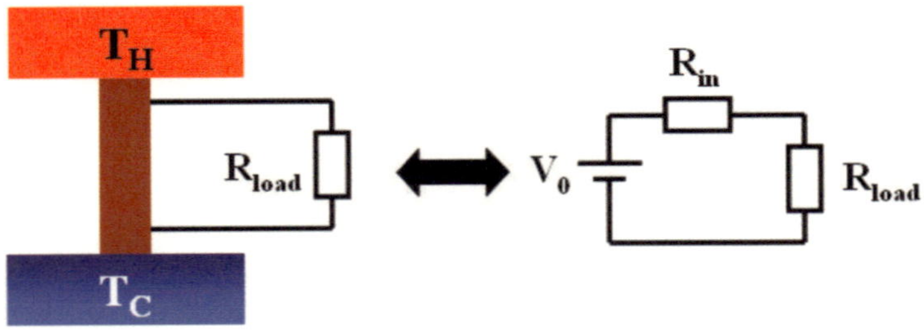

Fig. 2. Basic Thermogenerator.

$$\eta_C = \frac{T_h - T_c}{T_h} \tag{67}$$

Since the "thermoelectric engine" is not endoreversible the final fraction will be less than η_C. From the electrical point of view the system is equivalent to a voltage generator of open voltage V_0, internal resistance R_{in} and short-circuit current $I_{cc} = \frac{V_0}{R_{in}}$. The maximal electrical output power is obtained for $R_{load} = R_{in}$ giving

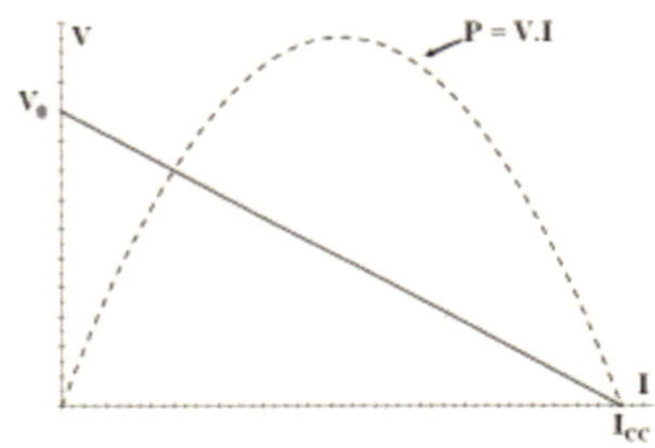

Fig. 3. IV Characteristic.

$$P_{\max} = \frac{V_0^2}{4R_{in}} = \frac{R_{in} I_{cc}^2}{4} \tag{68}$$

5.7 Figure of merit

As we have seen, if we force a heat flux to go through the thermoelectric material then an electrical current will flow. The efficiency of this process depends on two contributions:

- The material properties.
- The working conditions.

We will first consider the material properties, working conditions will be consider in a latter section.

The thermoelectric properties of a material are summarized in the so called "figure of merit" of a thermoelectric material, usually written "ZT". This scalar gives a direct measurement of the quality of the material, for practical applications. The ZT expression can be derived from technical arguments or from thermodynamical arguments. We propose here to derive the ZT expression from the thermal conductivity expressions. The illustration will be given on the basis of a thermogenerator description, but same conclusions could be drawn considering a Peltier heat pump configuration.

Let us consider the schematic thermoelectric engine described in the introduction. From the electrical point of view we can consider to extreme cases, open circuit and short-circuit.

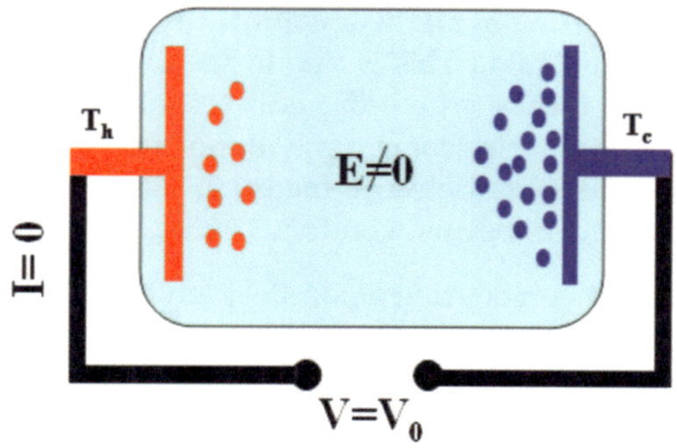

Fig. 4. Open circuit configuration.

These two configurations have strong consequences in term of thermal conductivity since it exists both an open circuit thermal conductivity κ_J and a short-circuit thermal conductivity κ_E. These two are not independent since from [47] we have, $\kappa_E = T\alpha^2\sigma_T + \kappa_J$
Again we see that the coupling parameter between electrical and thermal processes is α.

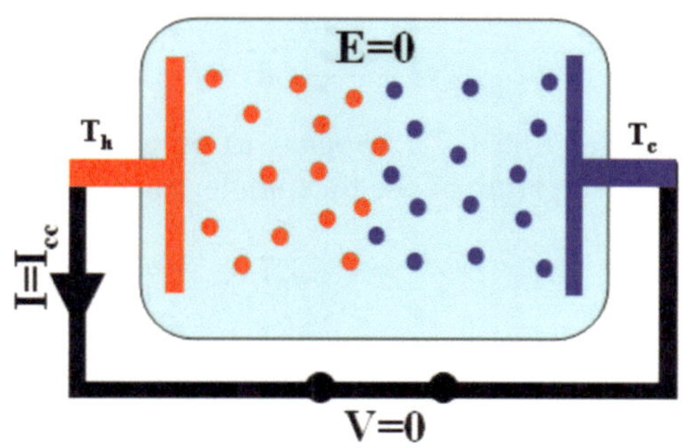

Fig. 5. Short circuit configuration.

From [33] we know that in order to get an efficient thermogenerator the open voltage should be as large as possible. As a consequence κ_J should be minimal.
From the IV curve we can notice that, for a given open voltage, the larger is the short-circuit current, the larger will be the output current. In other word the ratio of the two thermal conductivities $\frac{\kappa_E}{\kappa_J}$ should be as large as possible. Then we can write,

$$\frac{\kappa_E}{\kappa_J} = \left[\frac{\alpha^2\sigma_T}{\kappa_J}T + 1\right] \tag{69}$$

With

$$ZT = \frac{\alpha^2\sigma_T}{\kappa_J}T \tag{70}$$

The maximal efficiency, and output electrical power, can be obtained if ZT is large enough. Since only material parameters enter into ZT expression, the figure of merit is clearly the central term for material engineering research. In addition it should be noticed that the present description do not consider at all the anions of the cristal lattice of the thermoelectric material but only the electronic gas[2]. This is due to the Onsager description which follows the so called linear response theory where the lattice contribution is not included. So the present description reduces to the electronic gas with no contribution of the vibrations of the cristal lattice (phonons) of the thermoelectric material. Such a lattice contribution may be added to the L_{22} term as a lattice thermal conductivity κ_{Lat}. This latter can be added to the general conductance matrix by considering a parallel path for the heat flux, $\begin{bmatrix} 0 & 0 \\ 0 & \kappa_{Lat} \end{bmatrix}$. The complete matrix then takes the form, $\begin{bmatrix} \sigma_T & \alpha\sigma_T \\ T\alpha\sigma_T & \kappa_E + \kappa_{Lat} \end{bmatrix}$.

5.8 Isothermal and adiabatic electrical conductivity

As we have seen the electrical current density expression is given by,

$$\vec{J} = e\vec{J}_N = \sigma_T\left(\vec{E} - \alpha\vec{\nabla}T\right) \tag{71}$$

with the electrical field[3] $\vec{E} = -\frac{\vec{\nabla}\mu_e}{e}$. Let us now consider an adiabatic process instead of a isothermal one then the conductance matrix gives us the adiabatic electrical conductivity in the form

$$\vec{J}_{\vec{J}_Q=\vec{0}} = \frac{\sigma_T}{1+ZT}\vec{E} = \sigma_Q\vec{E} \tag{72}$$

Again we see that the figure of merit ZT appears to be a fundamental thermodynamical parameter in the core of the description of the thermoelectric process leading to,

$$\frac{\sigma_T}{\sigma_Q} = 1 + ZT$$

6. The thermoelectric potential

Until now we have not really take into account the working conditions point of view. As any working engine a thermoelectric device should be correctly driven in order to provide power in the best conditions. Then a precise control of the applied thermodynamic potentials is needed in order to get a correct use of the potentialities of the thermoelectric materials. Since the thermoelectric process implies the coupling of the heat and electrical fluxes, these two fluxes should both be driven optimally. This question has been addressed by Jeffrey Snyder in 2003, using the so called relative current (Snyder 2003).

6.1 Relative current and thermoelectric potential

The relative current density defined by the ratio of the electrical flux and the purely conductive fraction of the heat flux.

$$u = -\frac{J}{\kappa_J \nabla T} \tag{73}$$

[2]This would be the same as considering a steam engine without any boiling walls!

[3]$\vec{E} = \frac{\vec{J}}{\sigma_T} + \alpha\vec{\nabla}T$ so open voltage gives, $\vec{E}_{J=0} = \alpha\vec{\nabla}T$.

From [49] the heat flow becomes,

$$\vec{J}_Q = T\alpha\vec{J} + \frac{\vec{J}}{u} = \Phi\vec{J} \tag{74}$$

$$\Phi = \left[T\alpha + \frac{1}{u}\right] \tag{75}$$

where Φ is called the "Thermoelectric Potential".
The heat and carrier fluxes are now directly connected by the thermoelectric potential . This expression is fundamental since it allows to deriving the principle results of the Thermodynamics of thermoelectricity directly from this. According to the previous definitions the volumic heat production becomes,

$$\vec{\nabla}.\vec{J}_Q = \vec{J}.\vec{\nabla}\left[T\alpha + \frac{1}{u}\right] = \vec{J}.\vec{\nabla}\Phi \tag{76}$$

Since the heat production $\vec{J}.\vec{\nabla}\frac{1}{u}$ directly reduces the efficiency, it becomes now evident that the maximum efficiency coincides with the minimization of $\vec{\nabla}\frac{1}{u}$. This is currently obtained for a specific value $u = s$, where s is called "compatibility" (Snyder 2003).
Considering the entropy flux we get, $\vec{J}_S = \frac{1}{T}\left[\alpha + \frac{1}{u}\right]\vec{J} = \frac{\Phi}{T}\vec{J}$. The expression of the volumic entropy production becomes

$$\vec{\nabla}.\vec{J}_S = \vec{J}.\vec{\nabla}\left(\frac{\Phi}{T}\right) \tag{77}$$

We get the correct Onsager formulation of the entropy production as the summation of the Flux-Force products which here reduces to a single product. For a given material the thermoelectric potential give a direct measurement of the total volumic heat and entropy production by the respective degradation of Φ and $\frac{\Phi}{T}$. In other words, Φ can be considered to be the free energy of the electronic gas in the out of equilibrium process.

6.2 The irreversible factor

Due to the finite value of the ZT figure of merit, the maximal efficiency of a thermogenerator is only a fraction of the Carnot efficiency. In other word the thermoelectric engine is not endoreversible because ZT is not infinite. As we previously said the irreversiblities are partly due to finite value of ZT, but they are also due to the thermodynamic working conditions that can degrade strongly the overall efficiency. This is the subject of the present paragraph.

6.3 Thermoelectric potential and efficiency

Let us calculate the relative efficiency of a Thermoelectric Generator (TEG)(Goupil 2009). This latter is defined as the ratio between the electrical energy production and the entering heat flux. Using [42] and [1] we get $\vec{E}.\vec{J} = \vec{\nabla}.\vec{J}_Q = T\vec{\nabla}.\vec{J}_S + \vec{J}_S.\vec{\nabla}T$. Then we obtain the expression of the relative efficiency,

$$\eta_r = \frac{E.J}{J_S \nabla T} = \frac{E.J}{E.J + T\nabla J_S} \tag{78}$$

In the ideal case of a reversible process the entropy production is zero then the power production reduces to $\vec{E}.\vec{J} = \vec{J}_S.\vec{\nabla}T$ then the relative efficiency is $\eta_r = 1$ which means that the Carnot efficiency is reached. In an irreversible configuration the $T\vec{\nabla}.\vec{J}_S$ term contributes to reduce the efficiency then,

$$\eta_r = \frac{E_{out}.J}{J_S.\nabla T} = \frac{E_{out}.J}{E_{out}.J + T\nabla J_S} \tag{79}$$

where $E_{out}.J$ is the output electrical power. Let us now consider the introduction of the thermoelectric potential Φ into the calculation of the relative efficiency.

$$\eta_r = \frac{\vec{E}.\vec{J}}{\vec{J}_S.\vec{\nabla}T} = \frac{\vec{\nabla}.\vec{J}_Q}{\frac{\vec{J}_Q}{T}.\vec{\nabla}T} = \frac{\nabla\Phi}{\nabla T}\frac{T}{\Phi} \tag{80}$$

we see that η_r corresponds to the relative variation of the thermodynamical potential $\frac{\nabla\Phi}{\Phi}$ when changing the other potential $\frac{\nabla T}{T}$. This is coherent with a general definition of the efficiency of an out of equilibrium thermodynamical process as a coupled fluctuating system [4]. Using the expression of the thermoelectric potential Φ we can now rewrite the relative current in the form,

$$u = -\frac{J}{\kappa_J \nabla T} = -\frac{Z}{\alpha^2}\frac{\nabla\Phi}{\nabla T} + \frac{Z}{\alpha} \tag{81}$$

Introducing the expression of the relative efficiency we get,

$$u = \frac{Z}{\alpha}\left[-\left(1 + \frac{1}{\alpha T u}\right)\eta_r + 1\right] \tag{82}$$

The relative efficiency of the section of TEG becomes (Snyder 2003),

$$\eta_r = \frac{\frac{u\alpha}{Z}\left(1 - \frac{u\alpha}{Z}\right)}{\left(\frac{u\alpha}{Z} + \frac{1}{ZT}\right)} = \frac{1 - \frac{\alpha}{Z(\Phi - T\alpha)}}{1 + \frac{Z(\Phi - T\alpha)}{ZT\alpha}} \tag{83}$$

This classical expression of the reduce efficiency presents a maximum for the compatibility value $u_{opt} = s = \frac{\sqrt{1+ZT}-1}{\alpha T}$. In that case the reduce efficiency becomes,

$$\eta_r = \frac{\sqrt{1+ZT}-1}{\sqrt{1+ZT}+1} = \frac{1}{2\frac{\Phi_{opt}}{\alpha T} - 1} \tag{84}$$

where the equivalent optimal potential is,

$$\Phi_{opt} = \left[T\alpha + \frac{1}{s}\right] = T\alpha\left[\frac{\sqrt{1+ZT}}{\sqrt{1+ZT}-1}\right] \tag{85}$$

We can now plot the expression $\eta_r = f(\frac{\Phi}{T})$ for various ZT values. One can notice that this Φ dependence is sharper as ZT values increase. This is in agreement with the assumption that the proximity to reversibility implies a drastic control of the working contributions. This is obtained by a direct control of $\Phi = \Phi_{opt}$. As a consequence it becomes possible to control the maximum relative efficiency by keeping Φ to its optimal value $\Phi = \Phi_{opt}$. This approach extends the compatibility approach by adding a thermodynamical feedback by a direct measurement of the local entropy production $\vec{\nabla}.\vec{J}_S$. The expression of the irreversible

[4]We recover here another illustration of the equality $L_{ij} = Ł_{ji}$

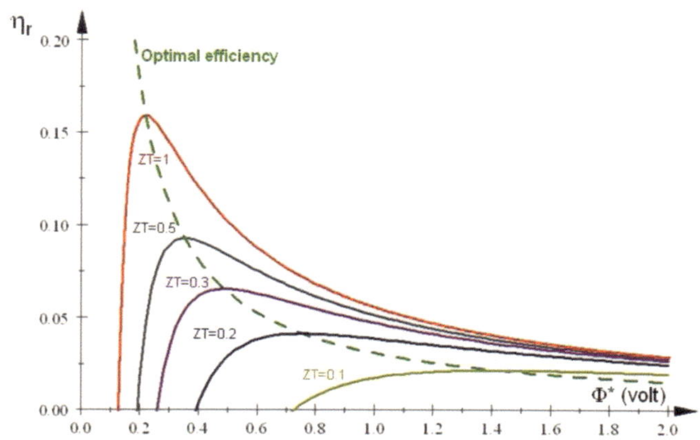

Fig. 6. Reduce Efficiency.

factor gives an insight of the sensibility of the thermoelectric system to varying working conditions. It shows that for large ZT this sensibility becomes very high and strong reduction of the efficiency can rapidly occur. For an optimal efficiency, obtained for $u = s$ it shows that Φ precises the intrinsic irreversibilities contribution for a given material.

7. Conclusion

The thermoelectric process has been described using a classical fluid approach. The "Fermi gas"' of electrons takes place of the traditional "steam" in the thermodynamical cycle, giving strong similarities in the description of the underlying mechanismes, which are based on the linear Onsager theory of "'out of equilibrium thermodynamics". It is shown that the so called "'figure of merit ZT"' of the thermoelectric material can be directly derived from this approach. Finally, the importance of the working conditions is demonstrated, leading to the concept of thermoelectric potential Φ which is an extension of the concept of free energy for the gas, under out of equilibrium conditions.

8. References

[Callen 1952] Herbert B. Callen and Richard F. Greene, Phys Rev **86** (5) 702-710 (1952).

[Callen 1954] Callen, H. B., Irreversible thermodynamics of thermoelectricity, Rev. Mod. Phys., 26, 237, 1954.

[Callen 1948] Herbert B. Callen, Phys Rev 73 (11) 1349-1358 (1948).

[Domenicali 1954] C.A. Domenicali, Rev. Mod. Phys. **26**, 237-275, (1954).

[Goupil 2009] C. Goupil, J. Appl. Phys. 106, 104907 (2009). 5

[Greene 1952] Richard F. Greene and Herbert B. Callen, Phys Rev **88** (6) 1387-1391 (1952).

[Ioffe 1960] A.F. Ioffe, Physics of Semiconductors (Infosearch, London, 1960)

[Landau 1984] L. D. Landau and E. M. Lifshitz, Electrodynamics of Continuous Media, 2nd Edition, Butterworth Heinemann (Oxford, 1984).

[Onsager 1931a] L. Onsager, Phys. Rev. 37, 405 - 426 (1931).

[Onsager 1931b] L. Onsager, Phys. Rev. 38, 2265 - 2279 (1931).

[Peltier 1834] Peltier, J. C. A., Nouvelles experiences sur la caloricité des courants electrique, Ann. Chem. Phys., 56, 371, 1834.

[Pottier 2007] Noëlle Pottier, Physique statisitique hors équilibre, processus irréversibles linéaires. Savoirs Actuels EDP Sciences/CNRS Editions, (2007).

[Rocard 1967] Y. Rocard ; Thermodynamique, Masson (2e édition-1967).

[Seebeck 1821] Seebeck, T. J., Ueber den magnetismus der galvenische kette, Abh. K. Akad. Wiss. Berlin, 289, 1821.

[Seebeck 1826] Seebeck, T. J., Ann. Phys. (Leipzig), 6, 1, 1826. 4. Seebeck, T. J., Methode, Platinatiegel auf ihr chemische reinheit durck thermomagnetismus zuprufen, Schweigger's J. Phys., 46, 101, 1826.

[Seebeck 1823] Seebeck, T. J., Magnetische polarisation der metalle und erze durck temperatur-differenz, Abh. K. Akad. Wiss. Berlin, 265, 1823.

[Snyder 2003] G. Jeffrey Snyder and Tristan S. Ursell,Phys. Rev. Lett. 91 148301 (2003).

[Thompson 1848] Thomson, W., On an absolute thermometric scale, Philos. Mag., 33, 313, 1848

[Thompson 1849] Thomson, W., An account of Carnot's theory of the motive power of heat, Proc. R. Soc. Edinburgh, 16, 541, 1849.

[Thompson 1852] Thomson, W., On a mechanical theory of thermo-electric currents, Philos. Mag. [5], 3, 529, 1852.

[Thompson 1854] Thomson, W., Account of researches in thermo-electricity, Philos. Mag. [5], 8,62, 1854.

[Thompson 1856] Thomson, W., On the electrodynamic qualities of metals, Philos. Trans. R. Soc. London, 146, 649, 1856.

[Vinning 1997] C. B. Vining. Materials Research Society Symposium Proceedings: Thermoelectric Materials - Mater. Res. Soc. 278 (1997) 3-13.

14

Application of the Continuum-Lattice Thermodynamics

Eun-Suok Oh
University of Ulsan
South Korea

1. Introduction

Through the continuum-lattice thermodynamic approach, the thermodynamic behaviors of two- and three-dimensional multicomponent, elastic, crystalline solids are developed. We begin with a discussion of non-equilibrium thermodynamics of an isolated body that is not undergoing a phase transformation.

Our analysis recognizes that the Helmholtz free energy, $\hat{A}$, is an explicit function of the deformed crystallographic or lattice vectors defining the deformed crystalline structure. Edelen (1975) gave a similar discussion with the assumption that $\hat{A}$ was an explicit function of the deformation gradient rather than a function of the deformed lattice vectors. For this reason, his analysis requires an additional step in which the implications of the isotropy group are observed (Truesdell & Noll, 1965, p. 310); the requirements of the isotropy group are automatically accounted for in our analysis through the use of the lattice vectors (Slattery & Lagoudas, 2005).

As applications, we obtain the stress-deformation behaviours of graphene, carbon nanotubes(CNTs), boron-nitride nanotubes(BNNTs) which are composed of a regular two-dimensional array of hexagonal lattices of atoms (Oh et al., 2005; Oh, 2010), and the stress-deformation behaviours of face-centred cubic crystals such as diamond, silicon, silicon-carbide, and boron-nitride (Oh & Slattery, 2008). Using an interatomic potential, the Tersoff (Tersoff, 1988; 1989) or Tersoff-like potential (Brenner, 1990; Albe & Moller, 1998; Brenner et al., 2002) to describe interaction between atoms, we compute the elastic properties for the crystals. These are compared with the available experimental and theoretical values.

2. Continuum-lattice thermodynamics

A simple two- or three-dimensional crystal is one in which two or three primitive lattice vectors, vectors drawn between immediate neighbor atoms, can express all of the lattice points as shown in Figs. 1(a) and 1(b) for two-dimensional crystals and in Figs. 2(a) and 2(b) for three-dimensional crystals. Generally, these primitive lattice vectors are not sufficient to describe more complicated structures such as those depicted in Figs. 1(c), 1(d), 2(c), and 2(d). The primitive lattice vectors $\mathbf{e}_{(1)}$ and $\mathbf{e}_{(2)}$ in Fig. 1 and $\mathbf{e}_{(1)}$, $\mathbf{e}_{(2)}$ and $\mathbf{e}_{(3)}$ in Fig. 2—so called external lattice vectors—determine the external structure of the unit cell. In order to describe the internal structure of the unit cell, one or more additional lattice vectors—so called internal lattice vectors—are required, such as $\mathbf{e}_{(3)}$ in Figs. 1(c) and 1(d) or $\mathbf{e}_{(4)}$ in Figs. 2(c) and 2(d).

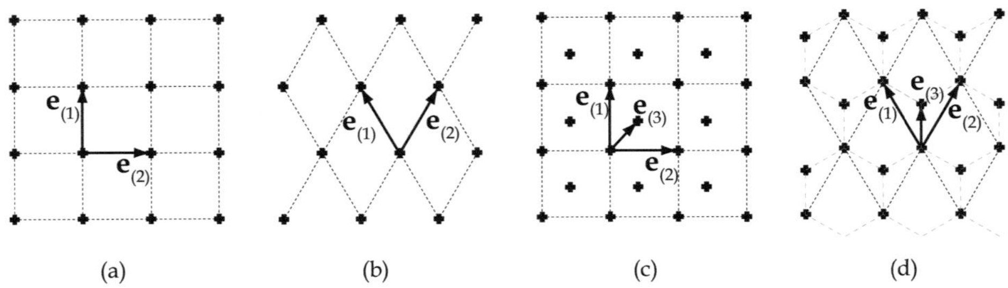

Fig. 1. Two-dimensional lattices: (a) a square Bravais lattice, (b) a diamond Bravais lattice, (c) a more general lattice having a square unit cell, and (d) a hexagonal lattice (Oh et al., 2005).

Our particular interest in what follows is complicated two- or three-dimensional crystals that are not undergoing a phase transformation. As seen in Fig. 3,
we will assume that the adjoining phases are not crystalline, and that they are not also undergoing a phase transformation.
For the adjoining non-crystalline phase (gas, liquid, or amorphous solid), we will assume that the Helmholtz free energy per unit mass is given by

$$\hat{A} = \hat{A}\left(T, \rho, \omega_{(1)}, \ldots, \omega_{(N-1)}\right). \tag{1}$$

Here T is the temperature, ρ is the total mass density, $\omega_{(A)} = \rho_{(A)}/\rho$ is the mass fraction of species A, and $\rho_{(A)}$ is the mass density of species A. For the two- or three-dimensional multicomponent crystal, let us initially assume that the surface Holmhotz free energy per mass is

$$\hat{A} = \hat{A}\left(T, \rho, \omega_{(1)}, \cdots, \omega_{(N-1)}, \mathbf{E}_{(1)}, \cdots, \mathbf{E}_{(k)}, \mathbf{e}_{(1)}, \cdots, \mathbf{e}_{(k)}\right). \tag{2}$$

Here $\mathbf{E}_{(i)}$ is a primitive lattice vector in the undeformed configuration and its length will be determined by the equilibrium bond length. Its corresponding lattice vector in the configuration deformed by an in-plane homogeneous deformation, $\mathbf{F}$, is denoted as $\mathbf{e}_{(i)}$. The deformation gradient is defined as

$$\mathbf{F} \equiv \text{grad}\, \mathbf{z}, \tag{3}$$

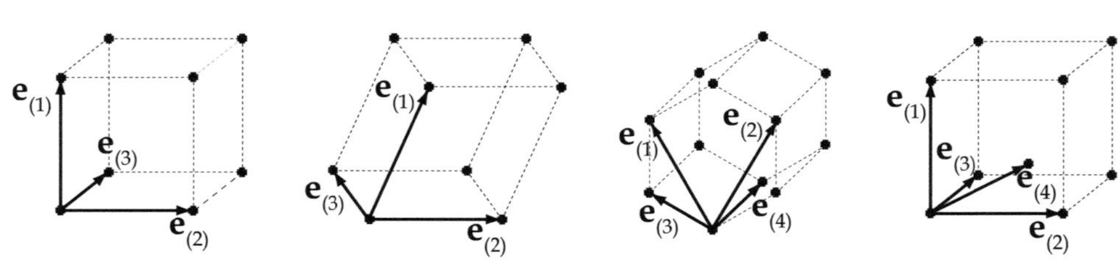

Fig. 2. Three-dimensional lattices: (a) a simple cubic lattice, (b) a rhombohedral lattice, (c) a hexagonal lattice, and (d) a more general cubic lattice having an atom inside unit cell (Oh & Slattery, 2008).

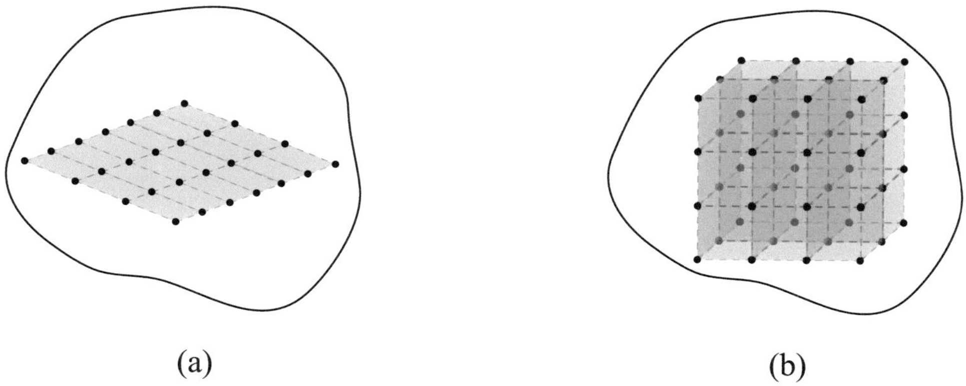

Fig. 3. An isolated body consisting of a crystalline solid and its adjoining phase: (a) two-dimensional crystal and (b) three-dimensional crystal.

where the gradient operation is performed in the undeformed configuration and $\mathbf{z}$ is a position vector. Taking a continuum point of view, we will regard these lattice vectors as being continuous functions of position on the space. Using the Born rule (Ericksen, 1992; Zanzotto, 1992; Klein, 199; James & Hane, 2000; Zhang et al., 2002), we can express the deformed external lattice vectors introduced to describe the external structure of the deformed unit cell as

$$\mathbf{e}_{(i)} \equiv \mathbf{F}\mathbf{E}_{(i)}, \qquad i=1,2, \text{ for 2-D crystals} \quad \text{or} \quad i=1,2,3 \text{ for 3-D crystals.} \tag{4}$$

It will be shown in equation (40) that the deformed internal lattice vectors $\mathbf{e}_{(i)}(i=3,\ldots,k$ for 2-D crystals or $i=4,\ldots,k$ for 3-D crystals) describing the internal structure of the unit cell are decided by minimizing the Helmholtz free energy at equilibrium.

The principle of frame indifference (Truesdell & Noll, 1965, p. 44) requires that the most general form of such a function is one in which all possible scalar products of the various lattice vectors appear (Truesdell & Noll, 1965, p. 29). We will eliminate scalar products of the form $\mathbf{E}_{(m)} \cdot \mathbf{e}_{(n)}$, since we will show later that they would lead to a non-symmetricstress tensor [see the discussion concluding with equation (38)], which we will not allow in this development. Scalar products of the form $\mathbf{E}_{(m)} \cdot \mathbf{E}_{(n)}$ may be retained in the background contributing to the behaviour of the surface. In view of this reasoning, we will write

$$\begin{aligned}\hat{A} = \hat{A}\big(&T,\rho,\omega_{(1)},\cdots,\omega_{(N-1)},\mathbf{e}_{(1)}\cdot\mathbf{e}_{(1)},\cdots,\mathbf{e}_{(1)}\cdot\mathbf{e}_{(k)},\mathbf{e}_{(2)}\cdot\mathbf{e}_{(2)},\cdots,\mathbf{e}_{(2)}\cdot\mathbf{e}_{(k)},\\ &\mathbf{e}_{(3)}\cdot\mathbf{e}_{(3)},\cdots,\mathbf{e}_{(3)}\cdot\mathbf{e}_{(k)}\big).\end{aligned} \tag{5}$$

Here all other scalar products are excluded from the dependence of $\hat{A}$ since they are not independent variables. It will be more convenient to represent the scalar products appearing in equation (5) as

$$I_{(mn)} \equiv \mathbf{e}_{(m)}\cdot\mathbf{e}_{(n)} - \mathbf{E}_{(m)}\cdot\mathbf{E}_{(n)} \qquad m=1,2,3 \text{ and } n=1,\cdots,k, \tag{6}$$

and equation (5) becomes

$$\hat{A} = \hat{A}\big(T,\rho,\omega_{(1)},\cdots,\omega_{(N-1)},I_{(11)},\cdots,I_{(1k)},I_{(22)},\cdots,I_{(2k)},I_{(33)},\cdots,I_{(3k)}\big). \tag{7}$$

Using equation (4) and the definition of the right Cauchy–Green strain tensor (Slattery et al., 2007, p. 7):

$$\mathbf{C} \equiv \mathbf{F}^{\mathrm{T}}\mathbf{F} \tag{8}$$

we can express the scalar products in equation (6) for three-dimensional crystals as

$$\begin{aligned} I_{(mn)} &= \mathbf{F}\mathbf{E}_{(m)} \cdot \mathbf{F}\mathbf{E}_{(n)} - \mathbf{E}_{(m)} \cdot \mathbf{E}_{(n)} \\ &= \mathbf{E}_{(m)} \cdot \mathbf{F}^{\mathrm{T}}\mathbf{F}\mathbf{E}_{(n)} - \mathbf{E}_{(m)} \cdot \mathbf{E}_{(n)} \\ &= \mathbf{E}_{(m)} \cdot (\mathbf{C} - \mathbf{I})\mathbf{E}_{(n)} \qquad m,n = 1,2,3. \end{aligned} \tag{9}$$

In addition, because the components of $\mathbf{C}$ are constrained by (Slattery, 1999, p. 49)

$$\sqrt{\det \mathbf{C}} = \frac{\rho_\kappa}{\rho}, \tag{10}$$

we see that ρ, $I_{(11)}$, $I_{(12)}$, $I_{(13)}$, $I_{(22)}$, $I_{(23)}$, and $I_{(33)}$ are not independent variables. For these reasons, we will write equation (7) as

$$\hat{A} = \hat{A}\big(T, \omega_{(1)}, \cdots, \omega_{(N-1)}, I_{(11)}, \cdots, I_{(1k)}, I_{(22)}, \cdots, I_{(2k)}, I_{(33)}, \cdots, I_{(3k)}\big), \tag{11}$$

for three-dimensional crystals.
In a very similar way, the Holmholtz free energy per unit mass for a two-dimensional multicomponent crystal can be expressed as

$$\hat{A} = \hat{A}\big(T, \omega_{(1)}, \cdots, \omega_{(N-1)}, I_{(11)}, I_{(12)}, I_{(22)}, I_{(13)}, \cdots, I_{(1k)}, I_{(23)}, \cdots, I_{(2k)}\big). \tag{12}$$

In order to simplify our theoretical approach, we will confine to three-dimensional crystals until we apply the results to some crystalline solids.

2.1 Euler, Gibbs, and Gibbs–Duhem equations

From the differential entropy inequality (Slattery, 1999, p. 438), we conclude

$$\hat{S} = -\left(\frac{\partial \hat{A}}{\partial T}\right)_{\omega_{(A)}, I_{(mn)}}, \tag{13}$$

where $\hat{S}$ is the entropy per unit mass. Slattery & Lagoudas (2005) have shown that

$$\left(\frac{\partial \hat{A}}{\partial \omega_{(A)}}\right)_{T, \omega_{(C \neq A)}, I_{(mn)}} = \mu_{(A)} - \mu_{(N)}, \tag{14}$$

where $\mu_{(A)}$ is the chemical potential of species A on a mass basis. We will also let (Slattery & Lagoudas, 2005)

$$\mu_{(I,mn)} \equiv \left(\frac{\partial \hat{A}}{\partial I_{(mn)}}\right)_{T, \omega_{(A)}, I_{(pq \neq mn)}}. \tag{15}$$

With these expressions, the differentiation of equation (11) can be expressed as

$$\begin{aligned} d\hat{A} &= -\hat{S}dT + \sum_{A=1}^{N-1} \left(\mu_{(A)} - \mu_{(N)}\right) d\omega_{(A)} + \sum_{m=1}^{3}\sum_{n=m}^{k} \mu_{(I,mn)} dI_{(mn)} \\ &= -\hat{S}dT + \sum_{A=1}^{N} \mu_{(A)} d\omega_{(A)} + \sum_{m=1}^{3}\sum_{n=m}^{k} \mu_{(I,mn)} dI_{(mn)}. \end{aligned} \tag{16}$$

This is referred to as the modified Gibbs equation.
In view of the Euler's equation (Slattery et al., 2007, p. 310)

$$\hat{A} = -\frac{P}{\rho} + \sum_{A=1}^{N} \mu_{(A)}\omega_{(A)}, \tag{17}$$

where the thermodynamic pressure is defined as (Slattery et al., 2007, p. 309)

$$P \equiv \rho^2 \left(\frac{\partial \hat{A}}{\partial \rho}\right)_{T,\omega_{(A)}}. \tag{18}$$

The modified Gibbs–Duhem equation follows immediately by subtracting equation (16) from the differentiation of equation (17):

$$\hat{S}dT + \frac{P}{\rho^2}d\rho - \frac{dP}{\rho} + \sum_{A=1}^{N} \omega_{(A)} d\mu_{(A)} - \sum_{m=1}^{3}\sum_{n=m}^{k} \mu_{(I,mn)} dI_{(mn)} = 0. \tag{19}$$

We would like to emphasize that the Euler's equation, the modified Gibbs equation, and the modified Gibbs–Duhem equation all apply to dynamic processes, so long as the underlying assumption about behaviour (11) is applicable.

3. Equilibrium: constraints on isolated systems

We define *equilibrium* to be achieved by an isolated body, when the entropy inequality becomes an equality. In the following sections, we wish to develop necessary and sufficient criteria for the achievement of equilibrium in the isolated body shown in Fig. 4. The following assumptions will be made

1. Once the body is isolated, it is totally enclosed by an impermeable, adiabatic boundary, the velocity of which is zero.
2. There is no mass transfer between the crystalline solid and its adjoining phase in the isolated body
3. No chemical reactions occur.

Let us begin by examining the constraints imposed upon the isolated body by the mass balance, by the momentum balance, by the energy balance, and by the entropy inequality.

3.1 Species mass balance

Since no chemical reactions occur in the isolated body, the mass balance for each species requires (Slattery et al., 2007, p. 269)

$$\frac{d}{dt}\left(\int_R \rho\omega_{(A)} dV\right) = 0. \tag{20}$$

Here dV indicates that a volume integration is to be performed. Applying the transport theorem for a multiphase body (Slattery, 1999, p. 433), we conclude

$$\begin{aligned} Z_{(A)} \equiv & \int_R \left(\rho \frac{\mathrm{d}_{(m)}\omega_{(A)}}{\mathrm{d}t} \right) \mathrm{d}V + \int_\Sigma \left[\rho\omega_{(A)}(\mathbf{v}-\mathbf{u})\cdot\boldsymbol{\xi} \right] \mathrm{d}A \\ = & \, 0. \end{aligned} \tag{21}$$

Here $\mathrm{d}_{(m)}/\mathrm{d}t$ is the derivative following a material particle within a phase (Slattery, 1999, p. 4), $\mathbf{v}$ is the mass average velocity, and $\mathbf{u}$ is the time rate of change of position following a surface point (Slattery, 1999, p. 23). The boldface brackets denote the jump quantity enclosed across the interface between phases α and β

$$[\mathbf{S}\boldsymbol{\xi}] \equiv \mathbf{S}^{(\alpha)}\boldsymbol{\xi}^{(\alpha)} + \mathbf{S}^{(\beta)}\boldsymbol{\xi}^{(\beta)}, \tag{22}$$

where $\mathbf{S}^{(\alpha)}$ is the value of the quantity $\mathbf{S}$ in phase α adjacent to the inteface Σ and $\boldsymbol{\xi}^{(\alpha)}$ is the unit normal to the interface pointing into phase α.

3.2 Momentum balance

Since the body is isolated, the sum of the forces exerted upon the body is zero, and the momentum balance requires (Slattery, 1999, p. 33):

$$\begin{aligned} \frac{\mathrm{d}}{\mathrm{d}t}\left(\int_R \rho\mathbf{v}\,\mathrm{d}V \right) &= \int_S \mathbf{T}\mathbf{n}\,\mathrm{d}A + \int_R \sum_{A=1}^{N} \rho_{(A)}\mathbf{b}_{(A)}\,\mathrm{d}V \\ &= \int_R \left(\mathrm{div}\,\mathbf{T} + \sum_{A=1}^{N} \rho_{(A)}\mathbf{b}_{(A)} \right) \mathrm{d}V + \int_\Sigma [\mathbf{T}\boldsymbol{\xi}]\,\mathrm{d}A, \end{aligned} \tag{23}$$

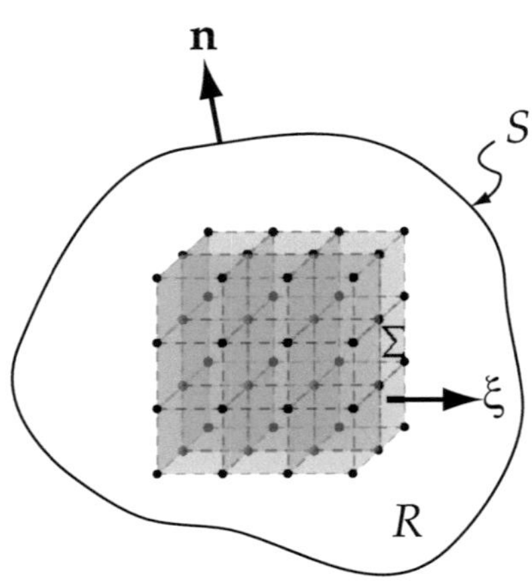

Fig. 4. An isolated body consisting of a three-dimensional crystalline solid and its adjoining phase. R indicates a region occupied by the body, S is a bounding surface of the body, and Σ denote a phase interface between the crystal and its adjoining phase. $\mathbf{n}$ and $\boldsymbol{\xi}$ are unit normal vectors to the surfaces S and Σ, respectively.

where $\mathbf{T}$ is the Cauchy stress tensor, and $\mathbf{b}_{(A)}$ is the body force per unit mass acting at each point within each phase. Here we have employed the divergence theorem (Slattery, 1999, p. 682). Again applying the transport theorem (Slattery, 1999, p. 433), we see

$$\begin{aligned}\mathbf{Z}_m \equiv & \int_R \left(\rho \frac{\mathrm{d}_{(m)}\mathbf{v}}{\mathrm{d}t} - \operatorname{div}\mathbf{T} - \sum_{A=1}^{N} \rho_{(A)}\mathbf{b}_{(A)}\right)\mathrm{d}V + \int_\Sigma [\rho\mathbf{v}(\mathbf{v}-\mathbf{u})\cdot\boldsymbol{\xi} - \mathbf{T}\boldsymbol{\xi}]\,\mathrm{d}A \\ & = 0.\end{aligned} \tag{24}$$

3.3 Energy balance

For this isolated body totally enclosed by an adiabatic boundary, the energy balance states that (Slattery et al., 2007, p. 288)

$$\begin{aligned}\frac{\mathrm{d}}{\mathrm{d}t}\left[\int_R \rho\left(\hat{U} + \frac{v^2}{2}\right)\mathrm{d}V\right] &= \int_S \mathbf{v}\cdot\mathbf{T}\mathbf{n}\,\mathrm{d}A + \int_R \mathbf{v}\cdot\sum_{A=1}^{N}\rho_{(A)}\mathbf{b}_{(A)}\,\mathrm{d}V \\ &= \int_R \left[\mathbf{v}\cdot\left(\operatorname{div}\mathbf{T} + \sum_{A=1}^{N}\rho_{(A)}\mathbf{b}_{(A)}\right) + \operatorname{tr}(\mathbf{T}\nabla\mathbf{v})\right]\mathrm{d}V + \int_\Sigma [\mathbf{v}\cdot\mathbf{T}\boldsymbol{\xi}]\,\mathrm{d}A.\end{aligned} \tag{25}$$

The contact forces do no work, since the boundary of the body is fixed in space. The boundary is adiabatic, which we interpret here as meaning that there is neither contact energy transmission with the surroundings nor external radiant energy transmission. We neglect the possibility of mutual radiant energy transmission. The time rate of change of the internal and kinetic energy of the body is the result only of work done by the body forces. We also neglect the possibility of mutual radiant energy transmission. In the second line we have once again used the transport theorem (Slattery, 1999, p. 433).

We can summarize equation (25) as

$$\begin{aligned}Z_e \equiv & \int_R \left[\rho\frac{\mathrm{d}_{(m)}(\hat{U} + v^2/2)}{\mathrm{d}t} - \mathbf{v}\cdot\left(\operatorname{div}\mathbf{T} + \sum_{A=1}^{N}\rho_{(A)}\mathbf{b}_{(A)}\right) - \operatorname{tr}(\mathbf{T}\nabla\mathbf{v})\right. \\ & \left. - \sum_{A=1}^{N}\rho_{(A)}(\mathbf{v}_{(A)} - \mathbf{v})\cdot\mathbf{b}_{(A)}\right]\mathrm{d}V + \int_\Sigma \left[\rho\left(\hat{U} + \frac{v^2}{2}\right)(\mathbf{v}-\mathbf{u})\cdot\boldsymbol{\xi} - \mathbf{v}\cdot\mathbf{T}\boldsymbol{\xi}\right]\mathrm{d}A \\ & = 0.\end{aligned} \tag{26}$$

3.4 Entropy inequality

For the isolated body under consideration here, the entropy inequality says that the time rate of change of the body's entropy must be greater than or equal to zero (Slattery et al., 2007, p. 295):

$$\frac{\mathrm{d}}{\mathrm{d}t}\left(\int_R \rho\hat{S}\,\mathrm{d}V\right) \geq 0. \tag{27}$$

Equilibrium is achieved, when this inequality becomes an equality. Applying the transport theorem again (Slattery, 1999, p. 433), we find that this may also be written as

$$\int_R \rho\frac{\mathrm{d}_{(m)}\hat{S}}{\mathrm{d}t}\,\mathrm{d}V + \int_\Sigma [\rho\hat{S}(\mathbf{v}-\mathbf{u})\cdot\boldsymbol{\xi}]\,\mathrm{d}A \geq 0. \tag{28}$$

4. Implications of equilibrium

As explained above, if equilibrium is to be achieved, the left side of equation (27) must be minimized and approaches to zero within the constraints imposed by conservation of mass for each species, by the momentum balance, and by the energy balance as developed in the prior section.
In view of equations (21), (24), and (26), there is no loss in generality in writing equation (28) as (Slattery et al., 2007, p. 391)

$$\int_R \rho \frac{\mathrm{d}_{(m)}\hat{S}}{\mathrm{d}t}\,\mathrm{d}V + \int_\Sigma \left[\rho\hat{S}(\mathbf{v}-\mathbf{u})\cdot\boldsymbol{\xi}\right]\mathrm{d}A + \sum_{A=1}^{N}\lambda_{(A)}Z_{(A)} + \boldsymbol{\lambda}_m\cdot\mathbf{Z}_m + \lambda_e Z_e \geq 0, \tag{29}$$

where $\lambda_{(A)}$ and λ_e are constants or Lagrangian multipliers, and $\boldsymbol{\lambda}_m$ is a constant spatial vector, the components of which are Lagrangian multipliers.
From the modified Gibbs equation (16) and the definition of $\hat{A} \equiv \hat{U} - T\hat{S}$ (Slattery et al., 2007, p. 305), we see that

$$\frac{\mathrm{d}_{(m)}\hat{S}}{\mathrm{d}t} = \frac{1}{T}\frac{\mathrm{d}_{(m)}\hat{U}}{\mathrm{d}t} - \frac{1}{T}\sum_{A=1}^{N}\mu_{(A)}\frac{\mathrm{d}\omega_{(A)}}{\mathrm{d}t} - \frac{1}{T}\sum_{m=1}^{3}\sum_{n=m}^{k}\mu_{(I,mn)}\frac{\mathrm{d}_{(m)}I_{(mn)}}{\mathrm{d}t}. \tag{30}$$

We will also need from equation (6)

$$\begin{aligned}\frac{\mathrm{d}_{(m)}I_{(mn)}}{\mathrm{d}t} &= \frac{\mathrm{d}_{(m)}\mathbf{e}_{(m)}}{\mathrm{d}t}\cdot\mathbf{e}_{(n)} + \mathbf{e}_{(m)}\cdot\frac{\mathrm{d}_{(m)}\mathbf{e}_{(n)}}{\mathrm{d}t}\\ &= \frac{\mathrm{d}_{(m)}\mathbf{F}}{\mathrm{d}t}\mathbf{E}_{(m)}\cdot\mathbf{e}_{(n)} + \mathbf{e}_{(m)}\cdot\frac{\mathrm{d}_{(m)}\mathbf{F}}{\mathrm{d}t}\mathbf{E}_{(n)}\\ &= \nabla\mathbf{v}\mathbf{e}_{(m)}\cdot\mathbf{e}_{(n)} + \mathbf{e}_{(m)}\cdot\nabla\mathbf{v}\mathbf{e}_{(n)}\\ &= \mathrm{tr}\left[\left(\mathbf{e}_{(m)}\otimes\mathbf{e}_{(n)} + \mathbf{e}_{(n)}\otimes\mathbf{e}_{(m)}\right)\nabla\mathbf{v}^{\mathrm{T}}\right] \qquad m,n=1,2,3.\end{aligned} \tag{31}$$

Here we used

$$\frac{\mathrm{d}_{(m)}\mathbf{F}}{\mathrm{d}t} = (\nabla\mathbf{v})\mathbf{F}, \tag{32}$$

and $\mathbf{e}_{(i)}\otimes\mathbf{e}_{(j)}$ is the tensor product or dyadic product of two vectors $\mathbf{e}_{(i)}$ and $\mathbf{e}_{(j)}$.
After rearranging equation (29) by means of equations (21) through (31), we have

$$\begin{aligned}&\int_R \Bigg\{\rho\left(\frac{1}{T}+\lambda_e\right)\frac{\mathrm{d}_{(m)}\hat{U}}{\mathrm{d}t} + \sum_{A=1}^{N}\rho\left(-\frac{\mu_{(A)}}{T}+\lambda_{(A)}\right)\frac{\mathrm{d}\omega_{(A)}}{\mathrm{d}t}\\ &\quad - \sum_{A=1}^{N}\rho_{(A)}\lambda_e(\mathbf{v}_{(A)}-\mathbf{v})\cdot\mathbf{b}_{(A)} - \frac{\rho}{T}\sum_{m=1}^{3}\sum_{n=4}^{k}\mu_{(I,mn)}\frac{\mathrm{d}_{(m)}I_{(mn)}}{\mathrm{d}t}\\ &\quad - \frac{1}{T}\mathrm{tr}\left[\left(T\lambda_e\mathbf{T} + \rho\sum_{m=1}^{3}\sum_{n=m}^{3}\mu_{(I,mn)}\left(\mathbf{e}_{(m)}\otimes\mathbf{e}_{(n)}+\mathbf{e}_{(n)}\otimes\mathbf{e}_{(m)}\right)\right)\nabla\mathbf{v}^{\mathrm{T}}\right]\Bigg\}\mathrm{d}V\\ &+ \int_\Sigma \Bigg[\rho\left\{\left(\frac{1}{T}+\lambda_e\right)\hat{U} + \frac{P}{T\rho} - \sum_{A=1}^{N}\left(\frac{\mu_{(A)}\omega_{(A)}}{T} - \lambda_{(A)}\omega_{(A)}\right) - \frac{\lambda_e}{2}v^2\right\}(\mathbf{v}-\mathbf{u})\cdot\boldsymbol{\xi}\\ &\quad + \rho(\boldsymbol{\lambda}_m + \lambda_e\mathbf{v})\cdot\left[\mathbf{v}\otimes(\mathbf{v}-\mathbf{u}) - \mathbf{T}\right]\boldsymbol{\xi}\Bigg]\mathrm{d}A\\ &\geq 0.\end{aligned} \tag{33}$$

In arriving at this result, we have recognized the differential momentum balance (Slattery, 1999, p. 434).

Necessary and sufficient conditions for equation (33) to be satisfied in the presence of small perturbations to the system are

$$T = -\frac{1}{\lambda_e} = \text{a constant}, \tag{34}$$

$$\mu_{(A)} = T\lambda_{(A)} \quad \text{for each species } A = 1,2,\cdots,N, \tag{35}$$

$$\mathbf{v}_{(A)} = \mathbf{v} \quad \text{for each species } A = 1,2,\cdots,N, \tag{36}$$

$$\mu_{(I,mn)} = 0 \quad \text{for } m = 1,2,3 \text{ and } n = 4,\cdots,k, \tag{37}$$

$$\mathbf{T} = \rho \sum_{m=1}^{3} \sum_{n=m}^{3} \mu_{(I,mn)} \left(\mathbf{e}_{(m)} \otimes \mathbf{e}_{(n)} + \mathbf{e}_{(n)} \otimes \mathbf{e}_{(m)} \right), \tag{38}$$

and

$$\text{on } \Sigma: \quad \mathbf{v} = \mathbf{u} = T\boldsymbol{\lambda}_m. \tag{39}$$

It is important to recognize that equation (38) is the description of stress-deformation behaviour of three-dimensional elastic crystalline solids. From equations (15) and (37), we can express that

$$\left(\frac{\partial \hat{A}}{\partial I_{(mn)}} \right)_{T,\omega_{(A)},I_{(pq \neq mn)}} = 0 \qquad m = 1,2,3 \text{ and } n = 4,\cdots,k. \tag{40}$$

These relations determine $\mathbf{e}_{(4)},\ldots,\mathbf{e}_{(k)}$ at equilibrium. Notice that, if we had retained scalar products of the form $\mathbf{E}_{(m)} \cdot \mathbf{e}_{(n)}$ in equation (5), the stress tensor would have been non-symmetric.

For a two-dimensional crystalline solid, similar results can be obtained (Oh et al., 2005) as follows:

$$\mathbf{T}^{(\sigma)} = \rho^{(\sigma)} \sum_{m=1}^{2} \sum_{n=m}^{2} \mu_{(I,mn)} \left(\mathbf{e}_{(m)} \otimes \mathbf{e}_{(n)} + \mathbf{e}_{(n)} \otimes \mathbf{e}_{(m)} \right), \tag{41}$$

and

$$\left(\frac{\partial \hat{A}}{\partial I_{(mn)}} \right)_{T,\omega_{(A)},I_{(pq \neq mn)}} = 0 \qquad m = 1,2 \text{ and } n = 3,\cdots,k. \tag{42}$$

Here the superscript (σ) denotes values assigned to a phase interphase.

5. Stress-deformation behaviour at equilibrium in the limit of infinitesimal deformations

Since it is common to consider infinitesimal deformations, let us consider how equation (38) reduces in this limit:

$$\begin{aligned}
\mathbf{T} &= \rho \sum_{m=1}^{3} \sum_{n=m}^{3} \mu_{(I,mn)} \left(\mathbf{e}_{(m)} \otimes \mathbf{e}_{(n)} + \mathbf{e}_{(n)} \otimes \mathbf{e}_{(m)} \right) \\
&= \rho \sum_{m=1}^{3} \sum_{n=m}^{3} \mu_{(I,mn)} \left(\mathbf{FE}_{(m)} \otimes \mathbf{FE}_{(n)} + \mathbf{FE}_{(n)} \otimes \mathbf{FE}_{(m)} \right) \\
&= \rho \sum_{m=1}^{3} \sum_{n=m}^{3} \mu_{(I,mn)} \mathbf{F} \left(\mathbf{E}_{(m)} \otimes \mathbf{E}_{(n)} + \mathbf{E}_{(n)} \otimes \mathbf{E}_{(m)} \right) \mathbf{F}^T.
\end{aligned} \tag{43}$$

Define

$$\mathbf{w} \equiv \mathbf{z} - \mathbf{z}_{\mathcal{K}}, \tag{44}$$

to be the displacement vector, and $\mathbf{z}$ and $\mathbf{z}_{\mathcal{K}}$ are the position vectors of a material particle in the deformed and undeformed configurations. It follows that the *displacement gradient* is

$$\mathbf{H} \equiv \operatorname{grad}\mathbf{w} = \mathbf{F} - \mathbf{I}. \tag{45}$$

Let ϵ be a very small dimensionless variable characterizing an infinitesimal deformation process. We will seek a solution of the form

$$\mathbf{w} = \mathbf{w}_{(0)} + \epsilon\mathbf{w}_{(1)} + \epsilon^2\mathbf{w}_{(2)} + \cdots. \tag{46}$$

Let us recognize that in the absence of an infinitesimal deformation or as $\epsilon \to 0$

$$\mathbf{w}_{(0)} = 0. \tag{47}$$

This means that

$$\mathbf{H} = \epsilon\mathbf{H}_{(1)} + \epsilon^2\mathbf{H}_{(2)} + \cdots, \quad \text{where } \mathbf{H}_{(i)} = \operatorname{grad}\mathbf{w}_{(i)}, \tag{48}$$

and thus the right Cauchy-Green strain tensor becomes

$$\begin{aligned}
\mathbf{C} &\equiv \mathbf{F}^T\mathbf{F} \\
&= (\mathbf{I} + \epsilon\mathbf{H}_{(1)} + \cdots)^T(\mathbf{I} + \epsilon\mathbf{H}_{(1)} + \cdots) \\
&= \mathbf{I} + \epsilon\left(\mathbf{H}_{(1)} + \mathbf{H}_{(1)}{}^T\right) + O\left(\epsilon^2\right).
\end{aligned} \tag{49}$$

From (10) and (49), we find that

$$\begin{aligned}
\rho &= \frac{\rho_{\mathcal{K}}}{\sqrt{\det\mathbf{C}}} \\
&= \frac{\rho_{\mathcal{K}}}{\sqrt{\det\left[\mathbf{I} + \epsilon(\mathbf{H}_{(1)} + \mathbf{H}_{(1)}{}^T)\right] + \cdots}} \\
&= \frac{\rho_{\mathcal{K}}}{\det\mathbf{I}} - \epsilon\frac{\rho_{\mathcal{K}}}{2\,(\det\mathbf{I})^2}\operatorname{tr}\left(\mathbf{H}_{(1)} + \mathbf{H}_{(1)}{}^T\right) + \cdots \\
&= \rho_{\mathcal{K}} - \epsilon\frac{\rho_{\mathcal{K}}}{2}\operatorname{tr}\left(\mathbf{H}_{(1)} + \mathbf{H}_{(1)}{}^T\right) + \cdots.
\end{aligned} \tag{50}$$

Since, in the limit of small deformations, the strain

$$\begin{aligned}
\boldsymbol{\varepsilon} &\equiv \frac{1}{2}\left[\operatorname{grad}\mathbf{w} + (\operatorname{grad}\mathbf{w})^T\right] = \epsilon\frac{1}{2}\left(\mathbf{H}_{(1)} + \mathbf{H}_{(1)}{}^T\right) + \cdots \\
&= \epsilon\boldsymbol{\varepsilon}_{(1)} + \cdots,
\end{aligned} \tag{51}$$

which allows us to also express equation (50) as

$$\rho = \rho_{\mathcal{K}}\left(1 - \epsilon\operatorname{tr}\boldsymbol{\varepsilon}_{(1)}\right) + \cdots. \tag{52}$$

Equations (45), (48), and (52) also permit us to rewrite equation (43) to the first order in ϵ as (for the moment not addressing the order of $\mu_{(I,mn)}$)

$$\mathbf{T} = \rho_\kappa \left(1 - \epsilon \operatorname{tr} \boldsymbol{\varepsilon}_{(1)}\right) \sum_{m=1}^{3} \sum_{n=m}^{3} \mu_{(I,mn)} \left(\mathbf{I} + \epsilon \mathbf{H}_{(1)}\right) \left(\mathbf{E}_{(m)} \otimes \mathbf{E}_{(n)} + \mathbf{E}_{(n)} \otimes \mathbf{E}_{(m)}\right) \left(\mathbf{I} + \epsilon \mathbf{H}_{(1)}\right)^T$$
$$= \rho_\kappa \left(1 - \epsilon \operatorname{tr} \boldsymbol{\varepsilon}_{(1)}\right) \sum_{m=1}^{3} \sum_{n=m}^{3} \mu_{(I,mn)} \left(\mathbf{E}_{(m)} \otimes \mathbf{E}_{(n)} + \mathbf{E}_{(n)} \otimes \mathbf{E}_{(m)}\right) + \rho_\kappa \epsilon \sum_{m=1}^{3} \sum_{n=m}^{3} \mu_{(I,mm)}$$
$$\times \left\{\mathbf{H}_{(1)} \left(\mathbf{E}_{(m)} \otimes \mathbf{E}_{(n)} + \mathbf{E}_{(n)} \otimes \mathbf{E}_{(m)}\right) + \left[\mathbf{H}_{(1)} \left(\mathbf{E}_{(m)} \otimes \mathbf{E}_{(n)} + \mathbf{E}_{(n)} \otimes \mathbf{E}_{(m)}\right)\right]^T\right\}. \tag{53}$$

If the reference(or undeformed) configuration is a stress-free configuration,

$$0 = \rho_\kappa \sum_{m=1}^{3} \sum_{n=m}^{3} \mu_{(I,mn)0} \left(\mathbf{E}_{(m)} \otimes \mathbf{E}_{(n)} + \mathbf{E}_{(n)} \otimes \mathbf{E}_{(m)}\right), \tag{54}$$

where the $\mu_{(I,mn)0}$ are evaluated in the reference configuration. This allows us to write equation (53) as

$$\begin{aligned}\mathbf{T} = \rho_\kappa &\sum_{m=1}^{3} \sum_{n=m}^{3} \left(\mu_{(I,mn)} - \mu_{(I,mn)0}\right) \left(\mathbf{E}_{(m)} \otimes \mathbf{E}_{(n)} + \mathbf{E}_{(n)} \otimes \mathbf{E}_{(m)}\right)\\ &- \rho_\kappa \epsilon \operatorname{tr} \boldsymbol{\varepsilon}_{(1)} \sum_{m=1}^{3} \sum_{n=m}^{3} \mu_{(I,mn)} \left(\mathbf{E}_{(m)} \otimes \mathbf{E}_{(n)} + \mathbf{E}_{(n)} \otimes \mathbf{E}_{(m)}\right)\\ &+ \rho_\kappa \epsilon \sum_{m=1}^{3} \sum_{n=m}^{3} \mu_{(I,mn)} \Big\{\mathbf{H}_{(1)} \left(\mathbf{E}_{(m)} \otimes \mathbf{E}_{(n)} + \mathbf{E}_{(n)} \otimes \mathbf{E}_{(m)}\right)\\ &\quad + \left[\mathbf{H}_{(1)} \left(\mathbf{E}_{(m)} \otimes \mathbf{E}_{(n)} + \mathbf{E}_{(n)} \otimes \mathbf{E}_{(m)}\right)\right]^T\Big\}.\end{aligned} \tag{55}$$

5.1 Implication of a quadratic equation of state

As a special case, let us assume that $\hat{A}$ can be represented as a quadratic function of $I_{(11)}$, $I_{(12)}$, $I_{(12)}$, $I_{(13)}$, $I_{(22)}$, $I_{(23)}$, and $I_{(33)}$ (Truesdell & Noll, 1965, pp. 311-312):

$$\hat{A} = a_{(0)} + \sum_{i=1}^{3} \sum_{j=i}^{3} a_{(ij)} I_{(ij)} + \frac{1}{2} \sum_{i=1}^{3} \sum_{j=i}^{3} \sum_{m=1}^{3} \sum_{n=m}^{3} a_{(ijmn)} I_{(ij)} I_{(mn)} + \cdots. \tag{56}$$

It should be understood here that $a_{(0)}$, $a_{(ij)}$, and $a_{(ijmn)}$ are functions of temperature and composition with the additional restrictions

$$\begin{aligned} a_{(ij)} &= a_{(ji)} \\ &= \left.\frac{\partial \hat{A}}{\partial I_{(ij)}}\right|_{I_{(pq)}=0}, \end{aligned} \tag{57}$$

and

$$\begin{aligned} a_{(ijmn)} &= a_{(mnij)} \\ &= \left.\frac{\partial^2 \hat{A}}{\partial I_{(ij)} \partial I_{(mn)}}\right|_{I_{(pq)}=0}. \end{aligned} \tag{58}$$

For this special case, from equation (15)

$$\mu_{(I,mn)} = a_{(mn)} + \sum_{i=1}^{3}\sum_{j=i}^{3} a_{(mnij)} I_{(ij)}, \tag{59}$$

and

$$\mu_{(I,mn)} - \mu_{(I,mn)0} = \sum_{i=1}^{3}\sum_{j=i}^{3} a_{(mnij)} I_{(ij)}. \tag{60}$$

From equations (9), (49), and (51), we see that

$$\begin{aligned} I_{(ij)} &= \mathbf{E}_{(i)} \cdot (\mathbf{C} - \mathbf{I})\,\mathbf{E}_{(j)} \\ &= 2\epsilon\,\mathbf{E}_{(i)} \cdot \boldsymbol{\varepsilon}_{(1)} \mathbf{E}_{(j)}. \end{aligned} \tag{61}$$

To the first order in ϵ, we find that equation (55) reduces to

$$\begin{aligned} \mathbf{T} = {} & 2\rho_\kappa \epsilon \sum_{m=1}^{3}\sum_{n=m}^{3}\sum_{i=1}^{3}\sum_{j=i}^{3} \left[a_{(ijmn)} \mathbf{E}_{(i)} \cdot \boldsymbol{\varepsilon}_{(1)} \mathbf{E}_{(j)} \left(\mathbf{E}_{(m)} \otimes \mathbf{E}_{(n)} + \mathbf{E}_{(n)} \otimes \mathbf{E}_{(m)} \right) \right] \\ & - \rho_\kappa \epsilon\, \mathrm{tr}\, \boldsymbol{\varepsilon}_{(1)} \sum_{m=1}^{3}\sum_{n=m}^{3} a_{(mn)} \left(\mathbf{E}_{(m)} \otimes \mathbf{E}_{(n)} + \mathbf{E}_{(n)} \otimes \mathbf{E}_{(m)} \right) \\ & + \rho_\kappa \epsilon \sum_{m=1}^{3}\sum_{n=m}^{3} a_{(mn)} \left\{ \mathbf{H}_{(1)} \left(\mathbf{E}_{(m)} \otimes \mathbf{E}_{(n)} + \mathbf{E}_{(n)} \otimes \mathbf{E}_{(m)} \right) \right. \\ & \left. + \left[\mathbf{H}_{(1)} \left(\mathbf{E}_{(m)} \otimes \mathbf{E}_{(n)} + \mathbf{E}_{(n)} \otimes \mathbf{E}_{(m)} \right) \right]^T \right\}. \end{aligned} \tag{62}$$

Using equations (54) and (59), we see that

$$a_{(mn)} = 0. \tag{63}$$

From this, equation (62) reduces to

$$\mathbf{T} = 2\rho_\kappa \epsilon \sum_{m=1}^{3}\sum_{n=m}^{3}\sum_{i=1}^{3}\sum_{j=i}^{3} a_{(ijmn)}\, \mathbf{E}_{(i)} \cdot \boldsymbol{\varepsilon}_{(1)} \mathbf{E}_{(j)} \left(\mathbf{E}_{(m)} \otimes \mathbf{E}_{(n)} + \mathbf{E}_{(n)} \otimes \mathbf{E}_{(m)} \right). \tag{64}$$

5.2 The helmholtz free energy

The internal energy is composed of the kinetic energy associated with molecular motions (translation, rotation, vibration) and the potential energy associated with intermolecular potentials between atoms or molecules (Tester & Modell, 1996, p. 33). Since the molecular kinetic energy is a constant for an isothermal system, the internal energy takes the form

$$\hat{U} = \hat{\Phi} + constant, \tag{65}$$

where $\hat{\Phi}$ is the intermolecular potential energy.
Since $\hat{A} = \hat{U} - T\hat{S}$ and at equilibrium no changes in temperature or entropy occur, the specfic Helmholtz free energy can be written as

$$\hat{A} = \hat{\Phi} + constant. \tag{66}$$

This represents the relationship between the Helmholtz free energy and the intermolecular potential energy at equilibrium.
We will employ two types of potentials, Tersoff (1988; 1989) and modified Tersoff potentials (Brenner, 1990; Albe & Moller, 1998; Brenner et al., 2002), to describe the interatomic potentials for the crystalline solids. However, the choice of the interatomic potential is rather arbitrary and is dependent of atoms consisting of the crystals.

5.2.1 Tersoff potential

The Tersoff (1988; 1989) potential between atoms i and j separated by the distance, r_{ij}, has the form

$$\Phi_{ij} = f_c(r_{ij})[\Phi_R(r_{ij}) + b_{ij}(r_{ik}, \theta_{ijk})\Phi_A(r_{ij})], \tag{67}$$

where θ_{ijk} is the angle between bonds i-j and i-k ($k \neq j$). The functions Φ_R, Φ_A, and b_{ij} represent repulsive and attractive pair potentials, and a multi-body coupling between bonds i-j and i-k:

$$\begin{aligned} \Phi_R(r) &= A\exp(-\lambda r), \\ \Phi_A(r) &= -B\exp(-\mu r), \end{aligned} \tag{68}$$

and

$$\begin{aligned} b_{ij} &= \chi\left\{1 + \beta^n \left[\sum_{k\neq i,j} f_c(r_{ik}) g(\theta_{ijk})\,\omega_{ijk}\right]^n\right\}^{-1/2n} \\ g(\theta_{ijk}) &= 1 + \frac{c^2}{d^2} - \frac{c^2}{d^2 + (h - \cos\theta_{ijk})^2}. \end{aligned} \tag{69}$$

The cutoff function is given by

$$f_c(r_{ij}) = \begin{cases} 1, & r < R_1, \\ \dfrac{1}{2} + \dfrac{1}{2}\cos\left[\dfrac{\pi(r - R_1)}{R_2 - R_1}\right], & R_1 \leq r \leq R_2, \\ 0, & r > R_2. \end{cases} \tag{70}$$

Tersoff (1988; 1989) has obtained the sets of parameters for diamond and silicon in Table 1 to fit data given by Yin & Cohen (1983a;b). The sets of the Tersoff potential parameters for boron and nitrogen have been reported by Matsunaga et al. (2000) and Kroll (1996), respectively.
Tersoff (1989) also proposed that the parameters for multicomponent systems such as silicon carbide could be estimated using the following rules:

$$\begin{aligned} &A_{\text{Si-C}} = \sqrt{A_{\text{Si}} A_{\text{C}}}, \quad B_{\text{Si-C}} = \sqrt{B_{\text{Si}} B_{\text{C}}}, \quad \lambda_{\text{Si-C}} = \frac{\lambda_{\text{Si}} + \lambda_{\text{C}}}{2}, \\ &\mu_{\text{Si-C}} = \frac{\mu_{\text{Si}} + \mu_{\text{C}}}{2}, \quad R_{1,\text{Si-C}} = \sqrt{R_{1,\text{Si}} R_{1,\text{C}}}, \quad R_{2,\text{Si-C}} = \sqrt{R_{2,\text{Si}} R_{2,\text{C}}}. \end{aligned} \tag{71}$$

The coefficients in b_{ij} for the multicomponent systems were chosen by the atom corresponding to the first index i. The fitting parameter χ and the C–Si bond length for silicon carbide were $d_0 = 0.9776$ and 1.87 Å, respectively. For boron nitride, Matsunaga et al. (2000) used $\chi = 1.1593$ and $d_0 = 1.584$ Å. The C–C bond length in diamond and the Si–Si bond length in silicon were 1.54 Å and 2.35 Å, respectively.

Parameters	Diamond	Silicon	Boron[a]	Nitrogen[b]
A (eV)	1393.6	1830.8	277.02	11000
B (eV)	346.7	471.2	83.49	219.45
λ (Å^{-1})	3.4879	2.4799	1.9922	5.7708
μ (Å^{-1})	2.2119	1.7322	1.5859	2.5115
β (10^{-7})	1.5724	11.0	16.00	1056200
n	0.72751	0.78734	3.9929	12.4498
c	38049	100390	0.52629	79934
d	4.384	16.217	0.001587	134.32
h	-0.57058	-0.59825	0.5	-0.9973
R_1 (Å)	1.8	2.7	1.8	2.0
R_2 (Å)	2.1	3.0	2.1	2.3
χ	1	1	1	1
ω_{ijk}	1	1	1	1

[a]parameters proposed by Matsunaga et al. (2000)
[b]parameters proposed by Kroll (1996)

Table 1. Tersoff potential parameters for C-C in diamond, Si-Si in silicon, B-B in boron, and N-N in nitrogen.

5.2.2 Modified tersoff potentials

A Tersoff-like potential proposed by Brenner (1990) has the same form as the Tersoff potential, except for repulsive and attractive pair potentials:

$$\begin{aligned}\Phi_R(r) &= \frac{D_0}{S-1}\exp\left[-\tau\sqrt{2S}\,(r-r_0)\right],\\ \Phi_A(r) &= -\frac{SD_0}{S-1}\exp\left[-\tau\sqrt{2/S}\,(r-r_0)\right],\end{aligned} \tag{72}$$

where D_0 and r_0 are the dimer energy and separation, respectively. The multi-body coupling function has been slightly changed to

$$\begin{aligned}b_{ij} &= \left[1+\sum_{k\neq i,j} f_c(r_{ik})g(\theta_{ijk})\right]^{-\delta},\\ g(\theta_{ijk}) &= a\left[1+\frac{c^2}{d^2}-\frac{c^2}{d^2+(1+\cos\theta_{ijk})^2}\right].\end{aligned} \tag{73}$$

This potential was employed to describe the interatomic interaction in CNTs and BNNTs including their plane sheets (Oh et al., 2005; Oh, 2010). Some of the potential parameters were adjusted from the original ones provided by Brenner (1990) to fit well the bond length and the cohesive energy of a boron nitride(BN) sheet given in the literature (Moon & Hwang, 2004; Verma et al., 2007). These are listed in Table 2.

Parameters	B. for C-C[c]	T.-B. for B-N[d]	T.-A.[e] for B-N
D_0 (eV)	10953.544	6.36	6.36
S	-	1.0769	1.0769
τ ($Å^{-1}$)	4.747	2.20	2.0431
r_0 (Å)	0.313	1.33	1.33
β (10^{-6})	-	-	11.134
δ	0.5	0.382	-
n	-	-	0.36415
a (10^{-4})	2.0813	2.0813	-
c	330	330	1092.93
d	3.5	3.5	12.38
h	-	-	-0.5413
λ_3	-	-	1.9925
ω	1	1	-
R_1 (Å)	1.7	1.9	1.9
R_2 (Å)	2.0	2.1	2.1

[c]parameters proposed by Brenner et al. (2002)
[d]parameters proposed by Oh (2010)
[e]parameters proposed by Albe & Moller (1998)

Table 2. Tersoff-like potential parameters for the interaction between atoms.

Albe & Moller (1998) proposed another Tersoff-like potential whose repulsive and attractive pair potentials are the same as (72). The parameter, ω_{ijk}, in the multi-body coupling function (69) was expressed as

$$\omega_{ijk} = \exp\left[{\lambda_3}^3 (r_{ij} - r_{ik})^3\right]. \tag{74}$$

Their potential parameters for boron-nitrogen interactions are also listed in Table 2. Using this potential, Koga et al. (2001) have determined the bond lengths for a cubic BN to be 1.555 Å. More recently, Brenner et al. (2002) proposed a similar potential to the Tersoff-Brenner potential with new repulsive and attractive pair potentials:

$$\begin{aligned} \Phi_R(r) &= (1 + r_0/r) D_0 e^{-\tau r} f_c(r), \\ \Phi_A(r) &= \sum_{n=1}^{3} B_n e^{-\beta_n r}, \end{aligned} \tag{75}$$

where

$$\begin{aligned} &B_1 = 12388.791 \text{ eV}, \quad B_2 = 17.567 \text{ eV}, \quad B_3 = 30.715 \text{ eV}, \\ &\beta_1 = 4.720 \text{ Å}^{-1}, \quad \beta_2 = 1.433 \text{ Å}^{-1}, \quad \beta_3 = 1.383 \text{ Å}^{-1}. \end{aligned} \tag{76}$$

The rest of parameters for C-C bond in graphite are described in Table 2.
We simply use B, TB, and TA as the three sets of Tersoff-like potential parameters, proposed by Brenner et al. (2002), Oh (2010). and Albe & Moller (1998).

6. Elastic properties

Now we will calculate the coefficients $a_{(ijmn)}$ in the stress-strain relation (64) for three-dimensional crystalline solids with a diamond-like structure, and for two-dimensional crystals with a graphene-like structure. Through these coefficients, the elastic properties of them will be obtained and compared to previous experimental and theoretical results.

6.1 Three-dimensional crystals– diamond, silicon and silicon-carbide

As shown in Fig. 5, diamond, silicon, silicon carbide, or boron nitride is composed of a face-centred cubic array with the centre of each tetrahedron filled by carbon or silicon, and four lattice vectors $\mathbf{E}_{(i)}$ $(i = 1, \cdots, 4)$ must be introduced to describe their structures. Here we consider cubic types of silicon carbide and boron nitride, not hexagonal types. The primitive lattice vectors in the undeformed diamond-like structure, $\mathbf{E}_{(i)}$, will be deformed into $\mathbf{e}_{(i)}$ $(i = 1, \cdots, 4)$ by an in-plane homogeneous deformation $\mathbf{F}$. The deformed lattice vectors $\mathbf{e}_{(1)}$, $\mathbf{e}_{(2)}$, and $\mathbf{e}_{(3)}$ determine the external structure of the deformed unit cell, while $\mathbf{e}_{(4)}$ determines its internal structure.

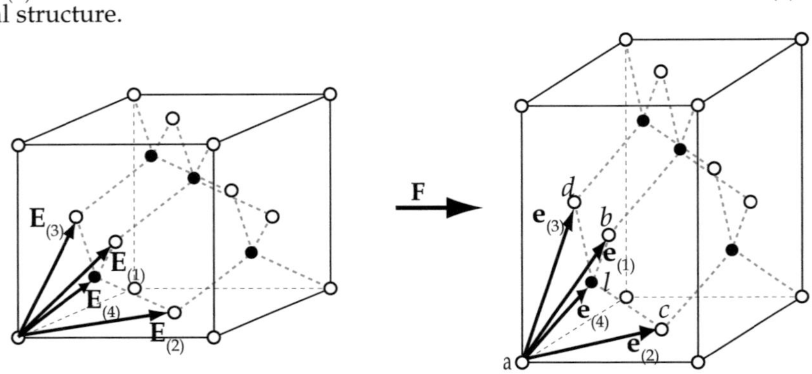

Fig. 5. Schematic of (a) an undeformed diamond-like structure and (b) its deformed structure by a homogeneous deformation $\mathbf{F}$: diamond (∘: C atom, •: C atom), silicon (∘: Si atom, •: Si atom), silicon carbide (∘: Si atom, •: C atom), or boron nitride (∘: B atom, •: N atom). The filled circles indicate centre atoms, while the open ones are atoms on each face.

The stress-deformation behaviour of the diamond-like crystals at constant temperature becomes

$$\mathbf{T} = 2\rho_\kappa \sum_{m=1}^{3} \sum_{n=m}^{3} \sum_{i=1}^{3} \sum_{j=i}^{3} a_{(ijmn)} \mathbf{E}_{(i)} \cdot \boldsymbol{\varepsilon} \mathbf{E}_{(j)} \left(\mathbf{E}_{(m)} \otimes \mathbf{E}_{(n)} + \mathbf{E}_{(n)} \otimes \mathbf{E}_{(m)} \right). \tag{77}$$

In view of equation (58) and (66), the coefficient $a_{(ijmn)}$ can be determined by

$$a_{(ijmn)} = \left. \frac{\partial^2 \hat{\Phi}}{\partial I_{(ij)} \partial I_{(mn)}} \right|_{I_{(pq)}=0}, \tag{78}$$

and we require at equilibrium by equations (40) and (66) that

$$\left(\frac{\partial \hat{\Phi}}{\partial I_{(14)}} \right)_{T, I_{(pq \neq 14)}} = \left(\frac{\partial \hat{\Phi}}{\partial I_{(24)}} \right)_{T, I_{(pq \neq 24)}} = \left(\frac{\partial \hat{\Phi}}{\partial I_{(34)}} \right)_{T, I_{(pq \neq 34)}} = 0. \tag{79}$$

	$a_{(1111)}$	$a_{(1122)}$	$a_{(1112)}$	$a_{(1123)}$	$a_{(1212)}$	$a_{(1213)}$
Diamond	1.282	0.146	-0.854	0.281	2.464	-0.378
Silicon	0.124	0.044	-0.083	-0.003	0.160	0.003
Silicon carbide	0.453	0.088	-0.302	0.063	0.745	-0.071
Boron nitride	0.598	0.295	-0.399	-0.096	0.654	0.072

$a_{(ijmn)} = a_{(mnij)}$, $a_{(1111)} = a_{(2222)} = a_{(3333)}$, $a_{(1122)} = a_{(1133)} = a_{(2233)}$,
$a_{(1112)} = a_{(1113)} = a_{(2212)} = a_{(2223)} = a_{(3313)} = a_{(3323)}$, $a_{(1123)} = a_{(2213)} = a_{(3312)}$,
$a_{(1212)} = a_{(1313)} = a_{(2323)}$, $a_{(1213)} = a_{(1223)} = a_{(1323)}$

Table 3. The coefficients $a_{(ijmn)}$ of diamond, silicon, and silicon carbide using the Tersoff potential. Here the unit of the coefficient is [eV Å^{-4}].

The internal lattice vector, $\mathbf{e}_{(4)}$, will be determined by equation (79).
Under the assumption that the deformation shown in Fig. 5 is infinitisimal, all of the distances between two atoms which are not covalently bonded are greater than the cutoff radius for the potentials described in Sect. 5.2. This means that the total stored energy in the diamond-like structure is just the sum of all covalent bond energies:

$$\begin{aligned}\hat{\Phi} &= 4(\Phi_{al} + \Phi_{bl} + \Phi_{cl} + \Phi_{dl}) \\ &= \hat{\Phi}\left(I_{(11)}, I_{(12)}, I_{(13)}, I_{(22)}, I_{(23)}, I_{(33)}, I_{(14)}, I_{(24)}, I_{(34)}\right),\end{aligned} \tag{80}$$

because the bond lengths and angles are determined by the invariants.
Using equations (78) and (80), we calculate the coefficients $a_{(ijmn)}$ for diamond, silicon, silicon carbide, and boron nitride and summarize them in Table 3. Here the Tersoff (1988; 1989) potential given in Sect. 5.2.1 is used.
For a cubic elastic material, all of the components of stress can be expressed in terms of strain components using the three independent elastic constants: C_{1111}, C_{1122}, and C_{1212} (Lovett, 1999, p. 66). Thus, the experiments are focusing on determining these three elastic constants with various techniques. These values for the cubic crystals are listed in Table 4 and compared with those calculated by our results, equation (82). In calculating the elastic constants, we express equation (64) in the index notation:

$$\begin{aligned}\mathrm{T}_{\alpha\beta} &= 2\rho_\kappa \sum_{m=1}^{3}\sum_{n=m}^{3}\sum_{i=1}^{3}\sum_{j=i}^{3} a_{(ijmn)}\, E_{(i)_\gamma} E_{(j)_\delta} \left(E_{(m)_\alpha} E_{(n)_\beta} + E_{(n)_\alpha} E_{(m)_\beta}\right) \varepsilon_{\gamma\delta} \\ &\equiv C_{\alpha\beta\delta\gamma}\, \varepsilon_{\gamma\delta},\end{aligned} \tag{81}$$

where

$$C_{\alpha\beta\gamma\delta} = 2\rho_\kappa \sum_{m=1}^{3}\sum_{n=m}^{3}\sum_{i=1}^{3}\sum_{j=i}^{3} a_{(ijmn)}\, \mathrm{E}_{(i)_\delta} \mathrm{E}_{(j)_\gamma} \left(\mathrm{E}_{(m)_\alpha} \mathrm{E}_{(n)_\beta} + \mathrm{E}_{(n)_\alpha} \mathrm{E}_{(m)_\beta}\right). \tag{82}$$

As seen in Table 4, our theory with the Tersoff potential could well estimate the elastic constants of diamond, silicon, and silicon carbide measured by various experiments: ultrasonic pulse technique (McSkimin & Andreatch, 1972), Brillouin light scattering (Grimsditch & Ramdas, 1975; Lee & Joannopoulos, 1982; Djemia et al., 2004), except for the elastic constants of boron nitride. A Tersoff-like potential, the Tersoff-Albe potential, give a more appropriate set of the elastic constants of boron nitride than the Tersoff potential.

	C_{1111} [GPa]	C_{1122} [GPa]	C_{1212} [GPa]
Diamond	1082	111	646
(exp.)[f]	1079	124	578
Silicon	144	76	69
(exp.)[g]	166	64	79
Silicon carbide	429	117	252
(exp.)[h]	363-395	132-154	149-236
Boron nitride by Tersoff	494	331	177
by Tersoff-Albe	946	131	568
(exp.)[i]	820	190	480

[f] measured by McSkimin & Andreatch (1972); Grimsditch & Ramdas (1975)
[g] measured by McSkimin & Andreatch (1964)
[h] measured by Lee & Joannopoulos (1982); Djemia et al. (2004)
[i] measured by Grimsditch & Zouboulis (1994)

Table 4. Elastic constants of diamond, silicon, and silicon carbide determined by our theory and measured by experiments.

6.2 Two–dimensional crystals–carbon nanotube and boron nitride nanotube

CNTs and BNNTs including their plane sheets are examples of what we will characterize as two-dimensional crystals. They are two-dimensional in the sense that atoms reside only in a surface or interface between two phases, typically two gases (air), a gas and a solid, or two solids. As illustrated in Fig. 6(a), graphene or BN sheet is composed of a regular two-dimensional array of hexagonal rings of atoms. The nanotubes depicted in Figs. 6(b) and 6(c) have structures of rolled-up sheets and are generally expressed as (m,n) by the method of rolling from the sheets. Here m and n are integers of the chiral vector of the tubes

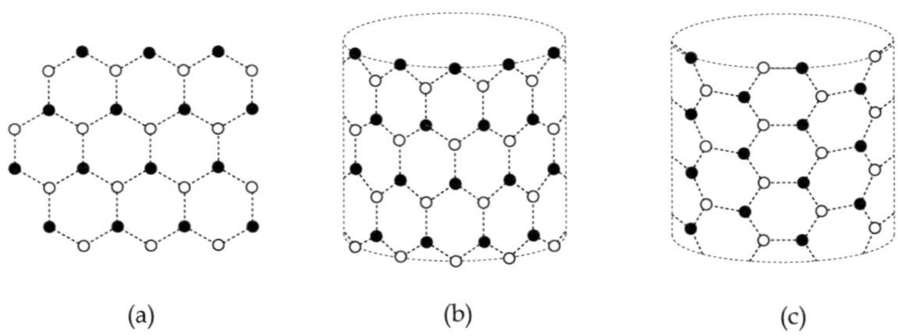

Fig. 6. Structures of two-dimensional crystals: (a) a plane sheet, (b) a zigzag nanotube, and (c) a armchair nanotube: graphene and CNT (∘: C atom, •: C atom), BN sheet or BNNT (∘: B atom, •: N atom) (Oh, 2010).

(Dresselhaus et al., 1995) and uniquely determine the diameter and the chiral angle of the tubes:

$$\text{Diameter} = \frac{\left[3(m^2 + n^2 + mn)\right]^{1/2}}{\pi} d_0, \quad \text{Angle} = \arctan\left[\frac{\sqrt{3}n}{2m+n}\right], \tag{83}$$

where d_0 is the bond length. The $(m,0)$ types of tubes are called *zigzag*, while (m,m) types are called *armchair*.
As shown in Fig. 7, the crystal structure deformed by a in-plane small deformation, $\mathbf{F}$, is determined by both two external lattice vectors, $\mathbf{e}_{(1)}$ and $\mathbf{e}_{(2)}$, and the internal lattice vector, $\mathbf{e}_{(3)}$.

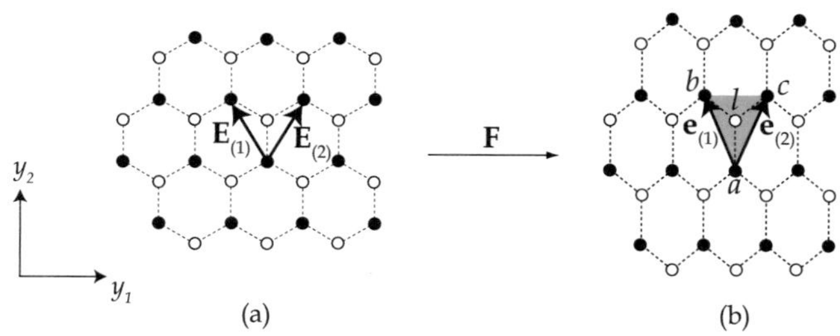

Fig. 7. Primitive lattice vectors (a) in an undeformed plane sheet and (b) in its deformed sheet by an in-plane deformation, **F**: graphene (○: C atom, •: C atom), BN sheet (○: B atom, •: N atom).

In view of equation (11) and (66), the intermolecular potential energy of the two-dimensional crystals at constant temperature is expressed in terms of five invariants:

$$\hat{\Phi} = \hat{\Phi}\left(I_{(11)}, I_{(12)}, I_{(22)}, I_{(13)}, I_{(23)}\right), \tag{84}$$

and the stress-deformation behaviour of the crystals becomes

$$\mathbf{T}^{(\sigma)} = 2\rho_{\kappa}^{(\sigma)} \sum_{m=1}^{2} \sum_{n=m}^{2} \sum_{i=1}^{2} \sum_{j=i}^{2} a_{(ijmn)} \mathbf{E}_{(i)} \cdot \boldsymbol{\varepsilon}^{(\sigma)} \mathbf{E}_{(j)} \left(\mathbf{E}_{(m)} \otimes \mathbf{E}_{(n)} + \mathbf{E}_{(n)} \otimes \mathbf{E}_{(m)}\right) \tag{85}$$

In addition, we also require at equilibrium by equations (40) and (66) that

$$\left(\frac{\partial \hat{\Phi}}{\partial I_{(13)}}\right)_{T, I_{(pq \neq 13)}} = \left(\frac{\partial \hat{\Phi}}{\partial I_{(23)}}\right)_{T, I_{(pq \neq 23)}} = 0. \tag{86}$$

The internal lattice vector $\mathbf{e}_{(3)}$ will be determined by equation (86).
In a very smilar way to the three-dimensional analysis, we can obtain the coefficient $a_{(ijmn)}$ as well as the elastic properties. As mensioned in previous section, all of the distances between two atoms which are not covalently bonded greater than the cutoff radius. Thus the intermolecular potential of the representative triangle lattice *a*-*b*-*c* in Fig. 7(b) is just the sum of three interactions and is also function of the five invarients:

$$\begin{aligned} \hat{\Phi} &= \Phi_{al} + \Phi_{bl} + \Phi_{cl} \\ &= \hat{\Phi}\left(I_{(11)}, I_{(12)}, I_{(22)}, I_{(13)}, I_{(23)}\right) \end{aligned} \tag{87}$$

because the bond lengths and angles are determined by the invariants.
Using equation (78) and the Tersoff-like potential described in Sect. 5.2.2, we calculate the coefficients $a_{(ijmn)}$ for CNT and BNNTs and finally determine their azimuthal and axial

Young's moduli as well as their Poisson's ratios:

$$E_1 = \left.\frac{T_{11}^{(\sigma)}}{\varepsilon_{11}^{(\sigma)}}\right|_{\boldsymbol{\varepsilon}^{(\sigma)}\to 0}, \quad E_2 = \left.\frac{T_{22}^{(\sigma)}}{\varepsilon_{22}^{(\sigma)}}\right|_{\boldsymbol{\varepsilon}^{(\sigma)}\to 0}, \quad \text{and} \quad \nu_{12} = -\left.\frac{\varepsilon_{11}^{(\sigma)}}{\varepsilon_{22}^{(\sigma)}}\right|_{\boldsymbol{\varepsilon}\to 0}, \tag{88}$$

with the help of equation (85).

In calculating the elastic properties we use the interlayer distance of 3.4 Å and 3.3 Å as the wall thicknesses of the CNTs and BNNTs (Krishnan et al., 1998; Salvetat et al., 1999; Yu et al., 2000; Verma et al., 2007). These results are displayed in Fig. 8.

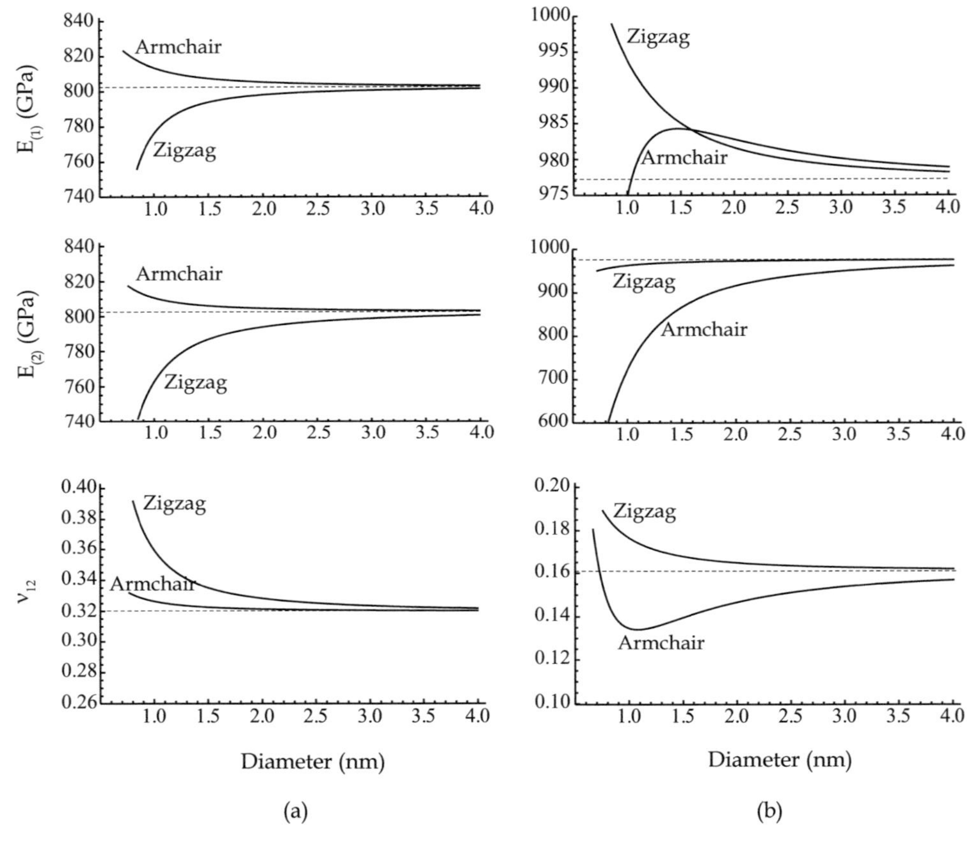

Fig. 8. Young's moduli and Poisson's ratio of (a) CNTs and (b) BNNTs. The dot lines indicate values corresponding to their plane sheets: graphene and BN sheets.

It is clear that the elastic properties of CNTs and BNNTs approach to those of their plane sheets, as the diameter of the nanotube increases. The Young's moduli of graphene and BN sheets are 802.4 and 977.2 GPa, and the Poisson's ratios of graphene and BN sheets are 0.32 and 0.161, respectively. It should be noticed that both azimuthal and axial Young's moduli of the plane sheet have to be the same due to isotropy in the plane of the sheet. While nanotubes are a little deviated from isotropy because of the effect of their curvatures and the anisotropy gradually disappears with decrease in curvature of the nanotubes.

As the diamenter increases, both of the Young's moduli of armchair CNTs decrease and ultimately approach to the Young's modulus of the graphene sheet, whereas both of the Young's moduli of zigzag CNTs increase. In the case of BNNTs, the axial Young's moduli, $E_{(2)}$, increase with the diameter of nanotubes. On the contrary, the azimuthal Young's moduli, $E_{(1)}$, of armchair BNNTs show a maximum value at a relatively small diameter and slightly decrease with increase in diameter. A similar phenomenon has been reported for the axial Young moduli of BNNTs (Verma et al., 2007). This may be closely related to the position of the equilibrium point corresponding to the third lattice vector in the deformed configuration, $\mathbf{e}_{(3)}$. As illustrated in Fig. 9, the equilibrium points for the zigzag nanotubes are located below the centroid, which is the equilibrium point for graphene or BN sheet. As the diameter increases, the equilibrium point is getting closed to the centroid. The opposite way happens to be the armchair types of nanotubes.

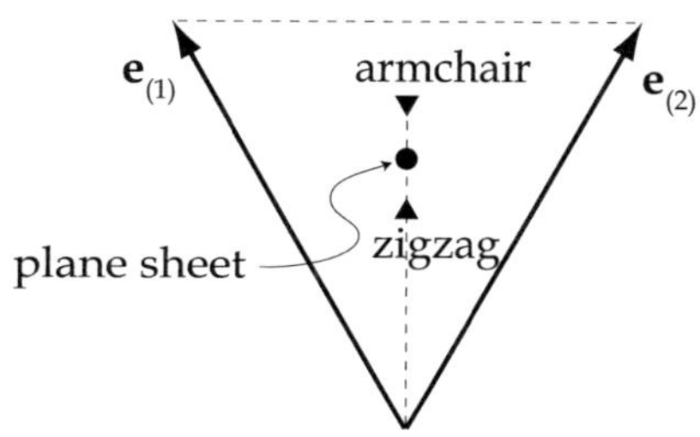

Fig. 9. The equilibrium position in the deformed configuration.

Experimentalists have reported values for only the axial Young's modulus, E_2. Salvetat et al. (1999) estimated 810±410 GPa of the Young's modulus by measuring the load–deflection of single-walled carbon nanotubes(SWCNTs). From tensile-loading experiments, Yu et al. (2000) measured the Young's modulus of SWCNTs in the range from 320 to 1471 GPa . Our result is also in good agreement with a theoretical result determined by atomistic calculations (Cornwell & Wille, 1997, 800 GPa). Besides, the Young's modulus of 1000 GPa has been reported by some theoretical calculations (Lu, 1997; Popov et al., 1999; Sanchez-Portal et al., 1999).

Using the electric-field-induced resonance method, Suryavanshi et al. (2004) obtained the axial Young's moduli of 18 BNNTs whose diameters are from 34 to 94 nm. The values were irregularly distributed in the range of 550∼1031 GPa with an average value of 722 GPa. Chopra & Zettl (1998) have measured the axial Young's modulus from a cantilevered BNNT equipped in transmission electron microscope. It was 1220±240 GPa. The elastic properties of BNNTs have been also theoretically calculated from quantumistic and atomistic simulation methods. Hernández et al. (1999) obtained 862∼940 GPa for the axial Young's moduli of BNNTs whose diameters range from 0.8 to 2.1 nm using the tight-binding method. In addition, they showed that the axial Young's moduli of both zigzag and armchair BNNTs increase as the diameter of the nanotubes increases. It corresponds to our results shown in Fig. 8. The Poisson's ratio of the BNNTs was from 0.232 to 0.268. Verma et al. (2007) have also calculated the axial Young's moduli and the Poisson's ratios of zigzag and armchair BNNTs: 982∼1110.6 GPa, 0.13∼0.16. These Poisson's ratios are close to our result as depicted in Fig. 8.

Consequently, the elastic properties of two-dimensional crystals obtained from the Tersoff-like potentials are in very good agreement with both experimental and theoretical ones.

7. References

Albe, K. & Möller, W. (1998). Modelling of boron nitride: Atomic scale simulations on thin film growth. *Comput. Mater. Sci.*, Vol. 10, 111-115.

Brenner, D. W. (1990). Empirical potential for hydrocarbons for use in simulation the chemical vapor deposition of diamond films. *Phys. Rev. B*, Vol. 42, 9458-9471.

Brenner, D. W.; Shenderova, O. A.; Harrison, J. A.; Stuart, S. J.; Ni, B. & Sinnott, S. B. (2002). A second-generation reactive empirical bond order (REBO) potential energy expression for hydrocarbons. *J. Phys.: Condens. Matter*, Vol. 14, 783-802.

Chopra, N. G. & Zettl, A. (1998). Measurement of the elastic modulus of a multi-wall boron nitride nanotube. *Solid State Commun.*, Vol. 105, 297-300.

Cornwell, C. F. & Wille, L. T. (1997). Elastic properties of single-walled carbon nanotubes in compression. *Solid State Commun.*, Vol. 101, 555-558.

Djemia, P.; Roussigne, Y.; Dirras, G. F. & Jackson, K. M. (2004). Elastic properties of β-SiC films by Brillouin light scattering. *J. Appl. Phys.*, Vol. 95, 2324-2340.

Dresselhause, M. S.; Dresselhaus, G. & Saito, R. (1995). Physics of carbon nanotubes. *Carbon*, Vol. 33, 883-891.

Edelen, D. G. B. (1975). The thermodynamics of evolving chemical systems and their approach to equilibrium. *Adv. Chem. Phys.*, Vol. 33, 399-422.

Ericksen, J. L. (1992). Bifurcation and martensitic transformations in Bravais lattices. *J. Elasticity*, Vol. 28, 55-78, ISSN

Grimsditch, M. H. & Ramdas, A. K. (1975). Brillouin scattering in diamond. *Phys. Rev. B*, Vol. 11, 3139-3148.

Grimsditch, M. H. & Zouboulis, E. S. (1994). Elastic constants of boron nitride. *J. Appl. Phys.*, Vol. 76, 832-834.

Hernández, E.; Goze, C.; Bernier, P. & Rubio, A. (1999). Elastic properties of single-wall nanotubes. *Appl. Phys. A*, Vol. 68, 287-292.

James, R. D. & Hane, K. F. (2000). Martensitic transformations and shape memory materials. *Acta Mater.*, Vol. 48, 197-222.

Klein, P. M. (1996). *A Virtual Internal Bond Approach to Modeling Crack Nucleation and Growth*, Stanford University, Ph. D. thesis.

Koga, H.; Nakamura, Y.; Watanabe, S. & Yoshida, T. (2001). Molecular dynamics study of deposition mechanism of cubic boron nitride. *Sci. Technol. Adv. Mater.*, Vol. 2, 349-356.

Krishnan, A.; Dujardin, E.; Ebbesen, T. W.; Yianilos, P. N. & Treacy, M. M. J. (1999). Young's modulus of single-walled nanotubes. *Phys. Rev. B*, Vol. 58, 14013-14019.

Kroll, P. M. (1996). *Computer Simulations and X-ray Absorption Near Edge Structure of Silicon Nitride and Silicon Carbonitride*, Technische Hochschule Darmstadt, Ph. D. thesis, Darmstadt, Germany.

Lee, D. H. & Joannopoulos, J. D. (1982). Simple Scheme for Deriving Atomic Force Constants: Application to SiC. *Phys. Rev. Lett.*, Vol. 48, 1846-1849.

Lovett, D. R. (1999). *Tensor Properties of Crystals*, Taylor & Francis, Second ed., ISBN 0-750-306262, Bristol and Philadelphia.

Lu, J. P. (1997). Elastic properties of carbon nanotubes and nanoropes. *Phys. Rev. Lett.*, Vol. 79, 1297-1300.

Matsunaga, K.; Fisher, C. & Matsubara, H. (2000). Tersoff potential parameters for simulating cubic boron carbonitrides. *Jpn. J. Appl. Phys.*, Vol. 39, L48-L51.

McSkimin, H. J. & Andreatch, P. (1964). Elastic moduli of silicon vs hydrostatic pressure at 25oC and -195.8oC. *J. Appl. Phys.*, Vol. 35, 2161-2165.

McSkimin, H. J. & Andreatch, P. (1972). Elastic Moduli of Diamond as a Function of Pressure and Temperature. *J. Appl. Phys.*, Vol. 43, 2944-2948.

Moon, W. H. & Hwang, H. J. (2004). Molecular-dynamics simulation of structure and thermal behaviour of boron nitride nanotubes. *Nanotechnology*, Vol. 15, 431-434.

Oh, E.-S. (2010). Elastic properties of boron-nitride nanotubes through the continiim lattice approach. *Mater. Lett.*, Vol. 64, 869-862.

Oh, E.-S.; Slattery, J. C. & Lagoudas, D. C. (2005). Thermodynamics of two-dimensional single-component elastic crystalline solids: single-wall carbon nanotubes. *Phil. Mag.*, Vol. 85, No. 20, 2249-2280.

Oh, E.-S. & Slattery, J. C. (2008). Nanoscale thermodynamics of multicomponent, elastic,crystalline solids: diamond, silicon, and silicon carbide. *Phil. Mag.*, Vol. 88, No. 3, 427-440.

Popov, V. N.; Van Doren, V. E. & Balkanski, M. (1999). Lattice dynamics of single-walled carbon nanotubes. *Phys. Rev. B*, Vol. 59, 8355-8358.

Salvetat, J. P.; Briggs, G. A. D.; Bonard, J. M.; Bacsa, R. R.; Kulik, A. J.; Stockli, T.; Burnham, N. A. & Forro, L. (1999). Elastic and shear moduli of single-walled carbon nanotube ropes. *Phys. Rev. Lett.*, Vol. 82, No. 3, 944-947.

Sanchez-Portal, D.; Artacho, E.; Solar, J. M.; Rubio, A. & Ordejon, P. (1999). Ab initio structural, elastic, and vibrational properties of carbon nanotubes. *Phys. Rev. B*, Vol. 59, 12678-12688.

Slattery, J. C. (1999). *Advanced Transport Phenomena*, Cambridge University Press, ISBN 0-521-63203X, Cambridge.

Slattery, J. C. & Lagoudas, D. C. (2005). Thermodynamics of multicomponent, elastic crystalline solids. *Mech. Mater.*, Vol. 37, 121-141.

Slattery, J. C.; Sagis, L. M. C. & Oh, E.-S. (2007). *Interfacial Transport Phenomena*, Springer, Second ed., ISBN 0-387-38438-3, New York.

Suryavanshi, A. P.; Yu, M. F.; Wen, J.; Tang, C. & Bando, Y. (2004). Elastic modulus and resonance behavior of boron nitride nanotubes. *Appl. Phys. Lett.*, Vol. 84, 2527-2529.

Tersoff, J. (1988). New empirical approach for the structure and energy of covalent systems. *Phys. Rev. B*, Vol. 37, 6991-7000.

Tersoff, J. (1989). Modeling solid-state chemistry: Interatomic potentials for multicomponent systems. *Phys. Rev. B*, Vol. 39, 5566-5568.

Tester, J. W. & Modell, M. (1996). *Thermodynamics and Its Applications*, Prentice Hall, Third ed., ISBN 0-139-15356X , New Jersey.

Truesdell, C. & Noll, W. (1965). The non-linear field theories of mechanics, In: *Handbuch der Physik*, Flügge, S. (Ed.), 1-602, Springer-Verlag, ISBN, Berlin.

Verma, V.; Jindal, V. K. & Dharamvir, K. (2007). Elastic moduli of a boron nitride nanotube. *Nanotechnology*, Vol. 18, 435711.

Yin, M. T. & Cohen, M. L. (1983). Structural theory of graphite and graphitic silicon. *Phys. Rev. B.*, Vol. 29, 6996-6998.

Yin, M. T. & Cohen, M. L. (1983). Will Diamond Transform under Megabar Pressures?. *Phys. Rev. Lett.*, Vol. 50, 2006-2009.

Yu, M. F.; Lourie, O.; Dyer, M. J.; Moloni, K.; Kelly, T. F. & Ruoff, R. S. (2000). Strength and breaking mechanism of multiwalled carbon nanotubes under tensile load. *Science*, Vol. 287, 637-640.

Zanzotto, G. (1992). On the material symmetry group of elastic crystals and the Born rule. *Arch. Rational Mech. Anal.*, Vol. 121, 1-36.

Zhang, P. ; Huang, Y.; Geubelle, P. H.; Klein, P. A. & Hwang, K. C. (2002). The Elastic Modulus of Single-Wall Carbon Nanotubes: A Continuum Analysis Incorporating Interatomic Potentials. *Int. J. Solids Struct.*, Vol. 39, 3893-3906.

15

Phonon Participation in Thermodynamics and Superconductive Properties of Thin Ceramic Films

Jovan P. Šetrajčić[1], Vojkan M. Zorić[1], Nenad V. Delić[1], Dragoljub Lj. Mirjanić[2] and Stevo K. Jaćimovski[3]
[1]*University of Novi Sad, Faculty of Sciences, Department of Physics, Novi Sad*
[2]*University of Banja Luka, Faculty of Medicine, Banja Luka, Academy of Sciences and Arts of the Republic of Srpska, Banja Luka*
[3]*Academy of Criminology and Police, Belgrade*
[1]*Vojvodina – Serbia,*
[2]*Republic of Srpska – B&H*
[3]*Serbia*

1. Introduction

The scope of our study in this paper is limited to the analysis of the phonons, i.e. phonon behavior in thin layered structures or crystalline films, which implies the existence of two boundary surfaces perpendicular to a preferred direction. Besides that, these film-structures could be doped by foreign atoms from one or both sides of the boundary surfaces in which way the internal configuration of the atom distribution is disturbed.

Since elastic constants and atomic masses define phonon spectra and states, we conclude that they must be different in the film-structures with respect to the corresponding ones in the ideal unbounded and translational invariant crystalline structures. The change of mass distribution along one direction and the existence of the finite structure width along that direction introduces additional boundary conditions into the analysis of the phonon behavior. We shall study the thin film "cut-off" from the ideal tetragonal crystalline structure with lattice constants $a_x = a_y = a$ and $a_z = 3a$. This structure has a finite width in the z-direction, while XY-planes are assumed to be infinite, meaning that the structure possesses two infinite boundary surfaces (parallel to the unbounded XY-planes) lying at $z = 0$ and $z = L$ (Fig. 1). The number of the atoms located along z-direction is assumed to be N_z, and it is also assumed that torsion constants $C_{\alpha\beta}$ $(\alpha \neq \beta)$ can be neglected with respect to the elongation constants $C_{\alpha\alpha}$ (Tošić et al., 1987). These structures will be titled the ideal crystalline films. The doping (by sputtering) of the ideal film with guest atoms (impurities) along z-direction, from one or mutually from both boundary surfaces, produces the film with disturbed internal distribution of atoms. Such structures will be entitled asymmetrically or symmetrically deformed crystalline films, respectively.

We have decided to study phonon behavior in the above mentioned film-structures for two reasons. Phonons are the basic elementary excitations in the condensed matter which have the decisive role in the creation of Cooper pairs of electrons in the low-temperature superconductivity. On the other hand, although the existence of phonons and Cooper pairs in the high-temperature superconductive ceramics is experimentally established (Chang & Esaki, 1992), the very mechanism of the superfluid charge transfer is not yet resolved (Bednorz & Müller, 1988). Taking into account the technical and technological treatments for the production of these high-temperature superconductors – they are small-grain crystalline structures (of small dimensions with pronounced boundaries) produced by doping, more precisely sputtering by guest atoms in certain stoichiometric ratio (Šetrajčić et al., 1990; Harshman & Mills, 1992), it is necessary to formulate the corresponding theoretical model. The simplest model for the bounded structures is the ideal crystalline film. Within the framework of this model we shall study only and exclusively the influence of the system boundaries onto spectra and states of phonons and their contribution to the basic physical properties of the system. In order to consider also the influence of the doping, we shall study the spectra and states of phonons in the deformed films and estimate what has the stronger influence to the change of the system behavior: the existence of the boundary surfaces or the disturbance of the internal distribution and type of atoms inside the system.

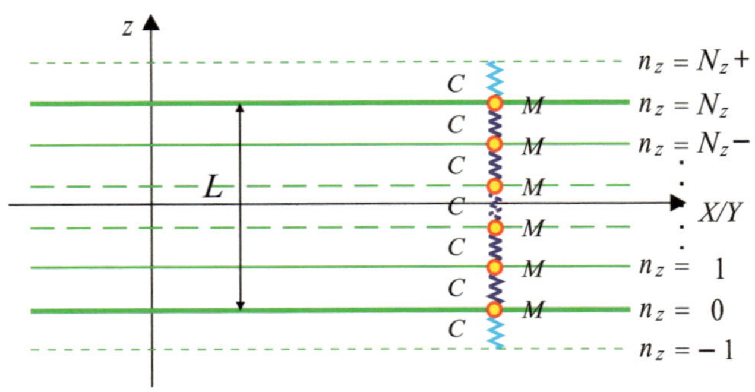

Fig. 1. Sight of crystalline film-structure model

The starting point of our study will be the standard Hamiltonian of the phonon system (Mahan, 1983; Jones & March, 1985) in the nearest neighbors' approximation:

$$H_{ID} = \frac{1}{2}\sum_{\vec{n}} \frac{p_{\vec{n}}^2}{M_{\vec{n}}} + \frac{1}{4}\sum_{\vec{n},\vec{\lambda}} C_{\vec{n},\vec{\lambda}}(\vec{u}_{\vec{n}} - \vec{u}_{\vec{n}+\vec{\lambda}})^2, \tag{1}$$

where: $\vec{p}_{\vec{n}}$ and $\vec{u}_{\vec{n}}$ – are the momentum and displacement of the atom of mass $M_{\vec{n}}$ at the crystal site $\vec{n} = a(n_x\vec{e}_x + n_y\vec{e}_y + 3n_z\vec{e}_z)$, while $C_{\vec{n},\vec{\lambda}} \equiv C_{\vec{\lambda},\vec{n}}$ – is Hooke's elastic constants between the atom at the site $\vec{n}$ and its neighboring atoms at the site $\vec{m} = \vec{n} + \vec{\lambda}$, $\vec{\lambda} = a(\vec{e}_x + \vec{e}_y + 3\vec{e}_z)$.

One of the most important aims is to study if the minimal frequencies of the atoms in the film are non-vanishing, i.e. does the phonon energy spectrum possesses the gap. In the structures where such gap exists, there can arise the damping or the elimination of the acoustical phonons (Tošić et al., 1987; Šetrajčić et al., 1990) so that there exist only the phonons of optical type. This would result in the film behavior like a "frozen" structure until certain corresponding activation temperature is reached – the temperature necessary for the creation of the phonons in the film (Djajić et al., 1987), since bellow that temperature, real (acoustical) phonons can not be present[1].

[1]For example, electrons would move in such a structure without friction up these temperatures, i.e. they would behave like ideal conductors.

Taking this into account, the presence of the phonon gap might represent the possible explanation of the fact (CRC HCP, 1989) that thin films represent better ordered thermodynamical systems and that they have higher critical superconductive temperature than the corresponding bulk structures. Since the change in the properties of the anisotropic structures, is caused by the change in the dispersion law, it is necessary to study the behavior of relevant physical quantities in order to obtain a more complete picture about these processes.
In Section 2 we first derive the dispersion law for phonons and calculate the possible phonon states in the above mentioned crystalline films, with separate study of the situation in the ideal (non-deformed), and in (symmetrically and asymmetrically) deformed films and the results are compared to the corresponding ones for unbounded structures with no breaking of the symmetry of the internal distribution of the atoms (ideal infinite structures). The Section 3 is devoted to the determination of the strength and the analysis of the diversity of the influences (the presence of boundaries, the type and manner of the doping, etc.) onto the changes of energy spectrum and specific moments in the population of the phonon states resulting from it, compared to the results with the known ones – for non-deformed bulk-structures. The conclusion gives the summary of the most important results and the discussion of the level of impact of boundary and deformation parameters.

2. Dispersion law and states of phonons in films

The Hamiltonian of the phonon subsystem of the model film-structure in the nearest neighbors approximation (Tošić et al., 1995) is given in the form expression 1, where:

$$-\frac{N_\beta}{2} \le n_\beta \le \frac{N_\beta}{2}; \;\; N_\beta \sim 10^8; \;\; \beta \in (x,y); \;\; 0 \le n_z \le N_z; \;\; N_z = \frac{L}{3a} \sim 20,$$

so that it can be written in the expanded form as:

$$\begin{aligned}
H_{FS} &= \frac{1}{2}\sum_\alpha \sum_{n_x,n_y} \sum_{n_z=0}^{N_z} \frac{(p^\alpha_{n_x,n_y,n_z})^2}{M_{n_x,n_y,n_z}} + \frac{1}{4}\sum_\alpha \sum_{n_x,n_y} \Big\{ C^\alpha_{-1}\left(u^\alpha_{n_x,n_y,0}\right)^2 + \\
&+ C^\alpha_0 \Big[\left(u^\alpha_{n_x+1,n_y,0} - u^\alpha_{n_x,n_y,0}\right)^2 + \left(u^\alpha_{n_x-1,n_y,0} - u^\alpha_{n_x,n_y,0}\right)^2 + \\
&+ \left(u^\alpha_{n_x,n_y+1,1} - u^\alpha_{n_x,n_y,0}\right)^2 + \left(u^\alpha_{n_x,n_y-1,0} - u^\alpha_{n_x,n_y,0}\right)^2 + \\
&+ \left(u^\alpha_{n_x,n_y,1} - u^\alpha_{n_x,n_y,0}\right)^2 + \left(u^\alpha_{n_x,n_y,0}\right)^2 \Big] + \\
&+ C^\alpha_1 \Big[\left(u^\alpha_{n_x,n_y,1} - u^\alpha_{n_x,n_y,2}\right)^2 + \left(u^\alpha_{n_x,n_y,1} - u^\alpha_{n_x,n_y,0}\right)^2 \Big] + \\
&+ \sum_{n_z=1}^{N_z-1} C^\alpha_{n_z} \Big[\left(u^\alpha_{n_x+1,n_y,n_z} - u^\alpha_{n_x,n_y,n_z}\right)^2 + \left(u^\alpha_{n_x-1,n_y,n_z} - u^\alpha_{n_x,n_y,n_z}\right)^2 +
\end{aligned}$$

$$
\begin{aligned}
+ & \left(u^{\alpha}_{n_x,n_y+1,n_z} - u^{\alpha}_{n_x,n_y,n_z}\right)^2 + \left(u^{\alpha}_{n_x,n_y-1,n_z} - u^{\alpha}_{n_x,n_y,n_z}\right)^2\Big] + \\
+ & \sum_{n_z=2}^{N_Z-2} C^{\alpha}_{n_z} \left[\left(u^{\alpha}_{n_x,n_y,n_z+1} - u^{\alpha}_{n_x,n_y,n_z}\right)^2 + \left(u^{\alpha}_{n_x,n_y,n_z-1} - u^{\alpha}_{n_x,n_y,n_z}\right)^2\right] + \\
+ & C^{\alpha}_{N_z-1} \left[\left(u^{\alpha}_{n_x,n_y,N_z-1} - u^{\alpha}_{n_x,n_y,N_z}\right)^2 + \left(u^{\alpha}_{n_x,n_y,N_z-1} - u^{\alpha}_{n_x,n_y,N_z-2}\right)^2\right] + \\
+ & C^{\alpha}_{N_z} \Big[\left(u^{\alpha}_{n_x+1,n_y,N_z} - u^{\alpha}_{n_x,n_y,N_z}\right)^2 + \left(u^{\alpha}_{n_x-1,n_y,N_z} - u^{\alpha}_{n_x,n_y,N_z}\right)^2 + \\
+ & \left(u^{\alpha}_{n_x,n_y+1,N_z} - u^{\alpha}_{n_x,n_y,N_z}\right)^2 + \left(u^{\alpha}_{n_x,n_y-1,N_z} - u^{\alpha}_{n_x,n_y,N_z}\right)^2 + \\
+ & \left(u^{\alpha}_{n_x,n_y,N_z-1} - u^{\alpha}_{n_x,n_y,N_z}\right)^2 + \left(u^{\alpha}_{n_x,n_y,N_z}\right)^2\Big] + C^{\alpha}_{N_z+1}\left(u^{\alpha}_{n_x,n_y,N_z}\right)^2\Big\} .
\end{aligned} \tag{2}
$$

This Hamiltonian describes the film-structure model presented at the Fig.1. It enables further theoretical analysis of the properties, specific effects and changes in the phonon behavior in above mentioned translational non-invariant systems. All changes and specific effects which can occur in the system, will be treated as a strict consequence of the contribution of the mechanical vibrations of the atoms of the crystal lattice under the influence of the presence of boundary surfaces and asymmetric distribution of the atoms along one preferred (z) crystallographic direction in that system.

2.1 Ideal film-structures

The concept of the ideal film means here the model of the crystal bounded by two parallel surfaces which can "breathe" (no rigid walls) along one crystallographic direction (which we choose for the positive direction of z-axis) perpendicular to the boundary surfaces and unbounded in the two other remaining directions. Furthermore, besides boundaries, there are no other defects in the ideal film, so inside the boundaries we encounter single atom tetragonal structure.

The Hamiltonian, expression 2, adapted to the above mentioned model can be separated into two parts: the first one H_S, which includes "surface" terms and the second one H_B, which includes "bulk" terms subject to the conditions (see Fig.1):

$$M_{n_x,n_y,n_z} \equiv M; \ \ C^{\alpha}_{n_xn_yn_z;n_x\pm1,n_yn_z} = C^{\alpha}_{n_xn_yn_z;n_xn_y\pm1,n_z} = C^{\alpha}_{n_xn_yn_z;n_xn_yn_z\pm1} \equiv C_{\alpha} .$$

Since there are no layers for $n_z \leq -1$ and for $n_z \geq N_z + 1$, we must include the following condition, too:

$$u^{\alpha}_{n_x,n_y,l} = 0; \ \ l \leq -1 \wedge l \geq N_z + 1 \ \ (\text{i.e. } l \notin [0, N_z]) .$$

If we would assign $C^{\alpha}_{-1} = C^{\alpha}_{N_z+1} = 0$, then the boundary atoms (for $n_z = 0$ and $n_z = N_z$) would be "frozen", i.e. we would have the effects of rigid walls (Maradudin, 1987). In this way, the expression for the total Hamiltonian of the ideal crystalline film obtains the following form:

$$H_{IF} = H_S + H_B , \tag{3}$$

where:

$$\begin{aligned}
H_S &= \frac{1}{2M}\sum_{\alpha}\sum_{n_x,n_y}\left[\left(p^{\alpha}_{n_x,n_y,0}\right)^2+\left(p^{\alpha}_{n_x,n_y,N_z}\right)^2\right]+\\
&+ \frac{1}{4}\sum_{\alpha}C_\alpha\sum_{n_x,n_y}\left[2\left(u^{\alpha}_{n_x,n_y,0}\right)^2+2\left(u^{\alpha}_{n_x,n_y,N_z}\right)^2+\right.\\
&+ \left(u^{\alpha}_{n_x,n_y,N_z-1}-u^{\alpha}_{n_x,n_y,N_z}\right)^2+\left(u^{\alpha}_{n_x,n_y,1}-u^{\alpha}_{n_x,n_y,0}\right)^2+\\
&+ \left(u^{\alpha}_{n_x,n_y,0}-u^{\alpha}_{n_x+1,n_y,0}\right)^2+\left(u^{\alpha}_{n_x,n_y,0}-u^{\alpha}_{n_x-1,n_y,0}\right)^2+\\
&+ \left(u^{\alpha}_{n_x,n_y,0}-u^{\alpha}_{n_x,n_y+1,0}\right)^2+\left(u^{\alpha}_{n_x,n_y,0}-u^{\alpha}_{n_x,n_y-1,0}\right)^2+\\
&+ \left(u^{\alpha}_{n_x,n_y,N_z}-u^{\alpha}_{n_x+1,n_y,N_z}\right)^2+\left(u^{\alpha}_{n_x,n_y,N_z}-u^{\alpha}_{n_x-1,n_y,N_z}\right)^2+\\
&+ \left.\left(u^{\alpha}_{n_x,n_y,N_z}-u^{\alpha}_{n_x,n_y+1,N_z}\right)^2+\left(u^{\alpha}_{n_x,n_y,N_z}-u^{\alpha}_{n_x,n_y-1,N_z}\right)^2\right];
\end{aligned} \tag{4}$$

$$\begin{aligned}
H_B &= \frac{1}{2M}\sum_{\alpha}\sum_{n_x,n_y}\left(p^{\alpha}_{n_x,n_y,n_z}\right)^2+\frac{1}{4}\sum_{\alpha}C_\alpha\times\\
&\times \sum_{n_x,n_y}\left\{\sum_{n_z=1}^{N_z-1}\left[\left(u^{\alpha}_{n_x+1,n_y,n_z}-u^{\alpha}_{n_x,n_y,n_z}\right)^2+\left(u^{\alpha}_{n_x-1,n_y,n_z}-u^{\alpha}_{n_x,n_y,n_z}\right)^2+\right.\right.\\
&+ \left.\left(u^{\alpha}_{n_x,n_y+1,n_z}-u^{\alpha}_{n_x,n_y,n_z}\right)^2+\left(u^{\alpha}_{n_x,n_y-1,n_z}-u^{\alpha}_{n_x,n_y,n_z}\right)^2\right]+\\
&+ \sum_{n_z=2}^{N_z-2}\left[\left(u^{\alpha}_{n_x,n_y,n_z+1}-u^{\alpha}_{n_x,n_y,n_z}\right)^2+\left(u^{\alpha}_{n_x,n_y,n_z-1}-u^{\alpha}_{n_x,n_y,n_z}\right)^2\right]+\\
&+ \left.\left(u^{\alpha}_{n_x,n_y,N_z-1}-u^{\alpha}_{n_x,n_y,N_z-2}\right)^2+\left(u^{\alpha}_{n_x,n_y,1}-u^{\alpha}_{n_x,n_y,2}\right)^2\right\}.
\end{aligned} \tag{5}$$

We have decided to use the approach of Heisenberg's equations of motion (Tošić et al., 1992; Šetrajčić et al., 1992; Šetrajčić & Pantić, 1994) for the determination of possible frequencies (energy spectrum) and the states of phonons. We start from the following system of the equations of motion for the phonon displacements:

– for $n_z = 0$

$$\begin{aligned}
\ddot{u}^{\alpha}_{n_x,n_y,0} &- \Omega^2_\alpha\left(u^{\alpha}_{n_x+1,n_y,0}+u^{\alpha}_{n_x-1,n_y,0}+u^{\alpha}_{n_x,n_y+1,0}+\right.\\
&+ \left.u^{\alpha}_{n_x,n_y-1,0}+u^{\alpha}_{n_x,n_y,1}-6u^{\alpha}_{n_x,n_y,0}\right)=0;
\end{aligned} \tag{6}$$

– for $1 \le n_z \le N_z-1$

$$\begin{aligned}
\ddot{u}^{\alpha}_{n_x,n_y,n_z} &- \Omega^2_\alpha\left(u^{\alpha}_{n_x+1,n_y,n_z}+u^{\alpha}_{n_x-1,n_y,n_z}+u^{\alpha}_{n_x,n_y+1,n_z}+\right.\\
&+ \left.u^{\alpha}_{n_x,n_y-1,n_z}+u^{\alpha}_{n_x,n_y,n_z+1}+u^{\alpha}_{n_x,n_y,n_z-1}-6u^{\alpha}_{n_x,n_y,n_z}\right)=0;
\end{aligned} \tag{7}$$

– for $n_z = N_z$

$$\begin{aligned} \ddot{u}^{\alpha}_{n_x,n_y,N_z} &- \Omega^2_{\alpha}\Big(u^{\alpha}_{n_x+1,n_y,N_z} + u^{\alpha}_{n_x-1,n_y,N_z} + u^{\alpha}_{n_x,n_y+1,N_z} + \\ &+ u^{\alpha}_{n_x,n_y-1,N_z} + u^{\alpha}_{n_x,n_y,N_z-1} - 6u^{\alpha}_{n_x,n_y,N_z}\Big) = 0. \end{aligned} \tag{8}$$

where $\Omega_{\alpha} = \sqrt{C_{\alpha}/M}$. The solution of this system of N_{z+1} homogeneous differential-difference equations for phonon displacements can be looked for in the form of the product of an unknown function (along z-axis) and harmonic function of the position (within XY-plane) known from the bulk solutions, i.e.

$$^{I}u^{\alpha}_{n_x,n_y,n_z}(t) = \sum_{k_x,k_y,k_z} \int_{-\infty}^{+\infty} d\omega\ e^{ia(k_x n_x + k_y n_y) - it\omega}\, \Phi^{\alpha}_{n_z};\quad \Phi^{\alpha}_{n_z} \equiv \Phi^{\alpha}_{n_z}(k_z,\omega)\,. \tag{9}$$

Substituting this expression into the equations 6–8 we obtain:

$$\begin{aligned} R\Phi^{\alpha}_0 + \Phi^{\alpha}_1 &= 0 \\ \Phi^{\alpha}_0 + R\Phi^{\alpha}_1 + \Phi^{\alpha}_2 &= 0 \\ \cdot\quad \cdot\quad \cdot\quad &\quad\ \cdot \\ \Phi^{\alpha}_{n_z-1} + R\Phi^{\alpha}_{n_z} + \Phi^{\alpha}_{n_z+1} &= 0 \\ \cdot\quad \cdot\quad \cdot\quad &\quad\ \cdot \\ \Phi^{\alpha}_{N_z-2} + R\Phi^{\alpha}_{N_z-1} + \Phi^{\alpha}_{N_z} &= 0 \\ \Phi^{\alpha}_{N_z-1} + R\Phi^{\alpha}_{N_z} &= 0 \end{aligned} \tag{10}$$

where:

$$R \equiv W^2_{\alpha} - 4\mathcal{F}_{k_x k_y} - 2;\quad W_{\alpha} \equiv \frac{\omega}{\Omega_{\alpha}};\quad \mathcal{F}_{k_x k_y} \equiv \sin^2\frac{ak_x}{2} + \sin^2\frac{ak_y}{2}\,. \tag{11}$$

In this way the system of $N_z + 1$ differential-difference equations 6–8 turns into a system of $N_z + 1$ homogeneous algebraic difference equations 10. In order that this system possesses nontrivial solutions, its determinant:

$$\mathcal{D}_{N_z+1}(R) = \begin{vmatrix} R & 1 & 0 & 0 & \cdots & 0 & 0 & 0 & 0 \\ 1 & R & 1 & 0 & \cdots & 0 & 0 & 0 & 0 \\ 0 & 1 & R & 1 & \cdots & 0 & 0 & 0 & 0 \\ \cdot & \cdot & \cdot & \cdot & \ddots & \cdot & \cdot & \cdot & \cdot \\ 0 & 0 & 0 & 0 & \cdots & 1 & R & 1 & 0 \\ 0 & 0 & 0 & 0 & \cdots & 0 & 1 & R & 1 \\ 0 & 0 & 0 & 0 & \cdots & 0 & 0 & 1 & R \end{vmatrix} \tag{12}$$

must vanish. The roots (poles) of this determinant represent one of the forms of Chebyshev's polynomials of the second order (Cottam & Tilley, 1989) and can be written in the form:

$$\mathcal{D}_{N_z+1}(R) \equiv \mathcal{D}_{N^I_z-1}(R) = \frac{\sin\left(N^I_z\zeta\right)}{\sin\zeta};\quad N^I_z = N_z + 2;\quad \zeta \neq 0,$$

where: $R = 2\cos\zeta$. Above mentioned condition ($\mathcal{D}_{N_z+1}(R_{\nu}) = 0$) is satisfied for:

$$\zeta_{\nu} = \frac{\pi\nu}{N^I_z};\quad \nu = 1,2,3,\ldots,N^I_z - 1, \tag{13}$$

whose substitution into expressions 11 leads to the expression for demanded (possible) unknown phonon frequencies:

$$ {}^{I}\omega^{\alpha}_{k_xk_y}(\mu) = 2\Omega_\alpha \sqrt{\mathcal{G}^{I}_{\mu} + \mathcal{F}_{k_xk_y}}\,, \tag{14}$$

where:

$$\mathcal{G}^{I}_{\mu} \equiv \sin^2 \frac{ak_z(\mu)}{2}; \quad k_z(\mu) = \frac{\pi}{a}\frac{\mu}{N^I_z}; \quad \mu \equiv N^I_z - \nu = 1,2,3,\dots,N^I_z - 1. \tag{15}$$

One must notice that contrary to k_x and k_y which range from 0 to π/a, one has:

$$\begin{aligned} k^{min}_z &\equiv k_z(1) = \frac{\pi}{a}\frac{1}{N^I_z} > 0; \\ k^{max}_z &\equiv k_z(N^I_z - 1) = \frac{\pi}{a}\frac{N^I_z - 1}{N^I_z} < \frac{\pi}{a}, \end{aligned} \tag{16}$$

because $N^I_z \ll (N_x, N_y)$.

If one divides the system of equations 10 by $\Phi^{\alpha}_0 \equiv \Phi^{\alpha}_0(k_z)$ and rejects the last equation, this system is obtained in the new form:

$$\begin{aligned} R_\nu + \varrho_1 &= 0, & \text{for} \quad & n_z = 0 \\ 1 + R_\nu\,\varrho_1 + \varrho_2 &= 0, & \text{for} \quad & n_z = 1 \\ \varrho_{n_z-1} + R_\nu\,\varrho_{n_z} + \varrho_{n_z+1} &= 0, & \text{for} \quad & 2 \le n_z \le N_z - 1 \end{aligned} \tag{17}$$

where $R_\nu \equiv 2\cos\zeta_\nu$ and:

$$\varrho_{n_z} \equiv \varrho^{\alpha}_{n_z} = (\Phi^{\alpha}_0)^{-1}\,\Phi^{\alpha}_{n_z} \quad \Longrightarrow \quad \Phi^{\alpha}_{n_z} = \Phi^{\alpha}_0 \nearrow^{\alpha}_{n_z}. \tag{18}$$

The last of the equations 18 is satisfied for:

$$\varrho_{n_z} = (-1)^{n_z}\left\{P\sin(n_z\zeta_\nu) + Q\sin[(n_z-1)\zeta_\nu]\right\}, \tag{19}$$

and using this and expression 13 it follows:

$$\varrho_1 = -P\sin(\zeta_\nu); \quad \varrho_2 = P\sin(2\zeta_\nu) + Q\sin(\zeta_\nu).$$

Substituting these expressions into the first and second equation in the system od difference equations 18 we arrive to the unknown coefficients $P \equiv P_\nu = R_\nu \sin^{-1}\zeta_\nu$ and $Q \equiv Q_\nu = -\sin^{-1}\zeta_\nu$, while returning them into expressions 19 and 18, it follows:

$$\Phi^{\alpha}_{n_z}(k_z) = (-1)^{n_z}\,\frac{\sin[(n_z+1)\zeta_\nu]}{\sin\zeta_\nu}\,\Phi^{\alpha}_0. \tag{20}$$

According to above calculations – combining 9, 20 and standard normalization (Callavay, 1974), one can easily obtain the final expression for phonon displacements in the form:

$$\begin{aligned} {}^{I}u^{\alpha}_{n_x,n_y,n_z}(t) &= \sum_{k_xk_y}\sum_{\mu=1}^{N^I_z-1} {}^{I}\mathcal{N}^{\alpha}_{n_z}(k_xk_y,\mu) \times \\ &\times \; e^{ia(k_xn_x+k_yn_y) - it\omega^{\alpha}_{k_xk_y}(\mu)} \sin[(n_z+1)\,ak_z(\mu)]; \\ {}^{I}\mathcal{N}^{\alpha}_{n_z}(k_xk_y,\mu) &= (-1)^{n_z}\sqrt{\frac{\hbar}{MN_xN_yN^I_z\,\omega^{\alpha}_{k_xk_y}(\mu)}}. \end{aligned} \tag{21}$$

Comparing the result obtained here with the corresponding one for ideal infinite structures, one can conclude that mechanical vibrations in the ideal unbounded structure are plane waves in all spatial directions, while in the thin film they represent the superposition of the standing waves in z-direction and plane waves in XY-planes. It is also evident that the displacement amplitude in the films is $\sim 10^4\sqrt{2/N_z^I}$ times larger[2] than the amplitude in corresponding unbounded structures.
Using expression 14 one can determine the dispersion law for phonons in thin undeformed ideal film[3] :

$$^{I}E^{\alpha}_{k_xk_y}(\mu) \equiv \hbar\, {}^{I}\omega^{\alpha}_{k_xk_y}(\mu) = E^{\alpha}_{I}\sqrt{\mathcal{G}^{I}_{\mu} + \mathcal{F}_{k_xk_y}}\,, \tag{22}$$

where $E^{\alpha}_{I} = 2\hbar\,\Omega_{\alpha}$ and which is valid together with expressions 11 and 15.
We represent graphically this energy spectrum at the Fig.2 vs XY-plane vector $k^2 = k_x^2 + k_y^2$:

$$^{I}\mathcal{E}^{z}_{\mu} \equiv \left({}^{I}E^{z}_{k_xk_y}(\mu)/E^{z}_{I}\right)^2 = {}^{I}\mathcal{E}^{z}_{k_z(\mu)}\left(\mathcal{F}_{k_xk_y}\right).$$

Fig.2 represent energy spectrum of phonons in the ideal (ultrathin $N_z = 4$) crystalline films vs. two-dimensional (XY planar) wave vector. Within the band of bulk energies with continual spectrum (bulk limits are denoted by solid dashed lines) one can notice five allowed discrete phonon energies in the film studied (thin solid lines). One can notice the narrowing of the energy band and the existence of the energy gap.

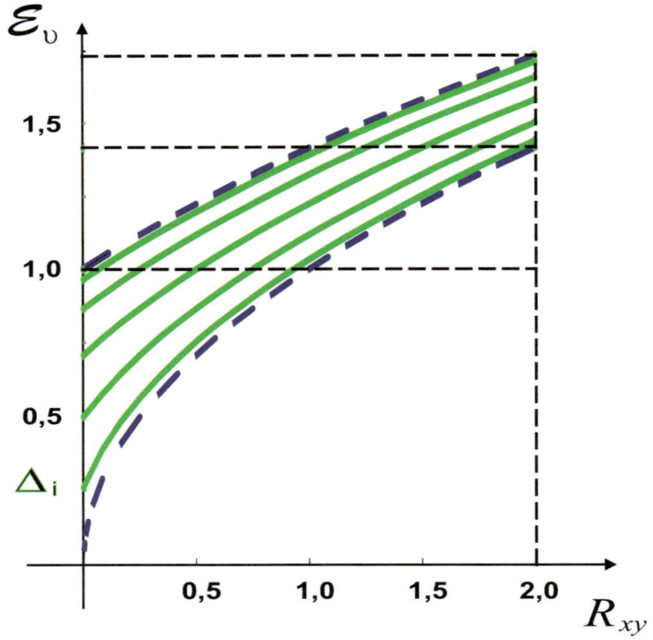

Fig. 2. Phonon spectra in the ideal ultrathin crystalline films

[2]For very thin films $N_z^I \sim 10$, so the factor of the amplitude increase can achieve even 2000.
[3]Most common treatment is that using classical procedure, for example, second quantization method (Callavay, 1974), on the basis of expressions 3–**??**, 14 and 20, the Hamiltonian H_{IF} is diagonalized, and then the energy spectrum in the form 22 is readily obtained.

One can clearly see from the plot explicate discreteness of the allowed energy levels of phonons in the ideal film with respect to the continuum of these values for the corresponding bulk-structures. All three acoustic frequencies in bulk-structures vanish when three-dimensional (spatial) vector $k = \left|\vec{k}\right|$ vanishes, while the minimal frequencies of phonons in the thin ideal film-structure are:

$$\Delta_i \equiv (\omega_{\alpha}^{I})_{min} \equiv \omega_{\alpha}^{I}(k_x = k_y = 0, k_z = k_z^{min}) \approx \Omega_{\alpha} \frac{\pi}{N_z^I} > 0. \tag{23}$$

On the other hand, maximal values of the frequencies of acoustic branches in the ideal infinite crystal tend to the value $(\omega_{\alpha}^{B})_{max} = 2\Omega_{\alpha}\sqrt{3}$ when $k_{\alpha} \to \pi/a, \alpha = x, y, z$, while in the studied ideal film they are:

$$(\omega_{\alpha}^{I})_{max} \equiv \omega_{\alpha}^{I}\left(k_x = k_y = \frac{\pi}{a}, k_z = k_z^{max}\right) \approx 2\Omega_{\alpha}\sqrt{3}\left[1 - \frac{\pi^2/12}{(N_z^I)^2}\right] < (\omega_{\alpha}^{B})_{max}\,. \tag{24}$$

It can be also seen from the same figure that the width of the energy band in the film is narrower. From expressions 23 and 24 we can determine the total narrowing of the band of allowed energies of the phonons in the film-structures with respect to the bulk band:

$$\mathcal{W}_{\alpha}^{I} \equiv \hbar\,\{(\omega_{\alpha}^{B})_{max} - [(\omega_{\alpha}^{I})_{max} - (\omega_{\alpha}^{I})_{min}]\} \approx \hbar\,\Omega_{\alpha}\,\frac{\pi\,(N_z^I + 1)}{(N_z^I)^2} > 0. \tag{25}$$

The functional behavior and the physical explanation, as well as the effects that might be caused by the existence of the frequency threshold 23 and the band narrowing 25 will be exposed in the next Section after the analysis of the phonon behavior in the deformed structures.

2.2 Deformed film-structures

During sputtering perpendicular to the boundary surfaces (Cava et al., 1987; Chu et al., 1987; Politis et al., 1987; Segre et al., 1987; Dietrich et al, 1987; Kuwahara, 1992; Notzel et al., 1992), the atoms that are introduced locate themselves along z-axis since this is energetically most convenient: $a_z = 3a$ and $a_x = a_y = a$. For that reason the effective mass of the atom (as well as the elongation constant), at some site in the crystalline lattice, will depend on its relative position (with respect to the origin of z-axis), i.e. on the lattice index in the z-direction, but not in x and y directions:

$$M_{\vec{n}} \equiv M_{n_x,n_y,n_z} = M_{n_z}\,.$$

Using the Heisenberg equations of motion for u and p with the Hamiltonian (given by expression 2), taking into account above mentioned conditions, one arrives to the system of $N_z + 1$ homogeneous differential-difference equations for the phonon displacements:

– for $n_z = 0$

$$\begin{aligned} \ddot{u}^{\alpha}_{n_x,n_y,0} &= \frac{C_0^{\alpha}}{M_0}\left(u^{\alpha}_{n_x+1,n_y,0} + u^{\alpha}_{n_x-1,n_y,0} + u^{\alpha}_{n_x,n_y+1,0}\right. \\ &+ \left. u^{\alpha}_{n_x,n_y-1,0} + u^{\alpha}_{n_x,n_y,1} - 5u^{\alpha}_{n_x,n_y,0}\right) + \\ &+ \frac{C_1^{\alpha}}{2M_0}\left(u^{\alpha}_{n_x,n_y,1} - u^{\alpha}_{n_x,n_y,0}\right) - \frac{C_{-1}^{\alpha}}{2M_0}\,u^{\alpha}_{n_x,n_y,0}, \end{aligned} \tag{26}$$

– for $1 \leq n_z \leq N_z - 1$

$$
\begin{aligned}
\ddot{u}^{\alpha}_{n_x,n_y,n_z} &= \frac{C^{\alpha}_{n_z}}{2M_{n_z}} \left[2 \left(u^{\alpha}_{n_x+1,n_y,n_z} + u^{\alpha}_{n_x-1,n_y,n_z} + u^{\alpha}_{n_x,n_y+1,n_z} + u^{\alpha}_{n_x,n_y-1,n_z} \right) + \right. \\
&+ \left. u^{\alpha}_{n_x,n_y,n_z+1} + u^{\alpha}_{n_x,n_y,n_z-1} - 10 u^{\alpha}_{n_x,n_y,n_z} \right] + \\
&+ \frac{C^{\alpha}_{n_z+1}}{2M_{n_z}} \left(u^{\alpha}_{n_x,n_y,n_z+1} - u^{\alpha}_{n_x,n_y,n_z} \right) + \frac{C^{\alpha}_{n_z-1}}{2M_{n_z}} \left(u^{\alpha}_{n_x,n_y,n_z-1} - u^{\alpha}_{n_x,n_y,n_z} \right),
\end{aligned} \tag{27}
$$

– for $n_z = N_z$

$$
\begin{aligned}
\ddot{u}^{\alpha}_{n_x,n_y,N_z} &= \frac{C^{\alpha}_{N_z}}{M_{N_z}} \left(u^{\alpha}_{n_x+1,n_y,N_z} + u^{\alpha}_{n_x-1,n_y,N_z} + u^{\alpha}_{n_x,n_y+1,N_z} + \right. \\
&+ \left. u^{\alpha}_{n_x,n_y-1,N_z} + u^{\alpha}_{n_x,n_y,N_z-1} - 5 u^{\alpha}_{n_x,n_y,N_z} \right) + \\
&+ \frac{C^{\alpha}_{N_z-1}}{2M_{N_z}} \left(u^{\alpha}_{n_x,n_y,N_z-1} - u^{\alpha}_{n_x,n_y,N_z} \right) - \frac{C^{\alpha}_{N_z+1}}{2M_{N_z}} u^{\alpha}_{n_x,n_y,N_z}.
\end{aligned} \tag{28}
$$

The solution of this system of equations can be looked for in the form of the product of an unknown function in z-direction and plane harmonic waves in XY-planes:

$$
{}^{D}u^{\alpha}_{n_x,n_y,n_z}(t) = \sum_{k_x,ky,k_z} \int_{-\infty}^{+\infty} dw\; e^{ia(k_x n_x + k_y n_y) - it\omega} \Psi^{\alpha}_{n_z}; \quad \Psi^{\alpha}_{n_z} \equiv \Psi^{\alpha}_{n_z}(k_z, \omega). \tag{29}
$$

Calculating corresponding derivatives and introducing them into equation 27 we obtain the difference equation for the unknown functions $\Psi^{\alpha}_{n_z}$:

$$
\begin{aligned}
2 \left(M_{n_z} \omega^2 - 4 C^{\alpha}_{n_z} \mathcal{F}_{k_x k_y} \right) \Psi^{\alpha}_{n_z} + C^{\alpha}_{n_z} \left(\Psi^{\alpha}_{n_z+1} + \Psi^{\alpha}_{n_z-1} - 2 \Psi^{\alpha}_{n_z} \right) &+ \\
+ C^{\alpha}_{n_z+1} \left(\Psi^{\alpha}_{n_z+1} - \Psi^{\alpha}_{n_z} \right) + C^{\alpha}_{n_z-1} \left(\Psi^{\alpha}_{n_z-1} - \Psi^{\alpha}_{n_z} \right) &= 0
\end{aligned} \tag{30}
$$

valid for $n_z = 1,2,3,\ldots, N_z - 1$. For $n_z = 0$, using the same procedure, one obtains from equation 26 the first (upper) boundary equation:

$$
\begin{aligned}
2 \left(M_0 \omega^2 - 4 C^{\alpha}_0 \mathcal{F}_{k_x k_y} \right) \Psi^{\alpha}_0 + C^{\alpha}_0 \left(\Psi^{\alpha}_1 - 2 \Psi^{\alpha}_0 \right) &+ \\
C^{\alpha}_1 \left(\Psi^{\alpha}_1 - \Psi^{\alpha}_0 \right) - C^{\alpha}_{-1} \Psi^{\alpha}_0 &= 0
\end{aligned} \tag{31}
$$

and for $n_z = N_z$, using equation 28, the second (lower) boundary equation:

$$
\begin{aligned}
2 \left(M_{N_z} \omega^2 - 4 C^{\alpha}_{N_z} \mathcal{F}_{k_x k_y} \right) \Psi^{\alpha}_{N_z} + C^{\alpha}_{N_z} \left(\Psi^{\alpha}_{N_z-1} - 2 \Psi^{\alpha}_{N_z} \right) &+ \\
+ C^{\alpha}_{N_z-1} \left(\Psi^{\alpha}_{N_z-1} - \Psi^{\alpha}_{N_z} \right) - C^{\alpha}_{N_z+1} \Psi^{\alpha}_{N_z} &= 0.
\end{aligned} \tag{32}
$$

The procedure for the determination of the allowed phonon states using the above equations is extremely complicated[4]. For that reason we are going to perform the transition to continuum

[4] In fact, it is not known or elaborated in the literature for this, completely general case.

(continuum approximation), i.e. transition from the discrete to continual variables, and expand the corresponding quantities into the Taylor's series:

$$n_z \longrightarrow z;\;\; \Psi^{\alpha}_{n_z} \longrightarrow \Psi_\alpha(z);\;\; M_{n_z} \longrightarrow M(z);\;\; a_z \longrightarrow a(z);\;\; C^{\alpha}_{n_z} \longrightarrow C_\alpha(z).$$

Besides that, as a consequence of sputtering, i.e. clustering of foreign atoms around the atoms of the basic matrix (Tošić et al., 1987; Šetrajčić et al., 1990; Tošić et al., 1992; Ristovski et al., 1989), the mass of the basic matrix must be substituted by the corresponding reduced mass:

$$M^{-1}(z) = \mathcal{M}_m^{-1} + n(z)\,\mathcal{M}_d^{-1}, \tag{33}$$

where: $\mathcal{M}_m$ – is the mass of the basic matrix, $\mathcal{M}_d$ – the mass of doping atoms and $n(z)$ – their number at the site z (measured from the lower boundary surface of the crystalline film).
After these transformations and introduction into the difference equation 30, it becomes a second order differential equation:

$$\frac{d^2\Psi_\alpha(z)}{dz^2} + C_\alpha^{-1}(z)\,\frac{dC_\alpha(z)}{dz}\,\frac{d\Psi_\alpha(z)}{dz} - \left[\frac{4\mathcal{F}_{k_xk_y}}{a^2(z)} - \frac{M(z)\,\omega^2}{C_\alpha(z)\,a^2(z)}\right]\Psi_\alpha(z) = 0. \tag{34}$$

Further solving of this differential equation demands the specification of the functional dependence of the quantities $M(z)$, $C_\alpha(z)$ and $a(z)$, and they depend not only on the procedure of the sputtering of the basic matrix – ideal crystalline film-structure, but also on the number, type and distribution of the sputtered atoms.

2.2.1 Asymmetrical deformation

Taking into account that the production of oxide superconductive ceramics includes the sputtering with foreign atoms (Cava et al., 1987; Chu et al., 1987; Politis et al., 1987; Segre et al., 1987; Dietrich et al, 1987; Kuwahara, 1992; Notzel et al., 1992; Johnson, 1995), we shall assume that it is performed perpendicularly to one (upper) of the boundary surfaces of the model film-structure. For this reason, doping atoms cluster along z-direction, from this upper surface towards lower boundary surface and let us assume the (approximate) parabolic distribution of such "weighted" atoms, i.e. their reduced masses:

$$M(z) \longrightarrow M_A(z) = A_M^A + B_M^A\,(z-L)^2.$$

Using boundary conditions:

$$M_A(0) = \mathcal{M}_m;\;\; M_A(L) = \frac{\mathcal{M}_m\mathcal{M}_d}{\mathcal{M}_d - n\,\mathcal{M}_m},$$

we determine the unknown coefficients A_M^A and B_M^A, so that we obtain:

$$M_A(z) = \frac{\mathcal{M}_m}{\mathcal{O}_M^A}\left[1 - \left(1-\mathcal{O}_M^A\right)\left(1-\frac{z}{L}\right)^2\right];\;\; \mathcal{O}_M^A = 1 - n\,\frac{\mathcal{M}_m}{\mathcal{M}_d}. \tag{35}$$

The sputtering of the basic matrix causes also (parabolic) change of the lattice constant:

$$a(z) \longrightarrow a_A(z) = A_a^A + B_a^A\,(z-L)^2,$$

with boundary conditions:

$$a_A(0) = a_Z; \;\; a_A(L) = \frac{a_Z}{n},$$

from which it follows:

$$a_A(z) = a_Z\left[1 - \frac{n-1}{n}\frac{z}{L}\left(2 - \frac{z}{L}\right)\right]. \tag{36}$$

Since Hooke's constants may be expressed as $C_\alpha(z) = \text{const}(\alpha)\, a^{-\gamma}(z)$, using expression 36 one can write:

$$C_\alpha^A(z) = C_\alpha\left[1 + \gamma\frac{n-1}{n}\frac{z}{L}\left(2 - \frac{z}{L}\right)\right]. \tag{37}$$

Furthermore, instead of $a_A(z)$ and $C_\alpha^A(z)$ we shall use their values averaged over the total film width (L):

$$\begin{aligned}
\overline{a_Z^A} &\equiv \overline{a_A(z)} = \frac{1}{L}\int_0^L a_A(z)\,dz = f_n^A\, a_Z; \;\; f_n^A \equiv \frac{n+2}{3n};\\
\overline{C_A^\alpha} &\equiv \overline{C_\alpha^A(z)} = \frac{1}{L}\int_0^L C_\alpha^A(z)\,dz = g_n^A\, C_Z^\alpha; \;\; g_n^A \equiv 1 + 2\gamma\frac{n-1}{3n}.
\end{aligned} \tag{38}$$

To simplify the solution of the last differential equation, besides 35 and 38, it is convenient to change variable $z \to \eta$: $1 - z/L = \Lambda\eta$, so that it becomes:

$$\frac{d^2\Psi_\alpha^A}{d\eta^2} + \mathcal{K}_\alpha^A\left\{\left[1 - \left(\Lambda_\alpha^A\right)^2\left(1 - \mathcal{O}_M^A\right)\eta^2\right]\omega^2 - 4\frac{\mathcal{O}_M^A}{\mathcal{M}_m}\overline{C_A^\alpha}\,\mathcal{F}_{k_x k_y}\right\}\Psi_\alpha^A = 0. \tag{39}$$

Introducing new notations:

$$\mathcal{K}_\alpha^A = \frac{L^2\left(\Lambda_\alpha^A\right)^2\mathcal{M}_m}{\overline{a_Z^A}^2\,\overline{C_A^\alpha}\,\mathcal{O}_M^A}; \;\; \Lambda_\alpha^A = \sqrt{\frac{\overline{a_Z^A}\,\Omega_\alpha^A}{L\omega}}; \;\; \Omega_\alpha^A = \Omega_\alpha\,\tau_\alpha^A(n); \;\; \tau_\alpha^A(n) = \sqrt{\frac{g_n^A\,\mathcal{O}_M^A}{1 - \mathcal{O}_M^A}} \tag{40}$$

and

$$\begin{aligned}
{}^A\mathcal{Q}_{k_x k_y}^\alpha(\omega) &= \frac{L}{\overline{a_Z^A}}\left[\frac{\omega}{\Omega_\alpha^A}\left(1 - \mathcal{O}_M^A\right)^{-1} - 4\frac{\Omega_\alpha^A}{\omega}\mathcal{F}_{k_x k_y}\right] \equiv\\
&\equiv 2s + 1; \;\; s = 0, 1, 2, \ldots,
\end{aligned} \tag{41}$$

the above differential equation can be turned into Hermit-Weber one (Callavay, 1974):

$$\frac{d^2\Psi_\alpha^A}{d\eta^2} + \left[{}^A\mathcal{Q}_{k_x k_y}^\alpha(\omega) - \eta^2\right]\Psi_\alpha^A = 0, \tag{42}$$

with the solution:

$$\Psi_\alpha^A \equiv \Psi_\alpha^A(\eta) \longrightarrow {}^A\Psi_s^\alpha(\eta) = {}^A\mathcal{N}_s^\alpha\,\mathcal{H}_s(\eta)\,\mathrm{e}^{-\eta^2/2}, \tag{43}$$

where $\mathcal{H}_s(\eta)$ – is Hermitian's polynomial of the order s. In order that atomic displacements remain finite, it is necessary that ${}^A\mathcal{Q}_{k_x k_y}^\alpha(\omega)$ satisfies the identity condition (expressed by 41) which, in fact, insures the physics-chemical (crystallographic) stability of the model film-structure. This identity allows the determination of the allowed vibrational frequencies of the system:

$$ {}^{A}\omega^{\alpha}_{k_xk_y}(s) = \frac{\Omega^{A}_{\alpha}}{2}\left[\mathcal{G}^{A}_{s} + \sqrt{\left(\mathcal{G}^{A}_{s}\right)^2 + \mathcal{F}_{k_xk_y}}\,\right], \tag{44} $$

where $\mathcal{F}_{k_xk_y}$ is defined in 11, and

$$ \mathcal{G}^{A}_{s} \equiv \frac{2s+1}{\overline{N^{A}_{z}}};\quad \overline{N^{A}_{z}} = \frac{L}{\overline{a^{A}_{z}}}\left(1 - \mathcal{O}^{A}_{M}\right)^{-1}. $$

It is clear from this expression that none of the possible frequencies ${}^{A}\omega^{\alpha}_{k_xk_y}(s)$ vanishes, neither for $s = 0$, nor for (dimensionless) twodimensional vector $q = a^{-1}\sqrt{k_x^2 + k_y^2} \to 0$.
Since we have solved Hermite-Weber's equation 42 without taking into account the boundary conditions, it must be supplied by two boundary equations 31 and 32, for $z = 0$ and $z = L$, i.e. its solution 43 must satisfy these supplementary conditions. The substitution of 43 into 31 for $z = 0 \Rightarrow \eta = \left(\Lambda^{A}_{\alpha}\right)^{-1}$ and $q = 0$ gives:

$$ \left\{2 - \mathcal{O}^{A}_{M}\left[\frac{{}^{A}\omega^{\alpha}_{0}(s)}{\Omega^{A}_{\alpha}}\right]^2\right\}\mathcal{H}_s\left(\frac{1}{\Lambda^{A}_{\alpha}}\right) = \exp\left\{\left(1 - \frac{\overline{a^{A}_{z}}}{2L}\right)\frac{{}^{A}\omega^{\alpha}_{0}(s)}{\Omega^{A}_{\alpha}}\right\}\mathcal{H}_s\left(\frac{L - \overline{a^{A}_{z}}}{\Lambda^{A}_{\alpha}L}\right). \tag{45} $$

By analogous procedure, solution expressed by 43 with boundary equation 32, for $z = L \Rightarrow \eta = 0$ and $q = 0$, gives:

$$ \left\{2 + \left(1 - \mathcal{O}^{A}_{M}\right)\left[\frac{{}^{A}\omega^{\alpha}_{0}(s)}{\Omega^{A}_{\alpha}}\right]^2\right\}\mathcal{H}_s(0) = \exp\left\{\frac{\overline{a^{A}_{z}}}{2L}\frac{{}^{A}\omega^{\alpha}_{0}(s)}{\Omega^{A}_{\alpha}}\right\}\mathcal{H}_s\left(\frac{\overline{a^{A}_{z}}}{\Lambda^{A}_{\alpha}L}\right). \tag{46} $$

Using the relation $\mathcal{H}_s(x + c) = \mathcal{H}_s(x) + (2c)^s$ one can write:

$$ \mathcal{H}_s\left(\frac{L - \overline{a^{A}_{z}}}{\Lambda^{A}_{\alpha}L}\right) = (-1)^s\,\mathcal{H}_s\left(\frac{\overline{a^{A}_{z}}}{\Lambda^{A}_{\alpha}L}\right) + \left(\frac{2}{\Lambda^{A}_{\alpha}}\right)^s, $$

after which the equations 45 and 46 turn into a single one:

$$ \left\{2 - \mathcal{O}^{A}_{M}\left[\frac{{}^{A}\omega^{\alpha}_{0}(s)}{\Omega^{A}_{\alpha}}\right]^2\right\}\left[\exp\left\{\frac{{}^{A}\omega^{\alpha}_{0}(s)}{\Omega^{A}_{\alpha}}\right\}\mathcal{H}_s\left(\frac{1}{\Lambda^{A}_{\alpha}}\right) + (-1)^{s+1}\mathcal{H}_s(0)\right] = $$

$$ = \left\{(-1)^s\left[\frac{{}^{A}\omega^{\alpha}_{0}(s)}{\Omega^{A}_{\alpha}}\right]^2\mathcal{H}_s(0) + \left(\Lambda^{A}_{\alpha}\right)^{-s}\exp\left\{\frac{\overline{a^{A}_{z}}}{2L}\frac{{}^{A}\omega^{\alpha}_{0}(s)}{\Omega^{A}_{\alpha}}\right\}\right\}. \tag{47} $$

It is obvious from here that the parameters $\mathcal{M}_m$, $\mathcal{M}_d$, n, L and quantum number s are not mutually independent. In fact, for given values, from expression 44 they define the conditions for the existence of phonon states with the energies $\hbar\,{}^{A}\omega^{\alpha}_{k_xk_y}(s)$. From this equation, one can determine for which value of quantum number s the function ${}^{A}\omega^{\alpha}_{0}(s)$ attains minimal value. A graphical-numerical solving method gives $s_{min} = 2$. Numerical calculations and estimates were performed for the compound $La(Ba_{2-\varepsilon}La_{\varepsilon})Cu_3O_{7+\delta}$, where it was taken $n = 3$, $\gamma = 12$, $\mathcal{O}^{A}_{M} = 1/0{,}83$, $\varepsilon = 0{,}125$, $\delta = 0{,}11$, $v_z \approx 3$, $a_z \approx 1{,}2$; all based on data from

(Šetrajčić et al., 1990; Cava et al., 1987; Chu et al., 1987; Politis et al., 1987; Segre et al., 1987; Dietrich et al, 1987; Kuwahara, 1992; Notzel et al., 1992; Johnson, 1995) and (Ristovski et al., 1989; Djajić et al., 1991; Šetrajčić et al., 1994). Due to the discreteness of the solutions (2.43) and the initial model, their total number must be equal to $N_z + 1$. It follows from here that the quantum number s is bounded also from above: $s_{max} = N_z + 2$, i.e. $s \in [2, N_z + 2]$.
Substituting of the solution expressed by 43 into difference equation 30, and normalizing it, the expression for the phonon displacements becomes:

$$\begin{aligned} {}^{A}u^{\alpha}_{n_x,n_y}(z,t) &= \sum_{k_xk_y}\sum_{s=2}^{N_z^A-2} {}^{A}\mathcal{N}^{\alpha}_{n_z}(k_xk_y,s)\,\mathcal{H}_s(z)\,\mathrm{e}^{-2(1-z/L)^2/(2\Lambda^{A}_{\alpha})^2}\times \\ &\times\ \mathrm{e}^{i\left[a(k_xn_x+k_yn_y)-t\,{}^{A}\omega^{\alpha}_{k_xk_y}(s)\right]}; \\ {}^{A}\mathcal{N}^{\alpha}_{n_z}(k_xk_y,s) &\equiv (-1)^{n_z}\sqrt{\frac{\hbar}{M_zN_xN_yN_z^A\,{}^{A}\omega^{\alpha}_{k_xk_y}(s)}}\ ;\quad N_z^A = N_z + 4. \end{aligned} \tag{48}$$

The analysis of this expression shows that, contrary to phonon displacements in ideal unbounded structures (plane waves in all three spatial directions), and similar to the ideal films (superposition of standing wave and plane waves), here they represent the superposition of the plane waves in XY-planes and collective vibrational harmonic motion along z-direction. The amplitude of the phonon displacements is here $\sim 10^4\sqrt{2/N_z^A}$ times larger than the corresponding one in the bulk structures, and approximately equal (in fact slightly smaller) than in the ideal films[5].
According to all above mentioned, it follows from expression 44 that the dispersion law for phonons in the asymmetrically deformed crystalline films has the following form:

$${}^{A}E^{\alpha}_{k_xk_y}(s) \equiv \hbar\,{}^{A}\omega^{\alpha}_{k_xk_y}(s) = E^{\alpha}_{A}\left[\mathcal{G}^{A}_{s} + \sqrt{\left(\mathcal{G}^{A}_{s}\right)^2 + \mathcal{F}_{k_xk_y}}\right], \tag{49}$$

where $E^{\alpha}_{A} \equiv \hbar\Omega^{A}_{\alpha}/2$, and $s = 2,3,4,\ldots, N_z^A - 2$. Graphical presentation of this dispersion law in the form ${}^{A}\mathcal{E}^{z}_{s} \equiv \left[{}^{A}E^{z}_{k_xk_y}(s)/E^{z}_{A}\right]^2 = {}^{A}\mathcal{E}^{z}_{s}\left(\mathcal{F}_{k_xk_y}\right)$ is given in the Fig.3.
Fig.3 represent the energy spectrum of phonons in the asymmetrically deformed (ultrathin $N_z = 4$) crystalline films vs. two-dimensional (XY planar) wave vector. Besides the narrowing of the energy band with five discrete levels and the presence of the energy gap (with respect to the bulk band denoted by solid dashed lines) a shift of this band outside bulk limits can be noticed, corresponding to the appearance of the localized phonon modes.
One can see from this plot that non of the allowed energies, i.e. possible frequencies ${}^{A}\omega^{\alpha}_{q}(s)$ does not vanish for $q \to 0$, implying that the presence of boundaries together with the deformation of the atom distribution of the parabolic type (expressed by 35–38) leads to the appearance of the energy gap in the phonon spectrum, i.e. to the possible creation of the phonons of only the optical type. Contrary to the dispersion law for phonons in unbounded and nondeformed structures, where minimal and maximal frequency of the acoustic phonon

[5] See the comment bellow the expression 22.

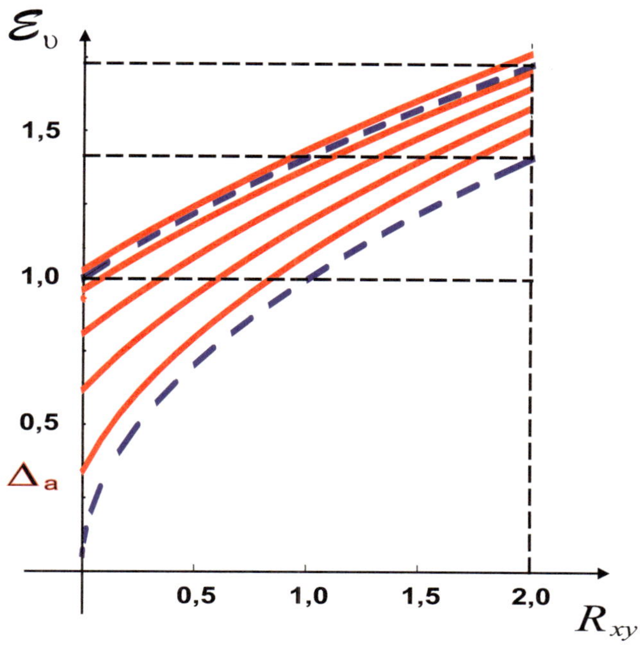

Fig. 3. Phonon spectra in the asymmetrically deformed ultrathin crystalline films

branches tend to 0, and $2\Omega_\alpha\sqrt{3}$, respectively, they have the following values here[6]:

$$(\omega_\alpha^A)_{min} \equiv {}^A\omega_0^\alpha(2) = \omega_\alpha^A\left(k_x = k_y = 0; s = 2\right) = \frac{5}{N_z^A}\Omega_\alpha^A > (\omega_\alpha^I)_{min}\,; \tag{50a}$$

$$(\omega_\alpha^A)_{max} \equiv \omega_\alpha^A\left(k_x = k_y = \frac{\pi}{a}; s = N_z^A - 2\right) \approx \frac{2(2N_z^A - 3)}{N_z^A}\Omega_\alpha^A < (\omega_\alpha^I)_{max}\,. \tag{50b}$$

It is evident from the same figure that the width of the allowed phonon energies band in the film is smaller[7]. Using expressions 49 and 50 one can determine the total narrowing of the energy band of phonons in the asymmetrically deformed film-structure with respect to the phonon bulk-band[8]:

$$\begin{aligned} \mathcal{W}_\alpha^A &\equiv \hbar\left\{(\omega_\alpha^B)_{max} - \left[(\omega_\alpha^A)_{max} - (\omega_\alpha^A)_{min}\right]\right\} \approx \\ &\approx \hbar\Omega_\alpha\left[2\sqrt{3} - \frac{4N_z^A - 11}{N_z^A}\tau_\alpha^A(n)\right] > \mathcal{W}_\alpha^I\,. \end{aligned} \tag{51}$$

More detailed analysis, functional behavior and physical interpretation as well as the possible effects caused by the existence of the frequency threshold in formula 49 and narrowing of the frequency range in expression 51 will be presented in the next Section.

[6]Compare with the expressions 23 and 24.

[7]It is also evident that this band is shifted so it leaves the bulk limits. This result which might mean the appearance of the localized phonon modes is not discussed here in particular, since it occurs for the higher values of the planar two-dimensional (XY) wave vector, for which the validity of the continual – long wavelength approximation might be questioned.

[8]See expression 25.

2.2.2 Symmetrical deformation

In the case of symmetrical sputtering (sputtering of the basic matrix – ideal crystalline film, by foreign atoms mutually perpendicular to both boundary surfaces of the film) within the framework of the parabolic approximation will be:

$$\begin{aligned} M(z) &\longrightarrow M_s(z) = A_M^s + B_M^s\left(z-\tfrac{L}{2}\right)^2; \\ a(z) &\longrightarrow a_s(z) = A_a^s + B_a^s\left(z-\tfrac{L}{2}\right)^2. \end{aligned} \tag{52}$$

The constants A and B are determined from the boundary conditions:

$$\begin{aligned} M_s(0) &= M_s(L) = n^{-1}\frac{\mathcal{M}_d}{\mathcal{O}_M^s}; \; \mathcal{O}_M^s = 1 + n^{-1}\frac{\mathcal{M}_d}{\mathcal{M}_m}; \\ M_s\left(\frac{L}{2}\right) &= \mathcal{M}_m; \; a_s(0) = a_s(L) = \frac{a_z}{n}; \; a_s\left(\frac{L}{2}\right) = a_z \end{aligned}$$

(n is the number of sputtered atoms of the mass $\mathcal{M}_d$ around the atom of the mass $\mathcal{M}_m$ of the basic – unsputtered matrix), so that we obtain:

$$M_s(z) = \frac{\mathcal{M}_m}{\mathcal{O}_M^s}\left[\mathcal{O}_M^s - \left(1-2\frac{z}{L}\right)^2\right]; \tag{53a}$$

$$a_s(z) = a_z\left[1 - \frac{n-1}{n}\left(1-2\frac{z}{L}\right)^2\right]. \tag{53b}$$

Due to $C_\alpha^s(z) = \text{const}(\alpha)\, a_s^{-\gamma}(z)$, where γ – is the decay exponent of the interatomic potentials with distance, sputtering will also cause the change of the Hooke's constants of elongation:

$$C_\alpha^s(z) = C_z^\alpha\left[1 + \gamma\frac{n-1}{n}\left(1-2\frac{z}{L}\right)^2\right]. \tag{54}$$

In order to simplify further analysis, just as in the previous case, instead of 53b and54 we shall use their values averaged over the total film width (L):

$$\begin{aligned} \overline{a_z^s} &\equiv \overline{a_s(z)} = f_n^s\, a_z; \quad f_n^s = \frac{2n+1}{3n}, \\ \overline{C_\alpha^s} &\equiv \overline{C_\alpha^s(z)} = g_n^s\, C_z^\alpha; \quad g_n^s = 1 + \gamma\frac{n-1}{3n}. \end{aligned} \tag{55}$$

The notations a_z and C_z^α in the expressions 53 and 54 are related to the corresponding quantities for the unsputtered matrix.
Now we can proceed to the solving of the equation 34. We introduce expressions 53a and 55 in it and perform the change of variable $z \to \zeta$: $(L-2z)/(2L) = \Lambda_\alpha^s \zeta$, after which it takes the form:

$$\frac{d^2\Psi_s^\alpha}{d\zeta^2} + \mathcal{K}_\alpha^s\left\{\left[\mathcal{O}_M^s - 4\left(\Lambda_\alpha^s\right)^2\zeta^2\right]\omega^2 - 4\frac{\mathcal{O}_M^s}{\mathcal{M}_m}\overline{C_s^\alpha}\,\mathcal{F}_{k_x k_y}\right\}\Psi_\alpha^s = 0, \tag{56}$$

$$\mathcal{K}_\alpha^s = \frac{L^2\left(\Lambda_\alpha^s\right)^2\mathcal{M}_m}{\overline{a_z^s}^2\,\overline{C_\alpha^s}\,\mathcal{O}_M^s}; \; \Lambda_\alpha^s = \sqrt{\frac{\overline{a_z^s}\Omega_\alpha^s}{2L\omega}}; \; \Omega_\alpha^s = \Omega_\alpha\, \tau_\alpha^s(n); \; \tau_\alpha^s(n) = \sqrt{g_n^s\, \mathcal{O}_M^s}.$$

Since the atoms in the studied film (along z-direction) represent the system of mutually coupled linear harmonic oscillators, above equation is reduced to the well-known (Callavay, 1974) Hermite-Weber's equation:

$$\frac{d^2 \overset{S}{\Psi}_\alpha}{d\zeta^2} + \left[{}^S\mathcal{Q}^\alpha_{k_x k_y}(\omega) - \zeta^2 \right] \overset{S}{\Psi}_\alpha = 0, \tag{57}$$

with:

$${}^S\mathcal{Q}^\alpha_{k_x k_y}(\omega) = \frac{L}{2\overline{a_z^S}} \left(\overset{S}{\mathcal{O}}_M \frac{\omega}{\overset{S}{\Omega}_\alpha} - 4\,\mathcal{F}_{k_x k_y} \frac{\overset{S}{\Omega}_\alpha}{\omega} \right). \tag{58}$$

In order to secure the crystal stability of the structure, the displacements of the atoms for any film width must remain finite, so we must introduce the restriction:

$${}^S\mathcal{Q}^\alpha_{k_x k_y}(\omega) = 2r+1; \quad r = 0,1,2,\ldots. \tag{59}$$

On the other hand, this is just the quantum-mechanical condition for the convergence of the solutions of the equation 57, which in this case can be expressed as:

$$\overset{S}{\Psi}_\alpha \equiv \overset{S}{\Psi}_\alpha(\zeta) \longrightarrow {}^S\Psi^\alpha_r(\zeta) = {}^S\mathcal{N}^\alpha_r\, \mathcal{H}_r(\eta)\, \mathrm{e}^{-\zeta^2/2}, \tag{60}$$

where $\mathcal{H}_r(\zeta)$ – is the Hermitian's polynomial of order r in terms of ζ.
Equating the condition equations 58 and 59, one obtains the expression for the possible phonon frequencies in the form:

$${}^S\omega^\alpha_{k_x k_y}(r) = \frac{\overset{S}{\Omega}_\alpha}{2} \left[\overset{S}{\mathcal{G}}_r + \sqrt{\left(\overset{S}{\mathcal{G}}_r\right)^2 + \mathcal{F}_{k_x k_y}} \right], \tag{61}$$

where the function $\mathcal{F}_{k_x k_y}$ is defined by formula 11, while

$$\overset{S}{\mathcal{G}}_r \equiv \frac{2r+1}{\overline{N_z^S}}; \quad \overline{N_z^S} = \frac{L}{\overline{a_z^S}} \overset{S}{\mathcal{O}}_M .$$

One can easily see by a simple analysis of this expression that allowed phonon frequencies express their discreteness, that they depend on all the parameters of the system (L, n, γ, $\mathcal{M}_m$, $\mathcal{M}_d$, a_z and $\mathcal{C}^\alpha_z$) and that their minimal values do not vanish neither for $r = 0$, nor for $(k_x, k_y) \to 0$.
Due to the presence of the boundaries, the solution 60 obtained without taking into account the presence of the boundaries of the studied system, must be additionally supplied by two boundary equations 31 and 32. Since we assume the symmetric conditions at the boundaries, these two equations are identical and after the substitution 60 they give:

$$\exp\left\{ \frac{{}^S\omega^\alpha_0(r)}{\overset{S}{\Omega}_\alpha} \left(1 - \frac{\overline{a_z^S}}{L} \right) \right\} \mathcal{H}_r\left(\frac{2\overline{a_z^S} - L}{2\overset{S}{\Lambda}_\alpha} \right) = \tag{62}$$
$$= \left\{ \frac{3}{2} - \left[\frac{{}^S\omega^\alpha_0(r)}{\overset{S}{\Omega}_\alpha} \right]^2 \frac{1}{n} \frac{\mathcal{M}_d}{\mathcal{M}_m} \right\} \mathcal{H}_r\left(-\frac{L}{2\overset{S}{\Lambda}_\alpha} \right).$$

One can see from here that the parameters of the studied system $\mathcal{M}_m$, $\mathcal{M}_d$, r, n and L (or N_z) are not mutually independent. In fact, for given values and through expression 61

they determine the conditions of the existence of phonon states with the energies $\hbar\,{}^{S}\omega_{q}^{\alpha}(r)$. Numerical solving and combination of the equations 61 and 62 allow us to determine the lowest possible energy state with $r = r_{min}$ and $q = 0$. These calculations[9] have shown that the value of the quantum number r_{min} can not be lower than 1. Due to the discreteness of the solutions (formula 61) and the initial model, their total number must be equal to $N_z + 1$. It follows from here that the quantum number r must be bounded from above, too: $r_{max} = N_z + 1$, i.e. $r \in [1, N_z + 1]$.
Substituting the solution expressed by 60 into 29 and normalizing it, the expression for the phonon displacements becomes:

$$\begin{aligned} {}^{S}u_{n_x,n_y}^{\alpha}(z,t) &= \sum_{k_x k_y}\sum_{r=1}^{N_z^S-2} {}^{S}\mathcal{N}_{n_z}^{\alpha}(k_x k_y, r)\,\mathcal{H}_r(z)\,\mathrm{e}^{-(1-2z/L)^2/(2\Lambda_{\alpha}^{S})^2}\times \\ &\times\ \mathrm{e}^{i\left[a(k_x n_x + k_y n_y) - t\,{}^{S}\omega_{k_x k_y}^{\alpha}(r)\right]}; \\ {}^{S}\mathcal{N}_{n_z}^{\alpha}(k_x k_y, r) &\equiv (-1)^{n_z}\sqrt{\frac{\hbar}{M_z N_x N_y N_z^S\,{}^{S}\omega_{k_x k_y}^{\alpha}(r)}};\quad N_z^S = N_z + 3. \end{aligned} \tag{63}$$

The analysis of this expression shows that contrary to the phonon displacements in the ideal unbounded structures (plane waves in all three spatial directions), similar to ideal films (superposition of the standing and plane wave) and in the same way as in asymmetrically deformed films, they represent here the superposition of the plane waves in XY-planes and the collective oscillatory harmonic motion along z-direction. The amplitude of the phonon displacements is of order $\sim 10^4\sqrt{2N_z^{-1}}$ times higher than the corresponding one in the bulk structures, and approximately the same as in the ideal crystalline films[10]. The largest difference between the bulk and film structures is for the thin films. We must mention also, that any relevant difference between the ideal and deformed film-structures appears only for ultrathin films, but the quantitative analysis of this dependence within their framework of of this analysis (continual approximation), can not be reliably performed[11].
According to all the above results, the solution expressed by 61 leads to the dispersion law of phonons in symmetrically deformed crystalline films:

$${}^{S}E_{k_x k_y}^{\alpha}(r) \equiv \hbar\,{}^{S}\omega_{k_x k_y}^{\alpha}(r) = E_{S}^{\alpha}\left[\mathcal{G}_r^{S} + \sqrt{\left(\mathcal{G}_r^{S}\right)^2 + \mathcal{F}_{k_x k_y}}\right], \tag{64}$$

where $E_S^{\alpha} = \hbar\Omega_{\alpha}^{S}/2$ and $r = 1,2,3, \dots, N_z^S - 2$. The plot of this dispersion law in the form:

$${}^{S}\mathcal{E}_{k}^{z} \equiv \left[\frac{{}^{S}E_{k_x k_y}^{z}(r)}{E_S^{z}}\right]^2 = {}^{S}\mathcal{E}_{k_x k_y}^{z}\left(\mathcal{G}_r^{S}\right)$$

is given at the Fig.4.
The Fig.4 represent the energy spectrum of phonons in the symmetrically deformed (ultrathin $N_z = 4$) crystalline films vs. two-dimensional (XY planar) wave vector. Besides the narrowing

[9]Estimates were based on the data from Refs. (Šetrajčić et al., 1990), (Cava et al., 1987; Chu et al., 1987; Politis et al., 1987; Segre et al., 1987; Dietrich et al, 1987; Kuwahara, 1992; Notzel et al., 1992; Johnson, 1995) and (Ristovski et al., 1989; Djajić et al., 1991; Šetrajčić et al., 1994) for the compound

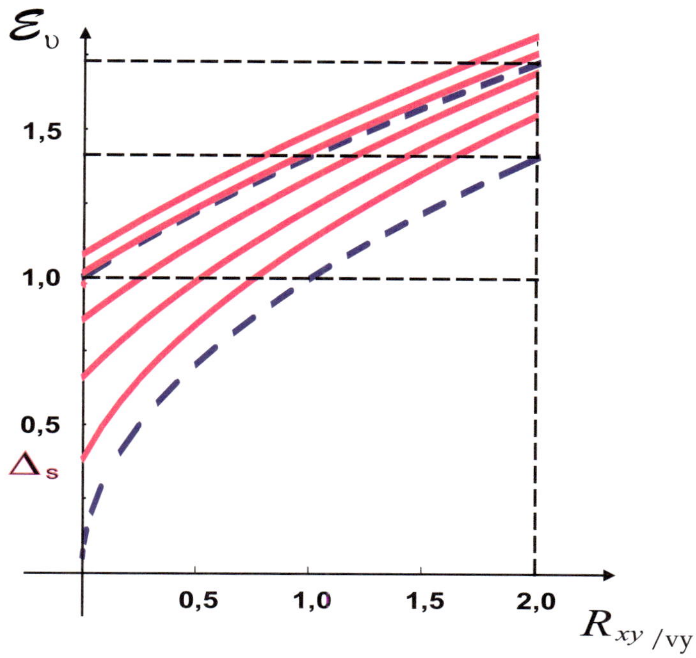

Fig. 4. Phonon spectra in the symmetrically deformed ultrathin crystalline films

of the energy band with five discrete levels and the presence of the energy gap (with respect to the bulk band denoted by thin dashed lines) a shift of this band outside bulk limits can be noticed, corresponding to the appearance of the localized phonon modes.
It is clear from this plot that none of the allowed energies $\overset{S}{E}{}^{\alpha}_{q}(r)$ does not vanish when two-dimensional vector $q \equiv a\sqrt{k_x^2 + k_y^2} \to 0$, meaning that the presence of boundaries together with the deformation of the parabolic type (expressions 52–55) of the atom distribution leads to the appearance of the energy gap in the phonon spectrum, i.e. to the possible creation of phonons of only optical type. As compared to the acoustic phonon energies in unbounded and nondeformed structures, here the minimal and maximal frequencies (compare with formula 50):

$$(\omega^S_{\vec{\alpha}})_{min} \equiv {}^S\omega^{\alpha}_{0}(1) = \omega^S_{\vec{\alpha}}\left(k_x = k_y = 0; r = 1\right) = \frac{3}{N^S_z}\overset{S}{\Omega}_{\alpha} > (\omega^A_{\vec{\alpha}})_{min}\,; \qquad (65a)$$

$$(\omega^S_{\vec{\alpha}})_{max} \equiv \omega^S_{\vec{\alpha}}\left(k_x = k_y = \frac{\pi}{a}; r = N^S_z - 2\right) \approx \frac{2\,(2\,N^S_z - 3)}{N^S_z}\overset{S}{\Omega}_{\alpha} < (\omega^A_{\vec{\alpha}})_{max}\,. \qquad (65b)$$

It is also evident from the same figure that there arises a narrowing of the band of allowed phonon energies in the studied film[12]. On the basis of formulas 64 and 65 one can

$La(Ba_{2-\varepsilon}La_{\varepsilon})Cu_3O_{7+\delta}$, where it was taken $n = 3$, $\gamma = 12$, $\varepsilon = 0{,}125$, $\delta = 0{,}11$, $v_z \approx 3$, $a_z \approx 1{,}2$.
[10]See the comment bellow the expression 21.
[11]The doubts that these results are the consequence of the applied (parabolic) approximation are rejected after testing on the simplest possible examples of three- and five-layered structures (Šetrajčić et al., 1990; Djajić et al., 1991), which can be also solved exactly.
[12]Besides that there is a visible shift of this band and its leaving the bulk limits. This results in the

determine the total narrowing of the energy band of phonons in the symmetrically deformed film-structures with respect to the phonon bulk-band (compare also with expression 51):

$$\begin{aligned} \mathcal{W}_{\alpha}^{S} &\equiv \hbar\left\{(\omega_{\alpha}^{B})_{max} - \left[(\omega_{\alpha}^{S})_{max} - (\omega_{\alpha}^{S})_{min}\right]\right\} \approx \\ &\approx \hbar\Omega_{\alpha}\left[2\sqrt{3} - \frac{4N_z^S - 9}{N_z^S}\,\tau_{\alpha}^{S}(n)\right] > \mathcal{W}_{\alpha}^{A}. \end{aligned} \tag{66}$$

The analysis of the functional behavior and physical interpretation, as well as the possible effects caused by the existence of the frequency threshold 65 and narrowing of the frequency range 66 will be presented in the next Section.

3. Energy gap and state density of phonons in ultrathin films

The basic property of the energy spectra of the phonons in the studied crystalline films is the narrowing of the allowed energy band[13]. Physically, the most interesting result of our calculations of the dispersion law is the presence of the energy gap, i.e. the existence of the acoustical phonons of the optical type. Analyzing the expressions 14, 44 and 61, and also the formulas 23, 50a and 65a, it is visible that:

$$\Delta_{\alpha}^{F} \equiv \hbar\,(\omega_{\alpha}^{F})_{min} \sim (N_z^F)^{-1}; \quad F \equiv \{I, A, S\}, \tag{67}$$

i.e. that the size of the gap decreases hyperbolically with the number of atomic XY layers (parallel to boundary surfaces) and that in the classical limit ($N_z^F \to \infty$) it vanishes. In Fig.5 we show this dependence for the studied types (nondeformed and deformed) crystalline film-structures.

Fig.5 shown the dependence of the size of energy gap or phonon activation temperature on the film width (i.e. the number of perpendicular layers) of particular film types: I – ideal crystalline film-structures, A – asymmetrically deformed crystalline film-structures, S – symmetrically deformed crystalline film-structures.

It can be clearly seen from the figure that for the same film width larger value of the gap belongs to deformed films (especially the symmetric ones) and that the effect is most expressed for ultrathin films ($N_z \leq 10$), meaning that the doping process can (in relatively small amount) to increase the effect (appearance of the gap) which is the consequence of the finiteness of (one of) the dimensions of the system[14].

Since the films studied are produced by "cutting them out" of the ideal infinite tetragonal structure with simple unit cell, it is clear that only acoustical phonon branches can appear. The existence of the gap, as the basic characteristic of the optical phonon branches, does not

---

phenomenon of localized phonon modes, which will not be discussed here in detail, since they appear for higher values of the perpendicular (z) wave-vector, for which the validity of the applied continual – long-wavelength approximation is questioned.

[13] See the expressions 25, 51 and 66 and notice that it is exclusive consequence of the boundness of the system – finite film width. Doping (within the framework of the parabolic model), can increase this effect only slightly, but for $L \to \infty$ (i.e. for $N_z \gg 10$) the narrowing magnitude $\mathcal{W} \to 0$.

[14] These, at first sight, puzzling conclusions become rather realistic on the basis of the general discussion based on the uncertainty relations. It is well-known that any micro-particle moving within the bounded region of space, can not possess zero energy. Applying the uncertainty relations $\Delta p_z \Delta z \geq \hbar/2$ and taking $\Delta z = N_z a_z = L_z$ and $\Delta p_z = \Delta E/v$, where v is the speed of sound, we obtain: $\Delta E \geq \hbar v/(2L_z)$. In bilk structures with $L_z \sim 1$ cm and $v = 4 \cdot 10^3$ m/s, it follows that $\Delta E \sim 10^{-10}$ eV, and this value is practically undetectable. On the other hand, in exceptionally thin films with the width $L \sim 10^{-7}$ cm, we find $\Delta E \sim 1$ meV, which is observable, experimentally accessible quantity.

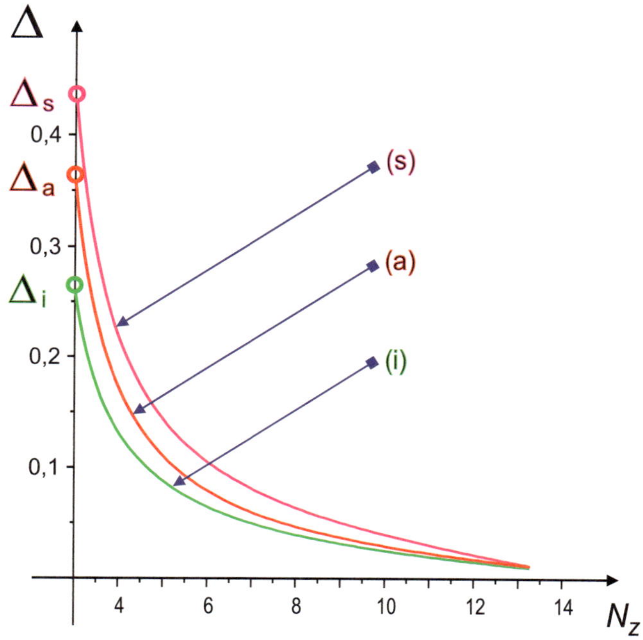

Fig. 5. Phonon energy gap vs. ultrathin film thickness

mean the elimination of the acoustical ones (as was the explanation in Refs. (Tošić et al., 1987; Šetrajčić et al., 1990; Djajić et al., 1987), but a completely new physical effect – the appearance of the acoustical phonons of the optical type. This agrees with the classical (thermodynamic) limit: for $N_z \to \infty$ it also follows that $T_{ac} \to 0$.
One can assign to the energy gap corresponding activation temperature:

$$T_{ac}^{F} = \frac{\Delta_F}{k_B} \equiv \frac{\Omega_\alpha}{k_B}\,\varphi_F(N_z) \approx$$
$$\approx \begin{cases} 35\ \mathrm{K}, & \text{for ideal } (F=I) \text{ nondeformed films;} \\ 38\ \mathrm{K}, & \text{for asymmetrically } (F=A) \text{ deformed films;} \\ 40\ \mathrm{K}, & \text{for symmetrically } (F=S) \text{ deformed films,} \end{cases} \tag{68}$$

for the typical values of the parameters[15] according to data from Refs. (Cava et al., 1987; Chu et al., 1987; Politis et al., 1987; Segre et al., 1987; Dietrich et al, 1987; Kuwahara, 1992) and (Notzel et al., 1992; Johnson, 1995; Ristovski et al., 1989; Djajić et al., 1991; Šetrajčić et al., 1994). According to al the above mentioned, physical interpretation which could be assigned to these temperatures might be the following: No phonons can appearing in the crystalline film-structure up to T_{ac},[16] and this structure behaves as completely "frozen". For phonon

[15]Following experimental data we have chosen: $\mathcal{M}_m = (2-x)\mathcal{M}_{\mathrm{Y}} + \mathcal{M}_{\mathrm{Cu}} + (4-\delta)\mathcal{M}_{\mathrm{O}}$, $\mathcal{M}_d = (x-\delta)\mathcal{M}_{\mathrm{Ba}}$ for the compound $\mathrm{Y}_{2-x}\mathrm{Ba}_{x-\delta}\mathrm{CuO}_{4-\delta}$, $x,\delta \in (0{,}5,\ 1{,}0)$, $n=4$, $\gamma=12$, $a_z \approx 1{,}2$ nm, $v \approx 3\cdot 10^3$ m/s; we gave a detailed explanation in our Refs. (Šetrajčić et al., 1992; 1994).
[16]This is the consequence of the existence of boundary surfaces in the system studied. The very effect of the sputtering, although not decisive, definitely contributes to this effect, especially for the symmetrical doping.

states to be created, energy must be brought to the system from outside (thermal, mechanical, etc) at least in the amount of the gap size. This could be also an acceptable explanation of the well-known effect (CRC HCP, 1989; Notzel et al., 1992) that superconductive critical temperature of the films is always higher than in the corresponding bulk structures.

From the plot (Fig.5) and from 68 one can notice two trends. The first one that the appearance of the activation temperature (energy gap) is the consequence of the existence of finite (small) film width, i.e the boundness of the structure. The sputtering which disturbers the symmetry of the distribution of atoms in the film along the direction in which the film is bounded (two parallel boundary surfaces) can only enhance this effect[17], and can not cause it in any way:

$$\Delta_S > \Delta_A > \Delta_I \gg \Delta_B \equiv 0 \tag{69}$$

(compare the expressions 23, 50a and 65a with ω_{min}).

The calculations of the energy bandwidth also support the statement that thin-layer, i.e. film-structures possess better superconductive characteristics. If we compare the results expressed by 25, 51 and 66, it is clear that:

$$\mathcal{W}_S > \mathcal{W}_A > \mathcal{W}_I \gg \mathcal{W}_B \equiv 0. \tag{70}$$

These expressions define the magnitude of the energy band narrowing in the films with respect to bulk band. We see that in the film-structures the allowed band of phonon energies is much smaller than the one in the corresponding massive sample. Taking into account the previous comment (the existence of the gap i.e. the rise of the ground state energy level) it is clear that the probability of the electron scattering on atomic vibrations is decreased (electric resistance decreases), but also that the effective – attractive – electron-phonon interaction "straightens" because virtual phonons possess more energy (phonons of the optical type). On the other hand, this can be used for the explanation of lower critical density of the electric current in the high-temperature superconductors (smaller number of phonons participates in the effective electron-phonon interaction), and less pronounced isotope effect (only higher energy phonons interact with electrons, so the role of the atoms is minority).

To support these statements, it is necessary to determine the density of states of the phonons at relevant energy levels, i.e. their distribution over the energies in film-structures. We have applied the well-known Debye's approach (Mahan, 1983; Jones & March, 1985), adapted for our model.

Following the standard definition (Maradudin, 1987; Cottam & Tilley, 1989; Callavay, 1974), the number of allowed values of the quasi-momentum $\vec{k} \equiv (k_x, k_y, k_z)$ per unit volume of $\vec{k}$-space is: $\frac{N_x}{k_x^{max}} \cdot \frac{N_y}{k_y^{max}} \cdot \frac{N_z}{k_z^{max}}$, where, according to the model, $k_x^{max} = k_y^{max} = \frac{2\pi}{a}$, and $k_z^{max} < \frac{2\pi}{a_z}$ (this value depends on the film kind). If we approximate the volume in the momentum space (in the long-wavelength approximation) by a cylinder of the basis πk^2 ($k^2 = k_x^2 + k_y^2$) and height k_z^{max}, in the frequency range ω, $\omega + d\omega$ the total number of the phonon states can

[17]This is not an abstract theoretical conclusion, since the parameters of the system were chosen from and according to experimental data, and not for some general cases. We are sure that for some other set of the parameters, the theoretical result of the sputtering might even oppose to this effect, i.e. to decrease T_{ac} with respect to the undoped film. Our leading idea was to cast at least some light in the shadows surrounding the mechanism of high-temperature superconductivity, so that is the reason for our choice of data.

be expressed as:

$$\mathcal{N}_F = \frac{N_F}{4\pi} a^2 k^2 ; \quad N_F = N_x N_y N_z^F . \tag{71}$$

The density of phonon states is then:

$$\mathcal{D}_F(\omega)|_{\omega=\omega_F} \equiv \frac{d\mathcal{N}_F}{d\omega_F} = \left(\frac{d\omega_F}{dk}\right)^{-1} \frac{N_F}{2\pi} a^2 k/,. \tag{72}$$

Long-wavelength approximation applied here allows us to choose and retain only the lowest possible value of the projection k_z: k_z^{min}. Taking this into account:

$$\omega_F \equiv \omega(k) \approx \Omega_F \left[\sqrt{a^2k^2 + \Delta_F^2} + \Delta_F (1 - \delta_{FI})\right] ; \quad \Delta_F = \frac{\kappa_F}{N_z^F} \tag{73}$$

(magnitudes and values for κ_F and N_z^F are given in the Table 1) and expressions 23, 24, 49, 50, 64 and 65 we can determine:

$$\mathcal{D}_F(\omega_F) = \frac{N_F}{2\pi\Omega_F^2} \left[\omega_F - \Omega_F \Delta_F (1 - \delta_{FI})\right] \tag{74}$$

(the values for Ω_F are also given in Table 1).
Table 1 shown the values of the parameters used for the calculation and comparison of the densities of phonon states and Debye's frequencies for the studied types of crystalline film-structures. The numerical values of the coefficients are calculated on the basis of the experimental data for high-temperature superconducting ceramics of the Y–Ba–Cu–O type from Refs. (Cava et al., 1987; Chu et al., 1987; Politis et al., 1987; Segre et al., 1987; Dietrich et al, 1987) and (Kuwahara, 1992; Notzel et al., 1992; Johnson, 1995); calculation procedure in details is given in our papers (Tošić et al., 1987; Šetrajčić et al., 1990; Djajić et al., 1987) and on conference talks (Ristovski et al., 1989; Djajić et al., 1991; Šetrajčić et al., 1994).
One can clearly see from the expression (3.8) that the density of phonon states for a given value of the frequency $\omega = \omega_F(k)$ is lower than in the bulk structures, and the lowest in the (symmetrically) deformed films:

$$\mathcal{D}_S < \mathcal{D}_A < \mathcal{D}_I \ll \mathcal{D}_B \tag{75}$$

and that they are posses linear dependence on the frequency.
If one would "abandon" the long-wavelength approximation then the expression 74 would have slightly different form: instead of a linear function of ω, the density of states would turn into a δ-function, since $\Delta_F \to \mathcal{G}_F$ which has a finite discrete series of possible values[18]. In this way, one would confirm the results of the Einstein's approach to the analyze of the density of phonon states (Jones & March, 1985).
At the end of this Section, let us determine the values of the Debye's frequencies for all three kinds of films. According to the model described, if the sample contains N_F unit cells, than the total number of states is also equal to N_F, so from:

$$N_F \equiv \int_0^{\omega_D^F} \mathcal{D}_F(\omega)\, d\omega = \frac{N_F \omega_D^F}{4\pi\Omega_F^2} \left[\omega_D^F - 2\Omega_F \Delta_F (1 - \delta_{FI})\right] \tag{76}$$

[18]It can be seen from the expressions 15, 16, 44 and 61.

	$F = I$	$F = S$	$F = A$
κ_F	π	4,2	4,7
N_z^F	$N_z + 2$	$N_z + 3$	$N_z + 4$
Ω_F/Ω	1,00	1,12	1,21

Table 1. Debye's parameters for different types ultrathin films

it follows:

$$\omega_D^F = 2\Omega_F \sqrt{\pi} \left[\sqrt{1 + \frac{\Delta_F^2}{4\pi}(1-\delta_{FI})^2} + \frac{\Delta_F}{2\sqrt{\pi}}(1-\delta_{FI}) \right]. \tag{77}$$

Simple analysis of this expression leads to another relation:

$$\omega_D^S < \omega_D^A < \omega_D^I \ll \omega_D^B. \tag{78}$$

For typical values of the parameters (already listed) one can estimate that Debye's frequencies in the film-structures are 10 – 15 % lower than in the corresponding bulk-structures. This means that in thinlayered structures there exist "softer" phonons.
It is interesting to compare the values of the density of phonon states at Debye's frequencies:

$$\frac{\mathcal{D}_F(\omega_D^F)}{\mathcal{D}_B(\omega_D^B)} = \frac{1}{\sqrt[3]{36\pi}} \frac{\Omega_B}{\Omega_F} \frac{N_z^F}{N_z^B}. \tag{79}$$

Since $N_z^F \sim 50$, and $N_z^B \sim 10^8$ it is clear that:

$$\mathcal{D}_F(\omega_D^F) \ll \mathcal{D}_B(\omega_D^B). \tag{80}$$

So, in the film-structures, i.e. the structures with broken translational symmetry, the population of phonons with Debye's frequencies is extremely small – much smaller than in the corresponding unbounded and nondeformed crystalline structures.
The phonons with precisely Debye's frequencies are responsible for the electrical and thermal transport properties of the materials, one can conclude that there will occur a large difference in these physical properties in bounded and deformed films with respect to the ideal and unbounded structures although there are no chemical or crystallographic differences. According to the results 78 and 80 – lower values of Debye's frequencies and lower densities of phonon states in the films studied – one can expect that film-structures will be poorer electric (lower conductivity) and heat (lower capacity) conductors. This is well known and well tabulated fact, for examples for metals and metallic alloys (CRC HCP, 1989).
Since the idea to study such film-structures arose after the discovery of the high-temperature superconductors and their specific properties differing them from the classical superconducting materials, we can not avoid turning our attention at the end to the possible consequences that these results could bear to the superconducting properties.
Due to smaller phonon population and the appearance of softer phonons in small-grain perovskite structures (Bednorz & Müller, 1988) of Y–Ba–Cu–O type and the similar ones, one can conclude that the probability of Cooper's pairs creation is smaller. This should result in the lower value of the (critical) density of (superconductive) current.
On the other hand, in the expression for matrix elements ($\mathcal{V}$) of the effective electron-electron interaction within the framework of the BCS theory there occur Debye's frequencies: $\mathcal{V} \sim$

$\omega_D^{-1/2}$, which, taking into account the results obtained, undoubtedly indicates that the attractive interaction of the electrons is more intense for strongly bounded and deformed structures, so they can be "more strongly" coupled into Cooper's pairs. For the destruction of such pairs one needs more energy so the critical temperature of these structures is much higher.

If we remember the fact that the materials with poor conductive properties (in the normal phase) possess better superconductive characteristics, then following above presented analyzes and estimates, we can state that the boundness of the structure and the deformation of its crystal symmetry is one of the important elements which explain the peculiarities of the phenomenon, and even the very mechanism of high-temperature superconductivity, in part.

At this stage we must notice that considerations presented here are only of qualitative nature and are the consequence of the incomplete[19] analysis of the behavior of a single subsystem (phonons) – one of the participants in the effects of high-temperature superconductivity. We are trying to establish if it is possible that maybe for copper-oxide ceramics we also deal with the same (BCS, i.e. Cooper's) mechanism of superconductivity – the same one as for the classical superconducting materials, only under substantially different conditions. The continuation of the research which should include the behavior of the electron system in the studied film-structures (these results are now in the stage of numerical treatment), and especially the formation and the analysis of the effective electron-(virtual phonon)-electron interaction will either confirm or deny our above exposed statements.

4. Phonon thermodynamics of thin film-structures

Forasmuch as the properties of anisotropic structures are conditioned by the change of dispersion law, it is necessary to observe behavior of certain thermodynamic properties towards obtainment of better understanding about those properties. Phonon participation in thermodynamic properties (or heat capacitance temperature behavior, i.e. generally – in heat transferring) in thin film was found in our previous paper (Lazarev et al., 2000; Jaćimovski et al., 2004; Šetrajčić et al., 2007; Ilić et al., 2007; Šetrajčić et al., 2009).

Getting that, when $k \to 0$ (in long-wave approximation: $4\left[\sin^2\left(ak_x/2\right)+\sin^2\left(ak_y/2\right)\right] \approx a^2k^2$, $k^2 = k_x^2 + k_y^2$), energies of all three phonon branches have non-zero values, it can be utilized dispersion relations 22, 49 and 64, in somewhat simplified form:

$$E(\vec{k}) = \sqrt{a^2k^2E_0^2 + \Delta_f^2}\,. \tag{81}$$

where

$$\Delta_f = ak_z^{min}E_0\,; \quad E_0 \equiv \hbar\sqrt{\frac{C_\alpha}{M}}\,. \tag{82}$$

It should be specifically emphasized that verification of phonon dispersion law at very low values of k is virtually impossible, so that verification of existence of phonon gap detects itself in measurement of low temperature thermal capacitances in film and corresponding ideal structure.

[19]For completing of these analyzes it is necessary to determine the phonon contribution in the heat capacity of the system.

The thermal capacitance is analyzed, whereby at first internal energy is calculated in terms of standard form (Mahan, 1983; Jones & March, 1985; Maradudin, 1987; Callavay, 1974):

$$U_f = 3 \sum_{k_x,k_y,k_z} E(\vec{k}) \left[\mathrm{e}^{E(\vec{k})/\theta} - 1 \right]^{-1} . \tag{83}$$

Going over from sum in last expression to integral in accordance with the formula[20]:

$$\sum_{k_x,k_y,k_z} \longrightarrow 3\,(N_z+1) \sum_{k_x,k_y} \longrightarrow \frac{3\,N_x N_y (N_z+1) a^2}{4\pi^2} \int_0^{2\pi} d\varphi \int_0^{k_{max}} k\,dk,$$

and taking $k_{max} \approx k_{_D} = \sqrt[3]{6\pi^2}$, after suitable notations:

$$\eta \equiv \sqrt{\frac{N_z^2}{3} + N_z + 1}\,; \quad \zeta \equiv \sqrt{1 + \left(\frac{N_z+2}{\pi} \sqrt[3]{6\pi^2} \right)^2}$$

and adequate operations, expression for internal energy has been obtained in form:

$$\begin{aligned} U_f(x) = \frac{3\,N_f}{4\pi^2} \frac{\Delta_f^4}{E_0^3}\, x^2 \Big\{ & \left[Z_2\left(\tfrac{1}{x}\right) - \eta^2 Z_2\left(\tfrac{\eta}{x}\right) + \eta^2\zeta^2 Z_2\left(\tfrac{\eta\zeta}{x}\right) - \zeta^2 Z_2\left(\tfrac{\zeta}{x}\right) \right] + \\ & +4x \left[Z_3\left(\tfrac{1}{x}\right) - \eta\, Z_3\left(\tfrac{\eta}{x}\right) + \eta\,\zeta\, Z_3\left(\tfrac{\eta\zeta}{x}\right) - \zeta\, Z_3\left(\tfrac{\zeta}{x}\right) \right] + \\ & + 6x^2 \left[Z_4\left(\tfrac{1}{x}\right) - Z_4\left(\tfrac{\eta}{x}\right) + Z_4\left(\tfrac{\eta\zeta}{x}\right) - Z_4\left(\tfrac{\zeta}{x}\right) \right] \Big\}, \end{aligned} \tag{84}$$

where the symbol x is introduced for reduced temperature: $x = \dfrac{\theta}{\Delta_f}$, $N_f = N_x N_y (N_z+1)$ and

$Z_r(X) = \sum_{j=1}^{\infty} j^{-r} \mathrm{e}^{-jX}$ – the functions are called Dyson's functions.

For finding of expression for the thermal capacitance per a unit cell (here: per an atom), the standard definitional form (Mahan, 1983; Jones & March, 1985; Maradudin, 1987; Callavay, 1974) is used:

$$C_f = \frac{1}{N_f} \frac{\partial U_f}{\partial T} \equiv \frac{k_B}{N_f} \frac{\partial U_f}{\partial \theta} = \frac{1}{\Delta_f} \frac{k_B}{N_f} \frac{\partial U_f}{\partial x} . \tag{85}$$

In accordance with that it is obtained:

$$\begin{aligned} C_f(x) = \frac{3k_B}{4\pi^2} \left(\frac{\Delta_f}{E_0} \right)^3 \Big\{ & \left[Z_1\left(\tfrac{1}{x}\right) - \eta^3 Z_1\left(\tfrac{\eta}{x}\right) + \eta^3\zeta^3 Z_1\left(\tfrac{\eta\zeta}{x}\right) - \zeta^3 Z_1\left(\tfrac{\zeta}{x}\right) \right] + \\ & +6x \left[Z_2\left(\tfrac{1}{x}\right) - \eta^2 Z_2\left(\tfrac{\eta}{x}\right) + \eta^2\zeta^2 Z_2\left(\tfrac{\eta\zeta}{x}\right) - \zeta^2 Z_2\left(\tfrac{\zeta}{x}\right) \right] + \\ & +18x^2 \left[Z_3\left(\tfrac{1}{x}\right) - \eta\, Z_3\left(\tfrac{\eta}{x}\right) + \eta\,\zeta\, Z_3\left(\tfrac{\eta\zeta}{x}\right) - \zeta\, Z_3\left(\tfrac{\zeta}{x}\right) \right] + \\ & + 24x^3 \left[Z_4\left(\tfrac{1}{x}\right) - Z_4\left(\tfrac{\eta}{x}\right) + Z_4\left(\tfrac{\eta\zeta}{x}\right) - Z_4\left(\tfrac{\zeta}{x}\right) \right] \Big\}. \end{aligned} \tag{86}$$

[20]The transition $\sum_{\vec{k}} \rightarrow \int d\vec{k} = \int d^3k$ of Descartes coordinates for film must to be carried out to cylindrical coordinates due to finite thickness

It is known that the phonon part in thermal capacitance of the system is described with cubic temperature dependence. By introducing nondimensional reduced temperature, this dependence amounts to: $C_b(x) = \frac{12}{5}\pi^4 N_b k_B \left(\frac{\Delta_f}{E_D}\right)^3 x^3$. For comparison of these dependencies, these and expression 86 are divided by the constant: $C_0 = \frac{k_B}{2}\left(\frac{\Delta_f}{E_D}\right)^3$, whose dimension is equal to dimension of thermal capacitance, and nondimensional properties are compared: $\mathcal{C}_{f/b} \equiv \frac{C_{f/b}}{C_0}$. On Fig.6 are shown relative (nondimensional) thermal capacitances of bulk (b) and film-structure (f) subject to the relative temperature x in low (a) and very low temperature region.

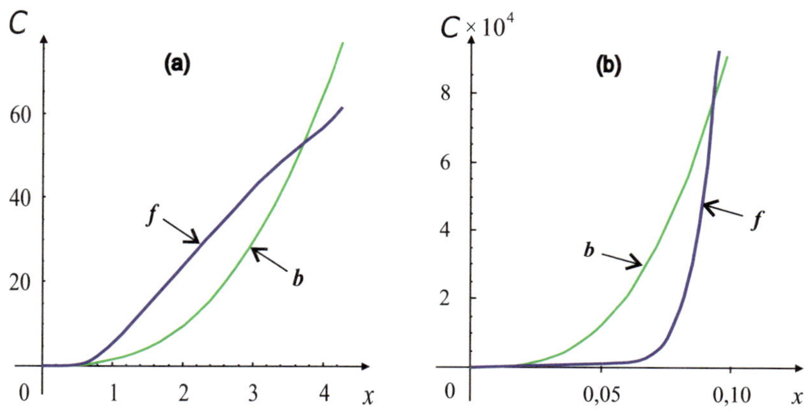

Fig. 6. Thermal capacitance vs. temperature at low and extremely low temperatures

On Fig.7 are shown relative (nondimensional) thermal capacitances of bulk (b) and film-structure (f) samples versus relative temperature x, for ultrathin – $N_z = 3$ (a), thin $N_z = 8$ (b) and thick $N_z = 48$ (c) film-structures, in comparatione with bulk ones.

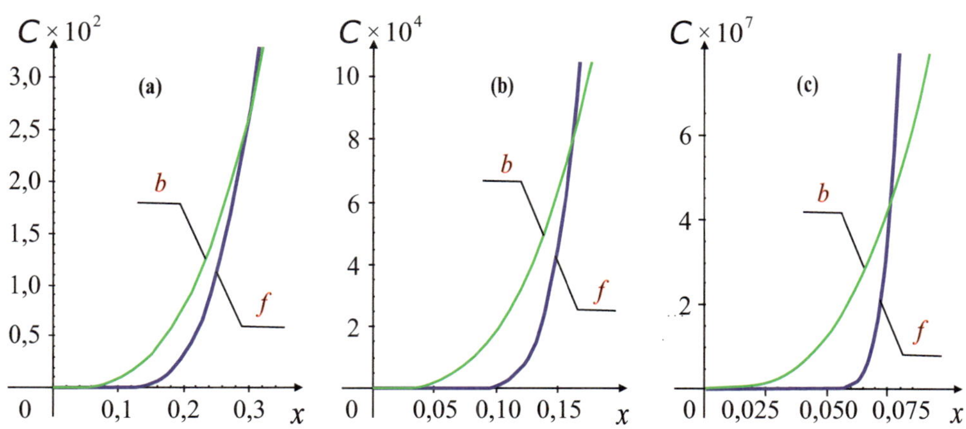

Fig. 7. Thermal capacitance in low temperature regime for ultrathin, thin, thick films

It can be seen that in low-temperature region (Fig.6) thermal capacitance of film is lower than that of massive specimens, whereas at the intermediate temperatures situation is reversed (Popov et al., 2003). Intersect point of two curves at low temperatures is moving – with increase of film-thickness – towards lower temperatures (Fig3 a –> c). Besides, it is noticeable that thermal capacitance of film with decrease of temperature declines faster than that of corresponding ideal structure, or slowly rises with the increase of temperature – to a certain upper temperature. Hence, for film heating from certain lower to a certain upper temperature, it is necessary to use more thermal energy per mass unit than for heating the same quantity of corresponding (with the identical crystallographical parameters) unbounded structure to the same temperature. It is in accordance with the fact that phonons in film have non-zero excitation energy.

5. Conclusion

Studying and comparing the phonon spectra and states in the ideal unbounded and nondeformed (bulk) structures and the structures with broken translational symmetry (films) we have reached the following conclusions.

1. Mechanical vibrations in bulk structures are plane waves in all directions, while in the films they represent the superposition of the standing waves in z-directions (perpendicular to the boundary surfaces) and plane waves in XY-planes (parallel to boundary surfaces).
2. The amplitude of phonon displacements in the films depends on the film width and it is $\sim 10^4\sqrt{2/N_z}$ times higher than in the ideal structures. This indicates their larger elastic "maneuvering space" without any negative effect to the mechanical properties of the given material (for example no breaking of interatomic bonds) which leads to higher resistance and higher melting point of the films with respect to bulk samples.
3. All three acoustic frequencies in bulk structures vanish for $\vec{k} \to 0$, while in the films they tend toward some minimal value depending on the film width. This means that phonons in the films possess the energy gap, that for their excitation (creation) one should spend certain energy, i.e. heat them up to certain – activation temperature, meaning that the system up to that temperature behaves as the "frozen" one, as if the phonons were not present.
4. Phonon gap, besides depending on the film width, depends also on the type of the atoms and their distribution along z-direction and also on the stoichiometric relation of the atoms injected in the films.
5. The densities of phonon states and Debye's frequencies have lower values in the films than in the corresponding bulk structures. This implies that in the films, phonon excitations are more "difficult" to appear, that they are less "present" and that created acoustic phonons of the optical type (above the activation temperature) are energetically "softer" than the classical ones which appear in the bulk structures.
6. Since phonons with Debye's frequencies define thermal and electrical properties of the materials, this means that films are worse thermal and electrical conductors.

These analyzes show that the films are better superconductors than the corresponding bulk samples, made from the same material with the same crystalline structure. This statement, which is an experimental fact is supported by the following of our results.

1. In the films there appear standing phonon waves along z-directions, the collective property specific for the macroscopic quantum-mechanical state which is the characteristics of the superconductors. In the ideal structures, where there exist only plane phonon waves, there is no such property.
2. Higher values of the amplitudes of phonon displacements in the films indicate to the possibility of the long-range phonon interaction with other excitations which can occur in the system (electrons or holes). This induces the higher possible coherence length, i.e. the larger radius of the Cooper's pairs which can be created here.
3. The appearance of the energy gap in the phonon spectrum of the films means that up to the activation temperature these systems behave as completely frozen, i.e. without any mechanical vibrations which would cause the real resistance to the electrical current conduction.
4. Lower values of Debye's frequencies in the films could result in the higher values of (BCS) matrix elements of the effective electron-electron interaction. The attraction between paired electrons is stronger, and one must spend more energy for their destruction, so the critical temperature of these systems is higher.

Comparing the results between the phonon behavior in the ideal and deformed films one can conclude that the strongest influence onto the change of phonon states and spectra has just the existence of finite boundaries of the system, while less important (although not negligible) influence has the disturbance of the atom distribution (due to doping) within the film – especially when we talk about symmetrical doping.

Since phonons with Debye's frequencies are responsible for electrically and thermically induced transport properties of material (Mahan, 1983; Jones & March, 1985), it follows that the nanofilm-structure will be inferior electrical and thermic conductor in contrast with the relative massive structures, providing there is no chemical and structural differences between them.

On the other hand, it is well known fact that the more inferior electrical conductor materials is (under normal conditions), the better superconductor it becomes (Maradudin, 1987; Callavay, 1974). Due to that, the experimental fact can be concluded and justified, that in spatially very confined structures more qualitative superconductive properties have been achieved.

In the region of low temperatures, the thermal capacitance of film is lower than in massive structures, while in medium temperatures it is reversed. Intersect point of two curves at low temperatures is moving – with increase of film-thickness – towards lower temperatures. Besides, it is noticeable that thermal capacitance of film with decrease of temperature declines faster than that of corresponding ideal structure, or slowly rises with the increase of temperature – to a certain upper temperature.

Hence, for film heating from certain lower to a certain upper temperature, it is necessary to use more thermal energy per mass unit than for heating the same quantity of corresponding (with the identical crystallographical parameters) unbounded structure to same temperature.

It is well known that poorer electric conductors are better superconductors, so that in ultrathin films it is possible to achieve much better superconducting properties!

All these conclusions, derived on the basis of the results of the analyzes of this work, are of more or less qualitative nature and deal with the change of phonon states and spectra under the influence of the presence of boundaries and the breaking of crystalline symmetry of the structure, and the possible influence of these changes onto the macroscopic physical characteristics of these systems. Since we have taken only phonon contributions into account, they could not be considered as definitive and final. The continuation of the research should study the influence of the boundaries onto spectra and states of other elementary charge carriers and their mutual interaction in the presence of the changed phonon field. On the basis of these results one might be able to say something more concrete about the order of magnitude of the superconductive characteristics of the films.

5.1 Acknowledgements

The greatest thanks the authors of this paper owe their teacher, mentor and friend, unfortunately – a recently deceased Academician, Prof. Dr. Bratislav S. Tošić! Its proper and quality oriented suggestions were crucial impact on the value and importance of this work.
Financial support was provided in part by the Ministry of Science and Technological Development of the Republic of Serbia and Ministry of Science and Technology of the Republic of Srpska.

6. References

Tošić B.S.; Šetrajčić J.P.; Djajić R.P. & Mirjanić D.Lj. (1987). Phonons in Broken-Symmetry Structures. *Physical Review* B (Cond.Matter), Vol. 36, No. 17, December 1987, 9094–9097, ISSN 1098-0121.

Chang L.L. & Esaki L. (1992). Semiconductor Quantum Heterostructures. *Physics Today*, Vol. 45, No. 10, October 1992, 36–43, ISSN 0031-9228.

Bednorz J.G. & Müller K.A. (1988). Perovskite-type oxides – The New Approach to High-Tc Superconductivity. *Reviews of Modern Physics*, Vol. 60, No. 3, July 1988, 585–600, ISSN 0034-6861.

Šetrajčić J.P.; Djajić R.P.; Mirjanić D.Lj. & Tošić B.S. (1990). Phonon Spectra in Superconducting Ceramics. *Physica Scripta*, Vol. 42, No. 6, December 1990, 732–736 ISSN 641-760.

Harshman D.R. & Mills, A.P.Jr. (1992). Concerning the Nature of High-Tc Superconductivity: Survey of Experimental Properties and Implications for Interlayer Coupling. *Physical Review* B (Cond.Matter), Vol. 45, No. 18, May 1992, 10684–10712, ISSN 1098-0121.

Mahan G. (1983). *Many-Particle Physics*, ISBN 0306404117, Plenum, London.

Jones W. & March N.H. (1985). *Theoretical Solid State Physics*, ISBN 0-486-65016-2, Dover, New York.

Djajić R.P.; Šetrajčić J.P.; Mirjanić D.Lj. & Tošić B.S. (1987). Elimination of Acoustical Phonons by Mass Deformation. *International Journal of Modern Physics* B, Vol. 1, No. 3-4, August 1987, 1001–1004, ISSN 0217-9792.

CRC HCP-70 (1989-1990). *CRC Handbook of Chemistry and Physics*, 70th Edition, ISBN 0147-6262, Chemical Rubber, Clevelend.

Tošić B.S.; Mirjanić D.Lj. & Šetrajčić J.P. (1995). *Spectra of Elementary Excitations in the Anisotropic Film-Structure*, ISBN 86-7622-099-9, Physical Society of the Republic of Srpska, Banja Luka (in serbian).

Maradudin A.A. (1987). *Physics of Phonons* – 33th Winter School of Theoretical Physics, Karpacz (Poland).
Proceedings of the XXIII Winter School of Theoretical Physics (Lecture Notes in Physics), held in Karpacz, February 1987, ISBN-13 978-3540182443.

Tošić B.S.; Šetrajčić J.P.; Mirjanić D.Lj. & Z.V.Bundalo (1992). Low-Temperature Properties of Thin Films. *Physica* A, Vol. 184, No. 3-4, July 1992, 354–366, ISSN 0378-4371.

Šetrajčić J.P.; Mirjanić D.Lj.; Sajfert V.D. & Tošić B.S. (1992). Perturbation Method in the Analysis of Thin Deformed Films and a Possible Application. *Physica* A, Vol. 190,

No. 3-4, December 1992, 363–374, ISSN 0378-4371.

Šetrajčić J.P. & Pantić M. (1994). Contribution to Thermodynamic Analysis of Thin Films. *Physics Letters* A, Vol. 192, No. 2-4, September 1994, 292–294, ISSN 0375-9601.

Cottam M.G. & Tilley D.R. (1989). *Introduction to Surface and Superlatice Excitations*, ISBN 0750305886, University Press, Cambridge.

Callavay J. (1974). *Quantum Theory of the Solid State*, ISBN 978-0121552565, Academic Press, New York.

Cava R.J.; van Dover R.B.; Batlogg B. & Rietman E. A. (1987). Bulk Superconductivity at 36 K in $La_{1.8}Sr_{0.2}CuO_4$. *Physical Review Letters*, Vol. 58, No. 4, January 1987, 408–410, ISSN 0031-9007.

Chu C.W.; Hou P.H.; Meng R.L.; Gao L.; Huang Z.J. & Wang Y.Q. (1987). Evidence for Superconductivity above 40 K in the La–Ba–Cu–O Compound System. *Physical Review Letters*, Vol. 58, No. 4, January 1987, 405–407, ISSN 0031-9007.

Politis C.; Geerk J.; Dietrich M. & Obst B. (1987). Superconductivity at 40 K in $La_{1.8}Sr_{0.2}CuO_4$. *Zeitschrift fur Physik* B (Condensed Matter), Vol. 66, No. 2, June 1987, 141–146, ISSN 0722-3277.

Segre C.U.; Dabrowski B.; HinksD.G.; Zhang K.; Jorgensen J.D.; BenoM.A. & Schuller I.K. (1987). Oxygen Ordering and Superconductivity in $La(Ba_{2-x}La_x)Cu_3O_{7+\delta}$. *Nature*, Vol. 329, No. 6136, September 1987, 227–229, ISSN 1476-1122.

Dietrich M.R.; Fietz W.H.; Ecke J.; Obst B. & Politis C. (1987). Structure and Superconductivity of La1.8Sr0.2cuo4 And Multi-Phase $Y_{1.2}Ba_{0.8}CuO_4$ Under High Pressure. *Zeitschrift fur Physik* B (Condensed Matter), Vol. 66, No. 3, September 1987, 283–287, ISSN 0722-3277.

Kuwahara M. (1992). Improvement of Superconducting Properties in $YBa_2Cu_3O_x$. *Ferroelectrics*, Vol. 128, April 1992, 237–241, ISSN 1563-5112.

Notzel R.; Daweritz L. & Ploog K. (1992). Topography of High- and Low-Index GaAs Surfaces. *Physical Review* B (Cond.Matter), Vol. 46, No. 8, August 1992, 4736–4743, ISSN 1098-0121.

Johnson N.F. (1995). Quantum Dots: Few-Body, Low-Dimensional Systems. *Journal of Physics* (Cond.Matter), Vol. 7, No. 6, February 1995. 965–989, ISSN 0953-8984.

Ristovski Lj.M.; Tošić B.S. & Davidović B.S. (1989). The Similarities between Thin Sputtered Films and High-Tc Superconductive Ceramics. *Physica* C: Superconductivity, Vol. 160, No. 5-6, October 1989, 548–566, ISSN 0921-4534.

Djajić R.P.; Tošić B.S.; Šetrajčić J.P. & Mirjanić D.Lj. (1991). High Tc as a Consequence of Structure Deformation, *International Conference on Materials and Mechanism of Superconductivity*, Kanazawa (Japan).

Šetrajčić J.P., M.Pantić, Tošić B.S. & Mirjanić D.Lj. (1994). Phonon States in Broken-Symmetry Thin Films and Some Possible Consequences on Superconductivity, *2nd General Conference of the Balkan Physics Union*, Izmir (Turkey).

Lazarev S.; Pantić M.R.; Stojković S.M.; Tošić B.S. & Šetrajčić J.P. (2000). Thermodynamical Properties of Ultrathin Layered Structures. *Journal of Physics & Chemisty of Solids*, Vol. 61, No. 6, June 2000, 931–936, ISSN 0022-3697.

Jaćimovski S.K.; Šetrajčić J.P.; Tošić B.S. & Sajfert V.D. (2004). Thermodynamics of Mechanical Oscillations in Superlattices. *Materials Science Forum*, Vol. 453–454, September 2004, 33–36, ISSN 0255-5476.

Šetrajčić J.P.; Zorić V.M.; Vučenović S.M.; Mirjanić D.Lj.; Sajfert V.D.; Jaćimovski S.K. & Ilić D.I. (2007). Phonon Thermodynamics in Crystalline Nanofilms. *Materials Science Forum*,

Vol. 555, July 2007, 291-296, ISSN 0255-5476.

Ilić D.I.; Vučenović S.M.; Jaćimovski S.K.; Zorić V.M. & Šetrajčić J.P. (2007). *Phonon Spectra and Thermodynamics of Crystalline Nanowires.* in Low-Dimensional Materials Synthesis, Assembly, Property Scaling, and Modeling, http://www.mrs.org/s_mrs/sec_subscribe.asp?CID=8761&DID=194295 Eds. Shim M.; Kuno M.; Lin X-M.; Pachter R. & Kumar S., *Materials Research Society*, ISBN 0-931837-15-4, San Francisco (USA) June 2007, Vol. 50, pp.1-6.

Šetrajčić J.P.; Mirjanić D.Lj.; Vučenović S.M.; Ilić D.I.; Markoski B.; Jaćimovski S.K.; Sajfert V.D. & Zorić V.M. (2009). Phonon Contribution in Thermodynamics of Nano-Crystalline Films and Wires. *Acta Physica Polonica* A, Vol. 115, No. 4, April 2009, 778-782, ISSN 0587-4246.

Popov D.; Jaćimovski S.K.; Tošić B.S. & Šetrajčić J.P. (2003). Kinetics of Thin Films Mechanical Oscillations. *Physica* A, Vol. 317, No. 1, January 2003, 129–139, ISSN 0378-4371.

16

Insight Into Adsorption Thermodynamics

Dr. Papita Saha and Shamik Chowdhury
Biotechnology Department, National Institute of Technology-Durgapur, Mahatma Gandhi Avenue, Durgapur (WB)-713209 India

1. Introduction

Saving the environment to save the Earth and to make the future of mankind safe is the need of the hour. Over the past several decades, the exponential population and social civilization change, affluent lifestyles and resources use, and continuing progress of the industrial and technologies has been accompanied by a sharp modernization and metropolitan growth. The world is reaching new horizons but the cost which we are paying or we will pay in near future is surely going to be high. Among the consequences of this rapid growth is environmental disorder with a big pollution problem. Rapid industrialization, unplanned urbanization and unskilled utilization of natural water resources have led to the destruction of water quality in many parts of the world. In many developing countries, groundwater provides drinking water for more than one-half of the nation's population, and is the sole source of drinking water for many rural communities and some large cities. However, due to industrial, agricultural and domestic activities, a variety of chemicals can pass through the soil and potentially contaminate natural water resources and reservoirs.

In recent years, the surge of industrial activities has led to tremendous increase in the use of heavy metals, synthetic dyes and other toxic chemicals, and inevitably resulted in an increased flux of these substances in the aquatic environment. Environmental contamination by toxic heavy metals and synthetic dyes is becoming a serious dilemma now days due to their negative ecotoxicological effects and bioaccumulation in wildlife. Contamination primarily result from industrial activities, such as iron and steel production, the non-ferrous metal industry, mining and mineral processing, pigment manufacturing, tanning, dyeing, painting, photographic and electroplating, gas exhaust, energy and fuel production, fertilizer, food, cosmetics, pharmaceuticals, pesticide applications, and generation of municipal wastes. The contamination of water due to toxic heavy metal ions and synthetic dye molecules is accountable for causing several damages to the environment and adverse effects on public health. Heavy metals form compounds that can be toxic, carcinogenic or mutagenic even in low quantities and due to their mobility in natural water ecosystems, they are prioritized as major inorganic contaminants of the environment. Furthermore, heavy metal ions are nonbiodegradable and can accumulate in living tissues, thus becoming concentrated throughout the food chain. Even, their minor content can bio-accumulate and enter the food chain causing mental retardation, reduction in hemoglobin production and interference with normal cellular metabolism and consequently may damage nervous system. Strong exposure may cause gastric pain, nausea, vomiting, severe diarrhea, hemorrhage and even cancer in the digestive tract and lungs. On the contrary, synthetic dyes contained in wastewaters may affect photosynthesis by preventing light penetration, thereby compromising aquatic life. Additionally, dye molecules can decompose into carcinogenic aromatic amines under aerobic conditions which can cause serious health problems to humans and animals. Also, dyes can cause allergy, dermatitis, skin irritation and cancer in humans.

With the rising awareness of the occurrences of industrial activities which has intensified numerous deteriorations on several ecosystems and seriously threatens the human health and environment, the enforcement of stringent rules and regulations concerning the emission of contaminants from industrial waste streams by various regulatory agencies has been promulgated. Simultaneously, a developing research by the invention of a wide range of treatment technologies (precipitation, coagulation–flocculation, sedimentation, flotation, filtration, membrane processes, electrochemical techniques, biological process, chemical reactions, adsorption, ion exchange, photocatalytic degradation, sonochemical degradation, micellar enhanced ultrafiltration, cation exchange membranes, electrochemical degradation, integrated chemical–biological degradation, integrated iron(III) photoassisted-biological treatment, solar photo-Fenton and biological processes, and Fenton-biological treatment scheme) with varying levels of successes has accelerated a dramatic progress in the scientific community. Of major interest, adsorption, a surface phenomenon by which a multi-component fluid (gas or liquid) mixture is attracted to the surface of a solid adsorbent and forms attachments via physical or chemical bonds, is recognized as the most efficient, promising and widely used fundamental approach in wastewater treatment processes, mainly hinges on its simplicity, economically viable, technically feasible and socially acceptable. Further, this process can remove/minimize different type of pollutants and thus it has a wider applicability in water pollution control. Adsorption separation in environmental engineering is now an aesthetic attention and consideration abroad the nations, owing to its low initial cost, simplicity of design, ease of operation, insensitivity to toxic substances and complete removal of pollutants even from dilute solutions. A large number of natural materials or the wastes/by-products of industries or synthetically prepared materials, which cost less and can be used as such or after some minor treatment have been tested and examined for their ability to remove various types of pollutants from water and wastewater.

At present, adsorption field has been enriched by a vast amount of studies published in different journals. Extensive research has been dedicated to sound understanding of adsorption isotherm and kinetics. However, a study on adsorption can only be meaningful and useful if it includes the structure as well as the dynamics of its different components, separately and interacting with each other. In this regard, it should be realized that an extensive study on adsorption thermodynamics is desirable. Confirming the assertion, this chapter presents a state of art review of adsorption thermodynamics.

In the design of adsorption systems, two types of thermodynamic properties, namely the directly measurable properties like temperature, equilibrium constant, and properties which cannot be measured directly such as activation energy, activation parameters, Gibb's free energy change, enthalpy, entropy, and isosteric heat of adsorption are required. These parameters are critical design variables in estimating the performance and predicting the mechanism of an adsorption separation process and are also one of the basic requirements for the characterization and optimization of an adsorption process. Thus, this chapter attempts to offer a better understanding of adsorption thermodynamics with special focus on its fundamental characteristics and mathematical derivations.

2.1 Activation energy

Activation energy is an important parameter in a thermodynamic study as it determines the temperature dependence of the reaction rate. In chemistry, activation energy is defined as the energy that must be overcome in order for a chemical reaction to occur. In adsorption separation, it is defined as the energy that must be overcome by the adsorbate ion/molecule to react/interact with the functional groups on the surface of the adsorbent. It is the minimum energy needed for a specific adsorbate-adsorbent interaction to take place, even though the process may already be thermodynamically possible. The activation energy of a reaction is usually denoted by E_a, and given in units of kJ mol^{-1}. The activation energy (E_a) for the adsorption of an adsorbate ion/molecule onto an adsorbent surface in an adsorption process can be determined from experimental measurements of the adsorption rate constant at different temperatures according to the Arrhenius equation as follows:

$$\ln k = \ln A - \frac{E_a}{RT} \tag{1}$$

where k is the adsorption rate constant, A is a constant called the frequency factor, E_a is the activation energy (kJ.mol^{-1}), R is the gas constant (8.314 J.$mol^{-1}K^{-1}$) and T is the temperature (K). By plotting ln k versus 1/T (Figure 1) and from the slope and the intercept, values of E_a and A can be obtained. The apparent activation energy of adsorption of heavy metal ions and synthetic dye molecules onto various low cost adsorbents is tabulated in Table 1.

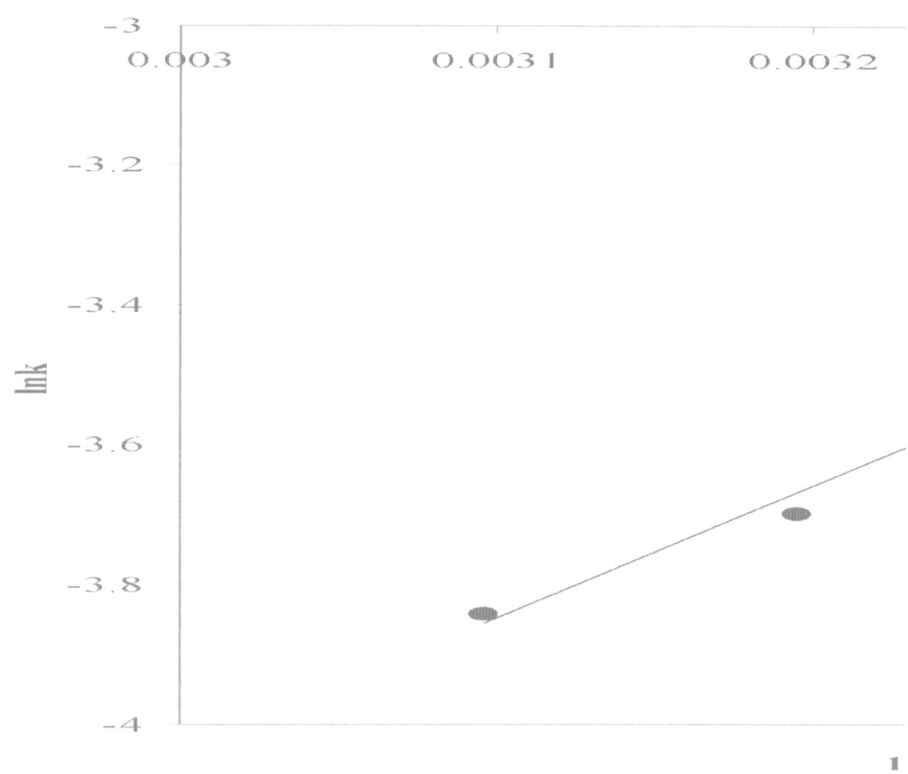

Fig. 1. A typical plot of ln k vs. 1/T (Arrhenius plot)

The magnitude of activation energy may give an idea about the type of adsorption. Two main types of adsorption may occur, physical and chemical. In physisorption, the equilibrium is usually rapidly attained and easily reversible, because the energy requirements are small. The activation energy for physisorption is usually no more than 4.2 kJ mol^{-1} since the forces involved in physisorption are weak. Chemisorption is specific and involves forces much stronger than in physisorption on. Therefore, the activation energy for chemisorption is of the same magnitude as the heat of chemical reactions. Two kinds of chemisorptions are encountered, activated and, less frequently, nonactivated. Activated chemisorption means that the rate varies with temperature according to finite activation energy (between 8.4 and 83.7 kJ/mol) in the Arrhenius equation (high E_a). However, in some systems the chemisorption occurs very rapidly, suggesting the activation energy is near zero. This is termed as a nonactivated chemisorption.

Adsorbent	**Adsorbate**	**E_a (kJ mol^{-1})**	**Reference**
Peanut hull	Cu(II)	17.02	Zhu et al., 2009
Laterite nickel ores	Pb(II)	7.6	Mohapatra et al., 2009
Cation exchanger derived from tamarind fruit shell	Cu(II)	10.84	Anirudhan & Radhakrishnan, 2008
Walnut hull	Cr(VI)	102.78	Wang et al., 2009
Wineyard pruning waste	Cr(III)	-15.65	Karaoglu et al., 2010
Sepioloite	Maxilon Blue 5G	19.25	Alkan et al., 2008
Chemically modified rice husk	Malachite Green	68.12	Chowdhury et al., 2010
Sea shell powder	Malachite Green	15.71	Chowdhury & Saha, 2010
Modified wheat straw	Methylene Blue	24.24	Han et al., 2010
Pinus sylvestris L.	Reactive Red 195	8.904	Aksakal & Ucun, 2010

Table 1. Activation energy for adsorption of heavy metal ions and dye molecules onto various low cost adsorbents

It is to be noted that in some cases rates of adsorption process decrease with increasing temperature. In order to follow an approximately exponential relationship so the rate constant can still be fit to the Arrhenius expression, results in a negative value of E_a. Sorption processes exhibiting negative activation energies are exothermic in nature and proceeds at lower temperatures. With the increase of temperature, the solubility of adsorbate species increases. Consequently, the interaction forces between the adsorbate and solvent are stronger than those between adsorbate and adsorbent. As a result, the adsorbate is more difficult to adsorb

2.2 Activation parameters

In order to get an insight whether the adsorption process follows an activated complex, it is absolutely necessary to consider the thermodynamic activation parameters of the process such as activation enthalpy (ΔH^*), activation entropy (ΔS^*) and free energy of activation (ΔG^*). The standard enthalpy of activation (ΔH^*), entropy of activation (ΔS^*), and free energy of activation (ΔG^*) in the adsorption process were calculated by the Eyring equation:

$$\frac{\ln k}{T} = \ln\frac{k_B}{h} + \frac{\Delta S^*}{R} - \frac{\Delta H^*}{RT} \tag{2}$$

where k is the adsorption rate constant, k_B is the Boltzman constant (1.3807×10^{-23} J K^{-1}), h is the Plank constant (6.6261×10^{-34} Js), R is the ideal gas constant (8.314 $J.mol^{-1}K^{-1}$), and T is temperature (K). The values of ΔH^* and ΔS^* can be determined from the slope and intercept of a plot of ln k/T versus 1/T (Figure 2). These values can be used to compute ΔG^* from the relation:

$$\Delta G^* = \Delta H^* - T\Delta S^* \tag{3}$$

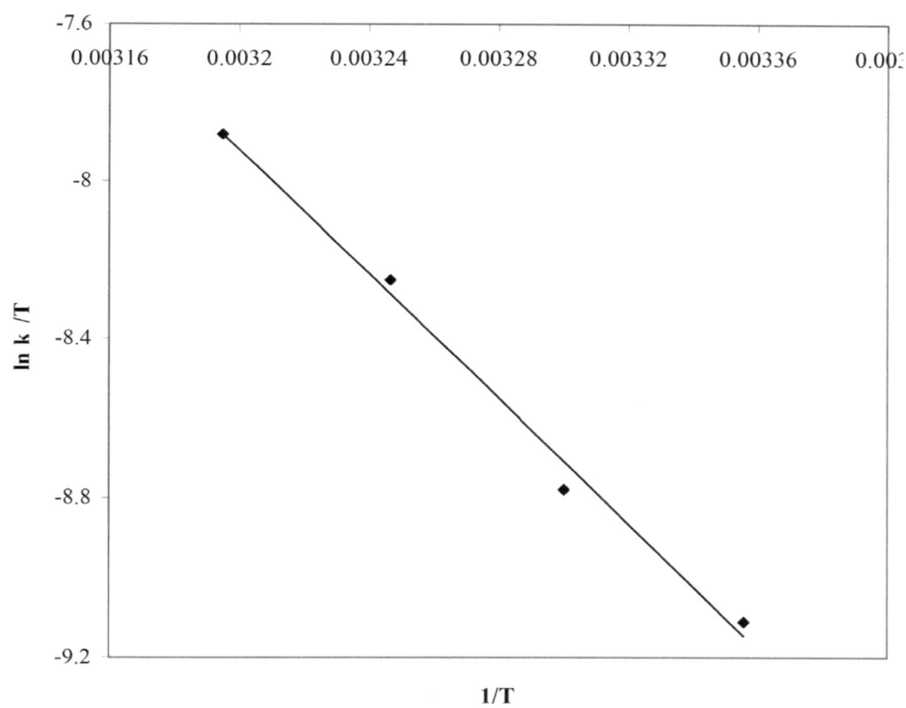

Fig. 2. A typical plot of ln k/T vs. 1/T (Eyring equation plot)

In general, the ΔG^* values are positive at all temperatures suggesting that adsorption reactions require some energy from an external source to convert reactants into products. A negative value of ΔH^* suggests that the adsorption phenomenon is exothermic while a positive value implies that the adsorption process is endothermic. The magnitude and sign of ΔS^* gives an indication whether the adsorption reaction is an associative or dissociative mechanism. A negative value of ΔS^* suggests that the adsorption process involves an associative mechanism. The adsorption leads to order through the formation of an activated complex between the adsorbate and adsorbent. Also a negative value of ΔS^* reflects that no significant change occurs in the internal structures of the adsorbent during the adsorption process. A positive value of ΔS^* suggests that the adsorption process involves a dissociative mechanism. Such adsorption phenomena are not favourable at high temperatures.

2.3 Thermodynamic parameters

Thermodynamic considerations of an adsorption process are necessary to conclude whether the process is spontaneous or not. The Gibb's free energy change, ΔG^0, is an indication of spontaneity of a chemical reaction and therefore is an important criterion for spontaneity. Both enthalpy (ΔH^0) and entropy (ΔS^0) factors must be considered in order to determine the Gibb's free energy of the process. Reactions occur spontaneously at a given temperature if ΔG^0 is a negative quantity. The free energy of an adsorption process is related to the equilibrium constant by the classical Van't Hoff equation:

$$\Delta G^0 = -RT \ln K_D \tag{4}$$

where, ΔG^0 is the Gibb's free energy change (kJ. mol^{-1}), R is the ideal gas constant (8.314 J.mol^{-1}K^{-1}), and T is temperature (K) and K_D is the single point or linear sorption distribution coefficient defined as:

$$K_D = \frac{C_a}{C_e} \tag{5}$$

where C_a is the equilibrium adsorbate concentration on the adsorbent (mg L^{-1}) and C_e is the equilibrium adsorbate concentration in solution (mg L^{-1}).
Considering the relationship between ΔG^0 and K_D, change in equilibrium constant with temperature can be obtained in the differential form as follows

$$\frac{d \ln K_D}{dT} = \frac{\Delta H^0}{RT^2} \tag{6}$$

After integration, the integrated form of Eq. (5) becomes:

$$\ln K_D = -\frac{\Delta H^0}{RT} + Y \tag{7}$$

where Y is a constant.
Eq (7) can be rearranged to obtain:

$$-RT \ln K_D = \Delta H^0 - TRY \tag{8}$$

Let $\Delta S^0 = RY$
Substituting Eqs. (4) and (8), ΔG^0 can be expressed as:

$$\Delta G^0 = \Delta H^0 - T\Delta S^0 \tag{9}$$

A plot of Gibb's free energy change, ΔG^0 versus temperature, T will be linear with the slope and intercept giving the values of ΔH^0 and ΔS^0 respectively.

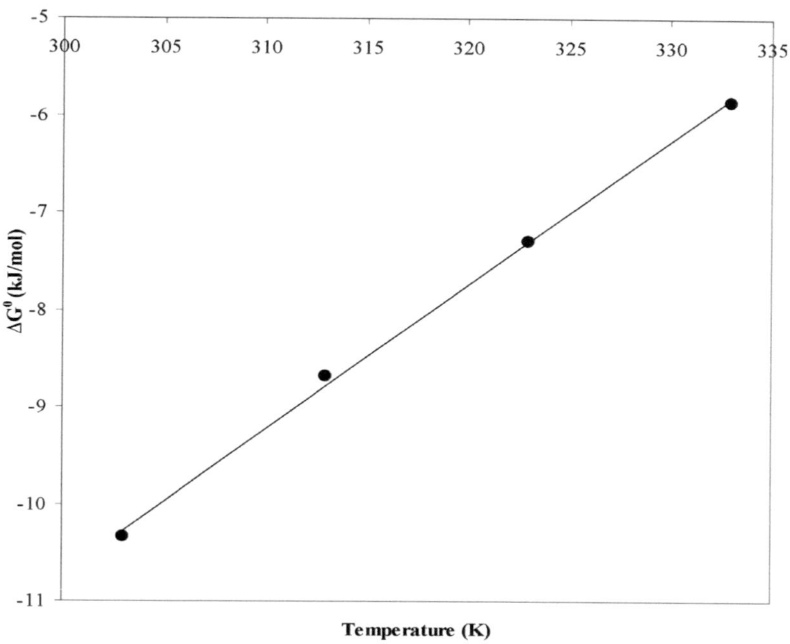

Fig. 3. Plot of Gibb's free energy change (ΔG^0) versus temperature for an exothermic process

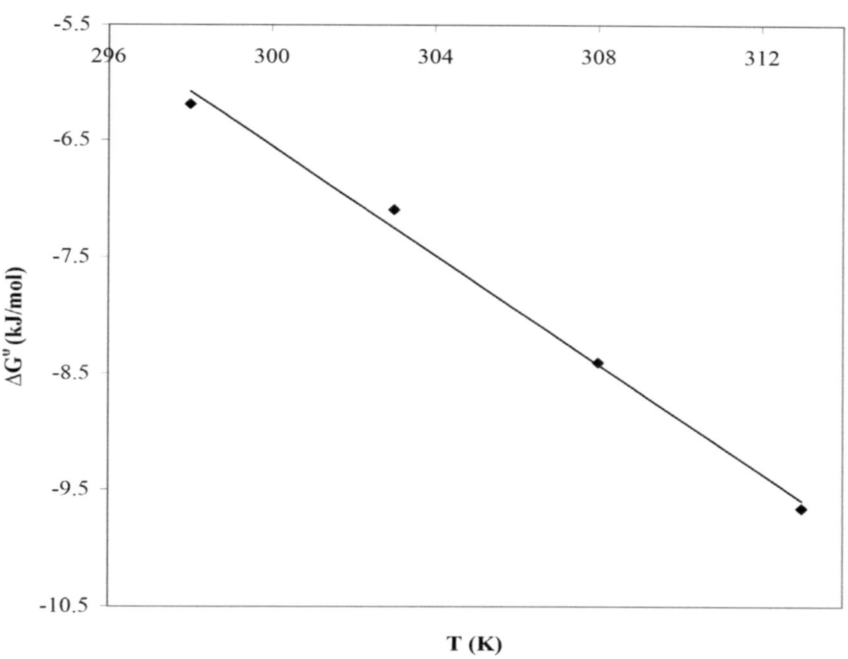

Fig. 4. Plot of Gibb's free energy change (ΔG^0) versus temperature for an endothermic process

The thermodynamic relation between ΔG^0, ΔH^0 and ΔS^0 suggests that either (i) ΔH^0 or ΔS^0 are positive and that the value of $T\Delta S^0$ is much larger than ΔH^0 (ii) ΔH^0 is negative and ΔS^0 is positive or (iii) ΔH^0 or ΔS^0 are negative and that the value of ΔH^0 is more than $T\Delta S^0$.
The typical value of the thermodynamic parameters for adsorption of heavy metal ions and synthetic dye molecules onto various low cost adsorbent are listed in Tables 2 and 3, respectively. For significant adsorption to occur, the Gibb's free energy change of adsorption, ΔG^0, must be negative. For example, as seen in Table 2, the Gibb's free energy change (ΔG^0) values were found to be negative below 313.15 K for adsorption of Cr(VI) onto chitosan, which indicates the feasibility and spontaneity of the adsorption process at temperatures below 313.15K. As a rule of thumb, a decrease in the negative value of ΔG^0 with an increase in temperature indicates that the adsorption process is more favourable at higher temperatures. This could be possible because the mobility of adsorbate ions/molecules in the solution increase with increase in temperature and that the affinity of adsorbate on the adsorbent is higher at high temperatures. On the contrary, an increase in the negative value of ΔG^0 with an increase in temperature implies that lower temperature makes the adsorption easier.

Adsorbent	**Adsorbate**	**T (K)**	**ΔG^0 (kJ mol^{-1})**	**ΔH^0 (kJ mol^{-1})**	**ΔS^0 (J mol^{-1})**	**Reference**
Rubber (*Hevea brasiliensis*) leaf powder	Cu(II)	300 310 320	-3.38 -2.17 -1.48	-31.96	-95.94	Ngah & Hanafiah, 2008
Modified oak sawdust	Cu(II)	293 303 313	-2.840 -3.064 -3.330	4.331	240	Argun et al., 2007
Mimosa tannin resin	Cu(II)	298 303 318 338 353	-2.47 -4.83 -8.51 -9.38 -11.45	42.09	153	Sengil & Ozacar, 2008
Hazelnut shell activated carbon	Cu(II)	293 303 313 323	-6.83 -6.66 -6.03 -5.71	18.77	40.4	Demribas et al., 2009
Penicillium simplicissimum	Cd(II)	293 303 313	-18.27 -19.81 -20.88	20.03	130.90	Fan et al., 2008
Red algae (*Ceramium virgatum*)	Cd(II)	293 303 313 323	-19.5 -19.0 -18.7 -18.2	-31.8	-42.4	Sari & Tuzen, 2008
Coconut copra meal	Cd(II)	299 311 323 333	-7.41 -7.15 -6.97 -6.66	-13.70	21.20	Ho & Ofomaja, 2006

Fennel biomass	Cd(II)	303 313 323	-5.017 -5.470 -6.016	10.34	51	Rao et al., 2010
Chitosan	Cr(VI)	303.15 313.15 323.15 333.15	-2.409 -1.326 0.178 2.429	-50.782	159	Aydin & Aksoy, 2009
Walnut hull	Cr(VI)	303 313 323	-23.03 -25.63 -28.77	64.14	287.4	Wang et al., 2009
Acacia leucocephala bark	Ni(II)	303 313 323	-6.147 -6.945 -7.847	10.389	55	Subbaiah et al., 2009
Baker's yeast	Ni(II)	300 313 323 333	-23.519 -23.408 -23.149 -22.708	-30.702	-23.658	Padmavathy, 2009
Oyster shell powder	Ni(II)	303 318 333	-20.0 -22.9 -26.4	44.90	127.7	Hsu, 2009
Lichen (*Cladonia furcata*) biomass	Ni(II)	293 303 313 323	-18.3 -14.4 -14.3 -14.4	-37.5	-71.5	Sari et al., 2009
Acacia leucocephala bark powder	Pb(II)	303 313 323	-3.876 -4.379 -4.997	-21.147	57	Munagapati et al., 2010
Penicillium simplicissimum	Pb(II)	293 303 313	-20.04 -22.60 -24.06	39.13	202.52	Fan et al., 2008
Lichen (*Cladonia furcata*) biomass	Pb(II)	293 303 313 323	-21.2 -17.4 -17.2 -17.1	-35.4	-57.6	Sari et al., 2009
Pine bark (*Pinus brutia* Ten.)	Pb(II)	273 283 293 303 313	-2.74 -2.89 -3.08 -3.25 -3.42	1.97	17.21	Gundogdu et al., 2009

Table 2. Thermodynamic parameters for adsorption of heavy metal ions on various low cost adsorbents

Adsorbent	Adsorbate	T (K)	ΔG⁰(kJ mol⁻¹)	ΔH⁰(kJ mol⁻¹)	ΔS⁰ (J mol⁻¹)	Reference
Treated ginger waste	Malachite Green	303 313 323	-1.515 -2.133 -3.016	47.491	167	Ahmad & Kumar, 2010
Degreased coffee bean	Malachite Green	298 308 318	-8.19 -10.0 -10.6	27.2	33.3	Baek et al., 2010
Neem sawdust	Malachite Green	298 308 318	-4.02 -2.33 -1.73	-54.56	-169.57	Khattri & Singh, 2009
Luffa cylindrical	Malachite Green	288 298 308	-6.1 -7.1 -8.7	32.1	132.2	Altınısık et al., 2010
Brazil nut shell	Methylene Blue	293 303 333	-2.27 -2.09 -1.97	-5.22	-112.23	Brito et al., 2010
Bentonite	Methylene Blue	283 293 303 308	-17.0 -17.7 -18.5 -19.4	9.21	92.2	Hong et al., 2009
Modified wheat straw	Methylene Blue	293 303 313	-9.96 -11.22 -12.14	21.92	108	Han et al., 2010
Cattail root	Congo Red	293 303 313	-7.871 -6.800 -4.702	-54.116	157	Hu et al., 2010
Ca-Bentonite	Congo Red	293 303 313 323	-6.4962 -6.7567 -7.1991 -11.179	5.1376	37.2	Lian et al., 2009
Non-living aerobic granular sludge	Acid Yellow 17	293 308 323	-5.14 -5.13 -4.65	-9.84	-15.79	Gao et al., 2010
P. vulgaris L. waste biomass	Reactive Red 198	293 303 313 323	-4.744 -4.573 -4.403 -4.232	-9.74	-17.04	Akar et al., 2009
Pinus sylvestris L. Biomass	Reactive Red 195	293 303 313 323	-13.253 -14.022 -15.723 -17.555	29.422	144.672	Aksakal & Ucun, 2010

Activated carbon from Brazilian-pine fruit shell	Reactive Orange 16	298 303 308 313 318 323	-32.9 -33.7 -34.6 -35.3 -36.2 -36.9	15.3	162	Calvete et al., 2010
Paulownia tomentosa Steud. leaf powder	Acid Orange 52	298 308 318	-0.85 -0.71 -0.51	-6.02	-17	Deniz & Saygideger, 2010
Brazil nut shell	Indigo carmine	293 303 333	-5.42 -5.71 -6.60	-3.20	-29.39	Brito et al., 2010
Activated carbon from bagasse pith	Rhodamine B	293 308 323 343	-7.939 -9.902 -12.361 -26.729	4.151	65.786	Gad & El-Sayed, 2009
Activated carbon from from *Euphorbia rigida*	Disperse Orange 25	283 288 293	-24.084 -25.736 -26.495	44.308	242.17	Gercel et al., 2008
Wheat bran	Astrazon Yellow 7GL	303 313 323	-14.472 -17.803 -22.552	46.81	175	Sulak et al., 2007

Table 3. Thermodynamic parameters for adsorption of synthetic dyes on various low cost adsorbents

A negative value of ΔH^0 implies that the adsorption phenomenon is exothermic while a positive value implies that the adsorption process is endothermic. The adsorption process in the solid–liquid system is a combination of two processes: (a) the desorption of the solvent (water) molecules previously adsorbed, and (b) the adsorption of the adsorbate species. In an endothermic process, the adsorbate species has to displace more than one water molecule for their adsorption and this result in the endothermicity of the adsorption process. Therefore ΔH^0 will be positive. In an exothermic process, the total energy absorbed in bond breaking is less than the total energy released in bond making between adsorbate and adsorbent, resulting in the release of extra energy in the form of heat. Therefore ΔH^0 will be negative. The magnitude of ΔH^0 may also give an idea about the type of sorption. The heat evolved during physical adsorption is of the same order of magnitude as the heats of condensation, i.e., 2.1–20.9 kJ mol^{-1}, while the heats of chemisorption generally falls into a range of 80–200 kJ mol^{-1}. Therefore, as seen from Tables 2 and 3, it seems that adsorption of most heavy metal ions and synthetic dye molecules by various low cost adsorbents can be attributed to a physico-chemical adsorption process rather than a pure physical or chemical adsorption process.

A positive value of ΔS^0 reflects the affinity of the adsorbent towards the adsorbate species. In addition, positive value of ΔS^0 suggests increased randomness at the solid/solution interface with some structural changes in the adsorbate and the adsorbent. The adsorbed solvent molecules, which are displaced by the adsorbate species, gain more translational entropy than is lost by the adsorbate ions/molecules, thus allowing for the prevalence of randomness in the system. The positive ΔS^0 value also corresponds to an increase in the degree of freedom of the adsorbed species. A negative value of ΔS^0 suggests that the adsorption process is enthalpy driven. A negative value of entropy change (ΔS^0) also implies a decreased disorder at the solid/liquid interface during the adsorption process causing the adsorbate ions/molecules to escape from the solid phase to the liquid phase. Therefore, the amount of adsorbate that can be adsorbed will decrease.

2.4 Isosteric heat of adsorption

The most relevant thermodynamic variable to describe the heat effects during the adsorption process is the isosteric heat of adsorption. Isosteric heat of adsorption (ΔH_x, kJ mol^{-1}) is defined as the heat of adsorption determined at constant amount of adsorbate adsorbed. The isosteric heat of adsorption is a specific combined property of an adsorbent–adsorbate combination. It is one of the basic requirements for the characterization and optimization of an adsorption process and is a critical design variable in estimating the performance of an adsorptive separation process. It also gives some indication about the surface energetic heterogeneity. Knowledge of the heats of sorption is very important for equipment and process design. However, the physical meaning of 'isosteric heat' is not clear and it is not even considered by some authors to be the most suitable way of understanding the adsorption phenomena

The isosteric heat of adsorption at constant surface coverage is calculated using the Clausius-Clapeyron equation:

$$\frac{d(\ln C_{e)}}{dT} = -\frac{\Delta H_X}{RT^2} \tag{10}$$

where, C_e is the equilibrium adsorbate concentration in the solution (mg.L^{-1}), ΔH_x is the isosteric heat of adsorption (kJ mol^{-1}), R is the ideal gas constant (8.314 J.mol^{-1}K^{-1}), and T is temperature (K).

Integrating the above equation, assuming that the isosteric heat of adsorption is temperature independent, gives the following equation:

$$\ln C_e = -\left(\frac{\Delta H_X}{R}\right)\frac{1}{T} + K \tag{11}$$

where K is a constant.

The isosteric heat of adsorption is calculated from the slope of the plot of ln C_e versus $1/T$ different amounts of adsorbate onto adsorbent. For this purpose, the equilibrium concentration (C_e) at constant amount of adsorbate adsorbed is obtained from the adsorption isotherm data at different temperatures. The isosteres corresponding to different equilibrium adsorption uptake of Cu(II) by tamarind fruit seed is shown in Fig. 5. Similar isosteres have been obtained for other systems as well.

The magnitude of ΔH_x value gives information about the adsorption mechanism as chemical ion-exchange or physical sorption. For physical adsorption, ΔH_x should be below 80 kJ mol^{-1} and for chemical adsorption it ranges between 80 and 400 kJ.mol^{-1}.

The isosteric heat of adsorption can also provide some information about the degree of heterogeneity of the adsorbent. Generally, the variation of ΔH_x with surface loading is indicative of the fact that the adsorbent is having energitically heterogeneous surfaces. If it were a homogeneous surface, the isosteric heat of adsorption would have been constant even with variation in surface loading. The ΔH_x is usually high at very low coverage and decreases steadily with an increase in q_e. The dependence of heat of adsorption with surface coverage is usually observed to display the adsorbent–adsorbate interaction followed by the adsorbate-adsorbate interaction. The decreasing of the heats of sorption indicates that the adsorbate-adsorbent interactions are strong in the range of lower q_e values and then they decrease with the increase in the surface coverage. It has been suggested that that the high values of the heats of sorption at low q_e values were due to the existence of highly active sites on the surface of the adsorbent. The adsorbent-adsorbate interaction takes place initially at lower q_e values resulting in high heats of adsorption. On the other hand, adsorbate-adsorbate interaction occurs with an increase in the surface coverage giving rise to lower heats of sorption. The variation in ΔH_x with surface loading can also be attributed to the possibility of having lateral interactions between the adsorbed adsorbate molecules.

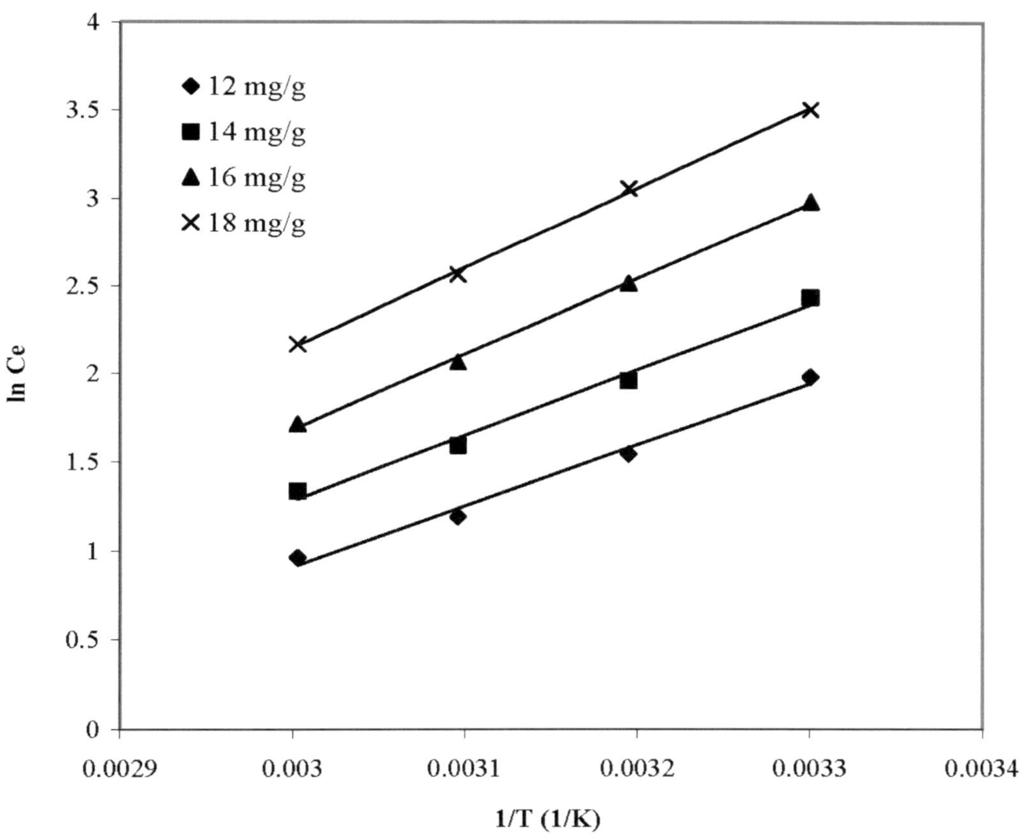

Fig. 5. Plots of ln C_e against 1/T for adsorption of Cu(II) onto tamarind seeds at constant surface coverage, q_e = 12, 14, 16, 18 mg g^{-1}

3. Conclusion

To date, adsorption has been regarded as an effective technology for the removal of soluble heavy metal ions, synthetic dye molecules and other toxic chemicals from aqueous solution.

In study of adsorption thermodynamics, it appears that determination of value of the thermodynamic quantities such as activation energy, activation parameters, Gibb's free energy change, enthalpy, entropy, and isosteric heat of adsorption are required. These parameters are critical design variables in estimating the performance and predicting the mechanism of an adsorption separation process and are also one of the basic requirements for the characterization and optimization of an adsorption process. So far, extensive research effort has been dedicated to a sound understanding of adsorption isotherm, kinetics and thermodynamics. Compared to adsorption isotherm and kinetics, there is lack of a theoretical basis behind the thermodynamic analysis of sorption data. In this regard, the next real challenge in the adsorption field is the identification and clarification of the underlying thermodynamics in various adsorption systems. Further explorations on developing in this area are recommended.

4. References

Ahmad, R. & Kumar, R. (2010). Adsorption studies of hazardous malachite green onto treated ginger waste. *Journal of Environmental Management*, 91, 1032-1038. ISSN: 0301-4797.

Akar, A.T.; Ozcan, A.S.; Akar, T.; Ozcan, A. & Kaynak, Z. (2009). Biosorption of a reactive textile dye from aqueous solutions utilizing an agro waste. *Desalination*, 249, 757-761. ISSN: 0011-9164.

Aksakal, O. & Ucun, H. (2010). Equilibrium, kinetic and thermodynamic studies of the biosorption of textile dye (Reactive Red 195) onto *Pinus sylvestris* L. *Journal of Hazardous Materials*, 181, 666-672. ISSN: 0304-3894.

Alkan, M.; Dogan, M.; Turhan, Y.; Demribas, O. & Turan, P. (2008). Adsorption kinetics and mechanism of maxilon blue 5G dye on sepiolite from aqueous solutions. *Chemical Engineering Journal*, 139, 213-223. ISSN: 1385-8947.

Altinisik, A.; Gur, E. & Seki, Y. (2010). A natyral sorbent, Luffa sylindrica for the removal of a model basic dye. *Journal of Hazardous Materials*, 179, 658-664.

Anirudhan, T.S. & Radhakrishnan, P.G. (2008). Thermodynamics and kinetics of adsorption of Cu(II) from aqueous solutions onto a new cation exchanger derived from tamarind fruit shell. *Journal of Chemical Thermodynamics*, 40, 702-709. ISSN: 0021-9614.

Argun, M.E.; Dursun, S.; Ozdemir, C. & Karatas, M. (2007). Heavy metal adsorption by modified oak sawdust: Thermodynamics and kinetics. *Journal of Hazardous Materials*, 141, 77-85. ISSN: 0304-3894.

Aydin, Y.A. & Aksoy, N.D. (2009). Adsorption of chromium on chitosan: Optimization, kinetics and thermodynamics. *Chemical Engineering Journal*, 151, 188-194. ISSN: 1385-8947.

Baek, M.-H.; Ijagbemi, C.O.; O, S.-J. & Kim, D.-S. (2010). Removal of Malachite Green from aqueous solution using degreased coffee bean. *Journal of Hazardous Materials*, 176, 820-828. ISSN: 0304-3894.

Brito, S.M.O.; Andrade, H.M.C.; Soares, L.F. & Azevedo, R.P. (2010). Brazil nut shells as a new biosorbent to remove methylene blue and indigo carmine from aqueous solutions, *Journal of Hazardous Materials*, 174, 84-92. ISSN: 0304-3894.

Chowdhury, S.; Mishra, R.; Saha, P. & Kushwaha, P. (2010). Adsorption thermodynamics, kinetics and isosteric heat of adsorption of malachite green onto chemically modified rice husk. *Desalination*, doi:10.1016/j.desal.2010.07.047. ISSN: 0011-9164.

Chowdhury, S. & Saha, P. (2010). Sea shell powder as a new adsorbent to remove Basic Green 4 (Malachite Green) from aqueous solutions: Equilibrium, kinetic and thermodynamic studies. *Chemical Engineering Journal*, doi:10.1016/j.cej.2010.08.050. ISSN: 1385-8947.

Calvete, T.; Lima, E.C.; Cardoso, N.F.; Vaghetti, J.C.P.; Dias, S.L.P & Pavan, F.A. (2010). Application of carbon adsorbents prepared from Brazailian-pine fruit shell for the removal of reactive orange 16 from aqueous solution: Kinetic, equilibrium, and thermodynamic studies. *Journal of Environmental Management*, 91, 1695-1706. ISSN: 0301-4797.

Demribas, E.; Dizge, N.; Sulak, M.T. & Kobya, M. (2009). Adsorption kinetics and equilibrium of copper from aqueous solutions using hazelnut shell activated carbon. *Chemical Engineering Journal*, 148, 480-487. ISSN: 1385-8947.

Deniz, F. & Saygideger, S.D. (2010). Corrigendum to "Equilibrium, kinetic and thermodynamic studies of Acid Orange 52 dye biosorption by *Paulownis tomentosa* Steud. Leaf powder as a low-cost natural biosorbent" [Bioresour. Technol. 101 (2010) 5137-5143]. *Bioresource Technology*, 101, 7688-7690. ISSN: 0960-8524.

Fan, T.; Liu, Y.; Feng, B.; Zeng, G.; Yang, C.; Zhou, M.; Zhou, H.; Tan, Z. & Wang, X. (2008). Biosorption of cadmium (II), zinc (II) and lead (II) by *Penicillium simplicissimum*: Isotherms, kinetics and thermodynamics. *Journal of Hazardous Materials*, 160, 655-661. ISSN: 0304-3894.

Gao, J.; Zhang, Q.; Su, K.;Chen, R. & Peng, Y. (2010). Biosorption of Acid Yellow 17 from aqueous solution by non-living aerobic granular sludge. *Journal of Hazardous Materials*, 174, 215-225. ISSN: 0304-3894.

Gad, H.M.H. & El-Sayed, A.A. (2009). Activated carbon from agricultural biproducts for the removal of Rhodamine B from aqueous solution. *Journal of Hazardous Materials*, 168, 1070-1081. ISSN: 0304-3894.

Gercel, O.; Gercel, H.F.; Koparal, A.S. & Ogutveren, U.B. (2008). Removal of disperse dye from aqueous solution by novel adsorbent prepared from biomass plant material. *Journal of Hazardous Materials*, 160, 668-674. ISSN: 0304-3894.

Gundogdu, A.; Ozdes, D.; Duran, C.; Bulut, V.N.; Soylak, M. & Senturk, H.B. (2009). Biosorption of Pb(II) ions from aqueous solution by pinr bark (*Pinus bruita* Ten.). *Chemical Engineering Journal*, 153, 62-69. ISSN: 1385-8947.

Han, R.; Zhang, L.; Song, C.; Zhang, M.; Zhu, H. & Zhang, L. (2010). Characterization of modified wheat straw, kinetic and equilibrium study about copper ion and methylene blue adsorption in batch mode. *Carbohydrate Polymers*. 79, 1140-1149. ISSN: 0144-8617.

Ho, Y.-S. & Ofomaja, A.E. (2006). Biosorption thermodynamics of cadmium on coconut copra meal as biosorbent. *Biochemical Engineering Journal*. 30, 117-123. ISSN: 1369-703X.

Hong, S.; Wen, C.; He, J.; Gan, F. & Ho, Y.-S. (2009). Adsorption thermodynamics of Methylene Blue onto bentonite. *Journal of Hazardous Materials*, 167,630-633. ISSN: 0304-3894.

Hsu, T.-C. (2009). Experimental assessment of adsorption of Cu^{2+} and Ni^{2+} from aqueous solution by oyster shell powder. *Journal of Hazardous Materials*, 171, 995-1000. ISSN: 0304-3894.

Hu, Z.; Chen, H.; Ji, F. & Yuan, S. (2010). Removal of Congo Red from aqueous solution by cattail root. *Journal of Hazardous Materials*, 173, 292-297. ISSN: 0304-3894.

Karaoglu, M.H.; Zor, S. & Ugurlu, M. (2010). Biosorption of Cr(III) from solutions using vineyard pruning waste. *Chemical Engineering Journal*, 159, 98-106. ISSN: 1385-8947.

Khattri, S.D. & Singh, M.K. (2009). Removal of malachite green from dye wastewater using neem sawdust by adsorption. *Journal of Hazardous Materials*, 167, 1089-1094. ISSN: 0304-3894.

Lian, L.; Guo, L. & Guo, C. (2009). Adsorption of Congo red fromaqueous solutions onto Ca-bentonite. *Journal of Hazardous Materials*, 161, 126-131. ISSN: 0304-3894.

Mohapatra, M.; Khatun, S. & Anand, S. (2009). Kinetics and thermodynamics of lead (II) adsorption on lateritic nickel ores of Indian origin. *Chemical Engineering Journal*, 155, 184-190. ISSN: 1385-8947.

Munagapati, V.S.; Yarramuthi, V.; Nadavala, S.K;, Alla, S.R. & Abburi, K. (2010). Biosorption of Cu(II), Cd(II) and Pb(II) by *Acacia leucocephala* bark powder: Kinetics, equilibrium and thermodynamics. *Chemical Engineering Journal*, 157, 357-365. ISSN: 1385-8947.

Ngah, W.S.W. & Hanafiah, M.A.K.M. (2008). Adsorption of copper on rubber (*Hevea brasiliensis*) leaf powder: Kinetic, equilibrium and thermodynamic studies. *Biochemical Engineering Journal*, 39, 521-530. ISSN: 1369-703X.

Padmavathy, V. (2008). Biosorption of nickel (II) ions by baker's yeast: Kinetic, thermodynamic and desorption studies. *Bioresource Technology*, 99, 3100-3109. ISSN: 0960-8524.

Rao, R.A.K.; Khan, M.A. & Rehman, F. (2010). Utilization of Fennel biomass (Foeniculum vulgari) a medicinal herb for the biosorption of Cd(II) from aqueous phase. *Chemical Engineering Journal*, 156, 106-113. ISSN: 1385-8947.

Sari, A. & Tuzen, M. (2008). Biosorption of cadmium (II) from aqueous solution by red algae (Ceramium virgatum): Equilibrium, kinetic and thermodynamic studies. *Journal of Hazardous Materials*, 157, 448-454. ISSN: 0304-3894.

Sari, A.; Tuzen, M.; Uluozlu, O.D. & Soylak, M. (2007). Biosorption of Pb(II) and Ni(II) from aqueous solution by lichen (*Cladonia furcata*) biomass. *Biochemical Engineering Journal*, 37,151-158. ISSN: 1369-703X.

Sengil, I. A. & Ozacar, M. (2008). Biosorption of Cu(II) from aqueous solutions by mimosa tannin gel. Journal of Hazardous Materials, 157, 277-285. ISSN: 0304-3894.

Subbaiah, M.V.; Vijaya, Y.; Kumar, N.S.; Reddy, A.S. & Krishnaiah, A. (2009). Biosorption of nickel from aqueous solutions by *Acacia leucocephala* bark: Kinetics and equilibrium studies. *Colloids and Surfaces B: Biointerfaces*, 74,260-265. ISSN: 0927-7765.

Sulak, M.T.; Demribas, E. & Kobya, M. (2007). Removal of Astrazon Yeloow 7GL from aqueous solutions by adsorption onto wheat bran. *Bioresource Technology*, 98, 2590-2598. ISSN: 0960-8524.

Wang, X.S.; Li, Z.Z. & Tao, S.R. (2009). Removal of chromium (VI) from aqueous solution using walnut hull. *Journal of Environmental Management*, 90, 721-729. ISSN: 0301-4797.

Zhu, C.-S.; Wang, L.-P. & Chen, W.-B. (2009). Removal of Cu(II) from aqueous solution by agricultural by-product: Peanut hull. *Journal of Hazardous Materials*, 168, 739-746. ISSN: 0304-3894.

17

Ion Exchanger as Gibbs Canonical Assembly

Heinrich Al'tshuler and Olga Al'tshuler
Institute of Solid State Chemistry and Mechanochemistry (Kemerovo division),
Siberian Branch of Russian Academy of Sciences
Russian Federation

1. Introduction

Ion exchange is one of the fundamental reversible processes occurring in nature. Predicting of the equilibrium compositions for ion exchange presents of considerable practical and scientific interest.

If two types of ions i and j with the same charge are exchanged in equivalent ratios, the ion exchange process can be presented by the following equation

$$i + \bar{j} \rightleftarrows \bar{i} + j, \tag{1}$$

where the overbar indicates that a species belongs to the polymer phase.

The reversible process (1) is determined by the change of Gibbs energy, enthalpy, entropy and constant of thermodynamic equilibrium. The thermodynamic constant of ion exchange equilibrium (1) is described by the formula

$$K_{i/j} = \frac{\bar{a}_i \cdot a_j}{\bar{a}_j \cdot a_i}, \tag{2}$$

where a_i is the activity of component i.

The value of $K_{i/j}$ is calculated by the formula (Reichenberg, 1966)

$$\ln K_{i/j} = \int_0^1 \ln k^a_{i/j} d\bar{x}_i, \tag{3}$$

based on the choice of the ion exchanger in the monoionic forms as the standard states for components i and j in the polymer phase. In formula (3), $k^a_{i/j}$ is the corrected selectivity coefficient of ion exchange, which can be presented as

$$k^a_{i/j} = \frac{\bar{x}_i \cdot a_j}{\bar{x}_j \cdot a_i}, \tag{4}$$

where $\bar{x}_i$ is the mole fraction of i- component in the ion exchanger phase; for ion exchange from dilute solution of 1-1 electrolytes with constant ionic strength, if the solution corresponds to Debye-Huckel theory, the corrected selectivity coefficient, $k^a_{i/j}$, equal to selectivity coefficient of ion exchange equilibrium, $k_{i/j}$; where $k_{i/j} = \frac{\bar{x}_i \cdot x_j}{\bar{x}_j \cdot x_i}$ and x_i is the mole fraction of i - component in the external solution.

Features of interaction of exchangeable ions with ionogenic groups can be reflect by the dependences of partial (differential) thermodynamic functions of process (1) from the content of exchangeable ions in polymer.

Partial Gibbs energy, $\Delta\bar{G}$, was calculated by the equation $\Delta\bar{G} = RT[(z_j - z_i) - \ln k^a_{i/j}]$ corresponding to (Gaines & Thomas, 1953).

Partial enthalpy of ion exchange was calculated as partial derivative of heat of ion exchange with respect to the quantity of sorbed ions, $\Delta\bar{H} = (\partial\Delta H / \partial\bar{x}_i)_T$.

For ion exchange in a multicomponent system containing *n* ions with the same charge, the phase composition is determined by solving a set of (n – 1) homogeneous equations for the selectivity coefficients of binary ion exchanges in conjunction with an equation of material balance (Tondeur & Klein, 1967). In addition, was postulated that the selectivity coefficients are constant in the entire range of compositions of the ion exchanger phase (Tondeur & Klein, 1967). A similar solution was obtained for an ion exchange complicated by the complexation (Al'tshuler et al., 1984).

In most cases, the selectivity coefficients of ion exchange are the functions of the ionic composition of the phases. The dependence of the selectivity coefficient from polymer composition was explained by an energetic inhomogeneity of the ion exchanger phase (Reichenberg, 1966). Thus, calculation results systematically deviate from experimental data (Soldatov & Bychkova, 1988). The known methods for calculating of ion exchanger phase compositions for multicomponent exchanges with consideration for the dependence of the selectivity coefficients on the compositions of the phases are applicable only within a limited region of the diagram of compositions (Tondeur & Klein, 1967). To extend region of the diagram of compositions correction coefficients were introduced (Soldatov & Bychkova, 1988). This provided satisfactory results only after several successive adjustments (Soldatov & Bychkova, 1988). How to determine the ion exchanger composition as a fraction of the structural conformers was discussed (Soldatov et al., 1994). Ion exchanger compositions for weakly acidic (Horst et al., 1990) and chelating ion exchangers (Horst et al., 1991) were calculated from the electrostatic potential by the Gouy-Chapman model, which is based on the Maxwell-Boltzmann distribution for an ideal gas. In this chapter we have shown, that it is possible in reality to calculate the phase compositions when the ion exchange with variable selectivity coefficients take a place from the multicomponent solutions if the ion exchanger represents Gibbs canonical assembly.

2. The theory. Ion exchanger as Gibbs canonical assembly

Consider an ion exchanger as the Gibbs canonical assembly. The Gibbs canonical distribution is

$$\rho = \exp\left(\frac{\Psi - \varepsilon}{\theta}\right), \quad (5)$$

where $\rho(\varepsilon)$ is the density of probability to find the phase having an energy ε; Ψ is a constant; θ is the modulus of distribution. If $\theta >> \psi\text{-}\varepsilon$, then ρ and ε are proportional:

$$\rho = 1 + (\psi - \varepsilon)/\theta\,. \quad (6)$$

For an ion exchanger containing two exchangeable ions $\Delta\bar{H}, \Delta\bar{G}$, $\ln k^{a}_{i/j}$ are continuous functions of energy, which vary along the ion exchange isotherm from $\Delta\bar{H}, \Delta\bar{G}, \ln k^{a}_{i/j}$ at $\bar{x}_i = 0$ to its values at $\bar{x}_i = 1$. In this case, the ion exchanger composition measured in mole fractions of the i-component corresponds to the probability density for the individual phases at statistical equilibrium. In other words, for a binary ion exchange, the canonical energy distribution results in a linear dependences both of the differential enthalpy and of differential Gibbs energy (or the logarithm of the corrected selectivity coefficient) of ion exchange from the mole fraction of counterions in the ion exchanger phase. For binary ion exchange systems, the relation between ρ and ε coincides with formulas (7), (8)

$$\Delta\bar{H} = a + b\bar{x}_i, \tag{7}$$

$$\Delta\bar{G} = c + g\bar{x}_i, \quad \text{or} \quad \ln k^{a}_{i/j} = A_{i/j} + B_{i/j}\bar{x}_i, \tag{8}$$

where a, b, c, g, $A_{i/j}$; $B_{i/j}$ are constants. Really, experimental results show, that at compositions $0 \le \bar{x}_i \le 1$, the $\Delta\bar{H}(\bar{x}_i)$, $\Delta\bar{G}(\bar{x}_i)$, $k^{a}_{i/j}(\bar{x}_i)$ dependences can be approximated by the linear equations (Al'tshuler et al., 1984), (Al'tshuler & Shkurenko, 1990). Apparently, both the partial enthalpy and partial Gibbs free energy of the individual components additively contribute to the total energy of ion exchange process (1). For this reason, we examined the ion exchanger phase as a system consisting of non-interacting elements (for example, electrically neutral resinates) within the region of validity of (5)-(8) formulas. This is also true for a multicomponent system consisting of non-interacting parts, i.e., a system described by a comprehensive canonical assembly.

3. Results and discussion

3.1 Ion exchange from the solutions containing two electrolytes

According to experimental data $\Delta\bar{H}(\bar{x}_i)$, $\ln k^{a}_{i/j}(\bar{x}_i)$ linear dependences are observed in next exchangers: strong - acid cross – linked polystyrene sulfonate cation exchangers (Boyd & Larson, 1967), (Soldatov et al., 1976), (Soldatov & Bychkova, 1988); cross-linked polystyrene phosphonic acid cation exchangers (Becker et al., 1966); zeolities (Al'tshuler & Shkurenko, 1990), calixarene-containing ion exchangers and strong - base anion exchangers. Short characteristics of ion exchangers and some of the specified dependences are summarized below.

Strong - acid cross – linked polystyrene sulfonate cation exchangers. Polystyrene sulfonate cation exchanger cross-linked with divinylbenzene (DVB). An ion exchanger contains only one type of ionogenic groups – sulfonic acid groups (SO_3H). The structure of repeating unit of cross-linked polystyrene sulfonate cation exchanger (1) is represented below.

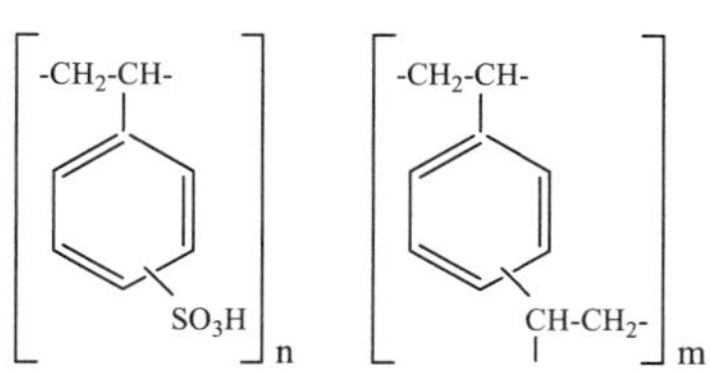

1

The ion exchanger contains nonequivalence ion exchange sites (Reichenberg, 1966) in spite of presence of only one type of ionogenic groups. The attempt to explain energy nonequivalence of ion exchange sites by the structure differences in cross-linking agent was made (Soldatov et al., 1994). The dependences (Soldatov et al. 1976) of corrected selectivity coefficients of ion exchange from mole fraction of sorbed potassium cations in variously cross-linked strong-acid sulfonate exchanger (KRS) were shown in fig.1.

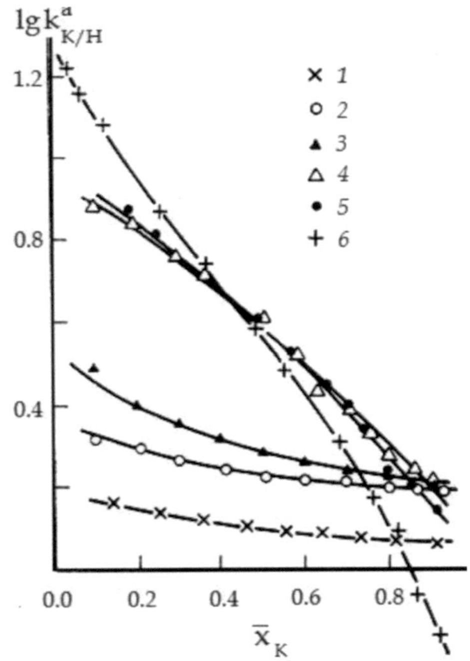

Fig. 1. Dependence of $\lg k^a_{K/H}$ from $\bar{x}_K$ of $H^+ \rightarrow K^+$ ion exchange in variously cross-linked strong-acid sulfonate exchanger (KRS). Nominal DVB, % : (*1*)-1, (*2*)-4, (*3*)-8, (*4*)-12, (*5*)-16, (*6*)-25. (Soldatov et al. 1976)

As we seen from fig.1, the energy nonequivalence of ion exchange sites of polymer rises with increasing of quantity of cross-linking agent. In that case any can to say that presented dependences (fig. 1) are described by the linear equations taking into account experimental errors (Soldatov & Suchover, 1968).

$\Delta\bar{H}(\bar{x}_i)$, $\ln k^a_{i/j}(\bar{x}_i)$ dependences of $Li^+ \rightarrow Cs^+$ ion exchange in middle-acid exchangers containing phosphonic acid groups (Becker et al., 1966) are shown in fig.2. Two types of phosphonic acid exchangers were used: a polystyrene-divinylbenzene cross-linked (5.5% DVB) preparation, $-C_6H_4-PO(OH)_2$, in which the phosphonate groups were attached to a benzene nucleus, and a polystyrene-divinylbenzene cross-linked (5.5% DVB) methylene phosphonic acid exchanger, $-C_6H_4-CH_2-PO(OH)_2$, in which the phosphonate was separated from the benzene ring by a methylene group.

The logarithms of selectivity coefficients are linear functions from Li^+ mole fraction in both phosphonic acid exchangers as can be seen from the fig. 2*a* and was noted (Becker et al., 1966). The measured heats of partial exchange are shown in fig. 2*b* by a chord for each experiment. The curves for the differential heat of exchange $\Delta\bar{H} = (\partial \Delta H / \partial \bar{x}_{Li})_T$ are the best least-squares fit through the midpoints of the chords. $\Delta\overline{H}$ are linear with respect to $\overline{x}_{Li}$ for both the nuclear phosphonic acid exchanger and ethylene phosphonic acid exchanger.

The energy nonequivalence of ion exchange sites was intensified on cation exchanger with methylene phosphonic acid groups. It was reflected in an evolution of $\Delta\overline{H}$ and $\lg k_{Li/Cs}$ of exchangers at $\overline{x}_{Li} = 0.0$ and $\overline{x}_{Li} = 1.0$. The magnitudes of partial enthalpy are in the range from 0 to 2000 cal per mole on the polymer in which the phosphonate groups were attached to a benzene nucleus. Partial enthalpy is increased from -500 to 3500 cal per mole on polystyrene-divinylbenzene cross-linked (5.5% DVB) methylene phosphonic acid exchanger.

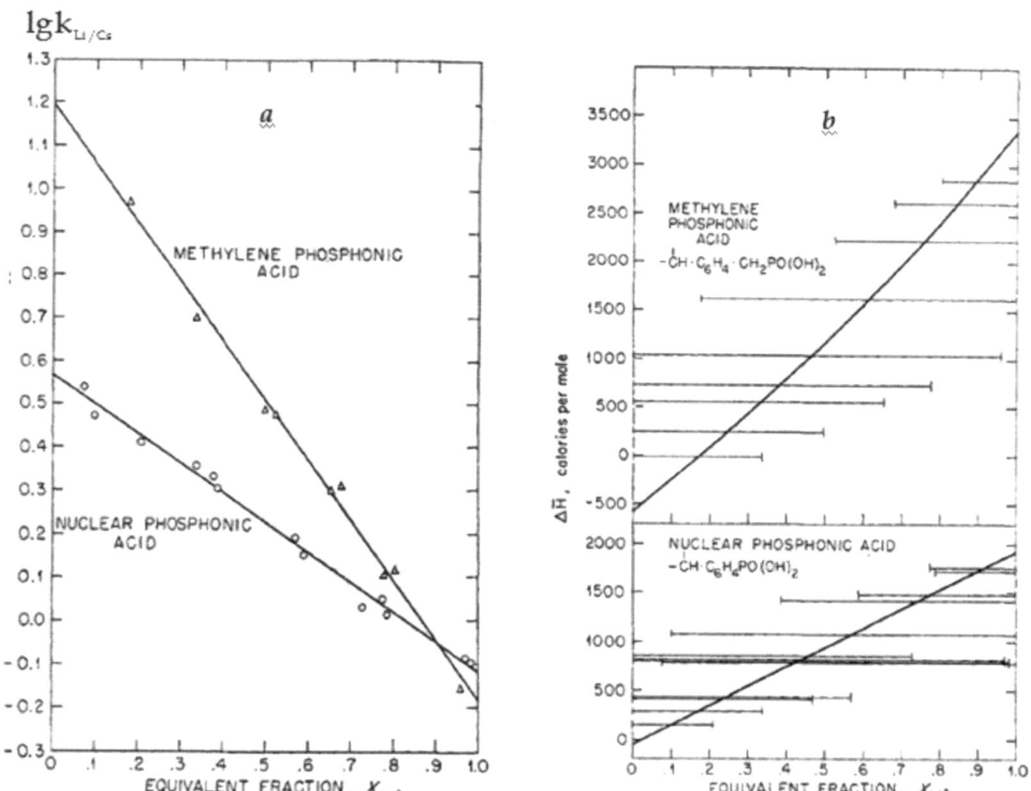

Fig. 2. Ion exchange of $Cs^+ \rightarrow Li^+$ in nominal 5.5% DVB cross-linked phosphonic acid ion exchangers: *a* –logarithmic function of selectivity coefficients, $\lg k_{Li/Cs}$ vs $\overline{x}_{Li}$; *b*-partial heat of exchange, $\Delta\overline{H}$ vs $\overline{x}_{Li}$ (Becker et al., 1966)

Zeolites (crystalline aluminium silicates) are weakly-acid cation exchangers. Particularity of ion exchange properties of zeolites is defined by crystal lattice structure and exchange cations distribution on the different position of the channels and cavities. In fig. 3 is shown a typical structure of zeolite repeating unit (2), presented in the templates of program Chem3D Ultra 8.0.

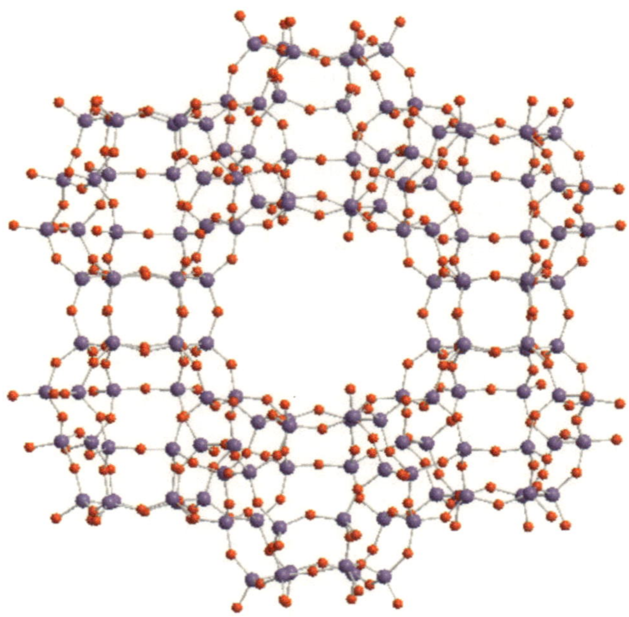

2

Fig. 3. Repeating unit of zeolite

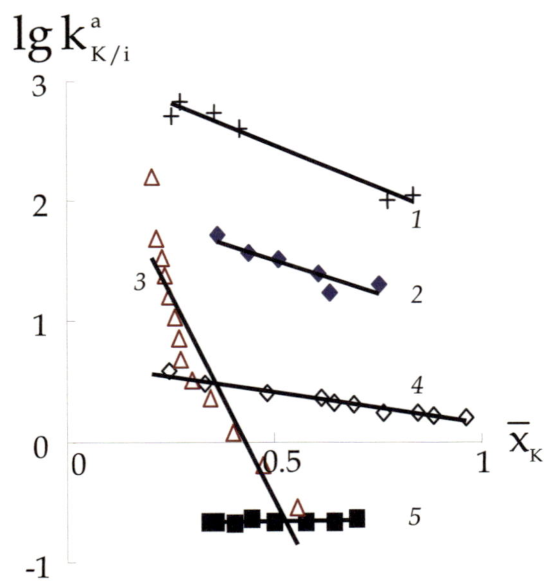

Fig. 4. Cation exchange in the natural heulandite (Al'tshuler & Shkurenko, 1990). $\lg^{a}_{K/Cat}$ vs $\overline{x}_K$. Cat: (*1*)-Li^+; (*2*)-Na^+; (*3*)-Rb^+; (*4*)-NH_4^+; (*5*)-Cs^+

The ion exchange thermodynamic functions in the heulandite – binary electrolyte solutions system were defined (Al'tshuler & Shkurenko, 1990). Natural heulandite (zeolite Sr-R), deposit of Kemerovo region, Russia. Idealized composition of elementary cell is $Ca_4[(AlO_2)_8(SiO_2)_{28}]\cdot 24H_2O$. Two – dimensional channel system was formed by rings, consisting from 8- and 10- members. Sizes of windows are: 4.0×5.5 Å, 4.1×4.7 Å, 4.4×7.2 Å (Breck, 1974). Cations of alkali metals and alkaline earth metals are the mobile exchangeable ions. The total ion exchange capacity of natural heulandite from Kemerovo region is 2.2 milliequivalents per gram of ion exchanger (Al'tshuler & Shkurenko, 1990). The isotherms of binary $Cat^+ \to K^+$ ion exchanges on natural heulandite are shown at the fig.4.

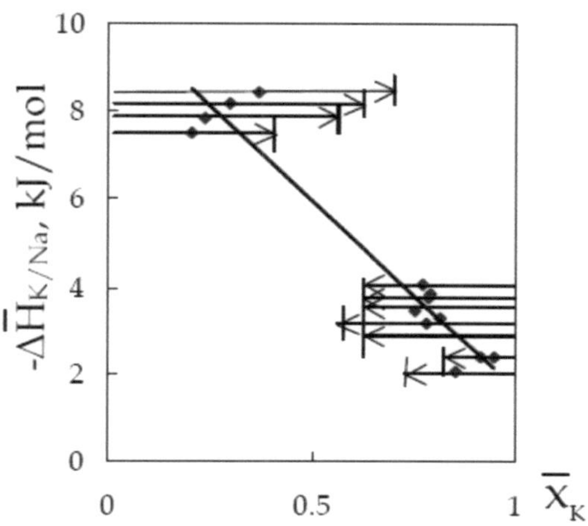

Fig. 5. Cation exchange of $Na^+ \to K^+$ in the natural heulandite (Al'tshuler & Shkurenko, 1990). $\Delta\bar{H}$ vs $\bar{x}_K$. Arrows reflect the change of heulandite composition

The dependence of differential enthalpy ($\Delta\bar{H}$) of $Na^+ \to K^+$ ion exchange from counterion composition in heulandite is shown in fig. 5. It is seen (fig. 4, fig. 5) partial thermodynamic functions of cation exchange linearly depend from counterion composition. The linear correlation coefficients are high, for example, its value equals to 0,97 for $\Delta\bar{H}(\bar{x}_i)$ function. The slope angle of the line depends on the energy nonequivalence of ion exchange sites.
Calixarenecontaining cationexchangers (3) have a common structure of polymer repeating unit

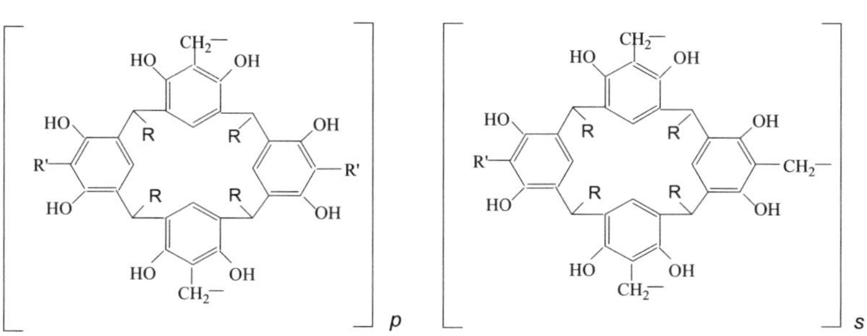

3a: $R=CH_3$, $R'=H$;
3b: $R=CH_3$, $R'=CH(OH)Fur$;
3c: $R=C_6H_4–PO(OH)_2$, $R'=H$;
3d: $R=C_6H_4–SO_3H$, $R'=H$

3

The experimental differential thermodynamic functions of ion exchange from aqueous solutions on weakly-acid cation exchangers 3a, 3b and strong-acid sulfonate cation exchanger 3d are shown in fig. 6-fig. 9.

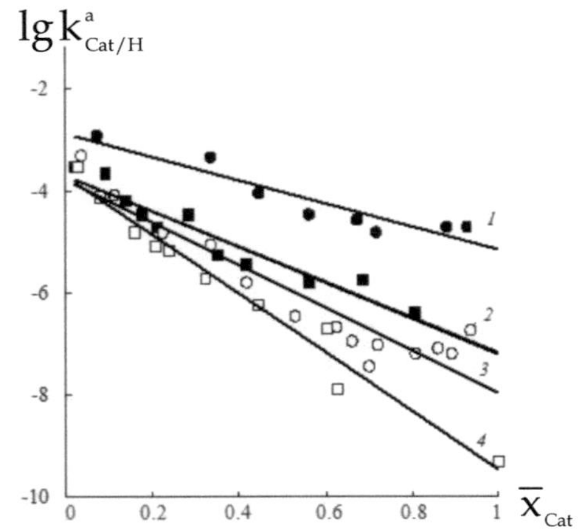

Fig. 6. Corrected selectivity coefficients of $H^+ \rightarrow NH_4^+$ ion exchange at 293 K in weakly – acid calixarenecontaining cation exchangers: (*1*) - in 3b, (*2*) - in 3a; $H^+ \rightarrow Na^+$ ion exchange: (*3*) - in 3b, (*4*) – in 3a (Al'tshuler et al., 2004)

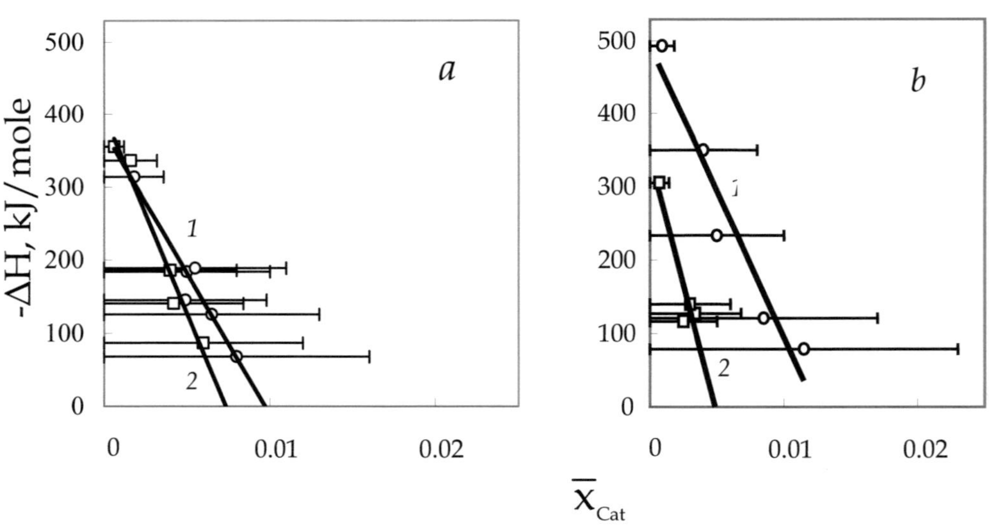

Fig. 7. Partial heat of $H^+ \rightarrow NH_4^+$ (*1*), $H^+ \rightarrow Na^+$ (*2*) cation exchange at 298 K in weakly – acid calixarenecontaining cation exchangers: (*a*) – in 3a, (*b*) – in 3b (Al'tshuler et al., 2004)

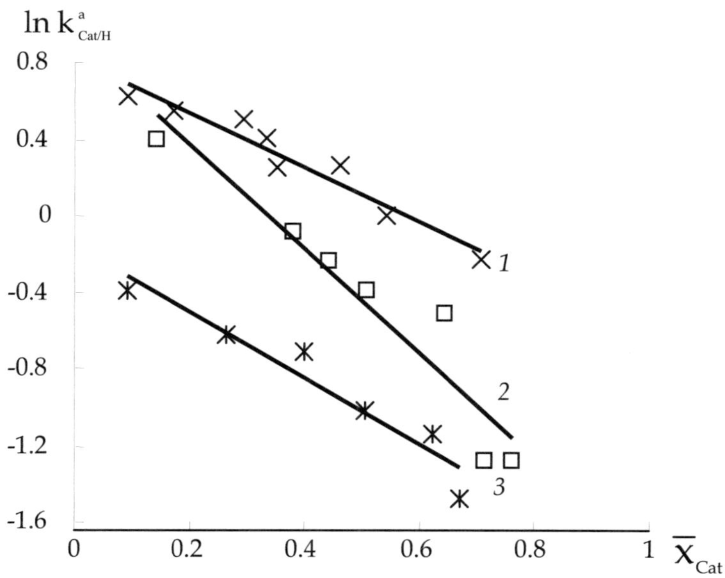

Fig. 8. Dependence of $\ln k^{a}_{Cat/H}$ from $\bar{x}_{Cat}$ of ion exchanges: (*1*) $H^+ \rightarrow Ag^+$, (*2*) $H^+ \rightarrow Na^+$, (*3*) $H^+ \rightarrow Li^+$ in strong – acid sulfonate cation exchanger 3d based on cis-tetraphenylcalix[4]-resorcinarene at 293 K (Al'tshuler et al., 2008

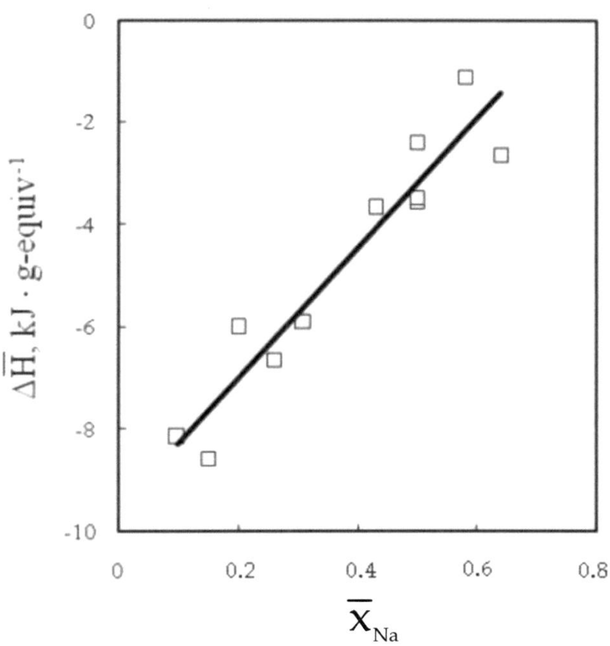

Fig. 9. Partial heat of $H^+ \rightarrow Na^+$ ion exchange in strong – acid sulfonate cation exchanger 3d based on cis-tetra-phenylcalix[4]resorcinarene, $\Delta\bar{H}$ vs $\bar{x}_{Na}$ (Al'tshuler et al., 2008)

The dependences of $\ln k^a_{Cat/H}$ vs $\bar{x}_{Cat}$ in the cation exchangers 3a-3d are shown in table 1.

Polymer	Cat	$\ln k^a_{Cat/H}$ ($\bar{x}_{Cat}$)	linear correlation coefficients
3a	Na^+	$\ln k^a_{Na/H} = -8.36 - 12.4\bar{x}_{Na}$	-0.96
3a	NH_4^+	$\ln k^a_{NH4/H} = -8.12 - 7.99\bar{x}_{NH4}$	-0.96
3b	Na^+	$\ln k^a_{Na/H} = -8.70 - 9.60\bar{x}_{Na}$	-0.94
3b	NH_4^+	$\ln k^a_{NH4/H} = -5.96 - 5.98\bar{x}_{NH4}$	-0.94
3c	K^+	$\ln k^a_{K/H} = -2.33 - 5.96\bar{x}_{K}$	-0.98
3d	Li^+	$\ln k^a_{Li/H} = -0.16 - 1.72\bar{x}_{Li}$	-0.99
3d	Na^+	$\ln k^a_{Na/H} = 0.91 - 2.73\bar{x}_{Na}$	-0.95
3d	Ag^+	$\ln k^a_{Ag/H} = 0.82 - 1.43\bar{x}_{Ag}$	-0.96

Table 1. Dependency of $\ln k^a_{Cat/H}$ vs $\bar{x}_{Cat}$ in calixarenecontaining cation exchangers

From the data about $\ln k^a_{i/j}(\bar{x}_i)$ and $\Delta\bar{H}(\bar{x}_i)$ linear dependences we can make a intermediate conclusion that investigated polymers 3a-3d in the form of single charge metal cations are Gibbs canonical assemblies in all intervals of counterion compositions.
From the experimental data (Soldatov& Bychkova, 1988) for Dowex-50W×12 strong - acid cation exchanger the following functions for independent binary exchanges were obtained:

$$\ln k^a_{NH_4^+/H^+} = 1.213 - 0.937\bar{x}_{NH_4^+}, \tag{9}$$

$$\ln k^a_{NH_4^+/Na^+} = 0.428 - 0.359\bar{x}_{NH_4^+}. \tag{10}$$

Strong-base anion exchangers. The ion exchange equilibria on strong-base anion exchanger of the Dowex-1 type were experimentally studied in aqueous solutions containing two types of anions: salicylate (Sal^-) - hydroxide (OH^-); nitrate (NO_3^-) - hydroxide; chloride (Cl^-) - hydroxide. The Dowex-1 is polystyrene cross-linked with divinylbenzene molecular network which carries the benzyltrimethylammonium ionogenic groups ($-CH_2C_6H_4-N^+(CH_3)_3$). The experimental data about the equilibrium phase composition for the binary ion exchanges on anion exchanger (Dowex-1) are presented here (fig. 10).
The equilibria for the binary exchanges on Dowex-1 can be described by the linear equations

$$\ln k^a_{Sal^-/OH^-} = 6.88 - 2.18\bar{x}_{Sal^-}, \tag{11}$$

$$\ln k^a_{NO_3^-/OH^-} = 4.815 - 1.17\bar{x}_{NO_3^-}, \tag{12}$$

$$\ln k^a_{Cl^-/OH^-} = 3.00 - 0.872\bar{x}_{Cl^-}. \tag{13}$$

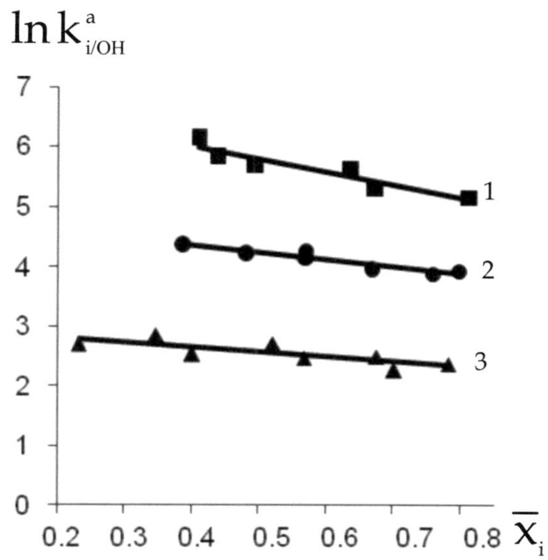

Fig. 10. Dependence of corrected selectivity coefficients of binary exchanges in Dowex-1 anion exchanger from mole fraction, $\overline{x}_i$, of sorbed anion: (*1*) - Sal^-, (*2*) - NO_3^-, (*3*) - Cl^-

3.2 Ion exchange from multicomponent solutions

For a multicomponent ion exchange the function $\ln k^a_{i/j}(\overline{x}_i)$ depends on the position of i and j ions in the series of selectivity. For any arbitrary point of interval of composition the following condition must be fulfilled

$$k^a_{i/k} = k^a_{i/j} / k^a_{k/j}. \tag{14}$$

Using the concept of reversible step reactions (Bjerrum, 1961) for a multicomponent ion exchange system containing n distinct ions with the same charge, the corrected selectivity coefficient is described by the equation (15)

$$\ln k^a_{1/n} = A_{1/n} + \sum_{i=1}^{n} B_{i/n} \overline{x}_i, \tag{15}$$

where n is the number (position) of the ion in the series of decreasing ion exchanger selectivity: $K_{1/n} > K_{2/n} > \ldots K_{i/n} \ldots > 1$; $A_{1/n}$, $B_{i/n}$ - are constants for binary exchanges of ions 1 and n; i and n. To calculate equilibria with participation of *n* exchangeable ions, we used a set of equations composed of (n - 1) equations of type (15), (n - 1) equations of type (4) for $k^a_{1/n}$, and the equation (16)

$$\sum_{i=1}^{n} \overline{x}_i = 1. \tag{16}$$

Thus, to calculate the phase composition of the ion exchanger for the system with n exchangeable ions, one should solve the following set of equations (17)

$$\begin{cases} \ln k^{a}_{1/2} = A_{1/2} + B_{1/2}\overline{x}_1 \\ \ln k^{a}_{1/3} = A_{1/3} + B_{1/3}\overline{x}_1 + B_{2/3}\overline{x}_2 \\ \ldots \\ \ln k^{a}_{1/n} = A_{1/n} + B_{1/n}\overline{x}_1 + B_{2/n}\overline{x}_2 + \ldots + B_{n-1/n}\overline{x}_{n-1} \\ k^{a}_{1/2} = \dfrac{\overline{x}_1 \cdot a_2}{\overline{x}_2 \cdot a_1} \\ k^{a}_{1/3} = \dfrac{\overline{x}_1 \cdot a_3}{\overline{x}_3 \cdot a_1} \\ \ldots \\ k^{a}_{1/n} = \dfrac{\overline{x}_1 \cdot a_n}{\overline{x}_n \cdot a_1} \\ \overline{x}_1 + \overline{x}_2 + \ldots + \overline{x}_n = 1. \end{cases} \quad (17)$$

The coefficients A and B were calculated from the experimental isotherms for independent binary exchanges by formulas (8) and (14).

We tested the proposed method for a vast array of experimental data on ion exchange equilibria in multicomponent systems in accordance with recommendations for testing new model (Soldatov & Bychkova, 1988). We imposed the following restrictions: ion exchange equilibria were considered at constant temperature, at constant ionic strengths of the solutions and only for monofunctional ion exchangers of specific chemical structure, for example, styrene-divinylbenzene sulfonate cation exchangers or anion exchangers containing only benzyltrimethylammonium groups. The equilibrium phase compositions were calculated for ion exchangers in which all exchange sites are accessible to exchangeable ions (the total exchange capacity of ion exchanger did not depend on the type of the sorbed ion). We considered ion exchange is not accompanied by complexation in the solution. Sorbate-sorbate interactions (Dmitryenko & Hale, 1965) don't present in ion exchanger phase. Exchange with participation of multicharged ions leads to nonlinear dependences of partial energies from counterion composition of calixarenecontaining polymers (Al'tshuler et al, 2008). High content of cross-linking agent creates additional difficulties (Vaslow & Boyd, 1966). Given these restrictions, a close agreement of calculated and experimental data was observed. All calculations were conducted on a computer with Intel (R) Core(TM)2 Duo T7300 2.00 GHz processor.

3.2.1 Three electrolytes in a solution

Consider thermodynamic system: sulfonate cation exchanger (Dowex-50W×12) – aqueous solution containing NH_4^+, Na^+ and H^+ cations. That system is recommended as a test (Soldatov & Bychkova, 1988) for calculating of the phase composition of an ion exchanger from data of binary exchanges.

$\bar{x}_{NH_4^+}$		$\bar{x}_{Na^+}$		$\bar{x}_{H^+}$	
exp.	calc.	exp.	calc.	exp.	calc.
0.123	0.117	0.142	0.159	0.735	0.724
0.132	0.125	0.066	0.085	0.802	0.79
0.115	0.105	0.268	0.285	0.617	0.61
0.104	0.097	0.384	0.391	0.512	0.512
0.096	0.090	0.483	0.486	0.421	0.424
0.085	0.085	0.573	0.572	0.342	0.343
0.082	0.081	0.654	0.654	0.264	0.265
0.079	0.078	0.727	0.732	0.194	0.190
0.072	0.074	0.804	0.810	0.124	0.116
0.224	0.218	0.062	0.076	0.714	0.706
0.215	0.207	0.132	0.145	0.653	0.648
0.201	0.189	0.250	0.264	0.549	0.547
0.186	0.176	0.358	0.367	0.456	0.457
0.165	0.166	0.461	0.459	0.374	0.375
0.438	0.427	0.201	0.216	0.361	0.357
0.404	0.408	0.308	0.308	0.288	0.284
0.388	0.394	0.392	0.394	0.220	0.212
0.374	0.381	0.474	0.476	0.152	0.143
0.360	0.372	0.555	0.555	0.085	0.073
0.558	0.553	0.048	0.055	0.394	0.392
0.540	0.540	0.104	0.106	0.356	0.354
0.521	0.517	0.193	0.203	0.286	0.280
0.498	0.498	0.282	0.291	0.220	0.211
0.486	0.484	0.362	0.375	0.152	0.141
0.625	0.632	0.055	0.052	0.320	0.316
0.611	0.620	0.104	0.101	0.285	0.279
0.154	0.158	0.549	0.544	0.297	0.298
0.146	0.151	0.629	0.626	0.225	0.223
0.139	0.146	0.704	0.704	0.157	0.150
0.131	0.142	0.781	0.782	0.088	0.076
0.361	0.357	0.059	0.065	0.580	0.578
0.347	0.343	0.119	0.126	0.534	0.531
0.328	0.321	0.224	0.235	0.448	0.444
0.304	0.304	0.331	0.332	0.365	0.364
0.285	0.290	0.422	0.421	0.293	0.289
0.468	0.472	0.448	0.456	0.084	0.072
0.269	0.279	0.509	0.504	0.222	0.217
0.261	0.271	0.585	0.584	0.154	0.145
0.250	0.264	0.657	0.662	0.093	0.074
0.456	0.463	0.063	0.060	0.481	0.477
0.447	0.450	0.115	0.115	0.438	0.435
0.592	0.598	0.191	0.193	0.217	0.209

Table 2. Experimental (exp.) and calculated (calc.) phase compositions of Dowex-50W×12

The series of selectivity of sulfonate cation exchanger (Dowex-50W×12) is: $NH_4^+ > Na^+ > H^+$. By using (9) and (10) the set of equations (17) is reduced to

$$\begin{cases} \ln k^a_{NH_4^+/Na^+} = 0.428 - 0.359\bar{x}_{NH_4^+} \\ \ln k^a_{NH_4^+/H^+} = 1.213 - 0.937\bar{x}_{NH_4^+} + 0.578\bar{x}_{Na^+} \\ k^a_{NH_4^+/Na^+} = \dfrac{\bar{x}_{NH_4^+} \cdot a_{Na^+}}{\bar{x}_{Na^+} \cdot a_{NH_4^+}} \\ k^a_{NH_4^+/H^+} = \dfrac{\bar{x}_{NH_4^+} \cdot a_{H^+}}{\bar{x}_{H^+} \cdot a_{NH_4^+}} \\ \bar{x}_{NH_4^+} + \bar{x}_{Na^+} + \bar{x}_{H^+} = 1. \end{cases}$$

In the table 2 are compared the experimental (Soldatov & Bychkova, 1988) and calculated in present work equilibrium phase compositions of Dowex-50W×12 cation exchanger for the $NH_4^+ - Na^+ - H^+$ three –ion exchange from aqueous solutions.

3.2.2 Four electrolytes in a solution

According to the obtained data (Fig. 10), the selectivity of the strong-base anion exchanger (Dowex-1) with respect to the studied anions decreases in the series: $Sal^- > NO_3^- > Cl^- > OH^-$. Equations (11)-(13) were used to calculate many – ion exchange equilibria by the set of equations (17). The experimental and calculated (by eqs. (17)) compositions of anion exchanger (Dowex-1) phase at equilibrium with aqueous solutions containing four competing anions (Sal^-, NO_3^-, Cl^- and OH^-) are presented in Table 3.

$\bar{x}_{Sal^-}$		$\bar{x}_{NO_3^-}$		$\bar{x}_{Cl^-}$		$\bar{x}_{OH^-}$	
exp.	calc.	exp.	calc.	exp.	calc.	exp.	calc.
0.208	0.205	0.592	0.603	0.149	0.147	0.050	0.050
0.341	0.337	-	-	0.289	0.298	-	-
0.432	0.410	-	-	0.339	0.341	-	-
0.519	0.526	0.239	0.229	0.100	0.102	0.142	0.143
0.598	0.604	0.102	0.101	0.202	0.198	0.098	0.098

Table 3. Experimental (exp.) and calculated (calc.) data of the $Sal^- - NO_3^- - Cl^- - OH^-$ tetra - ion exchange in strong-base anion exchanger (Dowex-1)

4. Conclusion

The predictive power of the calculation method was characterized by the mean absolute deviation for the entire array (Soldatov & Bychkova, 1988):

$$\Delta\bar{x} = \left(\sum_{i=1}^{n} \sum_{r=1}^{s} \left| \bar{x}_{ir\,exp} - \bar{x}_{ir\,calc} \right| \right) \Big/ ns, \tag{18}$$

where n is the number of exchangeable ions and s is the number of experimental points in many-ion equilibria. Processing the data of Table 2 by formula (18) gives $\Delta\bar{x}=0.006$, a value that corresponds to the most accurate among the calculation methods that were considered early (Soldatov & Bychkova, 1988). Processing the data from Table 3 by formula (18) yields $\Delta\bar{x}=0.005$. This results show that the proposed method has a high predictive power. The consideration of ion exchanger as a canonical assembly explains why the corrected ion exchange selectivity coefficient linearly depends on the ion exchanger composition. This circumstance makes it possible to predict many-ion exchange equilibria.

5. References

Al'tshuler, G.; Sapozhnikova, L. & Kirsanov, M. (1984) Division of unicharged cation mixtures on sulfoacidic and carboxylic ionites. *Russ. J. Phys. Chem,* Vol. 58, 162 - 166, ISSN: 0036-0244

Al'tshuler, G. & Shkurenko, G. (1990). Cation exchange equilibrium on natural heulandites. *Russ. Chem. Bull,* Vol. 39, 1331-1334, ISSN: 1066-5285

Al'tshuler, H.; Ostapova, E.; Sapozhnikova, L. & Altshuler O. (2004). Thermodynamics of sorption of sodium and ammonium cations by exchangers based on c-calix[4] resorcinarene, *Russ. Chem. Bull.,* Vol. 53, No. 12, 2670-2673, ISSN: 1066-5285

Altshuler, H.; Ostapova, E.; Sapoznikova, L. & Altshuler O. (2008). Thermodynamics of ion exchange in a sulfonated polymer based on cis-tetraphenylmetacyclophanoctol. *J. Therm. Anal. Calorim.,* Vol. 92, No. 3, 665-669, ISSN:1388-6150

Becker, K.; Lindenbaum, S. & Boyd G. (1966). Thermodynamic quantities in the exchange of lithium with cesium ions on cross-linked phosphonic acid cation exchangers. *J. Phys. Chem.* Vol. 70, No. 8, 3834-3837, ISSN: 0022-3654

Bjerrum, J. (1957). Metal Ammine Formation in Aqueous Solution. *Theory of Reversible Step Reactions,* Copenhagen

Boyd, G. & Larson, Q. The binding of quaternary ammonium ions by polystyrenesulfonic acid type cation exchangers. (1967). *J. Am. Chem. Soc.,* Vol. 89, P. 6038-6042, ISSN: 0002-7863

Breck, D. (1974). *Zeolite molecular sieves,* A Wiley – interscience publication, ISBN-10: 0471099856, ISBN-13: 9780471099857 ISBN-13: 9780471099857, New York – London-Sydney – Toronto

Dmitryenko, L. & Hale, D. (1965). The sorption of oxytetracycline and chlorotetracycline by sulfonated polystyrene resins. *J. Chem. Soc.,* (1965), 5570 – 5578, ISSN: 0368-1769

Gaines, G. & Thomas, H. (1953). Adsorption Studies on Clay Minerals. *J. Chem. Phys,* Vol. 21, No. 4, 714-718, ISSN (printed): 0021-9606

Horst, J.; Holl, W. & Eberle, S. (1990). Application of the surface complex formation model to exchange equilibria on ion exchange resins. Part I. Weak-acid resins. *Reactive polymers,* Vol. 13, 209-231, ISSN: 0923-1137

Horst, J.; Holl, W. & Wemet, M. (1991). Application of the surface complex formation model to exchange equilibria on ion exchange resins. Part II. Chelating resins. *Reactive Polymers,* Vol. 14, 251-261, ISSN: 0923-1137

Reichenberg, D. (1966). The ion exchanger selectivity, In: *Ion Exchange,* Marinsky, J.A., (Ed.), 104-173, Department of Chemistry State University of New York, New York

Soldatov, V. & Suhover, R. (1968). Identification of errors of experimental measurements and calculations of thermodynamic quantities in ion-exchange systems. *Russ. J. Phys. Chem*, Vol. 42., 1798-1801, ISSN: 0036-0244

Soldatov, V.; Marcinkiewicz, R.; Pokrovsky, A.; Kuvaev, Z.; Makarova, S.; Aptova, T. & Egorov E. (1976). Comparative characteristics sulphopolystyrene ion exchangers synthesized with different isomers of divinylbenzene. *Russ. J. Phys. Chem*, Vol. 50, No. 2, 480-484, ISSN: 0036-0244

Soldatov, V. & Bychkova, V. (1988). *Ion-Exchange Equilibrium in Multicomponent Systems*, Nauka i Tekhnika, Minsk

Soldatov, V.; Gogolinskii, V.; Zeienovskii, V. & Pushkarchuk, A. (1994). Structure calculation of network sulfopolystyrene ionite with meta- and para- divinylbenzene as cross agent. *Dokl. Ross. Akad. Nauk,* Vol. 337, No. 1, 63-67, ISSN: 0869-5652

Tondeur, D. & Klein, G. (1967). Multicomponent ion exchange in fixed beds. *Ind. Eng. Chem. Fundamentals,* Vol. 6, No. 3, 351-361, ISSN: 0019-7874

Vaslow, F. & Boyd E. (1966). Thermodynamic properties in exchange of silver with sodium ions in cross - linked polystyrene sulfonate cation exchangers. *J. Phys. Chem*, Vol. 70, 2295-2299, ISSN: 0022-3654

18

Microemulsions: Thermodynamic and Dynamic Properties

S.K. Mehta and Gurpreet Kaur
Department of Chemistry and Centre of Advanced Studies in Chemistry
Panjab University, Chandigarh
India

1. Introduction

Mixing two immiscible liquids (such as oil and water) using emulsifier and energy inputs has been the matter of study for decades. In early 1890's extensive work have been carried out on macroemulsions (i.e. oil dispersed in water in the form of fine droplets or vice versa) (Becher, 1977) and several theories and methods of their formation have been vastly explored (Lissant 1976 and 1984). However going along the line, microemulsion systems were well opted because of their stability and isotropic nature. Microemulsions are basically thermodynamically stable, isotropically clear dispersions of two immiscible liquids such as oil and water stabilized by the interfacial film of any surfactant and/or cosurfactant. Although a microemulsion is macroscopically homogeneous, or quasi-homogeneous but structured microscopically. Microemulsions in comparison to micellar systems are superior in terms of solubilization potential and thermodynamic stability and offers advantages over unstable dispersions, such as emulsions and suspensions, since they are manufactured with little energy input (heat, mixing) and have a long shelf life (Constantinides, 1995). The term "microemulsion" was first coined by Schulman group (Schulman et al., 1959). However, ambiguity in the microemulsion terminology persists till today as some authors differentiate them from swollen micelles (which either contain low volume fraction of water and oil) and transparent emulsions (Prince, 1977, Malcolmson et al., 1998).

One of the unique factors associated with microemulsions is the presence of different textures such as oil droplets in water, water droplets in oil, bicontinuous, lamellar mixtures etc., which are formed by altering the curvature of interface with the help of different factors such as salinity, temperature, etc. Such a variety in structure of microemulsion is a function of composition of the system. Phase study greatly helps to elucidate different phases that exist in the region depending upon the composition ratios. One peculiarity of microemulsions lies in the fact that these structures are interchangeable.

Construction of phase diagram enables determination of aqueous dilutability and range of compositions that form a monophasic region (Fig. 1). One of the unique factor associated with microemulsions is the presence of different structures as classified by Winsor (Winsor, 1948). Winsor I (o/w), Winsor II (w/o), Winsor III (bicontinuous or middle phase microemulsion) and Winsor IV systems are formed by altering the curvature of interface with the help of different factors such as salinity, temperature, etc. Where Type I indicates surfactant-rich water phase (lower phase) that coexists with surfactant-poor oil phase (Winsor I), Type II is surfactant-rich oil phase (the upper phase) that coexists with surfactant-poor water phase (Winsor II), Type III represents the surfactant rich middle-phase which coexists with both water (lower) and oil (upper) surfactant-poor phases (Winsor III) and Type IV is a single phase homogeneous mixture. Based upon the composition, these can be of various types *viz.*, water-in-oil (w/o) or oil-in-water (o/w) type or Lamellar or bicontinuous, hexagonal and reverse hexagonal, etc. (Fig. 2).

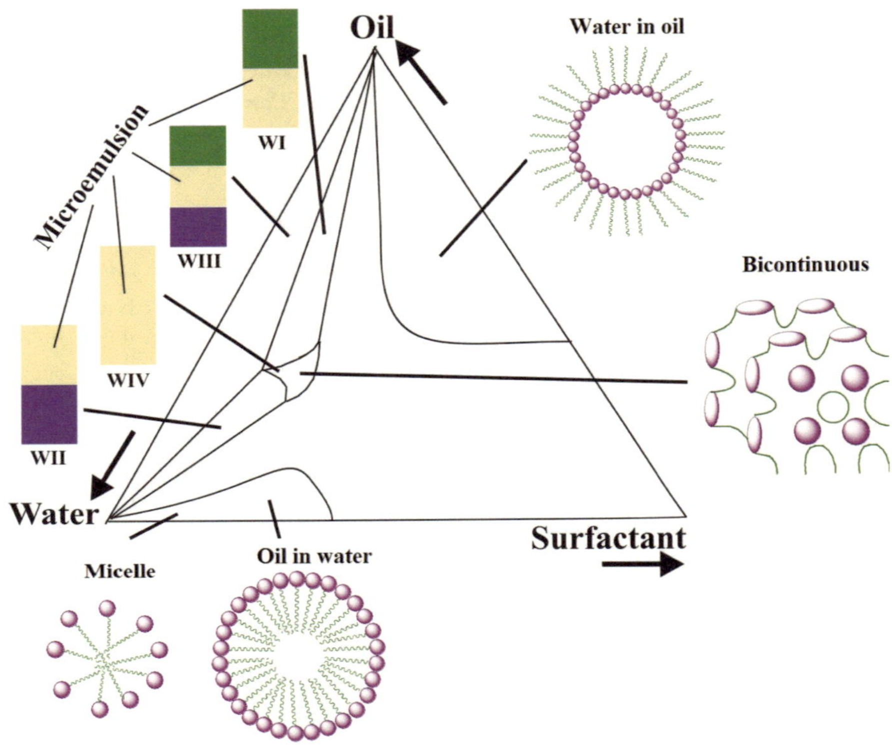

Fig. 1. A hypothetical ternary phase diagram representing three components of the system

Numerous attempts were made for predicting microemulsion types, the first was by Bancroft (later known as Bancroft's rule). It states that water-soluble emulsifiers tend to form o/w emulsions and oil-soluble emulsifiers tend to form w/o emulsions (Bancroft, 1913). Obviously, this is very qualitative, and therefore, it is of interest to put the area on a more quantitative footing. This section describes some of these concepts.

The preferred curvature of the interface is governed by the relative values of the head group area (a_o) and tail effective area (v/l_c) as described by Israelachvili et al., where v is the volume and l_c is the effective hydrocarbon chain (Fig. 3) (Israelachvili et al. 1976). By a simple geometrical consideration, the Critical Packing Parameter (CPP) is expressed as

$$CPP = v/l_c a_o \tag{1}$$

Theoretically, a CPP value less than $\frac{1}{3}$ corresponds to spherical micelles, between $\frac{4}{3}$ and $\frac{1}{2}$ corresponds to rod-like micelles and between $\frac{1}{2}$ and 1 to a planar structure (Fig. 4).

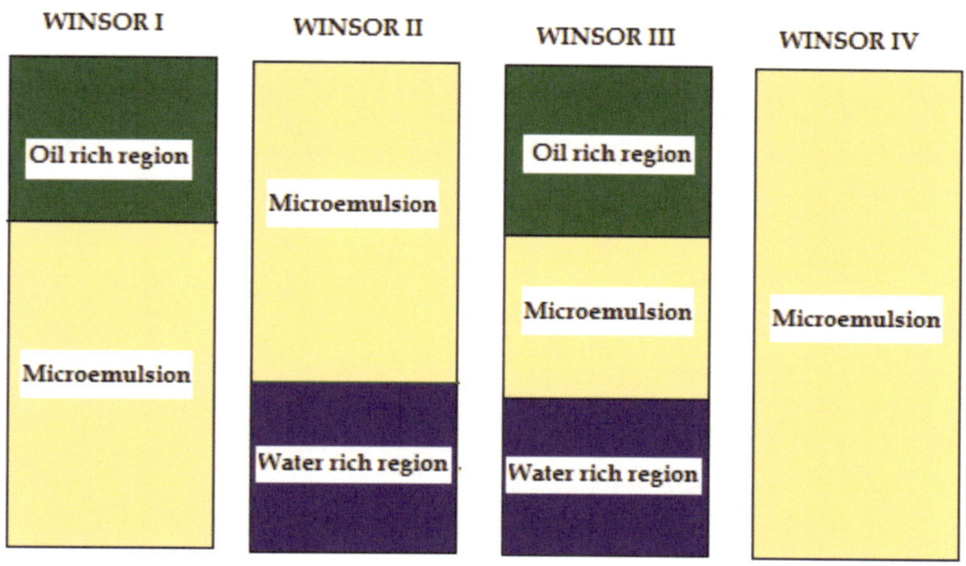

Fig. 2. Schematic presentation of most occurred surfactant associates

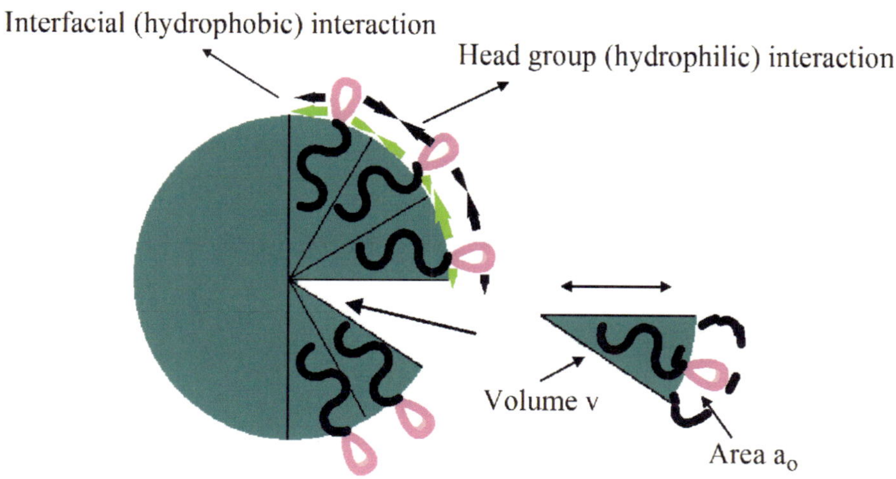

Fig. 3. The critical packing parameter relates the head group area, the length and the volume of the carbon chain into dimensionless number (eqn. 1)

The concept of HLB (Hydrophilic-Lipophilic Balance) was introduced by Griffin in 1949. As the name suggests, HLB is an empirical balance based on the relative percentage of the hydrophilic to the lipophilic moieties in the surfactant. Later, he (Griffin, 1954) defined an empirical equation that can be used to determine the HLB based on chemical composition. Davies et al. has offered a more general empirical equation (Davies et al., 1959) by assigning a number to the different hydrophilic and lipophilic chemical groups in a surfactant. The HLB number was calculated by the expression,

$$HLB = [(n_H xH) - (n_L xL)] + 7 \quad (2)$$

where H and L is the numbers assigned for the hydrophilic and lipophilic groups respectively, and n_H and n_L are the respective numbers of these groups per surfactant molecule. Both HLB and packing parameter numbers are closely correlated. However, it has been shown that, for a bicontinuous structure, which corresponds to a zero curvature, HLB ≈ 10. For HLB <10, negative curvature is favourable (i.e. w/o microemulsion), while for HLB >10, positive curvature results.

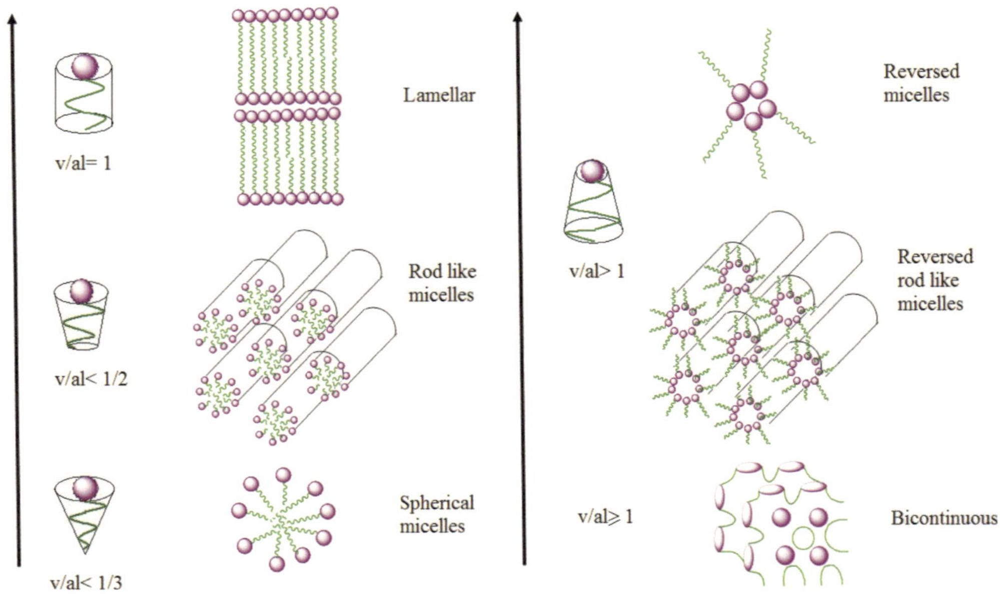

Fig. 4. The surfactant aggregate structure for critical packing parameters from < 1/3 (lower left) to >1 (upper right)

2. Basics of microemulsion formation

There are different theories relating to the formation of microemulsions i.e. interfacial, solubilization and thermodynamic theories, etc. The first theory known as mixed film theory considered the interfacial film as a duplux film. In 1955 (Bowcott & Schulman, 1955) it was postulated that the interface is a third phase, implying that such a monolayer is a duplex film, having diverse properties on the water side than oil side. Such a specialized liquid having two-dimensional region bounded by water on one side and oil on the other, has been based on the assumption that the spontaneous formation of a microemulsion is due to the interactions in the interphase and reducing the original oil/water interfacial tension to zero. However, zero interfacial tension does not ensure that a microemulsion is formed, as cylindrical and lamellar micelles are also believed to be formed. What differentiated an emulsion from other liquid crystalline phases is the kind of molecular interactions in the liquid interphase that produce an initial, transient tension or pressure gradient across the flat interphase, i.e., a duplux film, causing it to enclose one bulk phase in the other in the form of spheres. A liquid condensed film was considered essential to give the kind of flexibility to the interphase that would allow a tension gradient across it to produce curvature. Following the concept of mixed film theory, Robbins developed the theory of phase behavior of microemulsions using the concept that interactions in a mixed film are responsible for the direction and extent of curvature and thus can estimate the type and size of the droplets of microemulsions (Robbins, 1976). It is believed that kind and degree of curvature is imposed by the differential tendency of water to swell the heads and oil to swell the tails.

The stability of microemulsions has been the matter of interest for various research groups working in this area. The workers however feel that along with the depression in the interfacial tension due to surface pressure, a complex relationship between zero interfacial tension and thermodynamic stability holds the key for the formation of microemulsion systems. The thermodynamic factors include stress gradients, solubility parameters, interfacial compressibility, chemical potentials, enthalpy, entropy, bending and tensional components of interfacial free energy, osmotic pressure and concentrations of species present in the bulk and interphase, etc.

Based upon these facts, another theory i.e., the solubilization theory, was proposed which considers microemulsions as swollen micellar system. A model has been presented by Adamson (Adamson, 1969) in which the w/o emulsion is said to be formed because of the balance achieved in the Laplace and osmotic pressure. However, it has been emphasised that micellar emulsion phase can exist in equilibrium with non-colloidal aqueous phase. The model also concluded that the electrical double layer system of aqueous interior of the micelle is partially responsible for the interfacial energy. It was assumed that the interface has a positive free energy. However, this gave a contradiction to the concept of negative interfacial tension.

Considering the thermodynamic theory, the free energy of formation of microemulsion, ΔG_m, consists of different terms such as interfacial energy and energy of clustering droplets. Irrespective of the mechanism, the reduction of the interfacial free energy is critical in facilitating microemulsion formulation. Schulman and his co-workers have postulated that the negative interfacial tension is a transient phenomenon for the spontaneous uptake of water or oil in microemulsion (Schulman et al., 1959). It has been proved from thermodynamic consideration that a spontaneous formation of microemulsions take place where the interfacial tension is of order 10^{-4} to 10^{-5} dynes/cm (Garbacia & Rosano, 1973). However, the stability and the size of droplets in microemulsion can also be adjudged using the thermodynamic approach. This approach accounts for the free energy of the electric double layer along with the van der Waals and the electrical double layer interaction potentials among the droplets. It also takes into consideration the entropy of formation of microemulsion. Schulmen et al. also reported that the interfacial charge is responsible in controlling the phase continuity (Schulman et al., 1959).

Conversely from the thermodynamical point of view, it can also be said that microemulsions are rather complicated systems, mainly because of the existence of at least four components, and also because of the electric double layer surrounding the droplets, or the rods, or the layers which contribute noticeably to the free energy of the system. The role of the electrical double layer and molecular interactions in the formation and stability of microemulsions were well studied by Scriven (Scriven, 1977). Ruckenstein and Chi quantitatively explained the stability of microemulsions in terms of different free energy components and evaluated enthalpic and entropic components (Ruckenstein & Chi, 1975). For a dispersion to form spontaneously, the Gibbs free energy of mixing, ΔG_m must be negative. For the dispersion to be thermodynamically stable, ΔG_m must, furthermore, show a minimum.

When applying these conditions to microemulsions with an amphiphilic monolayer separating the polar and the nonpolar solvent, it has been customary to attribute a natural curvature as well as a bending energy to the saturated monolayer, thereby making the interfacial tension depend on the degree of dispersion. Kahlweit and Reissi have extensively worked on the stability of the microemulsions and paid attention to the reduction of amphiphile surface concentration below the saturation level, an effect that also makes the interfacial tension depend on the degree of dispersion (Kahlweit and Reissi, 1991).

Thermodynamic treatment of microemulsions provided by Ruckenstein and Chin not only provided the information about its stability but also estimated the size of the droplets (Ruckenstein & Chin, 1975). Treatment of their theory indicated that spontaneous formation of microemulsions occurs when the free energy change of mixing ΔG_m is negative. However, when ΔG_m is positive, a thermodynamically unstable and kinetically stable macroemulsions are produced. According to them, the model consists of monodisperse microdroplets, which are randomly distributed in the continuous phase. The theory postulated the factors which are responsible for the stability of these systems which includes van der Waals attraction potential between the dispersed droplet, the repulsive potential from the compression of the diffuse electric double layer, entropic contribution to the free energy from the space position combinations of the dispersed droplets along with the surface free energy. The van der Waals potential was calculated by Ninham-parsegian approach, however, the energy of the electric double layer was estimated from the Debye-Huckel distribution. Accordingly the highest and lowest limit of entropy has been estimated from the geometric considerations. The calculations depicted that the contribution from the van der Waals potential is negligible in comparison to the other factors contributing to the free energy.

It has been suggested (Rehbinder & Shchukin, 1972) that when the interfacial tension is low but positive, the interface may become unstable due to a sufficiently large increase in entropy by dispersion. The entropy change decreases the free energy and overpowers the increase caused by the formation of interfacial area and therefore net free energy change is negative. Along with this Murphy (Murphy, 1966) suggested that an interface having a low but positive interfacial tension could nevertheless be unstable with respect to bending if, the reduction in the interfacial free energy due to bending exceeds the increase in free energy due to the interfacial tension contribution. He also concluded that this bending instability might be responsible for spontaneous emulsification. Based upon these conclusions, Miller and Scriven interpreted the stability of interfaces with electric double layer (Miller & Scriven, 1970). According to them the total interfacial tension was divided into two components

$$\gamma_T = \gamma_p - \gamma_{dl} \tag{3}$$

where γ_T is total interfacial tension which is the excess tangential stress over the entire region between homogenous bulk fluids including the diffuse double layer. γ_p is the phase interphase tension which is that part of the excess tangential stress which does not arise in the region of the diffuse double layer and - γ_{dl} is the tension of the diffuse layer region. Equation 3 suggests that when γ_{dl} exceeds γ_p, the total interfacial tension becomes negative. For a plane interphase the destabilising effect of a diffuse layer is primarily that of a negative contribution to interfacial tension. Their results confirm that the double layer may indeed affect significantly the interfacial stability in low surface tension systems. However, the thermodynamic treatment used by Ruckenstein and Chi also included the facts that along with free energy of formation of double layers, double layer forces and London forces were also taken in consideration. For evaluating entropy of the system, the ideality of the system was not assumed.

The theory also predicts the phase inversion that can occur in a particular system. According to the calculations, the free energy change ΔG_m is the sum of changes in the interfacial free energy (ΔG_1), interaction energy among the droplets (ΔG_2) and the effect caused by the entropy of dispersion (ΔG_3=$T\Delta S_m$). The antagonism among these different factors mainly predicts the formation of microemulsions. The variation of ΔG_m with the radius of droplet, R, at constant value of water/oil ratio can be determined using $\Delta G_m(R) = \Delta G_1 + \Delta G_2 - T\Delta S_m$. R* is stable droplet size for a given volume fraction of the dispersed phase that leads to a minimum in G^*_m (Fig. 5). R* can be obtained from

$$\frac{d\Delta G_m}{dR_{R=R^*}} = 0 \tag{4}$$

if the condition, $(d^2\Delta G_m/dR^2)_{R=R^*}>0$ is satisfied. In order to determine the type of microemulsion and the phase inversion, the values of ΔG^*_m for both types of microemulsions have to be compared at same composition. The composition having more -ve value of Gibbs free energy will be favoured and the volume fraction for which the values of ΔG^*_m are same for both kinds of microemulsion are said to undergo phase inversion.

The quantitative outcome of the model for the given free energy as a function of droplet size has been shown in Fig. 5. The free energy change is positive for B and C i.e., only emulsions, the free energy change is positive. Hence, only emulsions which are thermodynamically unstable are formed in the case C. However, kinetically stable emulsions may be produced that too depending upon the energy barrier. The free energy change is negative within a certain range of radius (R) of dispersed phase in the case A, this means that droplets having a radius within this size range are stable towards phase separation and microemulsions formed are thermodynamically stable. The calculations show that at specific composition and surface potential, the transformation of curves from C to A can take place by decreasing the specific surface free energy.

As obvious from the above explanation, the assessment of the stability of microemulsions has been carried simultaneous by various groups after it was first triggered by Schulman and his collaborators. Simultaneously, Gerbacia and Rosano measured the interfacial tension at the interphase in the presence of an alcohol that is present in one of the phase (Gerbacia & Rosano, 1973). The presence of alcohol lowered the interfacial tension to zero as the alcohol diffuses to interface. This observation lead to the conclusion that for the formation of microemulsion, the diffusion of surface active molecules to the interface is mandatory and also the formation depends on the order in which the components are added.

The concept that interfacial tension becomes zero or negative for spontaneous formation of microemulsions has been later modified (Holmberg, 1998). It is believed that a monolayer is formed at the oil/water interphase which is responsible for a constant value of interfacial tension, which can be estimated using Gibbs equation.

$$\partial\gamma = -\Sigma\Gamma_i d\mu_i = -\Sigma\Gamma_i RT d\ln C_i \tag{5}$$

where Γ, , μ, C_i are Gibbs surface excess, chemical potential, and concentration of ith component. The presence of cosurfactant in the system further lowers this value.

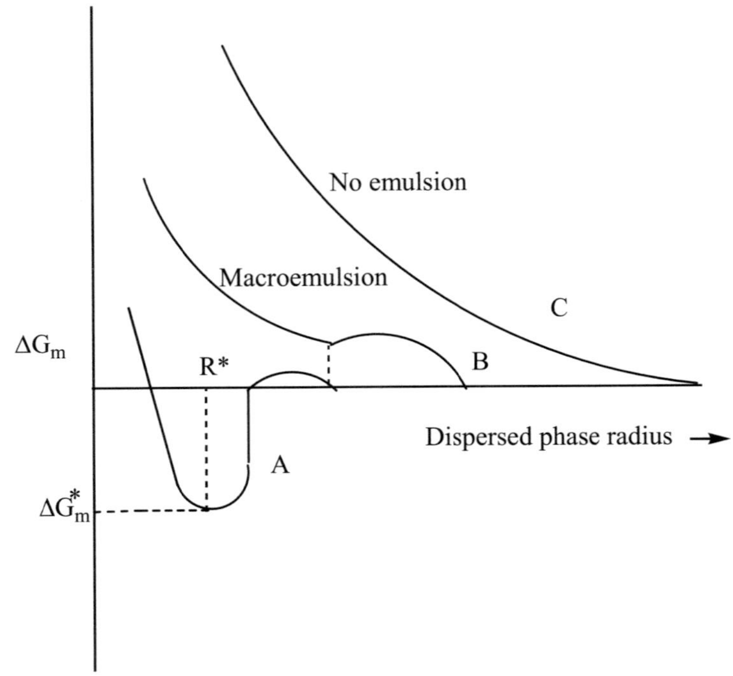

Fig. 5. Diagrammatic illustration of free energy as a function of droplet size

During 1980s the "dilution method" was well adopted by many, to extract energetic information for different combinations along with the understanding of their structural features (Bansal et al., 1979; Birdi, 1982; Singh et al., 1993; Bayrak, 2004; Zheng et al., 2006; Zheng et al., 2007), after it was first introduced by Schulman group (Bowcott & Schulman, 1955; Schulman et al., 1959). Basically, this method deals with the estimation of distribution of cosurfactant (k_d) and hence, determines the composition of interphase which is in turn responsible for the formation and stability of microemulsions. From the value of k_d (equilibrium constant), the Gibbs free energy of transfer of alcohol from the organic phase to the interphase can be obtained from the equation

$$\Delta G^o_{transfer} = -RT \ln K_d \tag{6}$$

Using this method, the different thermodynamic parameters such as entropy or/ and enthalpy of transfer can also be obtained. A polynomial fitting between $\Delta G^o_{transfer}$ and the temperature (T) was used to obtain $\Delta S^o_{transfer}$ from its first derivative

$$\frac{\Delta G^o_{transfer}}{\Delta T} = -\Delta S^o_{transfer} \tag{7}$$

From the knowledge of $\Delta G^o_{transfer}$ and $\Delta S^o_{transfer}$, the enthalpy of transfer was calculated according to the Gibbs Helmoltz equation

$$\Delta H^{o}_{transfer} = \Delta G^{o}_{transfer} + T\Delta S^{o}_{transfer} \tag{8}$$

Apart from the understanding of the composition of interphase, the "dilution method" also helps in the estimation and determination of structural aspects of the microemulsion system like droplet size, number of droplets, etc.

3. Percolation phenomenon

The integrity of the monolayer is often influenced by the events occurring upon collision between microemulsion droplets. One expects various changes of the properties of the microemulsions, when the volume fraction of the dispersed phase ϕ is increased. The electrical conductivity is especially sensitive to the aggregation of droplets. This is indeed observed in several reported studies (Lagues, 1978, 1979; Dvolaitzky et al., 1978; Lagues & Sauterey, 1980; Lagourette et al., 1979; Moha-Ouchane et al., 1987; Antalek et al., 1997) in aqueous microemulsions. The paper of 1978 by Lagues is the first to interpret the dramatic increase of the conductivity with droplet volume fraction for a water-in-oil microemulsion in terms of a percolation model and termed this physical situation as stirred percolation, referring to the Brownian motion of the medium. This was, however, soon followed by several investigations. According to most widely used theoretical model, which is based on the dynamic nature of the microemulsions (Grest et al., 1986; Bug et al., 1985; Safran et al., 1986), there are two pseudophases: one in which the charge is transported by the diffusion of the microemulsion globules and the other phase in which the change is conducted by diffusion of the charge carrier itself inside the reversed micelle clusters. According to this theory, two approaches (static and dynamic) have been proposed for the mechanism leading to percolation (Lagourette et al., 1979). These are being governed by scaling laws as given in equations 9 and 10.

$$\sigma = A(\phi_c - \phi)^{-s} \quad \text{pre-percolation} \tag{9}$$

$$\sigma = B(\phi - \phi_c)^{t} \quad \text{post-percolation} \tag{10}$$

where σ is the electrical conductivity, ϕ is the volume fraction, and ϕ_c is the critical volume fraction of the conducting phase (percolation threshold), and A and B are free parameters. These laws are only valid near percolation threshold (ϕ_c). It is impossible to use these laws at extremely small dilutions ($\phi \to 0$) or at limit concentration ($\phi \to 1$) and in the immediate vicinity of ϕ_c. The critical exponent *t* generally ranges between 1.5 and 2, whereas the exponent *s* allows the assignment of the time dependent percolation regime. Thus, $s > 1$ (generally around 1.3) identifies a dynamic percolation (Cametti et al., 1992, Pitzalis et al., 2000; Mehta et al., 2005). The static percolation is related to the appearance of bicontinuous microemulsions, where a sharp increase in conductivity, due to both counter - ions and to lesser extent, surfactant ions, can be justified by a connected water path in the system. The dynamic percolation is related to rapid process of fusion- fission among the droplets. Transient water channels form when the surfactant interface breaks down during collisions or through the merging of droplets (Fig. 6). In this latter case, conductivity is mainly due to the motion of counter ions along the water channels. For dynamic percolation model, the overall process involves the diffusional approach of two droplets, leading to an encounter pair (Fletcher et al., 1987).

In a small fraction of the encounters, the interpenetration proceeds to a degree where the aqueous pools become directly exposed to each other through a large open channel between the two compartments, created by the rearrangement of surfactant molecules at the area of mutual impact of the droplets. The channel is probably a wide constriction of the monolayer shell between the two interconnected compartments. Due to the geometry of the constriction, the monolayer at that site has an energetically unfavorable positive curvature, which contributes to the instability of the droplet dimmer. The short lived droplet dimmer than decoalesces with a concomitant randomization of the occupancy of all the constituents and the droplets re-separate. Thus, during the transient exchange of channels, solubilizate can exchange between the two compartments. This offers an elegant approach to study the dynamic percolation phenomenon. However, another approach depicts a static percolation picture which attributes percolation to the appearance of a "bicontinuous structure" i.e formation of open water channels (Bhattacharaya et al., 1985).
The conductivity of the microemulsion system is very sensitive to their structure (Eicke et al., 1989; Kallay and Chittofrati, 1990; Giustini et al., 1996 ; D'Aprano et al., 1993, Feldman et al., 1996). The occurrence of percolation conductance reveals the increase in droplet size, attractive interactions and the exchange of materials between the droplets. The percolation threshold corresponds to the formation of first infinite cluster of droplets (Kallay and Chittofrati, 1990). Even before the occurrence of percolation transition the change in conductivity indicates the variation of reverse micellar microstructure. The conductivity is closely related to the radius of the droplet but other factors like the composition of the microemulsions system, presence of external entity, temperatures etc. Under normal conditions, water in oil microemulsions represent very low specific conductivity (ca.10^{-9} – 10^{-7} $\Omega^{-1}cm^{-1}$). This conductivity is significantly greater than it would be if we consider the alkane, which constitutes the continuous medium and is the main component of the water in oil microemulsions (~10^{-14} Ω^{-1} cm^{-1}). This increase in the electrical conductivity of the microemulsions by comparison with that of the pure continuous medium is due to the fact that microemulsions are able to transport charges.

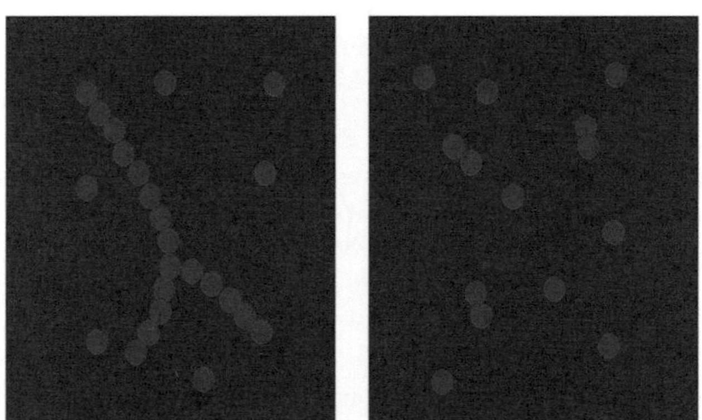

(a) Static droplet fusion (b) Dynamic droplet fusion

Fig. 6. Dynamics of droplet fusion

When we reach a certain volume of the disperse phase, the conductivity abruptly increases to give values of up to four orders of magnitude, which is greater than typical conductivity of water in oil microemulsions. This increase remains invariable after reaching the maximum value, which is much higher than that for the microemulsion present before this transition occurs. Similar behavior is observed with variation in water content, temperature or volume fraction for the fixed composition of the microemulsion. This phenomenon is known as *electric percolation*, (Hamilton et al., 1990, Garcia-Rio et al., 1997, Hait et al., 2001, Borkovec et al., 1988, Papadimitriou et al., 1993, Mehta et al., 1995, 1998, 1999, 2000, 2002) the moment at which an abrupt transition occurs from poor electric conductor, system (`$10^{-7}\ \Omega^{-1}$ cm^{-1}) to the system with fluid electric circulation ($10^{-3}\ \Omega^{-1}\ cm^{-1}$). As a consequence of ion transfer it yields a sigmoidal $\sigma-\theta$ and $\sigma-\phi$ profile. The point of maximum gradient of the $d\log\sigma/d\theta$ or $d\log\sigma/d\phi$ profile corresponds to the transition of the percolation process and is designated as the threshold volume fraction (ϕ_c) or the threshold temperature (θ_c), characteristic feature of a percolating system.

Moulik group (Hait et al., 2001) has proposed the sigmoidal Boltzmann equation (SBE) to determine the threshold characteristics of microemulsion systems. In conductance percolation, the equivalent equation can be written as

$$\log\sigma = \log\sigma_f\left[1+\left(\frac{\log\sigma_i-\log\sigma_f}{\log\sigma_f}\right)\{1+\exp(\theta-\theta_c)/\Delta\theta\}^{-1}\right] \tag{11}$$

where i, f and c are the initial, final and percolative stages. The composition of the system and other environmental conditions such as pressure and presence of additives control the threshold values.

The specific conductance of the system is calculated with the help of *droplet charge fluctuation model* (Eicke et al., 1989). In this model which is valid for the free diffusing species, it is assumed that the net charges of a droplet around an average net zero charge and its transport is associated with the free diffusive process of single droplets. Using this model the final result of the conductivity of a dilute microemulsion is

$$\sigma = \frac{\varepsilon_o\varepsilon k_B T}{2\pi\eta}\frac{\phi}{r_n^3} \tag{12}$$

However, in order to arrive at this expression, consider a nanodroplet composed of N_1 negatively charged surfactant molecules and N_2 positively charged counterions. For electroneutrality the average values are equal $\langle N_1\rangle=\langle N_2\rangle=N.$ However, due to spontaneous fluctuations in these numbers the droplet will carry can excess charge

$$z = N_2 - N_1 \tag{13}$$

in units of the elementary charge e. Even though the valency of a droplet z will fluctuate in time, the conductivity of microemulsion and a dilute electrolyte solution containing different ions can be evaluated in an entirely equivalent manner. This is because only the mean square valency of the ions (or droplets) determines the conductivity. The conductivity σ of a dilute electrolyte solution of different ions i, with a valency z_i, radius r_n (taken as independent of i for simplicity), number density ρ_i, is given (Berry et al., 1980) by

$$\sigma = \frac{e^2}{6\pi\eta r_n}\sum_i z_i^2\rho_i \tag{14}$$

where η is the viscosity of the solvent and i runs over all different ionic species in the solution. In case of microemulsion droplets, it is more convenient to write equation 14 as

$$\sigma = \frac{\rho e^2}{6\pi\eta r_n}\langle z^2\rangle \quad (15)$$

where ρ is the number of droplets per unit volume and $\langle .....\rangle$ is the canonical average over all droplets. Note that due to electroneutrality $\langle z\rangle = 0$.
The quantity of interest is the mean square charge $\langle z^2\rangle$ of a droplet. It can be expressed in the terms of the mean squared fluctuations of the number of ions residing on a droplet $\delta N_i = N_i - \langle N_i\rangle$ by

$$\langle z^2\rangle = \langle \delta N_2^2\rangle - 2\langle \delta N_2 \delta N_1\rangle + \langle \delta N_1^2\rangle \quad (16)$$

Such averages are related to derivatives with respect to the conjugated thermodynamic forces (Callen, 1960)

$$\langle \delta N_i \delta N_j\rangle = k_B T\left(\frac{\partial N_i}{\partial \mu_j}\right)_{T,\mu_{k\neq j}} \quad (17)$$

where μ_j is the chemical potential of the jth component (j=1,2), T is the absolute temperature and k_B is the Boltzman constant.
To evaluate $\langle z^2\rangle$ explicitly, one need a model for the chemical potential μ_i of the ion residing on a droplet. One may write

$$\mu_i = \mu_i^o + k_B T \log N_i + \mu_i^{(ex)} \quad (18)$$

where first two terms on the right hand side represents the chemical potential for an ideal solution while $\mu_i^{(ex)}$ is the excess chemical potential

$$\mu_i^{(ex)} = \left(\frac{\partial G^{(ex)}}{\partial N_i}\right)_{T,N_{k\neq j}} \quad (19)$$

Here a simple model is adopted to identify the electrostatic work required to charge a droplet in the solvent with the excess Gibbs free energy, i.e.

$$G^{(ex)} = \left(\frac{z^2 e^2}{8\pi\varepsilon_o \varepsilon r_n}\right) \quad (20)$$

where z is given in equation 13, ε_o is the dielectric permittivity of the vacuum, and ε is the dielectric constant of the solvent. The excess Gibbs free energy (eqn no. 20) is also the basis of Born's theory of ionic solvation (Berry et al., 1980).
Now the calculation of $\langle z^2\rangle$ explicitly, using equations 18-20, the 2X2 matrix with the elements $(\partial\mu_i / \partial N_j)_{N_{k\neq j},T}$ was calculated and it was found

$$k_B T\begin{bmatrix} 1/N_1+\alpha & -\alpha \\ -\alpha & 1/N_2+\alpha \end{bmatrix} \quad (21)$$

The abbreviation $\alpha = e^2/(4\pi k_B T \varepsilon_o \varepsilon r_n)$ has been introduced. The derivatives $(\partial N_i / \partial \mu_j)_{\mu_{k\neq j},T}$ required in equation 15 can be obtained most easily by noting the fact that the matrices with the elements $(\partial\mu_i / \partial N_j)_{N_{k\neq j}, T}$ and $(\partial N_i / \partial\mu_j)_{\mu_{k\neq j}, T}$ are related by simple matrix inversion (Callen, 1960). Inverting the matrix in equation in 21 and using equation 16 and 17, it gives

$$\langle z^2 \rangle = \frac{2N}{1+2N\alpha} \tag{22}$$

There are two limiting cases to consider.
For $\alpha << 1$ the second term in the denominator of equation 22 is negligible, and therefore $\langle z^2 \rangle = 2N$. This is the limiting case of ideal behavior where the mean-square fluctuations are essentially given by the number of ions residing on the droplet. As N>>1, the realistic case corresponds to the second limit where $\langle z^2 \rangle = 1/\alpha$. This means that it is determined by the ration of coulomb and thermal energies. Inserting $\langle z^2 \rangle = 1/\alpha$ into equation 15, we obtain the final result for the conductivity of a dilute microemulsion i.e. equation no. 12.
ρ has been replaced with the volume fraction of the droplet ϕ by using the relation $\phi = 4\pi r^3 \rho / 3$. Equation 12 predicts that the specific conductivity of microemulsion (σ/ϕ) should be constant and for a given solvent and temperature, depend on the radius of the droplet only $\propto r_n^3$. this result is independent of the charge of the ions in question.
However the phenomenon of percolation process (Peyrelasse et al., 1988) was also assessed using permittivity studies by Peyrelasse et al. From a very general point of view the complex permittivity ε* of a heterogeneous binary system may be represented by a relationship of the form $\varepsilon^* = G(\varepsilon_1^*, \varepsilon_2^*, \phi, p_k)$ in which ε_1^* and ε_2^* are the complex permittivities of the constituents 1 and 2, ϕ is the volume fraction and p_k represents the parameter, which enable the function G to contain all the information on the geometry of the dispersion and on the interactions, which takes place within the system. The models, which enable a approximation of the function G, are often based on effective and mean field theories. Satisfactory results are generally obtained when the interactions within the mixtures are weak, which is often the case when the volume fraction of one of the constituents is small, and as long as the dispersion can be considered macroscopically homogenous. But when the dispersed particles are no longer isolated from each other, in other words, when clusters of varying sizes form, the conventional models no longer apply and the concept of percolation can be successfully used. The general relationship for complex permittivity (Efros & Shklovskii, 1976; Stroud & Bergman, 1982) is

$$\frac{\varepsilon^*}{\varepsilon_1^*} = |\phi - \phi_c|^t f\left[\frac{\varepsilon_2^* / \varepsilon_1^*}{|\phi - \phi_c|^{t+s}}\right] \tag{23}$$

where ε^*, ε^*_1, ε^*_2 are the complex permittivities of microemulsion and the components, ϕ is the volume fraction of dispersed phase ϕ_c is the percolation threshold. Similar to the conductivity, permittivity also obeys the characteristic scaling laws. The function f(z) in which z is a complex variable satisfies the following asymptotic forms

$$\phi > \phi_c, |z| << 1, f(z) = C_1 + C_1' z \tag{24}$$

$$\phi < \phi_c, |z| << 1, f(z) = C_2 z \tag{25}$$

$$z >> 1 \forall \phi, f(z) = Cz^{t(t+s)} \tag{26}$$

If it is assumed that the two components are dielectric conductors of static permittivities ε_{1s} and ε_{2s} and the conductivities σ_1 and σ_2, then one obtains

$$\varepsilon_1^* = \varepsilon_{1s} - j\frac{\sigma_1}{\varepsilon_o \omega},\ \varepsilon_2^* = \varepsilon_{2s} - j\frac{\sigma_2}{\varepsilon_o \omega},\ \omega = 2\pi\upsilon \tag{27}$$

in which υ is the frequency of the electric field applied, ε_o i.e. the dielectric permittivity of a vacuum and $j^2 = -1$. The heterogeneous system built from components 1 and 2 presents static permittivity ε_s and conductivity σ. The following relations are obtained when $\phi < \phi_c, |z| << 1$,

$$\varepsilon_s = C_2 \varepsilon_{2s} (\phi_c - \phi)^{-s} \tag{28}$$

$$\sigma = C_2 \sigma_2 (\phi_c - \phi)^{-s} \tag{29}$$

But when $\phi > \phi_c, |z| << 1$

$$\varepsilon_s = C_1' \varepsilon_{2s} (\phi - \phi_c)^{-s} [1 + \frac{C_1 \varepsilon_{1s}}{C_1' \varepsilon_{2s}} (\phi - \phi_c)^{(t+s)}] \tag{30}$$

$$\sigma = C_1 \sigma_1 (\phi - \phi_c)^{t} [1 + \frac{C_1' \sigma_2}{C_1 \sigma_1} (\phi - \phi_c)^{-(t+s)}] \tag{31}$$

As (*t*+*s*) are positive $\varepsilon_{1s} / \varepsilon_{2s}$ does not tends towards infinity, while ϕ is close to ϕ_c we have

$$\varepsilon_s = C_1' \varepsilon_{2s} (\phi - \phi_c)^{-s} \tag{32}$$

So when ϕ is close to ϕ_c, $\varepsilon \alpha |\phi - \phi_c|^{-s}$ on either side of ϕ_c. Moreover, of $\sigma_2 / \sigma_1 << 1$ (for example, with a perfect insulator $\sigma_2 = 0$), close to the percolation threshold one can still have

$$\frac{\sigma_2}{\sigma_1} << |\phi - \phi_c|^{(t+s)} \tag{33}$$

One consequently obtains

$$\sigma = C_1 \sigma_1 (\phi - \phi_c)^{t} \tag{34}$$

It will be noted that equations 29 and 34 indicate that the derivative $(1/\sigma)(d\sigma / d\phi)$ tends towards infinity at percolation threshold. It should also be observed that equations 28-34 are only valid if z <<1. For conductivity this implies that $\frac{\sigma_2}{\sigma_1} << (\phi - \phi_c)^{(t+s)}$. These are not valid in the neighbourhood of ϕ_c in which $z \to \infty$. In fact, experimentally, there is a continuous transition with in an interval of width (Δ) in the immediate neighbourhood of ϕ_c. The width of this transition interval (the cross over regime) is of the order of

$$\Delta = s_1 + s_2 = (\frac{\sigma_2}{\sigma_1})^{1/(t+s)} \quad (35)$$

The above discussion leads to asymptotic relationships for conductance as

$$\sigma \alpha (\phi - \phi_c)^t \text{if} \phi > \phi_c + s_1 \quad (36)$$

$$\sigma \alpha (\phi_c - \phi)^{-s} \text{if} \phi < \phi_c - s_2 \quad (37)$$

Finally one obtain the equations in the form of equation 9 and 10.

In terms of permittivity, when in equation (23) $\left[\frac{\varepsilon_2^* / \varepsilon_1^*}{|\phi - \phi_c|^{t+s}} \right]$ is much smaller than 1,

$$\varepsilon_s = A'(\phi_c - \phi)^{-s} \qquad \text{pre-percolation} \quad (38)$$

$$\varepsilon_s = B'(\phi - \phi_c)^t \qquad \text{post-percolation} \quad (39)$$

where ϕ is the volume fraction, and ϕ_c is the critical volume fraction of the conducting phase (percolation threshold), and A and B are free parameters. These laws are only valid near percolation threshold (ϕ_c).

The study of microemulsions provides a characteristic insight in to its structural features. However, in the dilute limit and for Newtonian behavior, the microemulsions are said to obey well-known Einstein relation

$$\eta_r = \frac{\eta}{\eta_o} = 1 + 2.5\phi \quad (40)$$

According to the relation, the dispersed particles in the liquid are in the form of rigid spheres, which are larger than the solvent molecules. However, on account of complex interaction between the particles and solvent, the relation no longer remains linear when the concentration is increased (ϕ volume fraction >0.05). Ward and Whitemore have found that η_r is a function of size distribution and is independent of the viscosity of the suspending liquid and the absolute size of the spheres at a given concentration (Ward & Whitemore, 1950). According to Roscoe and Brinkman the viscosity of solutions and suspensions of finite concentration with spherical particles of equal size (Roscoe, 1952; Brinkman, 1952) is given by

$$\eta_r = (1 - 1.35\phi)^{-2.5} \quad (41)$$

For large volume fractions one must, on the one hand, account for hydrodynamic interactions between the spheres and on the other hand, for direct interactions between the particles that are, e.g., of a thermodynamic origin. However, there is no accurate theory to explain the viscosity of microemulsions but several semiempirical relations are used for the purpose (Saidi et al., 1990). When the composition of microemulsion is subjected to change it subsequently changes its viscosity profile yielding different structural organisations. Further, in order to understand the percolation phenomena viscosity studies are successfully employed. With time the viscosity increases and at time t_c a molecule of infinite length is formed. t_c corresponds to the gelation point and scaling laws can be written as follows

$$\eta \alpha (t_c - t)^{-s} \text{ pre percolation} \tag{42}$$

$$E \alpha (t - t_c)^{\mu} \text{ post percolation} \tag{43}$$

where E is the elastic modulus. However, the critical exponent obtained for η can have various values. It depends on the experimental conditions which suggest the existence of different growth mechanism.

4. Energetics of droplet clustering

Various studies (Peyrelasse et al., 1989; Peyrelasse & Boned, 1990; Mathew et al., 1991) have shown that conductivity percolation in microemulsions is the result of droplet clustering. This is being depicted in Fig. 7. At low ϕ, individual droplets maintain a low conductivity. As ϕ increases, the conductivity also increases as droplets assemble in the clusters and ion diffusion or exchange of droplet content is facilitated and an infinite percolating droplet network is eventually formed at ϕ_c. The droplets clustering process has been estimated utilizing the values of dispersed phase volume fraction (ϕ_c) at the percolation threshold (Ray et al., 1993; Alexandridis et al., 1995). It has been postulated that the threshold of electrical percolation corresponds to the formation of first open structure of an infinite cluster (Eicke et al., 1989; Riddick & Bunger, 1970).

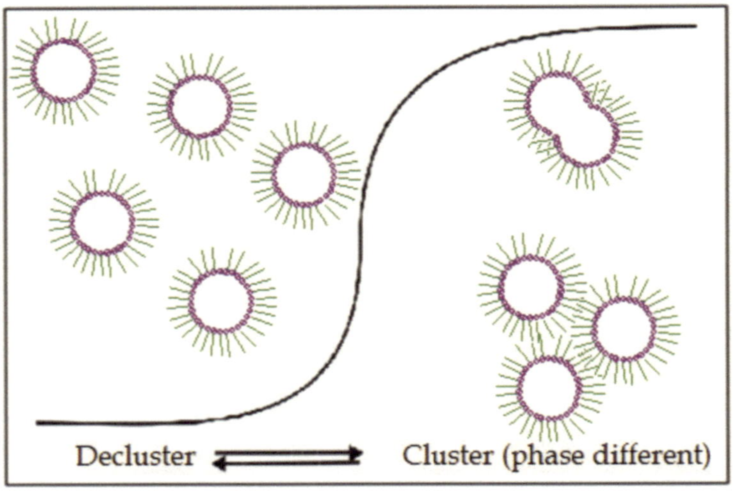

Fig. 7. Schematic representation of droplet cluster formation and the course of electrical conductivity with varying temperature and composition

Organodiselenide	S=[Oil]/[AOT]	$-\Delta G^o_{cl}$ (kJ mol^{-1})	ΔH^o_{cl} (kJ mol^{-1})	ΔS^o_{cl} (JK^{-1}mol^{-1})
Without	5	21.67,24.36	64.34, 67.49	130.80, 131.24
	7	22.30,25.41		
	9	23.14,26.35		
	15	24.93,28.33		
$C_{12}H_{16}N_6Se_2$	5	21.25, 24.48	74.59, 82.36	162.63,175.34
	7	22.39, 25.66		
	9	23.23, 26.60		
	15	25.04, 28.40		
$C_8H_4N_4Se_2Cl_2$	5	25.84,24.21	61.33,68.82	133.95,134.24
	7	26.84, 25.26		
	9	27.77, 27.84		
	15	29.74, 28.84		
Nap_2Se_2	5	25.36,24.57	68.62,69.22	131.50,132.24
	7	26.36,25.87		
	9	27.25,27.62		
	15	29.23,30.59		
$(Ph_2CHSe)_2$	5	24.36,24.30	67.29, 69.68	133.36,137.48
	7	25.41,25.47		
	9	23.26,26.34		
	15	24.95,28.95		
$C_{10}H_6N_2Se_2Br_2$	5	28.70, -----		
	15	32.70, 33.02		
$C_{10}H_4N_2Se_2Br_4$	5	28.90, 35.40		
	15	-------, 42.50		

Table 1. Energetics of droplet clustering, ΔG^o_{cl} for water-induced percolation for AOT/isooctane/water and Lecithin+AOT/isooctane/water for different organodiselenide at various concentrations

The microemulsion droplet above the percolation threshold, aggregated in clusters, are considered to be phase different than that of nonpercolating droplets, with distinct physical properties such as conductivity. This is comparable to psuedophase concept used for modeling the formation of micelles in solutions of amphiphillic molecules (association model). Dilution of clustered (percolating) microemulsion system upon addition of an apolar solvent at constant droplet size lowers the conductivity rapidly until the clusters dissociate into individual droplets below percolation threshold ϕ_c. This phenomenon is comparable to the process of demicellization occurring when the surfactant concentration is lowered below the critical micellar concentration. In the light of the concept of droplet association, the Gibb's free energy of droplet clustering (standard free energy change for the transfer of 1 mole of droplets from an infinitely diluted solution to the percolating cluster), ΔG^o_{cl} is calculated from the relationship

$$\Delta G^o_{cl} = RT\ln X_p \tag{44}$$

where R is the gas law constant, T is the absolute temperature, X_p is the mole fraction of the microemulsion droplets corresponding to percolation threshold (ϕ_c) at constant temperature T. The Xp value is obtained from the relation

$$X_p = \frac{n_d}{(n_d + n_o)} = A_t R_w M_o / [A_t R_w M_o + 3V_o \rho_o N] \tag{45}$$

where, n_d and n_o are the number of moles of droplet and oil, respectively. A_t is the total cross-sectional area of the surfactant, V_o, ρ_o, M_o are the volume, density and molar mass of the oil, respectively. N is the Avogadro number, R_w is the water pool radius is obtained from

$$R_w = 3[M_s + rM_{cs} + \{(W_w + W_{cs}) / W_s\}M_s / \{N\rho_d(A_s + rA_{cs}\}] \tag{46}$$

where M stands for molar mass, W is the weight fraction and subscript c and cs represents for surfactant and cosurfactant respectively. ρ_d, A and N are the density of the dispersed phase, cross-sectional area and Avogadro number, respectively. $r = (N_{cs} / N_s)$ is called structural ratio (ratio between the number of molecules of s and cs at the droplet interface) However, in the absence of cosurfactant equation 46 becomes

$$R_w = 3[M_s + (W_w + W_s)] / N\rho_d a_s \tag{47}$$

Further, the enthalpy of clustering ΔH^o_{cl} can also be obtained from Gibbs Helmoltz equation

$$d(\Delta G^o_{cl} / \theta_{cl}) / d(1 / \theta_{cl}) = \Delta H^o_{cl} \tag{48}$$

The entropy change of the process is obtained from

$$\Delta S^o_{cl} = (\Delta H^o_{cl} - \Delta G^o_{cl}) / \theta_t \tag{49}$$

The energetics of the clustering have been well utilized during all these years for different microemulsions with and without additives. The dynamics of percolation and thereafter energetics of the clustering have been extensively investigated in our group for various cationic, anionic and non-ionic microemulsions (Mehta et al., 2006, 2007, 2009). The data for energetics of the clustering of AOT/isooctane/water and Lecithin+AOT/isooctane/water microemulsion in presence of different additive i.e., glycols, aminopyridines, modified aminoacids and organodiselenide have been tabulated in Tables (1-3) (Mehta et al., 2007, 2009).

Anti-tuberculosis drug	$-\Delta G^{o}_{cl}$ (kJ mol^{-1})
Tween 80/ oleic acid/ ethanol/phosphate buffer	
without	4.35
rifampicin	5.17
isoniazid	5.17
pyrazinamide	5.17
Brij 96/Ethyl Oleate/butanol/water	
without	48.8
rifampicin	46.04
isoniazid	48.80
pyrazinamide	42.97

Table 2. Energetics of droplet clustering, ΔG^{o}_{cl} for water-induced percolation for different non-ionic microemulsion with and without the presence of anti-tuberculosis drugs (rifampicin, isoniazid and pyrazinamide)

	S = [Oil]/[AOT]	Conc. (mM)	$-\Delta G^{o}_{cl}$ (kJ mol^{-1})	ΔH^{o}_{cl} (kJ mol^{-1})	ΔS^{o}_{cl} (JK^{-1}mol^{-1})
Glycols					
Without	5	15	20.0	61.2±7.5	264.6
	7	15	21.0		
	9	15	22.0		
PG	5	15	20.0	69.4±5.9	291.4
	7	15	21.2		
	9	15	22.0		
BG	5	15	19.8	55.1±5.8	226.4
	7	15	21.1		
	9	15	21.9		

PMG	5	15	19.8	49.0±5.8	226.4
	7	15	20.9		
	9	15	21.8		
			Aminopyridines		
Without		-	21.68	55.40±5.2	243.00
2-Ampy	9	50 30 15 5	20.86 21.13 21.55 21.74	48.45±4.8	220.00
3-Ampy	9	50 30 15 5	20.65 21.00 21.47 21.55	48.31±3.6	219.70
2-Am-4-mpy	9	50 30 15 5	21.20 21.40 21.54 21.61	54.53±4.2	239.69
2-Am-6-mpy	9	50 30 15 5	21.13 21.48 21.55 21.61	53.74±1.5	237.33
			Modified Aminoacids		
Without	9	5 7 9 15	20.08 21.54 22.35 24.37	47.99	220.4
NAG	9	5 7 9 15	20.33 21.68 22.41 24.81	45.20	209.27
NAA	9	5 7 9 15	20.27 21.61 22.49 24.66	45.64	211.75
NAC	9	5 7 9 15	20.43 21.82 22.62 22.44	43.16	203.6

Table 3. Standard free energy values (kJ mol^{-1}) for volume induced percolation for Glycol, Aminopyridines and Modified Aminoacids (MAA) in AOT/isooctane/water microemulsion

5. Conclusions

Microemulsions are thermodynamically stable, transparent, low viscosity mixtures of oil and water stabilized by a monolayer of surfactant. The chemistry of microemulsions is at an incredibly exciting stage of development. The advent of systems that are easy to handle allows those without specialist knowledge of the field to use them for the first time. Because of its versatility and thermodynamic stability, the microemulsion systems find potential applications in pharmaceutical, oil recovery, as food additives and as reaction media, etc. There are different theories which have been developed over the course of time relating to the formation of microemulsions. Most acceptable among them are interfacial, solubilization and thermodynamic theories, etc. The spontaneity of formation of microemulsion involves change in the free energy of formation of microemulsion, ΔG_m, which mainly consists of interfacial energy and energy of clustering droplets. Irrespective of the mechanism, the reduction of the interfacial free energy is critical in facilitating microemulsion formulation. Alongwith the stability, the size of droplets in microemulsions can also be adjudged using the thermodynamic approach.

The unique solvent properties of microemulsion compared to conventional organic solvents, indicate that the novel micro-heterogeneous media can be efficiently used as reaction media for various biocatalytic reactions. Presence of different domains of variable polarity due to compartmentalization in microemulsions makes them particularly useful in the area of drug delivery. Microemulsion media finds several applications ranging from drug delivery to drug nanoparticles templating due to its ability to enhance solubility, stability and bioavailability. To thoroughly understand the drug delivery potential of microemulsion, it is necessary to know the possible phase transitions occurring in the system and the influence of drug on its microstructure. Apart from being used as a carrier, microemulsions also act as template for synthesis of nanoparticles. In its pharmaceutical applications, attempts are being made to synthesize drug nanoparticles for controlled release.

This chapter deals with different theories and models developed during the course of time using various approaches to interpret the formation, stability and structure of microemulsion systems which offers application and adaptability in multiple fields.

6. References

Adamson, A.W. (1969). A model for micellar emulsion. *J. Colloid Interface Sci.*, 29, 261-267. ISSN 0021-9797

Alexandridis, P.; Holzwarth, J.F. & Hatton T.A. (1995).Thermodynamics of Droplet Clustering in Percolating AOT Water-in-Oil Microemulsions *J. Phys. Chem., 99*, 8222–8232. ISSN 0022-3654

Antalek, B.; Williams, J.; Texter, J., Feldman, Y.; & Garti, N. (1997). Microstructure analysis at the percolation threshold in reverse microemulsions. *Colloids surf. A*, 128, 1-11. ISSN 0927-7757

Bancroft, W. D. (1913). The Theory of Emulsification, V, *J. Phys. Chem. 17*, 501-519. ISSN 0022-3654

Bansal, V.K.; Chinnaswamy, K.; Ramachandran, C. & Shah, D.O. (1979). Structural aspects of microemulsions using dielectric relaxation and spin label techniques. *J. Colloid Interface Sci.*, 72, 524-537. ISSN 0021-9797

Bayrak, Y. (2004). Interfacial composition and formation of w/o microemulsion with different amphiphiles and oils. *Colloids Surf. A*, 247, 99-103. ISSN 0927-7757

Becher, P. (1977). *Emulsions: theory and Practice*, Kriegar Publishers, New York. ISBN-10 0882755897

Berry, R.S.; Rice, S.A. & Ross, J. (1980). *Physical Chemistry*, Wiley, New York.

Bhattacharya, S.; Stokes, J. P.; Kim, M. W. & Huang, J. S. (1985). Percolation in an Oil-Continuous Microemulsion. *Phys. Rev. Lett.*, 55, 1884 - 1857. ISSN 0031-9007 (print)

Birdi, K.S. (1982). Microemulsions: effect of alkyl chain length of alcohol and alkane. *Colloid Polym. Sci.*, 26, 628-631. ISSN *0303-402X*

Borkovec, M.; Eicke, H. F.; Hammerich, H. & Gupta, B. D. (1988). Two Percolation processes in Microemulsions. *J. Phys. Chem.*, 92, 206–211. ISSN 0022-3654

Bowcott, J.E. & Schulman J.H. (1955). Emulsions. *Z Elektrochem.*, 59, 283-290.

Brinkman, R. (1952). The Viscosity of Concentrated Suspensions and Solutions. *J. Chem. Phys.*, 20, 571. ISSN *0021-9606*

Bug, A. L. R.; Safran, S.A.; Grest, G. S. & Webman, I. (1985). Do Interactions Raise or Lower a Percolation Threshold. *Phys. Rev. Lett.*, 55, 1896-1899. ISSN 0031-9007 (print)

Callen, H.B. (1960). *Thermodynamics*, Wiley, New York.

Cametti, C.; Codastefno, P.; Tartaglia, P.; Chen, S. & Rouch, J. (1992). Electrical conductivity and percolation phenomena in water-in-oil microemulsions. *Phys. Rev. A.* 45, R5358–R5361. ISSN 1050-2947 (print)

Constantinides, P. (1995). Lipid Microemulsions for Improving Drug Dissolution and Oral Absorption: Physical and Biopharmaceutical Aspects. *Pharm. Res.*, 12, 1561-1572. ISSN *0724-8741*

D'Aprano, A.; D'Arrigo, G.; Paparelli, A.; Goffredi, M. & Liveri, V. T. (1993). Volumetric and transport properties of water/AOT/n-heptane microemulsions. J. Phys. Chem., 97, 3614-3618. ISSN 0022-3654

Davis, J.T. (1959). A quantitative kinetic theory of emulsion type, I. Physical chemistry of the emulsifying agent, Gas/Liquid and Liquid/Liquid Interface. *Proceedings 2. Intern. Cong. Surface Activity*, London, 1, 426-438.

Efros A.L. & Shklovskii B. L. (1976). Structure and Conductivity of Bilinear Organic Compounds. *Phys. Stat. Sol.* 76B, 475-485. ISSN 1521-3951.

Eicke, H. F.; Borkovec, M. & Gupta, B. D. (1989). Conductivity of water-in-oil microemulsions: a quantitative charge fluctuation model. J. Phys. Chem., 93, 314-317. ISSN 0022-3654

Feldman, Y.; Kozlovich, N.; Nir, I.; Garti, N.; Archipov, V.; Idiyatullin, Z.; Zuev, Y. & Fedotov, V. (1996). Mechanism of Transport of Charge Carriers in the Sodium Bis(2-ethylhexyl) Sulfosuccinate-Water-Decane Microemulsion near the Percolation Temperature Threshold. J. Phys. Chem., 100, 3745-3748. ISSN 0022-3654

Fletcher, P. D. I.; Howe, A. M. & Robinson, B. R. (1987). The Kinetics of solubilizate exchange between water droplets in water in oil microemulsions. *Faraday Trans. I*, 83, 985-1006. ISSN 0300-9599

Garbacia, W. & Rosano, H.L. (1973). Microemulsion: Formation and stabilization. *J. Colloid Interface Sci.*, 44, 242-248. ISSN 0021-9797

Garcia- Rio, L.; Herves, P.; Leis, J. R. & Mujeto, J. C. (1997). Influence of Crow ethers and Macrocyclic Kryptands upon the Percolation phenomenon in AOT/ Isooctane/ H2O Microemulsions. Langmuir, 13, 6083–6088. ISSN 0743-7463

Giustini, M. ; Palazzo, G. ; Colafemmina, G.; Monica ; Marcello Giomini, M. D. & Ceglie, C. (1996). Microstructure and Dynamics of the Water-in-Oil CTAB/n-Pentanol/n-Hexane/Water Microemulsion: A Spectroscopic and Conductivity Study. J. Phys. Chem., 100, 3190-3198. ISSN 0022-3654

Grest, G. S.; Webman, I.; Safran, S. A. & Bug, A. L. R. (1986). Dynamic percolation in microemulsions. *Phys. Rev. A*, 33, 2842-2845. ISSN 1050-2947 (print)

Griffin, W. C., (1949). Classification of Surface-Active Agents by 'HLB'. *J. Soc. Cosmetics Chemists*, 1, 311-326. ISSN 1525-7886

Griffin, W. C., (1954). Calculation of HLB Values of Non- Ionic Surfactants. *J. Soc. Cosmetics Chemists* 5, 249-256. ISSN 1525-7886

Hait, S. K.; Moulik, S. P.; Rodgers, M. P.; Burke, S. E. & Palepu, R. (2001). Physiochemical Studies on Microemulsions of dynamics of percolation and Energetics of Clustering in Water/AOT/Isooctane and Water/AOT/ Decane w/o Microemulsions in the presence of Hydrotopes (Sodium Salicylate, α-Napthol, β-Napthol, Resorcinol, Catechol, Hydroquinone, Pyrogallol and Urea) and Bile Salt (Sodium Cholate). *J. Phys. Chem.*, 105, 7145–7154. ISSN 0022-3654

Hamilton, R. T.; Billman, J. F. & Kaler, E. W. (1990). Measurements of Interdroplet Attractions and the onset of Percolation in Water- in –Oil Microemulsions. Langmuir, 6, 1696–1700. ISSN 0743-7463

Holmberg, K. (1998). *Quater century progress and new horizons in microemulsions in Micelles, microemulsions and monolayer science and technology*, Marcel Dekker, Inc. New York, 161-192.

Israelachvili, J.N.; Mitchell, D.J. and Ninham, B.W. (1976). Theory of self-assembly of hydrocarbon amphiphiles into micelles and bilayers. *J. Chem. Soc. Faraday Trans II*, 72, 1525-1568. ISSN 0300-9238

Kallay, N. & Chittofrati, A. (1990). Conductivity of microemulsions: refinement of charge fluctuation model. J. Phys. Chem., 94, 4755-4756. ISSN 0022-3654

Kahlweit, M. & Reissi, H. (1991). On the Stability of Microemulsions. *Langmuir* 7, 2928-2933. ISSN 0743-7463

Lagourette, B; Peyrelasse, J. & Boned, C. (1979). Bicontinuous structure zones in microemulsions, *Nature, 5726, 60-62.* ISSN 0028-0836

Lagues, M. (1978). Study of structure and electrical conductivity in microemulsions: Evidence for percolation mechanism and phase inversion. *J. Phys. (France) Lett.*, 39, L487-L491. ISSN 0003-6951 (print)

Lagues, M. (1979). Electrical conductivity of microemulsions: a case of stirred percolation. *J. Phys. (France) Lett*, 40, L331- L333. ISSN 0003-6951 (print)

Lagues, M. & Sauterey, C. (1980). Percolation transition in water in oil microemulsions. Electrical conductivity measurements. *J. Phys. Chem.*, 84, 3503-3508. ISSN 0022-3654

Lissant, K.L. (1976, 1984). *Emulsions and emulsion technology*, Part II and III, Dekker, New York. ISBN 0824760980

Malcolmson, C.; Satra, C.; Kantaria , S.; Sidhu, A. & Lawrence, M.J. (1998). Effect of oil on the level of solubilization of testosterone propionate into non-ionic oil-in-water microemulsions. *J. Pharm. Sci.*, 87, 109-116. ISSN 1520-6017

Mathew, C.; Saidi, Z.; Peyrelasse, J. & Boned C. (1991). Viscosity, conductivity, and dielectric relaxation of waterless glycerol–sodium bis(2-ethylhexyl) sulfosuccinate–isooctane microemulsions: The percolation effect. *Phys. Rev. A*, 43, 873- 882. ISSN 1050-2947 (print)

Mehta, S.K. & Bala K. (1995)Volumetric and transport properties in microemulsions and the point of view of percolation theory, *Physical Review E*, 51, 5732-5737 ISSN 1539-3755 (print)

Mehta, S.K. & Kawaljit. (1998). Isentropic compressibility and transport properties of CTAB-Alkanol-Hydrocarbon-Water microemulsion systems, *Colloids and Surfaces A*, 136, 35-41. ISSN 0927-7757

Mehta, S.K. & Kawaljit & Bala K. (1999). Phase Behaviour, Structural Effects, Volumetric and Transport Properties in Non-aqueous Microemulsion, *Physical Review E.*, 59, 4317-4325 ISSN 1539-3755 (print)

Mehta, S.K. & Bala K. (2000).Tween-based microemulsions: a percolation view, *Fluid Phase Equilibria*, 172, 197-209. ISSN 0378-3812

Mehta, S.K. & Kawaljit. (2002). Phase diagram and physical properties of a waterless sodium bis.2-ethylhexylsulfosuccinate.- ethylbenzene-ethyleneglycol microemulsion: An insight into percolation. *Phys. Rev. E*, 65, 021502. ISSN 1539-3755 (print)

Mehta, S. K.; Sharma, S. & Bhasin, K. K. (2005). On the Temperature Percolation in w/o Microemulsions in the Presence of Organic Derivatives of Chalcogens. *J. Phys. Chem. B*, 109, 9751-9759. ISSN 1520-6106

Mehta, S.K. & Sharma, S. (2006). Temperature-induced percolation behavior of AOT reverse micelles affected by poly(ethylene glycol)s. *J. Colloid Interface Sci.*, 296, 690-699 ISSN 0021-9797

Mehta, S. K.; Kaur, K. K.; Sharma, S. & Bhasin, K. K. (2007). Incorporation of aromatic heterocyclic compounds in reverse micelles: A physicochemical and spectroscopic approach. *Colloids Surf.* A, 298, 252-261. ISSN 0927-7757

Mehta, S. K.; Kaur, K. K.; Sharma, S. & Bhasin, K. K. (2007). Behavior of acetyl modified amino acids in reverse micelles: A non-invasive and physicochemical approach. *J. Colloid Interface Sci.*, 314, 689-698. ISSN 0021-9797

Mehta, S. K.; Kaur, K. K.; Arora, E. & Bhasin, K. K. (2009) Mixed Surfactant Based Microemulsions as Vehicles for Enhanced Solubilization and Synthesis of Organoselenium Compounds. *J. Phys. Chem. B*, 113, 10686–10692. ISSN 1520-6106

Mehta, S.K.; Kaur, G. & Bhasin K.K. (2010). Tween embedded microemulsions-Physicochemical and spectroscopic analysis for antitubercular Drugs. *AAPS Pharm. Sci. Tech.*, 11,143-153. ISSN 1530-9932

Miller, C.A. & Scriven, L.E. (1970a). Interfacial instability due to electrical forces double layer I: General consideration. *J. Colloid Interface Sci.*, 33, 360-370. ISSN 0021-9797

Moha-Ouchane, M.; Peyrelasse, J. & Boned C. (1987). Percolation transition in microemulsions: Effect of water-surfactant ratio, temperature, and salinity. *Phys. Rev. A*, 35, 3027-3032. ISSN 1050-2947 (print)

Murphy, C.L. (1966).*PhD. Thesis*, University of Minnesota, Minnea- polis, Minnesota.

Papadimitriou, V.; Xenakis, A. & Lianos, P. (1993). Electric Percolation in Enzyme-Containing Microemulsions. *Langmuir*, 9, 912-915. ISSN 0743-7463

Peyrelasse, J.; Moha-Ouchane, M. & Boned, C. (1988). Viscosity and the phenomenon of percolation in microemulsions *Phys. Rev. A*, 38, 4155-4161. ISSN 1050-2947 (print)

Peyrelasse, J. & Boned, C. (1990). Conductivity, dielectric relaxation, and viscosity of ternary microemulsions: The role of the experimental path and the point of view of percolation theory. *Phys. Rev. A*, 41, 938–953. ISSN 1050-2947

Peyrelasse, J.; Boned, C. & Saidi, Z. (1993). Quantitative determination of the percolation threshold in waterless microemulsions. *Phys. Rev. E*, 47, 3412-3417. ISSN 1539-3755 (print)

Pitzalis, P.; Angelico, A.; Soderman, O. & Monduzzi, M. (2000). A Structural Investigation of Ca AOT/ water/Oil Microemulsions. *Langmuir*, 16, 442–450. ISSN 0743-7463

Prince, L.M. (1977). *Microemulsions*, Academic Press, New York, ISBN0-12-565750-1.

Ray, S.; Bisal, S. R. & Moulik, S.P. (1993). Structure and dynamics of microemulsions. Part 1—Effect of additives on percolation of conductance and energetics of clustering in water–AOT–heptane microemulsions. *J. Chem. Soc. Faraday Trans.*, 89, 3277-3282. ISSN 0956-5000

Rehbinder, P. & Shchukin, E.D. (1972). Surface phenomena in solids during deformation and fracture processes. *Prog. Surface Sci.*, 3, 97-188. *IS*SN 0079-6816

Riddick, J.A. & Bunger, W.B. (1970). *Organic solvents: Physical properties and methods of purification.* Wiley Interscience, New York. ISBN 0-7506-7571-3

Robbins, M.L. (1976). *Theory of the phase behaviour of microemulsions*, paper no. 5839, presented at the improved oil recovery symposium of the society of petroleum engineers of AIME, Oklahoma, March 22-24.

Roscoe, R. (1952). The viscosity of suspensions of rigid spheres. *Br. J. Appl. Phys.*, 3, 267-269. ISSN 0508-3443(print)

Ruckenstein, E., & Chi, J.C. (1975). Stability of microemulsion. *J. Chem. Soc. Faraday Trans.*, 71, 1690-1707. ISSN 0956-5000

Safran, S.A.; Webman, I. & Grest, G.S. (1986). Percolation in interacting colloids. *Phys. Rev. A*, 32, 506-511. ISSN 1050-2947 (print)

Saidi, Z.; Mathew, C.; Peyrelasse, J. & Boned C. (1990). Percolation and critical exponents for the viscosity of microemulsions. *Phys. Rev. A*, 42, 872- 876. ISSN 1050-2947 (print)

Singh, H.N.; Prasad, C.D. & Kumar, S. (1993). Water solubilization in microemulsions containing amines as cosurfactant. *J. Am. Oil Chem. Soc.*, 70, 69-73. ISSN 0003-021X

Schulman, J.H.; Staeckenius, W. & Prince, L.M. (1959). Mechanism of Formation and Structure of Micro Emulsions by Electron Microscopy. *J. Phys. Chem.*, 63, 1677- 1680. ISSN 0022-3654

Scriven, L.E. (1977). *Micellization, solubilization and microemulsion*, Plenum Press, vol 2, 877-893, ISBN 0-306-31023-6.

Stroud, D. & Bergman, J. (1982). Frequency Dependence of the Polarization Catastrophe at a Metal-Insulator Transition and Related Problems. *Phys. Rev. B*, 25, 2061-2064. ISSN 1098-0121 (print)

Ward S.G. & Whitemore, R.L. (1950). Studies of the viscosity and sedimentation of suspensions Part 1. - The viscosity of suspension of spherical particles. *Br. J. Appl. Phys.* 1, 286-290. ISSN 0508-3443(print)

Winsor, P.A. (1948). Hydrotropy, solubilization, and related emulsification processes I. *J. Chem. Soc. Faraday Trans.*, 44, 376-382. ISSN 0956-5000

Zheng, O.; Zhao, J.-X. & Fu, X.-M. (2006). Interfacial Composition and Structural Parameters of Water/C12-*s*-C12·2Br/*n*-Hexanol/*n*-Heptane Microemulsions Studied by the Dilution Method. *Langmuir*, 22, 3528-3532. ISSN 0743-7463

Zheng, O.; Zhao, J.-X.; Yan, H. & Gao, S.-K. (2007). Dilution method study on the interfacial composition and structural parameters of water/C_{12}–EO_x–C_{12}.2Br/*n*-hexanol/*n*-heptane microemulsions: Effect of the oxyethylene groups in the spacer. *J. Colloid Interface Sci.*, 310, 331-336. ISSN 0021-9797

19

The Atmosphere and Internal Structure of Saturn's Moon Titan, a Thermodynamic Study

Andreas Heintz and Eckard Bich
University of Rostock
Germany

1. Introduction

The atmosphere and internal structure of the icy satellites of Jupiter and Saturn offer a most interesting subject of studying their thermodynamic properties. The treatment provides a motivating example of applying thermodynamic and physical principles beyond the well known classical areas of chemical engineering and chemical separation or reaction processes. Particularly Saturn's largest moon Titan is most suitable for that purpose for several reasons. There exist numerous data of Titan obtained by observations from Earth, from the Voyager 1 and 2 missions in 1977 and more recently from the Cassini-Huygens mission which in particular provided most spectacular results in the years 2005 to 2008:

- Titan has a dense atmosphere consisting mainly of N_2 (93 – 98 %) and methane (7 – 2 %) with a pressure of 1.5 bar and a temperature of $-180\,^{\circ}\mathrm{C} = 93$ K at the bottom.
- Methane plays a similar role on Titan as water on Earth causing similar meteorological processes.
- Liquid lakes consisting of the atmospheric components cover at least 4 % of Titan's surface, maybe even more, because only parts of Titan's surface have been investigated so far by the Cassini probe.
- Solid water ice plays a similar role as rocky material and sediments on Earth's surface.
- Due to its low density ($1.88\ \mathrm{g \cdot cm^{-3}}$) Titan must contain remarkable amounts of water in its mantle while its inner core consists of rocky material.

Important questions arise which can be answered by thermodynamic methods as shown in the following sections:

- What is the composition of the condensed liquid phase on Titan's surface?
- What does the temperature profile of the atmosphere look like and what role does methane play in the formation of clouds and rainfall?
- How have the atmosphere and surface developed in the past and what will be their future fate?
- What can be said about Titan's internal structure?

Table 1 gives some information on physical properties and interplanetary data of Titan.
Titan always turns the same side to Saturn, i. e., its rotational period is identical with its revolution time around Saturn. Much more detailed information than presented in this

article can be found in the literature (Atreya et al., 2006; Barth & Toon, 2003; Flasar et al., 2005; Fulchignoni et al., 2005; Griffith et al., 2006; Lewis, 1997; Lorenz et al., 1997; 2008; Mitri et al., 2007; Mousis & Schmitt, 2008; Niemann et al., 2005; Rannou et al., 2006; Stofan et al., 2007; Yung & DeMore, 1999; Yung et al., 1984).

2. Photochemistry in Titan's atmosphere

The most important gaseous components detected and determined quantitatively in Titan's atmosphere are presented in Table 2.
The atmosphere is characterized by clouds in the troposphere and a hazy upper atmosphere containing aerosol particles of $\sim 1\mu$ m size which conceals a free view to Titan's surface in the visible region of light. However, UV, IR and mainly radar imaging technique is able to reveal details of Titan's atmosphere and surface. This situation indicates an intensive photochemical activity in the atmosphere. A simplified version of the photochemical destruction processes caused by solar UV-irradiation reads as follows (Atreya et al., 2006; Lorenz et al., 1997; Yung et al., 1984; Yung & DeMore, 1999)

$$\begin{array}{lcll}
2CH_4 + h\nu & \rightarrow & C_2H_6 + H_2 & \text{(reaction I)} \\
2CH_4 + h\nu' & \rightarrow & C_2H_4 + 2H_2 & \text{(reaction II)} \\
C_2H_4 + h\nu'' & \rightarrow & C_2H_2 + H_2 & \text{(reaction IIb)} \\
2CH_4 & \rightarrow & C_2H_2 + 3H_2 & \text{(reaction II} = \text{IIa} + \text{IIb)} \\
2CH_4 + h\nu''' & \rightarrow & \frac{2}{n}C_nH_2 + \left(4 - \frac{2}{n}\right) H_2 & \text{(reaction III)}
\end{array}$$

This reaction scheme shows that the appearance of C_2H_6 and C_2H_2 in the atmosphere (s. Table 2) has its origin in the photochemistry of CH_4. C_2H_4 is a photochemically unstable intermediate product. The haze in the upper atmosphere is mainly caused by the production of polyenes (reaction III). H_2 escapes rapidly from the atmosphere into the space due to Titan's low escape velocity. It is assumed that ca. 80 % of the destructed CH_4 is transformed into C_2H_6, ca. 12 % into C_2H_2 and 8 % into polyenes (Yung & DeMore, 1999; Griffith et al., 2006). C_2H_6 is assumed to be dissolved partly in the liquid lakes on the surface. Most of C_2H_6 and all other photochemical products remain as small particles in the atmosphere (polyenes) or are adsorbed on the icy surface on the bottom.

3. The composition of lakes on Titan's surface

The discovery of liquid lakes with areas up to the size of Lake Superior (USA) was one of the most spectacular results of the Cassini mission (Stofan et al., 2007; Mitri et al., 2007; Lorenz et al., 2008). What is the composition of these lakes covering at least several percent of the surface? The most simple assumption is that these lakes consist of a liquid mixture of N_2 and CH_4, which are also the dominant components in the gaseous atmosphere. It is an easy task to calculate the liquid composition treating the binary system as an ideal liquid mixture. We only have to know the temperature of the lakes and the saturation pressure.

mass/kg	radius/km	gravity acceleration/ $m \cdot s^{-2}$	average density/ $g \cdot cm^{-3}$	rotational period/days
$1.344 \cdot 10^{23}$	2575	1.354	1.88	16
surface temperature/K	surface pressure/bar	surface area/m^2	distance to sun/km	distance to Saturn/km
93	1.49	$8.332 \cdot 10^{13}$	$1.428 \cdot 10^9$	$1.222 \cdot 10^6$

Table 1. Physical parameters of Titan.

N_2	CH_4	H_2	CO	C_2H_6	C_2H_2	C_2H_4	HCN
$93-98$	$7-2$	0.1	$6 \cdot 10^{-3}$	$2 \cdot 10^{-3}$	$2 \cdot 10^{-4}$	$4 \cdot 10^{-5}$	$2 \cdot 10^{-5}$

Table 2. Atmospheric composition in mol % (Yung & DeMore, 1999)

Obvious choices are the experimentally determined surface temperature of $T = 93$ K and the total surface pressure $p = 1.49$ bar. According to Raoult's ideal law it follows

$$p \cdot y_{CH_4} = p_{CH_4}^{sat} \cdot x_{CH_4} \quad \text{and} \quad p \cdot y_{N_2} = p_{N_2}^{sat} (1 - x_{CH_4}) \tag{1}$$

with the molefractions y_i and x_i in the gaseous atmosphere and in the liquid phase respectively ($i = CH_4, N_2$).
The saturation pressures of methane and nitrogen at 93 K are (Prydz & Goodwin, 1972; Wagner & de Reuck K. M., 1996; Span et al., 2001):

$$p_{CH_4}^{sat} = 0.1598 \text{ bar}, \; p_{N_2}^{sat} = 4.625 \text{ bar}$$

Using these data we obtain

$$x_{CH_4} = \frac{p - p_{N_2}^{sat}}{p_{CH_4}^{sat} - p_{N_2}^{sat}} = 0.698 \text{ and } x_{N_2} = 1 - x_{CH_4} = 0.302 \tag{2}$$

The predicted composition of the atmosphere is therefore

$$\begin{aligned} y_{CH_4} &= p_{CH_4}^{sat} \cdot x_{CH_4} / p = 0.074 \\ y_{N_2} &= p_{N_2}^{sat} \cdot x_{N_2} / p = 0.926 \end{aligned} \tag{3}$$

This is close to the experimental data of y_{N_2} (0.93 – 0,98) and y_{CH_4} (0.02 – 0.07) and suggests that the predicted composition of the lakes with 70 mole percent CH_4 and 30 mole percent N_2 is no unreasonable result.
However, there exist serious arguments in the literature that the liquid phase of the lakes and also humidity in the pores of the solid water ice on the bottom may contain also considerable amounts of ethane (C_2H_6) (Lorenz et al., 2008; Barth & Toon, 2003; Rannou et al., 2006; Griffith et al., 2006; Mousis & Schmitt, 2008) since C_2H_6 is soluble in liquid CH_4 or liquid $CH_4 + N_2$ mixtures. Therefore a more refined model of the ternary system $CH_4 + N_2 + C_2H_6$ has to be applied accounting also for non-ideality of the three mixture components in the liquid as well as in the gaseous phase.
We start with the following expression for the total pressure p of a real liquid ternary mixture being in thermodynamic equilibrium with its real gaseous phase ($C = CH_4, N = N_2, E = \text{Ethan} = C_2H_6$).

$$p = (x_C \gamma_C) \cdot (\pi_C \cdot p_C^{sat}) + (x_N \gamma_N) \cdot (\pi_N \cdot p_N^{sat}) + (x_E \gamma_E) \cdot (\pi_E \cdot p_E^{sat}) \tag{4}$$

$\gamma_i (i = C,N,E)$ and $\pi_i (i = C,N,E)$ are the activity coefficients of i in the liquid phase and the so-called Poynting correction factors for component i respectively. Eq. (4) exhibits a sufficiently correct description of ternary vapor-liquid phase equilibria provided the reality of the gaseous phase can be described by accounting for second virial coefficients B_i and the approximation is acceptable that all activity coefficients in the vapor phase are unity (Lewis-Randall rule in the vapor phase).
γ_i and π_i are given by the following equations:

$a_{CN}/\mathrm{J\cdot mol^{-1}}$	$a_{CE}/\mathrm{J\cdot mol^{-1}}$	$a_{NE}/\mathrm{J\cdot mol^{-1}}$
720	440	800
$\bar{V}^l_{CH_4}/\mathrm{cm^3\cdot mol^{-1}}$	$\bar{V}^l_{N_2}/\mathrm{cm^3\cdot mol^{-1}}$	$\bar{V}^l_{C_2H_6}/\mathrm{cm^3\cdot mol^{-1}}$
35.65	38.42	67.83
$B_{CH_4}/\mathrm{cm^3\cdot mol^{-1}}$	$B_{N_2}/\mathrm{cm^3\cdot mol^{-1}}$	$B_{C_2H_6}/\mathrm{cm^3\cdot mol^{-1}}$
- 455	- 186	- 2500

Table 3. Thermodynamic parameters of the ternary mixture components at 93 K

$$\begin{aligned} RT\ln\gamma_C &= a_{CN}\cdot x_N(1-x_C) + a_{CE}(1-x_C)x_E - a_{NE}x_N\cdot x_E \\ RT\ln\gamma_N &= a_{CN}\cdot x_C(1-x_N) + a_{NE}(1-x_N)x_E - a_{CE}x_C\cdot x_E \\ RT\ln\gamma_E &= a_{CE}\cdot x_C(1-x_E) + a_{NE}(1-x_E)x_N - a_{CN}x_C\cdot x_N \end{aligned} \tag{5}$$

with the balance for mole fractions

$$x_C + x_N + x_E = 1$$

and the Poynting factors

$$\pi_i = \exp\left[\frac{(\bar{V}_{i,l} - B_i)(p - p_i^{sat})}{RT}\right] \qquad (i = \mathrm{C,N,E}) \tag{6}$$

The parameters a_{ij} characterize the difference of intermolecular interactions in binary liquid mixtures $i + j$ and are based on the simple expression for the molar Gibbs excess enthalpy G^E_{ij} of the binary mixture $i + j$:

$$G^E_{ij} = a_{ij}\, x_i\, x_j$$

The Gibbs excess enthalpy of the ternary mixture is

$$G^E_{ter} = RT(x_C\ln\gamma_C + x_N + \ln\gamma_N + x_E\ln\gamma_E) \tag{7}$$

with $\ln\gamma_i$ from eq. (5).
Parameters a_{CN}, a_{CE} and a_{NE} can be obtained from experimental data of binary G^E_{ij} available from the literature (Miller et al., 1973; McClure et al., 1976; Ponte et al., 1978) by fitting eq. (7) to the experiments. Values obtained are listed in Table 3. Calculation of π_i in eq. (6) requires data of the liquid molar volume $\bar{V}^l_i$ (Liu & Miller, 1972; Massengill & Miller, 1973) and the second virial coefficient B_i (Span et al., 2001; Bücker & Wagner, 2006; Wagner & de Reuck K. M., 1996) for each component at $T = 93$. Values are also listed in Table 3.
At given liquid composition x_C, x_N and $x_E = 1 - x_E - x_N, \gamma_C, \gamma_N$ and γ_E can be calculated using eq. (5). π_C, π_N and π_E can be calculated by eq. (6) using $p = 1.49\cdot 10^5$ Pa and the known values of p_i^{sat}. Compositions $y_i (i = \mathrm{C,N,E})$ are than obtained by

$$y_i = x_i\gamma_i p_i^{sat}\cdot\pi_i \qquad (i = \mathrm{C,N,E}) \tag{8}$$

It turns out that $y_E \approx 0$ for almost all compositions x_C and x_N due to the negligible vapor pressure $p_E^{sat} = 2.3\cdot 10^{-5}$ bar (Bücker & Wagner, 2006) compared to p_C^{sat} and p_N^{sat} at 93 K. As a result the vapor phase can be treated as a binary mixture of CH_4 and N_2. For presenting

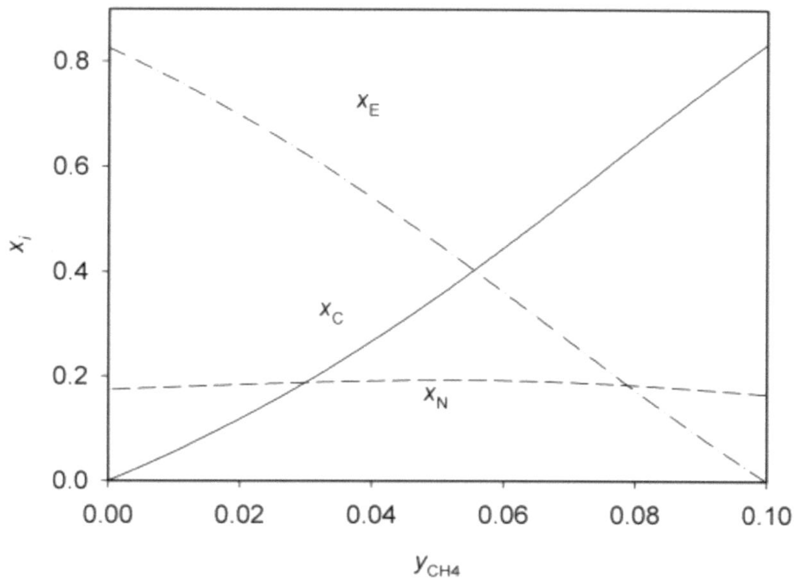

Fig. 1. Liquid composition $x_i (i = \mathrm{C,N,E})$ of the ternary system $CH_4 + N_2 + C_2H_6$ as function of molfraction of methane, y_C, in a saturated atmosphere

our results of the ternary liquid mixture in Fig. 1 the mole fractions in the liquid state x_C, x_N and $x_E = 1 - x_C - x_N$ are plotted as function of the composition of $y_{CH_4} \cong 1 - y_{N_2}$ up to $y_{CH_4} = 0.1$. Fig. 1 shows that x_E is continuously decreasing with increasing y_C while x_C is continously increasing. At $y_{CH_4} \cong 0.10, x_E = 0$, i. e. values of $y_{CH_4} \geq 0.1$ would indicate that no ethane can be present in the liquid phase. Interestingly x_N is almost independent of y_C with values close to 0.18. The following conclusions can be drawn from Fig. 1. Since experimental values of y_{CH_4} lie between 0.02 and 0.07 the results suggest that ethane would be present in the liquid phase with values of x_E in the range of 0.7 to 0.2. However, there is no evidence that the measured values of y_{CH_4} are really values being in thermodynamic equilibrium with a saturated liquid mixtures at the places where these values have been measured in Titan's atmosphere. Therefore equilibrium values of y_{CH_4} which are representative for the liquid composition of the lakes may reach or even exceed 0.1. In this case most likely no ethane would be present in the lakes and the liquid composition would be $x_{CH_4} \cong 0.834$ and $x_{N_2} \cong 0.166$ which is different from the result obtained by eq. (2) under the assumption of an ideal binary mixture ($x_{CH_4} = 0.698, x_{N_2} = 0.302$).

4. Cloud formation and rainfall in Titan's troposphere

Early attempts to describe quantitatively the situation of a saturated atmosphere of Titan can be found in the literature (Kouvaris & Flasar, 1991; Thompson et al., 1992). We provide here a simple and straight forward procedure based on the most recent results of the temperature profile of the lower atmosphere.

Fig. 2 shows the temperature profile of Titan's atmosphere as measured by the landing probe Huygens (Fulchignoni et al., 2005). The relatively high temperatures in the thermosphere

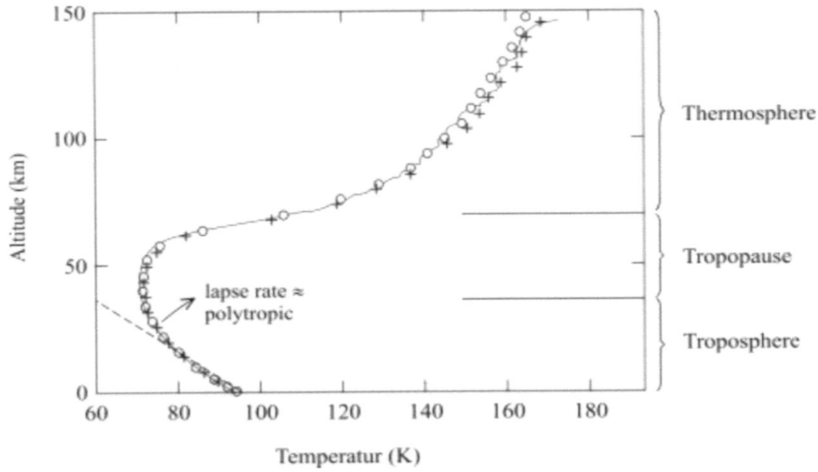

Fig. 2. Temperature profile in Titan's atmosphere (see text).

are caused by the absorption of solar radiation. This is the region where the photochemical processes take place. At ca. 40 km the temperature reaches a minimum value of ca. 73 K increasing again below this altitude. The nearly linear temperature profile below 20 km is called the polytropic lapse rate. Its slope (dT/dh)) is negative $(-0.92\ \mathrm{K}\cdot\mathrm{km}^{-1})$. This is the part of the troposphere where cloud formation of $CH_4 + N_2$-mixtures can take place as well as rainfall. Such negative lapse rates of temperature are also observed in other dense atmospheres, e. g. on the Earth or on the Venus and can be explained by the convection of gases in a gravitational field which corresponds approximately to an isentropic process which is given by the following differential relationship valid for ideal gases:

$$\frac{dT}{T} = \frac{\gamma - 1}{\gamma}\frac{dp}{p} \quad \text{with} \quad \gamma = C_p/C_V \tag{9}$$

C_p and C_V are the molar heat capacities at constant pressure and volume respectively. Real processes are often better described by ε instead of γ with

$$1 \le \varepsilon \le \gamma \tag{10}$$

ε is called the polytropic coefficient.
Considering hydrostatic equilibrium as a necessary condition in any atmosphere we have for ideal gases:

$$dp = -p\frac{\bar{M}\cdot g}{RT}\,dh \tag{11}$$

where $\bar{M}$ is the average molar mass of the gas and h is the altitude.
Combining eq. (9) with eq. (11) and using ε instead of γ integration gives the temperature profile in the atmosphere

$$T(h) = T_0\left(1 - \frac{\bar{M}\cdot g}{R}\frac{\varepsilon - 1}{\varepsilon}\frac{h}{T_0}\right) \tag{12}$$

with $\bar{M} = 0.028\ \mathrm{kg}\cdot\mathrm{mol}^{-1}, T_0 = 93$ K and $g = 1.354\ \mathrm{m}\cdot\mathrm{s}^{-2}$ (see Table 1).
Substituting eq. (12) into eq. (9) integration gives the pressure profile in the atmosphere:

$$p(h) = p_0 \left(1 - \frac{\bar{M} \cdot g}{R} \frac{\varepsilon - 1}{\varepsilon} \frac{h}{T_0}\right)^{\frac{\varepsilon}{\varepsilon-1}} = p_0 \left(\frac{T(h)}{T_0}\right)^{\frac{\varepsilon}{\varepsilon-1}} \quad (13)$$

with $p_0 = 1.49 \cdot 10^5$ Pa.
The experimental lapse rate $(dT/dp) = -0.92\ \mathrm{K} \cdot \mathrm{km}^{-1}$ is best described by eq. (12) with $\varepsilon = 1.25$. It is worth to note that eq. (13) gives the barometric formula for an isothermic atmosphere with $T = T_0$ in the limiting case of $\lim_{\varepsilon \to 1}$ [eq. (13)].
We are now prepared to develop a straightforward procedure for calculating cloud formation based on the assumption of a real binary mixture in the liquid state consisting of CH_4 and N_2. By equating eq. (4) with $x_{\mathrm{Ethane}} = 0$ and eq. (13) we obtain:

$$\begin{aligned} p(h) &= x_{CH_4} \cdot \gamma_{CH_4} \cdot \pi_{CH_4} \cdot p^{\mathrm{sat}}_{CH_4}(T(h)) + (1 - x_{CH_4}) \cdot \gamma_{N_2} \pi_{N_2} \cdot p^{\mathrm{sat}}_{N_2}(T(h)) \\ &= p_0 \left[1 - \frac{\overline{M} \cdot g}{R} \cdot \frac{\varepsilon - 1}{\varepsilon} \frac{h}{T_0}\right]^{\varepsilon/(\varepsilon-1)} \end{aligned} \quad (14)$$

Substituting now $T(h)$ from eq. (12) into the left hand side of eq. (14) gives x_{CH_4} of the saturated $CH_4 + N_2$ mixture as function of the altitude h. The corresponding mole fraction y_{CH_4} is calculated by

$$y_{CH_4} = \frac{x_{CH_4} \cdot \gamma_{CH_4} \cdot p^{\mathrm{sat}}_{CH_4}(T(h)) \cdot \pi_{CH_4}}{p(h)} \quad (15)$$

where $p(h)$ is eq. (14).
Results are shown in Fig. 3 which also shows the temperature profile (eq. (12)) and the solid-liquid equilibrium of methane. Fig. 3 demonstrates that only values of y_{CH_4} **below** the $y_{CH_4}(h)$ curve represent a dry atmosphere. For $y_{CH_4} > y_{CH_4}(h)$ phase splitting, i. e. condensation occurs, e. g. for $y = 0.049$ above 8700 m or for $y = 0.036$ above 12000 km. These are the cloud heights where we also can expect rain fall provided there is no supersaturation. Fig. 3 also shows that "methane snow" will never occur in Titan's atmosphere since the solid-liquid line of CH_4 does not intersect the $x_{CH_4}(h)$ curve above the temperature minimum of 73 K (s. Fig. 2) due to freezing point depression of the $CH_4 + N_2$ mixture. $y_{CH_4} = 0.0975$ and $x_{CH_4} = 0.834$ are the saturation values at the bottom as already calculated for the real model in section 3.

5. Approximative scenario of Titan's atmosphere in the past and in the future

To our knowledge no attempts have been made so far to develop a thermodynamically consistent procedure of a time dependent scenario of Titan's atmosphere.
The simplified scenario presented here is based on the assumption that the gaseous atmosphere as well as the liquid reservoirs on Titan's surface consist of binary $CH_4 + N_2$ mixtures which behave as ideal gases in the vapor phase and obey Raoult's ideal law. Further we assume that the total amount of N_2 remains unchanged over the time, only CH_4 underlies a photochemical destruction process occurring exclusively in the gaseous phase, i. e. in the atmosphere, with a known destruction rate constant. The photokinetic process is assumed to be slow compared to the rate for establishing the thermodynamic phase equilibrium. Starting with the mole numbers of $CH_4, n^g_{CH_4}$ and $N_2, n^g_{N_2}$ in the atmospheric (gaseous) phase given by the force balances between gravitational forces and pressure forces at $h = 0$

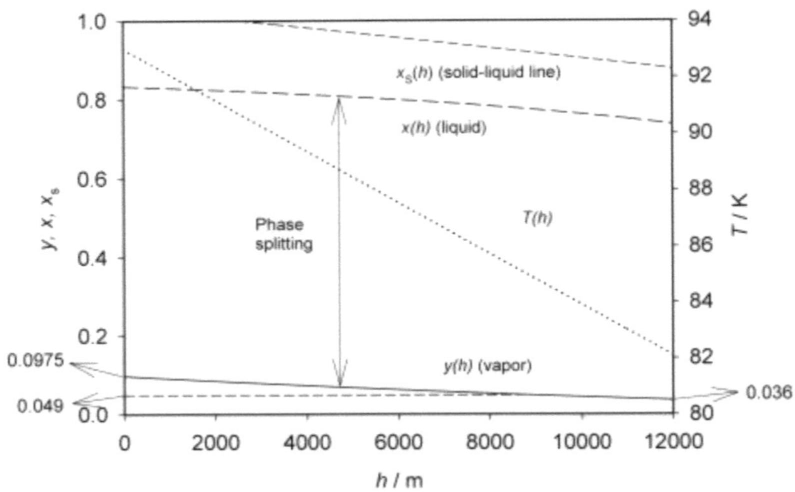

Fig. 3. Composition profile in a polytropic atmosphere (see text)

$$n^g_{\mathrm{CH_4}} = \frac{p^{\mathrm{sat}}_{\mathrm{CH_4}} \cdot A}{M_{\mathrm{CH_4}} \cdot g} x_{\mathrm{CH_4}} \quad \text{and} \quad n^g_{\mathrm{N_2}} = \frac{p^{\mathrm{sat}}_{\mathrm{N_2}} \cdot A}{M_{\mathrm{N_2}} \cdot g}(1 - x_{\mathrm{CH_4}}) \tag{16}$$

the total mole fraction of methane $\bar{x}_{\mathrm{CH_4}}$ being an averaged value of **both** phases is

$$\bar{x}_{\mathrm{CH_4}} = \frac{n^l_{\mathrm{CH_4}} + x_{\mathrm{CH_4}} \cdot \frac{p^{\mathrm{sat}}_{\mathrm{CH_4}} \cdot A}{M_{\mathrm{CH_4}} \cdot g}}{n^l_{\mathrm{CH_4}} + x_{\mathrm{CH_4}} \cdot \frac{p^{\mathrm{sat}}_{\mathrm{CH_4}} \cdot A}{M_{r\mathrm{CH_4}} \cdot g} + n^l_{\mathrm{N_2}} + (1 - x_{\mathrm{CH_4}}) \frac{p^{\mathrm{sat}}_{\mathrm{N_2}} \cdot A}{M_{\mathrm{CH_4}} \cdot g}} \tag{17}$$

where $n^l_{\mathrm{CH_4}}$ and $n^l_{\mathrm{N_2}}$ are the mole numbers of CH_4 and N_2 in the liquid phase respectively. A is the surface area of Titan (s. Table 1).
Since the total mole number of N_2, $n^{\mathrm{tot}}_{\mathrm{N_2}}$, is given by

$$n^{\mathrm{tot}}_{\mathrm{N_2}} = n^l_{\mathrm{N_2}} + n^g_{\mathrm{N_2}} = (1 - x_{\mathrm{CH_4}})(n^l_{\mathrm{N_2}} + n^l_{\mathrm{CH_4}}) + (1 - x_{\mathrm{CH_4}}) \frac{p^{\mathrm{sat}}_{\mathrm{N_2}} \cdot A}{M_{\mathrm{N_2}} \cdot g} \tag{18}$$

the sum of mole numbers of CH_4 and N_2 in the liquid phase is

$$n^l_{\mathrm{N_2}} + n^l_{\mathrm{CH_4}} = \frac{n^{\mathrm{tot}}_{\mathrm{N_2}}}{1 - x_{\mathrm{CH_4}}} - \frac{p^{\mathrm{sat}}_{\mathrm{N_2}} \cdot A}{M_{\mathrm{N_2}} \cdot g}$$

With $x_{\mathrm{CH_4}} = n^l_{\mathrm{CH_4}} / (n^l_{\mathrm{CH_4}} + n^l_{\mathrm{N_2}})$ eq. (17) can now be rewritten as

$$\bar{x}_{\mathrm{CH_4}} = \frac{x_{\mathrm{CH_4}} \left[\frac{n^{\mathrm{tot}}_{\mathrm{N_2}}}{1 - x_{\mathrm{CH_4}}} - \frac{p^{\mathrm{sat}}_{\mathrm{N_2}} \cdot A}{M_{\mathrm{N_2}} \cdot g} + \frac{p^{\mathrm{sat}}_{\mathrm{CH_4}} \cdot A}{M_{\mathrm{CH_4}} \cdot g} \right]}{x_{\mathrm{CH_4}} \left[\frac{n^{\mathrm{tot}}_{\mathrm{N_2}}}{1 - x_{\mathrm{CH_4}}} - \frac{p^{\mathrm{sat}}_{\mathrm{N_2}} \cdot A}{M_{\mathrm{N_2}} \cdot g} + \frac{p^{\mathrm{sat}}_{\mathrm{CH_4}} \cdot A}{M_{\mathrm{CH_4}} \cdot g} \right] + n^{\mathrm{tot}}_{\mathrm{N_2}}} \tag{19}$$

$\bar{x}_{CH_4}$ in eq. (19) depends on time through x_{CH_4}, the mole fraction of CH_4 in the liquid phase, all other parameters $p^{sat}_{CH_4}, p^{sat}_{N_2}, M_{N_2}, M_{CH_4}, g$ and in particular $n^{tot}_{N_2}$ are constant, i. e. they do not depend on time provided the temperature is also independent on time.
Below the wavelength $\lambda = 1650\mathring{A}$ methane dissociates according to the reaction scheme presented in the introductory section with the destruction rate of $4 \cdot 10^{-12}$ kg m^{-2} s^{-1} = $2.5 \cdot 10^{-10}$ mol m$^{-2} \cdot$ s^{-1} (Lorenz et al., 1997; Yung & DeMore, 1999). The total destruction rate on Titan is therefore $2.5 \cdot 10^{-10} \cdot 4\pi R_T^2 = 2.1 \cdot 10^4$ mol $\cdot$ s^{-1}. The loss of CH_4 is proportional to the sunlight intensity I_S and the mole number of CH_4 in the atmosphere $n^g_{CH_4}$ (see eq. (16)).

$$\frac{dn^t_{CH_4}}{dt} = 2.1 \cdot 10^4 = -n^g_{CH_4} \cdot I_S \cdot k'$$
$$= -k \cdot x_{CH_4}(t) \cdot p^{sat}_{CH_4} \cdot A/(M_{CH_4} \cdot g) \text{ mol} \cdot \text{s}^{-1} \quad (20)$$

with $I_s \cdot k' = k$ where $x_{CH_4}(t)$ is the mole fraction of CH_4 in the liquid phase at time t. From eq. (2) $x_{CH_4}(t = 0) = 0.698$ is the mole fraction of CH_4 in the liquid phase at present, and it follows:

$$k = \frac{2.1 \cdot 10^4 \cdot M_{CH_4} \cdot g}{0.698 \cdot p^{sat}_{CH_4} \cdot A}$$

We now have to integrate eq. (20):

$$\int_0^t \frac{dn^{tot}_{CH_4}}{x_{CH_4}(t)} = -K \cdot t \quad (21)$$

with

$$K = k \cdot \frac{p^{sat}_{CH_4} \cdot A}{M_{CH_4} \cdot g} = 2.1 \cdot 10^4/0.698 = 3.0 \cdot 10^4 \text{ mol} \cdot \text{s}^{-1}$$

where the time t can be positive (future) or negative (past). Eq. (21) with $K =$ const implies that the luminosity of the sun has been constant all the time, which is a realistic assumption with exception of the early time of the solar system Yung & DeMore (1999).
To solve the integral in eq. (21) we write:

$$dn^{tot}_{CH_4} = \left(\frac{dn^{tot}_{CH_4}}{d\bar{x}}\right) \cdot \left(\frac{d\bar{x}}{dx_{CH_4}}\right) dx_{CH_4} \quad (22)$$

Considering that $n^{tot}_{N_2} =$ const we obtain with $\bar{x} = n^{tot}_{CH_4} / \left(n^{tot}_{CH_4} + n^{tot}_{N_2}\right)$:

$$\frac{dn^{tot}_{CH_4}}{d\bar{x}} = \frac{\left(n^{tot}_{CH_4} + n^{tot}_{N_2}\right)^2}{n^{tot}_{N_2}} \quad (23)$$

and from differentiating eq. (19):

$$\frac{d\bar{x}_{CH_4}}{dx_{CH_4}} = \frac{n^{tot}_{N_2}\left(\frac{n^{tot}_{N_2}}{(1-x_{CH_4})^2} - \frac{p^{sat}_{N_2} \cdot A}{M_{N_2} \cdot g} + \frac{p^{sat}_{CH_4} \cdot A}{M_{CH_4} \cdot g}\right)}{\left[x_{CH_4}\left(\frac{n^{tot}_{N_2}}{1-x_{CH_4}} - \frac{p^{sat}_{N_2} \cdot A}{M_{N_2} \cdot g} + \frac{p^{sat}_{CH_4} \cdot A}{M_{CH_4} \cdot g}\right) + n^{tot}_{N_2}\right]^2} \quad (24)$$

Since the denominator of eq. (24) is equal to $\left(n_{N_2}^{tot}+n_{CH_4}^{tot}\right)^2$ substituting eq. (23) and eq. (24) into eq. (22) and then in eq: (21) gives:

$$-Kt=\int\limits_{x_{CH_4}(t=0)}^{x_{CH_4}(t)}\left(\frac{n_{N_2}^{tot}}{(1-x_{CH_4})^2}-\frac{p_{N_2}^{sat}\cdot A}{M_{N_2}\cdot g}+\frac{p_{CH_4}^{sat}\cdot A}{M_{CH_4}\cdot g}\right)\cdot\frac{dx_{CH_4}}{x_{CH_4}} \quad (25)$$

The integral in eq. (25) can be solved analytically and the result is:

$$\begin{aligned}-Kt &= n_{N_2}^{tot}\left[\frac{1}{1-x_{CH_4}(t)}-\frac{1}{1-x_{CH_4}(t=0)}\right.\\ &\left.-\ln\left(\frac{1-x_{CH_4}(t)}{1-x_{CH_4}(t=0)}\cdot\frac{x_{CH_4}(t=0)}{x_{CH_4}(t)}\right)\right]+\frac{A}{g}\left(\frac{p_{CH_4}^{sat}}{M_{CH_4}}-\frac{p_{N_2}^{sat}}{M_{N_2}}\right)\\ &\cdot\ln\left(\frac{x_{CH_4}(t)}{x_{CH_4}(t=0)}\right)\end{aligned} \quad (26)$$

Eq. (19) and eq. (26) are the basis for discussing the scenario.
The pressure of the CH_4+N_2 mixture is given by

$$p(x_{CH_4})=x_{CH_4}\left(p_{CH_4}^{sat}-p_{N_2}^{sat}\right)+p_{N_2}^{sat} \quad (27)$$

or

$$p(y_{CH_4})=\frac{y_{CH_4}\cdot p_{N_2}^{sat}}{p_{CH_4}^{sat}+y_{CH_4}\left(p_{N_2}^{sat}-p_{CH_4}^{sat}\right)}\cdot\left(p_{CH_4}^{sat}-p_{N_2}^{sat}\right)+p_{N_2}^{sat} \quad (28)$$

where $p(x_{CH_4})=p(y_{CH_4})$ is the total pressure as function of x_{CH_4} or y_{CH_4} respectively.
In Fig. 4 $p(x_{CH_4}), p(y_{CH_4})$ and $p(\bar{x}_{CH_4})$ with $\bar{x}_{CH_4}$ taken from eq. (19) are plotted in a common diagram at 93 K. Three different values of $n_{N_2}^{tot}$ have been chosen for calculating $p(\bar{x}_{CH_4})$: $n_{N_2}^{tot}=3.08\cdot10^{20}$ mol corresponds to a surface of Titan which is covered by 4 % of lakes with a depth of 100 m, $n_{N_2}^{tot}=3.39\cdot10^{20}$ mol has the same coverage but a depth of 1000 m. $n_{N_2}^{tot}=10.16\cdot10^{20}$ mol corresponds to a coverage of 28 percent and a depth of 600 m. This is exactly the value where $n_{N_2}^{g}=n_{N}^{tot}$ when x_{CH_4} becomes zero according to eq. (16).
Fig. 4 illustrates the "lever rule" of binary phase diagrams. For the present situation with $y_{CH_4}=0.074$ and $x_{CH_4}=0.698$ the dashed horizontal line indicates the 2-phase region with $\bar{x}_{CH_4}$-values corresponding to their $n_{N_2}^{tot}$-values. The higher $n_{N_2}^{tot}$ is the closer is $\bar{x}_{CH_4}$ to the value of x_{CH_4}. The $\bar{x}_{CH_4}$-trajectories indicated by arrows show how $\bar{x}_{CH_4}$ is changed with decreasing values of x_{CH_4}, i. e. with increasing time. It is interesting to note that the trajectories with $n_{N_2}^{tot}=3.03\cdot10^{20}$ mol and $n_{N_2}^{tot}=3.39\cdot10^{20}$ mol end on the $p(y_{CH_4})$-curve. This means that after a certain time $\bar{x}_{CH_4}$ becomes equal to y_{CH_4}, and all liquid reservoirs on Titan would have disappeared. The surface has dried out and methane being now exclusively present in the atmosphere will be photochemically destructed according to a first order kinetics. Finally a dry atmosphere containing pure N_2 will survive. In case of $n_{N_2}^{tot}=10.16\cdot10^{20}$ mol the trajectory ends at $y_{CH_4}=\bar{x}_{CH_4}=x_{CH_4}=0$ and at $p=p_{N_2}^{sat}$ which means that the atmosphere consist of pure N_2. For $n_{N_2}^{tot}>10.16\cdot10^{20}$ mol the final situation

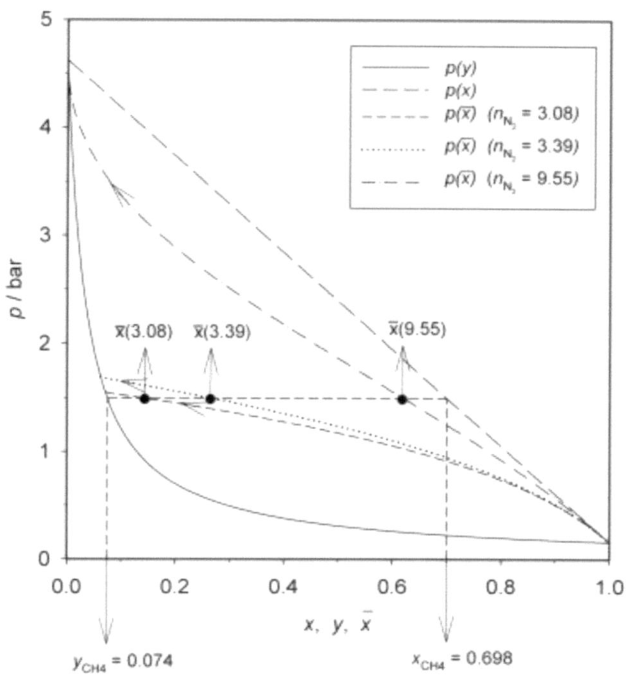

Fig. 4. Equilibrium x,g diagram of an ideal binary $N_2 + CH_4$ mixture with trajectories of the total molefraction of CH_4 $\bar{x}$ at different total mole numbers of N_2 $n_{N_2}^{tot}$

will be an atmosphere of pure N_2 with $p = p_{sat}^{N_2}$ and liquid reservoirs with pure N_2 on the surface.

The explicit time evolution of x_{CH_4} obtained by eq. (26) is shown in Fig. 5 and Fig. 6. According to these results $x_{CH_4} > 0.9$ can be expected at the time of Titan's formation ($4 \cdot 10^9$ years ago) for all values of $n_{N_2}^{tot}$ considered here.

According to Fig. 5 the lakes will have disappeared in $10 \cdot 10^6$ years ($n_{N_2}^{tot} = 3.08 \cdot 10^{20}$ mol) or in $23 \cdot 10^6$ years ($n_{N_2}^{tot} = 3.39 \cdot 10^{20}$ mol) with the "last drop" of a liquid composition $x_{CH_4} \approx 0.685$ or 0.655 respectively. At higher values of $n_{N_2}^{tot}$ the liquid mixtures phase would exist much longer since $n_{CH_4}^{tot}$ is also larger at the same composition. In case of $n_{N_2}^{tot} = 10.16 \cdot 10^{20}$ mol and $n_{N_2}^{tot} = 25 \cdot 10^{20}$ mol (see Fig. 6) CH_4 would have disappeared at Titan in ca. $350 \cdot 10^6$ years and $2.0 \cdot 10^9$ years respectively resulting in a pure N_2-atmosphere with liquid N_2 reservoirs ($n_{N_2}^{tot} > 10 \cdot 16 \cdot 10^{20}$ mol).

Fig. 7 shows the change of the total amount of CH_4 $n_{CH_4}^{tot} = n_{CH_4}^{g} + n_{CH_4}^{l}$ on Titan as function of time for the past and the future also based on the present values of $x_{CH_4} = 0.658$ and $y_{CH_4} = 0.074$ for different values of $n_{N_2}^{tot}$ as indicated. $n_{CH_4}^{tot}$ has been calculated by

$$n_{CH_4}^{tot}(t) = \frac{\bar{x}_{CH_4}(t)}{1 - \bar{x}_{CH_4}(t)} \cdot n_{N_2}^{tot}$$

with $\bar{x}_{CH_4} = \bar{x}_{CH_4}(x_{CH_4}(t))$ using eq. (19) with $x_{CH_4}(t)$ from eq. (26). According to the model

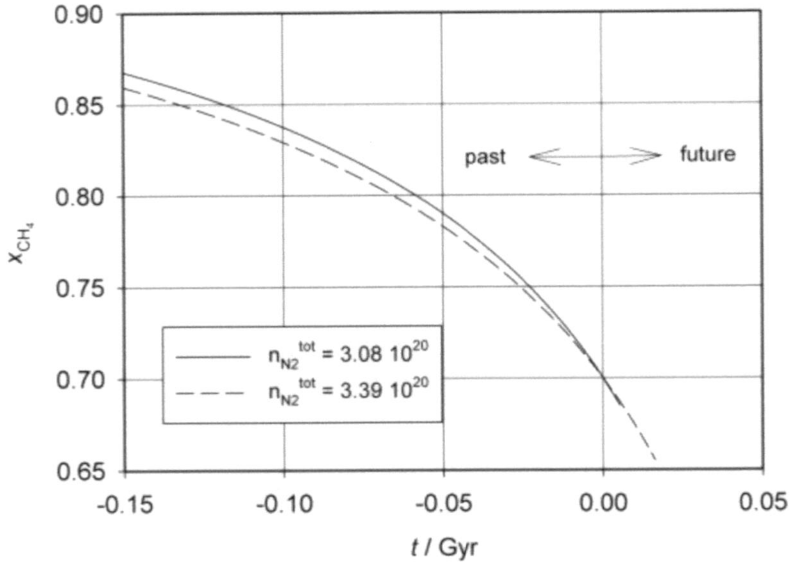

Fig. 5. Molefraction x_{CH_4} in the liquid phase as function of time at fixed total mole numbers of N_2 on Titan's surface $n_{N_2}^{tot} = 3.08 \cdot 10^{20}$ mol and $n_{N_2}^{tot} = 3.39 \cdot 10^{20}$ mol.

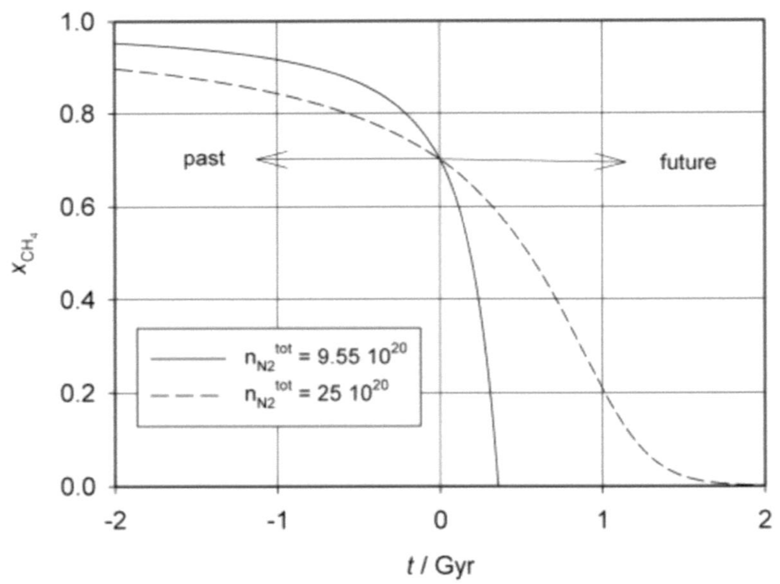

Fig. 6. Molefraction x_{CH_4} in the liquid phase as function of time at fixed total mole numbers of N_2 on Titan's surface: $n_{N_2}^{tot} = 9.55 \cdot 10^{20}$ mol and $n_{N_2}^{tot} = 25 \cdot 10^{20}$ mol.

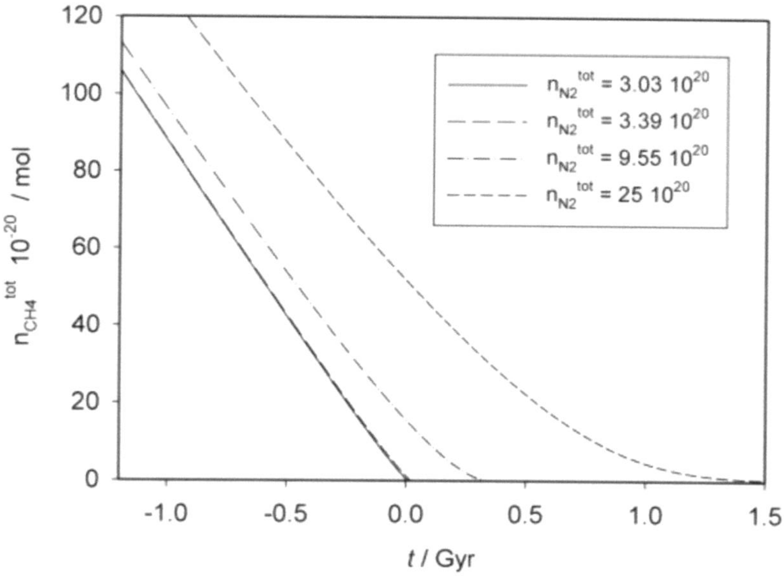

Fig. 7. Total mole number of CH_4 on Titan's surface as function of time at fixed total model numbers of N_2 as indicated

much more methane must have existed on Titan in the past than today, i. e., Titan has possibly been covered by a deep liquid ocean consisting of a $N_2 + CH_4$ mixture with distinctly higher concentrations of methane than today. It might also be possible that the amount of CH_4 and N_2 is already much higher at present than the detectable lakes on Titan's surface suggest due to hidden reservoirs such as "humidity" in micropores of Titan's icy crust.
The calculations of this simple scenario show that the fate of Titan's atmosphere and lakes sensitively depends on the amount and on the composition of liquid present today on Titan's surface. As long as there are no certain values available neither the future nor the past can be predicted with acceptable reliability. If this situation is changed the model of the scenario can be extended to real ternary mixture also including ethane. At this point the purpose of this section was to demonstrate how scenario calculations can be performed.

6. Titan's internal structure

As already mentioned the low density indicates that Titan's interior must contain considerable amounts of water beside rocky material. Assuming that these different chemical components are well separated we expect that Titan has an inner core consisting of rock, i. e. silicates, with an averaged density $\rho_{Rock} = 3\ g \cdot cm^{-3}$, the mantel and the crust will mainly consist of liquid water and ice with an averaged density $\rho_{H_2O} = 1.1\ g \cdot cm^{-3}$. Accepting these figures we are able to determine roughly the amount of water in the outer shell as well as the amount of silicate in the core. We also can calculate the central pressure and the pressure as function of radius r with $r = 0$ at the center. First the radius r_1 of transition from the rocky material to the water phase is determined from the following mass balance:

$$m_{H_2O} + m_{Rock} = \frac{4}{3}\pi < \rho > \cdot R_T^3 = \frac{4}{3}\pi\rho_{H_2O}\left[R_T^3 - r_1^3\right] + \frac{4}{3}\pi\rho_{Rock}\cdot r_1^3 \tag{29}$$

where R_T is Titan's radius (see Table 1).
Solving eq. (29) for r_1^3 gives:

$$r_1^3 = R_T^3\,\frac{<\rho> - \rho_{H_2O}}{\rho_{Rock} - \rho_{H_2O}} \tag{30}$$

Using the known data of $<\rho>= 1.88\ \mathrm{g\cdot cm^{-3}}$ (s. Table 1), ρ_{H_2O}, ρ_{Rock}, and R_T the result is

$$r_1 = 1913\ \mathrm{km} \quad \text{or} \quad r_1/R_T = 0.743$$

This is a rough estimate because ρ_{H_2O} and ρ_{Rock} are averaged values over the pressure and temperature profile of Titan's interior. Improved results can be obtained by considering compressibilities and thermal expansion coefficients if these profiles would be known.
The mass fraction w_{H_2O} of H_2O in Titan is:

$$w_{H_2O} = \rho_{H_2O}\left(R_T^3 - r_1^3\right) \Big/ \left[\rho_{H_2O}\left(R_T^3 - r_1^3\right) + \rho_{Rock}\cdot r_1^3\right] = 0.345$$

and the corresponding mass fraction of rock is $w_{Rock} = 1 - w_{H_2O} = 0.655$.
Using the hydrostatic equilibrium condition

$$dp = -\rho\cdot g\cdot dr \tag{31}$$

the pressure profile of Titan's interior can easily be calculated. Using the local gravity acceleration $g(r)$

$$g(r) = G\cdot\frac{m(r)}{r^2} = \frac{4}{3}\pi G\cdot\rho(r)$$

eq. (31) can be written:

$$dp = -G\rho^2(r)\cdot\frac{4}{3}\pi r\cdot dr \tag{32}$$

with the gravitational constant $G = 6.673\cdot 10^{-11}\ [\mathrm{J\cdot m\cdot kg^{-2}}]$ and ρ either ρ_{H_2O} or ρ_{Rock}. Integration of eq. (32) gives the central pressure p_0 assuming that ρ_{H_2O} and ρ_{Rock} are independent of the pressure p with $p(r = R_T) \approx 0$:

$$\begin{aligned} p_0 &= \frac{2}{3}\pi G\left[\rho_{H_2O}^2(1-(0.743)^2) + \rho_{Rock}\cdot(0.743)^2\right]\cdot R_T^2 = 5.106\cdot 10^9\ \mathrm{Pa} \\ &= 51\ \mathrm{kbar} \end{aligned} \tag{33}$$

where $r_1/R_T = 0.743$ has been used.
The pressure p_1 at r_1 is:

$$p_1 = p_0 - \rho_{Rock}^2\cdot\frac{2}{3}G\cdot r_1^2 = 0.502\cdot 10^9\ \mathrm{Pa} \cong 5\ \mathrm{kbar} \tag{34}$$

The dependence of the pressure p on r is given by:

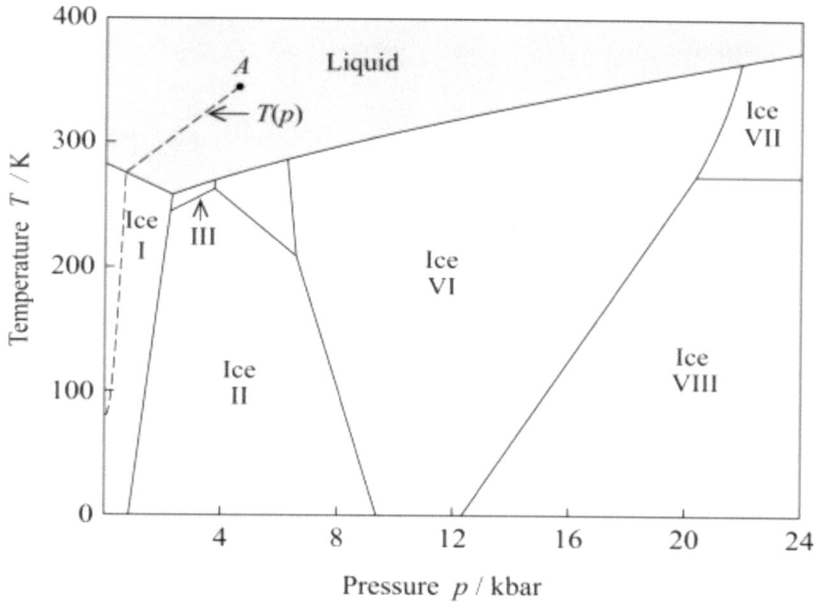

Fig. 8. Phase diagram of water. – – – – estimated $T(p)$ curve inside Titan. A = assumed transition point from aqueous phase to the rocky material

$$
\begin{aligned}
p(r) &= p_0 - \rho^2_{\mathrm{Rock}} \cdot \frac{2}{3}\pi G \cdot r^2 \qquad (\text{for } r \le 0.743\ R_T) \\
p(r) &= p_0 + \left[\left(\rho^2_{\mathrm{H_2O}} - \rho^2_{\mathrm{Rock}}\right) \cdot (0.743 \cdot R_T)^2 - \rho^2_{\mathrm{H_2O}} \cdot r^2\right] \frac{2}{3}\pi G \ (\text{for } r \ge 0.743 \cdot R_T)
\end{aligned}
\tag{35}
$$

The upper layer of the aqueous system is the crust consisting of water ice up to a depth of ca. 200 km followed by a layer of liquid water up to the depth where the solid phase of rocky materials begins. The existence of solid and liquid water layers can be understood by inspecting the phase diagram of water shown in Fig. 8.
Figure 8 shows that the tentative $T(p)$ curve most probably intersects the solid-liquid equilibrium line of water at ca. 0.5 kbar and r = 2300 km. In the literature (Lunine & Stevenson, 1987) there are also aqueous NH_3 solutions discussed instead of pure water leading to a second solid phase of pure water which is likely to exist between r = 2100 km and r = 1913 km. The dependence of pressure p on r calculated according to eq. (35) is shown in Fig. 9.

7. Conclusions

- Using thermodynamic methods and comparatively simple theoretical tools the composition of liquid lakes on Titan's surface can be predicted being in acceptable agreement with the known atmospheric composition.

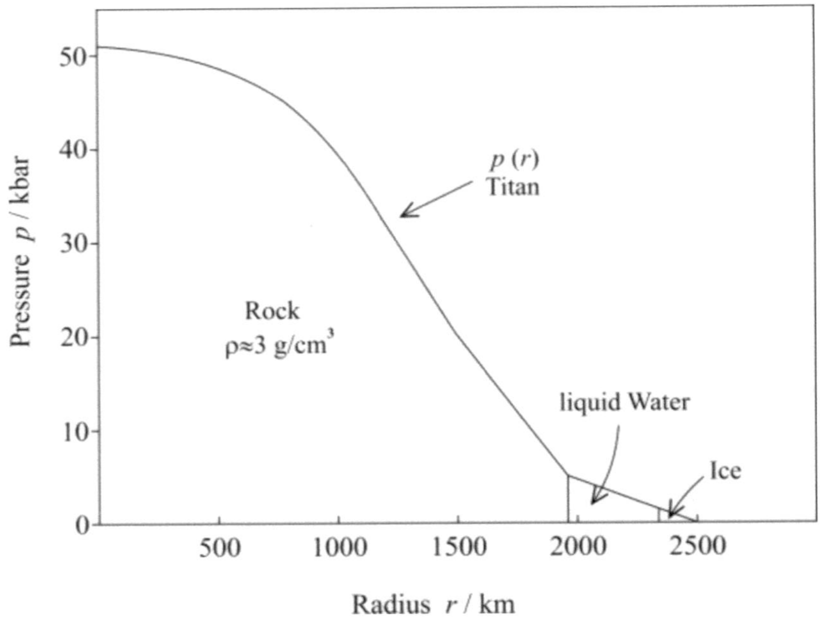

Fig. 9. Pressure as function of the distance r from Titan's center

- The cloud ceiling of $CH_4 + N_2$ mixtures in the troposphere can be estimated. Depending on the degree of saturation on the bottom altitudes between 8 and 12 km are predicted provided the vapor is not supersaturated. In the model calculations presented here a real mixture without ethane has been treated.
- A rough estimation of atmospheric scenarios in the past and the future can be made. The results depend essentially on the total amount of liquid on Titan's surface present today.
- An imagination of the internal structure of Titan can be provided predicting an aqueous mantle of ca. 700 km thickness and a hard core consisting of rocky material with a central pressure of ca. 51 kbar.

8. References

Atreya, S. K., Adams, E. Y., Niemann, H. B., Demick-Montelara, J. E., Owen, T. C., Fulchignoni, M., Ferri, F. & Wilson, E. H. (2006). Titan's methane cycle, *Planetary and Space Science* 54(12): 1177–1187.

Barth, E. L. & Toon, O. B. (2003). Microphysical modeling of ethane ice clouds in titan's atmosphere, *Icarus* 162(1): 94–113.

Bücker, D. & Wagner, W. (2006). A reference equation of state for the thermodynamic properties of ethane for temperatures from the melting line to 675 K and pressures up to 900 MPa, *J. Phys. Chem. Ref. Data* 35: 205–266.

Flasar, F. M., Achterberg, R. K., Conrath, B. J., Gierasch, P. J., Kunde, V. G., Nixon, C. A., Bjoraker, G. L., Jennings, D. E., Romani, P. N., Simon-Miller, A. A., Bezard, B., Coustenis, A., Irwin, P. G. J., Teanby, N. A., Brasunas, J., Pearl, J. C., Segura, M. E., Carlson, R. C., Mamoutkine, A., Schinder, P. J., Barucci, A., Courtin, R., Fouchet, T., Gautier, D., Lellouch, E., Marten, A., Prange, R., Vinatier, S., Strobel, D. F., Calcutt, S. B., Read, P. L., Taylor, F. W., Bowles, N., Samuelson, R. E., Orton, G. S., Spilker, L. J.,

Owen, T. C., Spencer, J. R., Showalter, M. R., Ferrari, C., Abbas, M. M., Raulin, F., Edgington, S., Ade, P. & Wishnow, E. H. (2005). Titan's Atmospheric Temperatures, Winds, and Composition, *Science* 308(5724): 975–978.

Fulchignoni, M., Ferri, F., Angrilli, F., Ball, A. J., Bar-Nun, A., Barucci, M. A., Bettanini, C., Bianchini, G., Borucki, W., Colombatti, G., Coradini, M., Coustenis, A., Debei, S., Falkner, P., Fanti, G., Flamini, E., Gaborit, V., Grard, R., Hamelin, M., Harri, A. M., Hathi, B., Jernej, I., Leese, M. R., Lehto, A., Lion Stoppato, P. F., Lopez-Moreno, J. J., Makinen, T., McDonnell, J. A. M., McKay, C. P., Molina-Cuberos, G., Neubauer, F. M., Pirronello, V., Rodrigo, R., Saggin, B., Schwingenschuh, K., Seiff, A., Simoes, F., Svedhem, H., Tokano, T., Towner, M. C., Trautner, R., Withers, P. & Zarnecki, J. C. (2005). In situ measurements of the physical characteristics of titan's environment, *Nature* 438(7069): 785–791.

Griffith, C. A., Penteado, P., Rannou, P., Brown, R., Boudon, V., Baines, K. H., Clark, R., Drossart, P., Buratti, B., Nicholson, P., McKay, C. P., Coustenis, A., Negrao, A. & Jaumann, R. (2006). Evidence for a Polar Ethane Cloud on Titan, *Science* 313(5793): 1620–1622.

Kouvaris, L. C. & Flasar, F. M. (1991). Phase equilibrium of methane and nitrogen at low temperatures: Application to titan, *Icarus* 91(1): 112–124.

Lewis, J. S. (1997). *Physics and Chemistry of the Solar System*, Academic Press, New York.

Liu, Y.-P. & Miller, R. (1972). Temperature dependence of excess volumes for simple liquid mixtures: Ar + CH_4, N_2 + CH_4, *The Journal of Chemical Thermodynamics* 4(1): 85–98.

Lorenz, R. D., McKay, C. P. & Lunine, J. I. (1997). Photochemically Driven Collapse of Titan's Atmosphere, *Science* 275(5300): 642–644.

Lorenz, R. D., Mitchell, K. L., Kirk, R. L., Hayes, A. G., Aharonson, O., Zebker, H. A., Paillou, P., Radebaugh, J., Lunine, J. I., Janssen, M. A., Wall, S. D., Lopes, R. M., Stiles, B., Ostro, S., Mitri, G. & Stofan, E. R. (2008). Titan's inventory of organic surface materials, *Geophys. Res. Lett.* 35(2): L02206.

Lunine, J. I. & Stevenson, D. J. (1987). Clathrate and ammonia hydrates at high pressure: Application to the origin of methane on titan, *Icarus* 70(1): 61–77.

Massengill, D. & Miller, R. (1973). Temperature dependence of excess volumes for simple liquid mixtures: N_2 + Ar, N_2 + Ar + CH_4, *The Journal of Chemical Thermodynamics* 5(2): 207–217.

McClure, D. W., Lewis, K. L., Miller, R. C. & Staveley, L. A. K. (1976). Excess enthalpies and gibbs free energies for nitrogen + methane at temperatures below the critical point of nitrogen, *The Journal of Chemical Thermodynamics* 8(8): 785–792.

Miller, R. C., Kidnay, A. J. & Hiza, M. J. (1973). Liquid-vapor equilibria at 112.00 k for systems containing nitrogen, argon, and methane, *AIChE Journal* 19(1): 145–151.

Mitri, G., Showman, A. P., Lunine, J. I. & Lorenz, R. D. (2007). Hydrocarbon lakes on titan, *Icarus* 186(2): 385–394.

Mousis, O. & Schmitt, B. (2008). Sequestration of ethane in the cryovolcanic subsurface of titan, *The Astrophysical Journal Letters* 677(1): L67.

Niemann, H. B., Atreya, S. K., Bauer, S. J., Carignan, G. R., Demick, J. E., Frost, R. L., Gautier, D., Haberman, J. A., Harpold, D. N., Hunten, D. M., Israel, G., Lunine, J. I., Kasprzak, W. T., Owen, T. C., Paulkovich, M., Raulin, F., Raaen, E. & Way, S. H. (2005). The abundances of constituents of titan's atmosphere from the gcms instrument on the huygens probe, *Nature* 438(7069): 779–784.

Ponte, M. N. D., Streett, W. B. & Staveley, L. A. K. (1978). An experimental study of the equation of state of liquid mixtures of nitrogen and methane, and the effect of pressure on their excess thermodynamic functions, *The Journal of Chemical Thermodynamics* 10(2): 151–168.

Prydz, R. & Goodwin, R. D. (1972). Experimental melting and vapor pressures of methane, *The Journal of Chemical Thermodynamics* 4(1): 127–133.

Rannou, P., Montmessin, F., Hourdin, F. & Lebonnois, S. (2006). The Latitudinal Distribution of Clouds on Titan, *Science* 311(5758): 201–205.

Span, R., Lemmon, E. W., Jacobsen, R. T., Wagner, W. & Yokozeki, A. (2001). A reference equation of state for the thermodynamic properties of nitrogen for temperatures from 63.151 to 1000 K and pressures to 2200 MPa, *J. Phys. Chem. Ref. Data* 29: 1361–1433.

Stofan, E. R., Elachi, C., Lunine, J. I., Lorenz, R. D., Stiles, B., Mitchell, K. L., Ostro, S., Soderblom, L., Wood, C., Zebker, H., Wall, S., Janssen, M., Kirk, R., Lopes, R., Paganelli, F., Radebaugh, J., Wye, L., Anderson, Y., Allison, M., Boehmer, R., Callahan, P., Encrenaz, P., Flamini, E., Francescetti, G., Gim, Y., Hamilton, G., Hensley, S., Johnson, W. T. K., Kelleher, K., Muhleman, D., Paillou, P., Picardi, G., Posa, F., Roth, L., Seu, R., Shaffer, S., Vetrella, S. & West, R. (2007). The lakes of titan, *Nature* 445(7123): 61–64.

Thompson, W. R., Zollweg, J. A. & Gabis, D. H. (1992). Vapor-liquid equilibrium thermodynamics of $N_2 + CH_4$: Model and titan applications, *Icarus* 97(2): 187–199.

Wagner, W. & de Reuck K. M. (1996). *Methane*, Vol. 13 of *International thermodynamic tables of the fluid state*, Blackwell Science, Oxford.

Yung, Y. L., Allen, M. & Pinto, J. P. (1984). Photochemistry of the atmosphere of Titan - Comparison between model and observations, *Astophys. J. Suppl. Ser.* 55: 465–506.

Yung, Y. L. & DeMore, W. B. (1999). *Photochemistry of Planetary Atmospheres*, Oxford University Press, Oxford.

20

Interoperability between Modelling Tools (MoT) with Thermodynamic Property Prediction Packages (Simulis® Thermodynamics) and Process Simulators (ProSimPlus) Via CAPE-OPEN Standards

Ricardo Morales-Rodriguez[1], Rafiqul Gani[1], Stéphane Déchelotte[2], Alain Vacher[2] and Olivier Baudouin[2]
[1]CAPEC, Technical University of Denmark
[2]ProSim S.A.
[1]Denmark
[2]France

1. Introduction

Simulation and modelling continues to play a very important role for chemical engineers in the study, evaluation, development, etc., of chemical processes by producing different alternatives in the design/production of product/process of chemicals and thereby, avoiding expenses in experimentation.
The CAPE-OPEN effort is a standardisation process for achieving true plug and play of process industry simulation software components and environments, where, CAPE-OPEN Laboratorties Network (CO-LaN) consortium is in charge of managing the lifecycle of the CAPE-OPEN standard (Belaud, 2002). The objective of CAPE-OPEN project was to clarify user priorities for process modelling software component/environment interoperability and promote the use of CAPE-OPEN standards to create commercially-valuable interoperability (Pons, 2005a).
The follow-up of the CAPE-OPEN project, called the Global CAPE-OPEN project, focused on the development of standards in new subfields of process modelling and simulation addressing complex physical properties, kinetic models, new numerical algorithms and distributed models. Also, future support for the development of simulation software in the CAPE-OPEN-compliant interface components was established through the creation of the CO-LaN. The CO-LaN promotes the integration of open process simulation technology in the work process, and use of CAPE-OPEN compliant interoperated software for taking real industrial case studies and in assessing the use of CAPE-OPEN technology. In addition, CO-LaN provides support and user training; definition of open standards for new technologies beyond process modelling and simulation, developing prototypes for on-line systems, discrete and mixer batch-continuous processes, finer granularity interfaces, and scheduling and planning systems. Further dissemination of the technical results of CAPE-OPEN using

both traditional and internet-based mechanism; assessment of the use and benefits of CAPE-OPEN standards for educations and training (Braunschweig et al., 2000).

Currently, several commercial simulator vendors, modelling tools developers, etc. have incorporated CAPE-OPEN standard in their products, allowing the user a better and easier manner for implementation/combination among process modelling components (PMC) and process modelling environments (PME) (Pons, 2003).

CAPE-OPEN is an abstract specification that can be subsequently implemented in COM, CORBA and .NET for bridging PMCs and PMEs. Recently, .NET framework has been introduced as a new alternative to provide the interoperability among different platforms. This new technology has been presented by Microsoft and it seems to be visualized as the future of the connections between different platforms. CO-LaN has published guidelines on how to use .NET with CAPE-OPEN, this is available in CO-LaN website at http://www.colan.org/News/Y06/news-0616.htm.

In this chapter, a connection between a modelling tool (ICAS-MoT) representing a PMC and external simulators (ProSimPlus and Simulis® Thermodynamics) representing a PME is established and implemented through the CAPE-OPEN standards and highlighted through two case studies. The interoperability issues of ICAS-MoT and Simulis® Thermodynamics are highlighted through case study number one. Here, a thermodynamic property model is generated by ICAS-MoT and wrapped to satisfy the CAPE-OPEN standard. A second case study highlights the interaction issues between ICAS-MoT and ProSimPlus regarding the use of a non-conventional unit operation (model generated in ICAS-MoT) plugged to the ProSimPlus simulator environment. Furthermore, as this unit operation model is employing a multiscale modelling approach, these issues are also highlighted through the case study.

2. What do we need for the Interoperability between modelling tool (MoT) and external simulators (ProSim)?

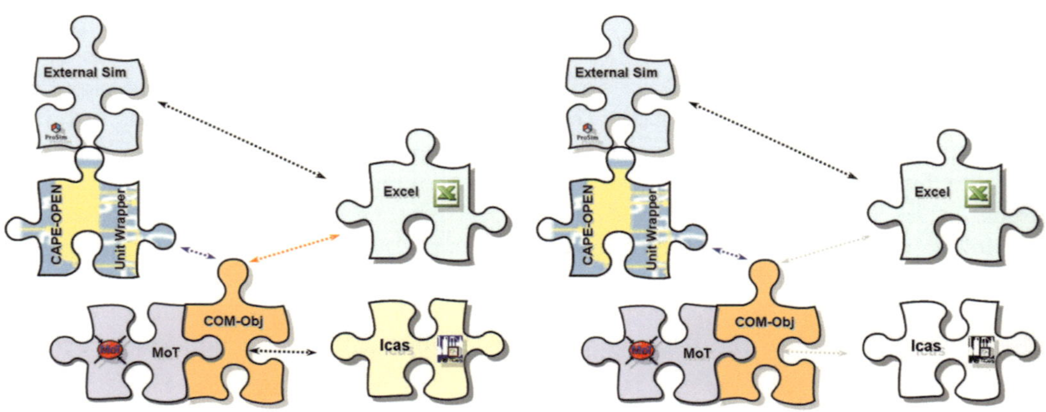

Fig. 1. Interoperability between the PMC and PME

Process modelling component (represented by ICAS-MoT) is able to achieve interoperability and be wrapped for use within a PME (represented by a simulation engine, external software and external simulator) through CAPE-OPEN interfaces. Fig. 1 illustrates the interaction that can be done between a PMC and a PME. This interaction has been applicated in this chapter for two case studies. In both cases, the ICAS-MoT is representing a

PMC with different models. In the first case, the PMC is integrated with a PME (represented by Simulis® Thermodynamics) while in the second case, the PMC is integrated with a PME (represented by one external simulator as ProSimPlus). To establish the integration, here, the following components are needed:

- PME (Process Modelling Environment)
- PMC (Process Modelling Component)
- Middleware (COM)

These are briefly explained below.

2.1 Process modelling component (PMC)

Process modelling components are objects that can be added to the flowsheet to represent unit operations or mathematical/information/energy flows within the flowsheet, thermodynamic property models, reaction models or mathematical model solvers (Barrett, 2005). PMCs basically are pieces of software that are defined for a specific function. Most of the applications are for: physical properties, unit operation modules, numerical solver and flowsheet analysis tools. This work focuses mostly on the connection of PMCs related to physical properties and unit operation to an appropriate PME.

As far as physical properties are concerned, they are an important part in the evaluation of chemical processes that involve some kind of phase equilibrium calculations (vapor-liquid, liquid-liquid, solid-liquid and so on) or some transport properties or the use of some derivatives of some properties with respect to temperature or pressure (*i.e.;* fugacity coefficients, enthalpy, etc.) using some equations of states or special correlations (depending on the property that it is being calculated) that cannot be easily found in some commercial or free software.

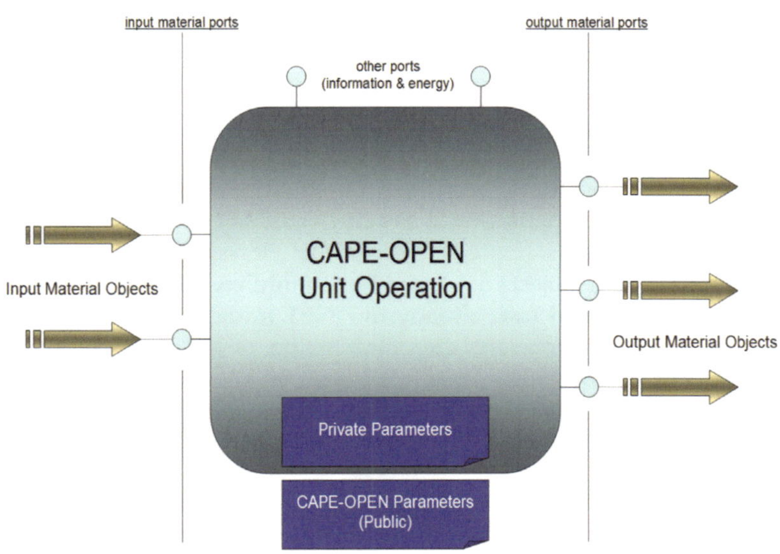

Fig. 2. CAPE-OPEN Unit operation

In CAPE-OPEN compliant PMCs for unit operations, the interface includes ports that can be classified as material, energy and information ports. In this work, the notion of Material Object is used to represent a material stream. In fact, a CAPE-OPEN Material Object is a container of properties describing the material stream. Also, this object allows property and equilibrium calculations by its associated CAPE-OPEN Thermo Property Package. This association is performed by the simulator (PME). The CAPE-OPEN Material Object can also provide pure component properties, constant or temperature dependent. CAPE-OPEN Material Objects are connected to the Unit Operation using inlet or outlet material ports. It is not necessary to connect all ports but, in many cases, a minimum number of connections (inlet and outlet) must be respected. Parameters are another important part in CAPE-OPEN. They can be classified into private and public parameters. On the one hand, private parameters are only modifiable from the Unit Operation itself using, in general, a graphical user interface. On the other hand, we find public parameters, also called CAPE-OPEN parameters, which are exposed to the outside of the Unit Operation. Private and public parameters can be input or output parameters of the model. In some cases, the values of these parameters can be imported or exported using CAPE-OPEN Information Objects connected to information ports. Fig. 2 is showing a scheme of a CAPE-OPEN Unit Operation. Some PMCs that can be found are: ICAS-MoT, Aspen Properties, ChemSep, CPA Property Package and so on.

2.2 Process modelling environment (PME)

The process modelling environment supports the construction of process models from first-principles and/or library of unit operation models; number of model-based applications as, simulation, optimization; and they may use of one or more PMCs (Pons, 2005b); furthermore, PMEs allow process engineers to use software from heterogeneous sources operating together to carry out complex mode-based tasks (Braunschweig et al., 2000). PMEs also consist of the graphical interface and functionality required to create the flow network being modeled; input, review and modify values for parameters of components; input, review and modify material or energy flows, and calculate the conditions of the flowsheet based on the inputs (Barrett, 2005). Some examples of PMEs that can be found are: ProSimPlus, Simulis® Thermodynamics, Aspen Plus, COFE, gPROMS and so on.

2.3 Middleware

One of the most important parts to carry out this interoperability between the PMCs and PMEs is the part that allows connecting those entities. CAPE-OPEN has chosen to adopt a component software and object-oriented approach that views each PMC as a separate object. All communication between objects is handled by "middleware" such as CORBA or COM (Braunschweig et al., 2000) and now already the .NET framework. Fig. 3 is illustrating the three main components to carry out the combination of software components through CAPE-OPEN standards.

PMC offers a "CAPE-OPEN Plug" when it can be used within different PMEs using CAPE-OPEN interfaces. CAPE-OPEN Unit Operations offers a CAPE-OPEN Unit Plug, while, CAPE-OPEN Thermo Property Packages offers a CAPE-OPEN Thermo Plug. For PMEs, the "socket" term will be used to express the capacity to employ CAPE-OPEN PMCs using CAPE-OPEN interfaces. It will be found on one hand, a CAPE-OPEN Thermo Socket allowing the use of CAPE-OPEN Thermo Property Packages, and on another hand a CAPE-OPEN Unit Socket allowing the use of CAPE-OPEN Units. In this chapter, the term PMC will be used to indicate a CAPE-OPEN thermo plug or CAPE-OPEN unit plug, while the term PME will be used to indicate a CAPE-OPEN Thermo socket or CAPE-OPEN Unit socket, respectively.

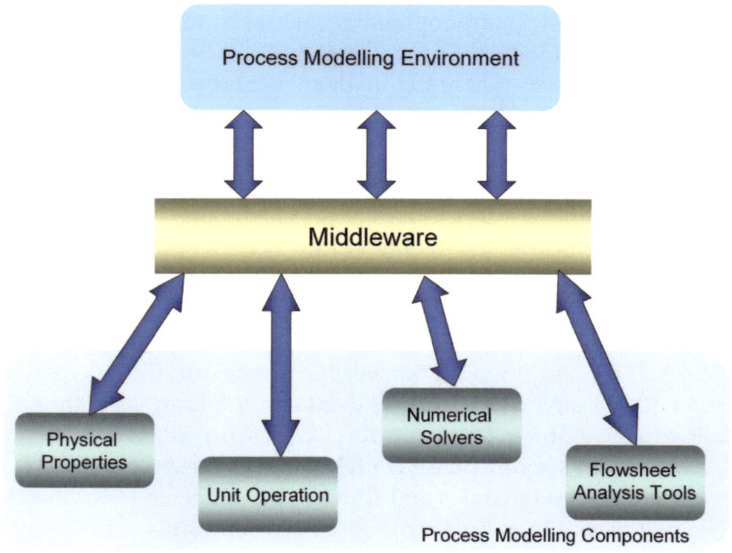

Fig. 3. CAPE-OPEN components

3. Case studies

3.1 Simulis® thermodynamic – ICAS-MoT

The first case study is demonstrating the integration between ICAS-MoT and Simulis® Thermodynamics, which is carried out through the use of a DLL file as the middleware. Fig. 4 is illustrating the general structure of the combination of these different computational tools. The introduction of data is carried out through Simulis® Thermodynamics that provides a graphic interface for this purpose. Information (data) is transferred through the DLL file where variables that are shared between the integrated computational tools are specified. This information is transferred to ICAS-MoT to carry out the calculations using the ICAS-MoT solver. Afterwards, results are returned through the DLL file again

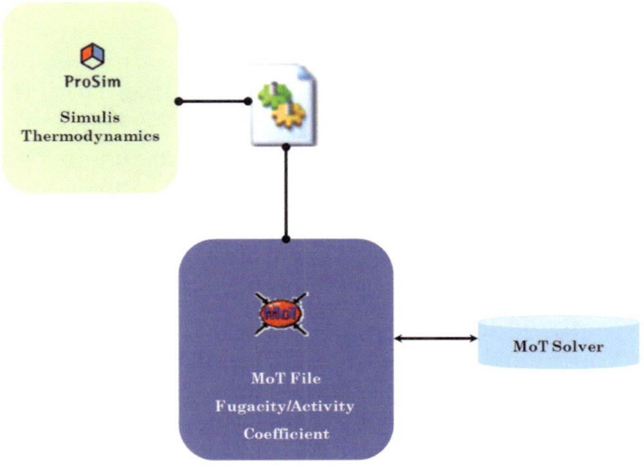

Fig. 4. ICAS-MoT interoperability with Simulis® Thermodynamics

and presented in the Simulis® Thermodynamics dialogs as the same way than calculations carried out using its native models. This case study is illustrated using the calculation of fugacity and activity coefficients where the mathematical (thermodynamic property) model is specified in the ICAS-MoT file.

Fugacity and Activity Coefficients Calculation

In order to show the reliability of this integration, calculations of fugacity coefficients have been performed with three different mathematical models equations of state: SRK, SAFT and PC-SAFT. Note that the same middleware has been used for the different equations of state models (see Fig. 5). For instance, whether one wants to perform the calculations using the SAFT model and afterwards with the PC-SAFT model, the only "work" needed is to change the ICAS-MoT file and it would be possible to perform the calculations in Simulis® Thermodynamics without any extra effort. The data-flow taking place through the DLL file involves the following variables: temperature (T), pressure (P), vapour composition (Yv) and fugacity coefficients in vapour phase (Phiv). Among these variables, (T, P and Yv) are specified in Simulis® Thermodynamics and through the DLL file to ICAS-MoT, which then employs the specified model to calculate the fugacity coefficients (PhiV) sent it to Simulis® Thermodynamics through the DLL file.

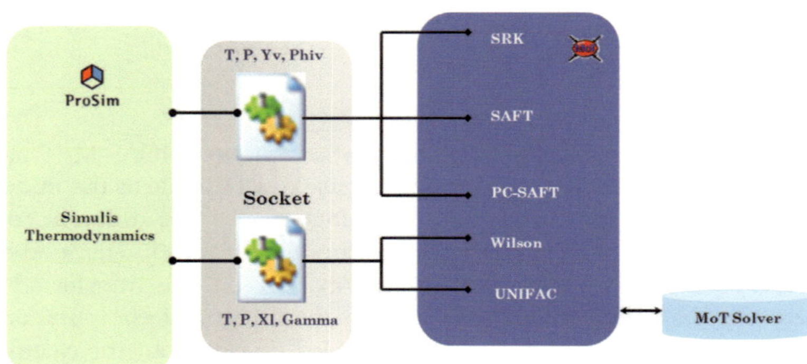

Fig. 5. Structure for the calculation of fugacity and activity coefficients though the synergy between ICAS-MoT and Simulis® thermodynamic

As far as activity coefficient calculation is concerned, two different mathematical models have been performed: Wilson and UNIFAC models solution (see Fig. 5). The data-flow through DLL file involves: temperature (T), pressure (P), liquid composition (Xl) and activity coefficient (Gamma). T, P and Xl are also specified in Simulis® thermodynamics and sent them through the DLL file to ICAS-MoT, which employs them to calculate the activity coefficients (Gamma) and send it to Simulis® Thermodynamics through the DLL file.

Testing of Interoperability

The interoperability of the PME-PMC integration is tested through the calculation of the component fugacity coefficients by three different equations of state for the binary mixture of methanol-methane, using the same middleware. Note that the model parameters have not been adjusted. The SAFT and PC-SAFT EOS give similar values while the SRK EOS give very different values. For the chemical system already used in this case study, at high pressures, the calculations with the PC-SAFT and the SAFT EOS are supposed to give more accurate values. Since the calculated values with the SRK EOS are very different, it suggests that there is a need for parameter regression to obtain more accurate results. Activity coefficient calculations for methanol-water mixture are also performed in order to show the interoperability of the PME and PMC employing a different thermodynamic property. Results for fugacity and activity coefficient calculation are illustrated in Fig. 6.

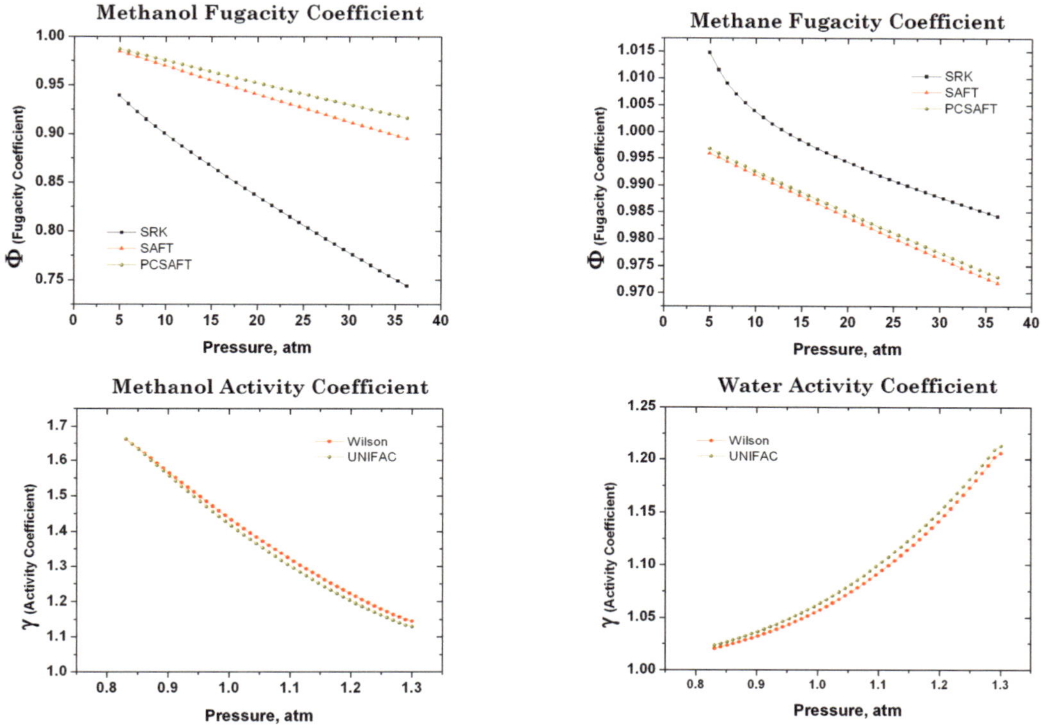

Fig. 6. Results for fugacity and activity coefficient calculations using ICAS-MoT and Simulis® Thermodynamics

3.2 ProSimPlus – ICAS-MoT

The interoperability between ProSimPlus and ICAS-MoT is shown in this case study. A new unit operation (Direct Methanol Fuel Cell) model employing a multiscale modelling approach is combined with unit operations that can be found in the model library of ProSimPlus; other examples of this interoperability between ProSimPlus and MoT can also be found at this reference Morales-Rodriguez et al. (2006). The first step is to understand the interoperability between the PMC and PME.

Fig. 7 illustrates a "Generic CAPE-OPEN Unit Operation" where objects are wrapped by an ICAS-MoT object (COM object) representing a model generated through ICAS-MoT (a ICAS-MoT file) and where all the necessary interfaces for connection to other tools are CAPE-OPEN compliant. An XML configuration file describes the mapping between variables of the ICAS-MoT model and variables required by CAPE-OPEN specifications.

The input Material Objects must, at least, provide the following variables: temperature, pressure, composition (either total flowrate and molar fractions, either partial flowrates) through the COM interfaces. More generally, the wrapper will provide/ask to MoT Objects for each property described in the XML configuration file (containing the mapping between ICAS-MoT Objects variable and wrapper variables). The values of the variables are obtained using the ICAS-MoT model and calculated by the ICAS-MoT Solver. As far as output Material Objects concerns, the same variables should be described and returned plus enthalpy of the stream (in this case calculated by ProSimPlus).

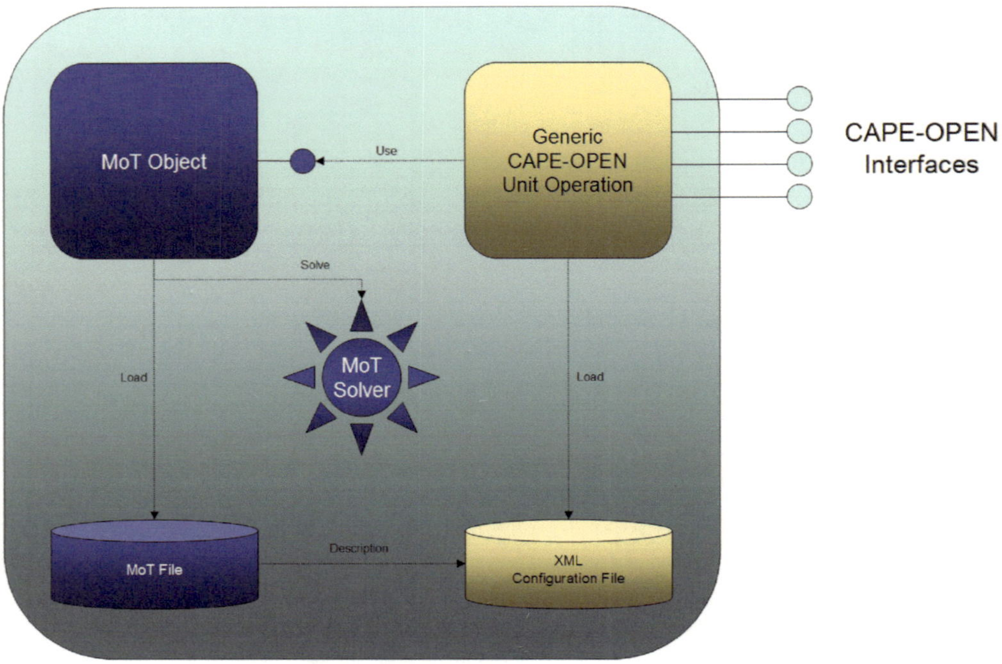

Fig. 7. ICAS-MoT CAPE-OPEN Unit Operation

The generic CAPE-OPEN MoT Unit Operation uses configuration files to determine the number and the definition of ports, variables and constants. These files are text files (".MoTUO" filename extension) with a specific syntax and can be written with any text editor. They are based on the well known XML and XSD standard. The structure of the XML configuration file consists of three main sections: "Globals", "UserParameters" and "MaterialPorts" as shown in Fig. 8. First of all, encoding type used in the XML file (in our case "ISO-8859-1") and schema used in the XLM file are specified. The schema describes the grammar of the XML file. The "Globals" block consists of general information about: the unit operation, minimum and maximum number of material ports, compounds, etc. "UserParameters" block describes the values of parameters that the user can modify through the GUI (Graphical User Interface) of the unit operation in the simulator. Here, it is possible to specify the type of parameter, default initial value, lower and upper bound for the parameters and so on. "MaterialPorts" block consists of a list of ports. These ports describe the connections available for the simulator and the values to copy from/to CAPE-OPEN Material objects connected to the port. Each "MaterialPort" section describes the connection port, variables and constants used to communicate with the CAPE-OPEN Material Objects connected to these ports. "MaterialPorts" block also contents the

```xml
<?xml version="1.0" encoding="ISO-8859-1"?>
<UnitOperation
            xmlns:xsi="http://www.w3.org/2001/XMLSchema-instance"
            xsi:noNamespaceSchemaLocation="GenericMoTUnit.xsd">

  <Globals>
    <MOTFile>C:\Program Files\COMoTUO\FuelGen.mot</MOTFile>
    <Name>FuelCell</Name>
    <Description>Test for CAPE-OPEN ProSim-Mot</Description>
    <MinimumCompoundsNumber>1</MinimumCompoundsNumber>
    <MaximumCompoundsNumber>2</MaximumCompoundsNumber>
    <MinimumInputMaterialPortsNumber>1</MinimumInputMaterialPortsNumber>
    <MaximumInputMaterialPortsNumber>1</MaximumInputMaterialPortsNumber>
    <MinimumOutputMaterialPortsNumber>1</MinimumOutputMaterialPortsNumber>
    <MaximumOutputMaterialPortsNumber>2</MaximumOutputMaterialPortsNumber>
  </Globals>

  <UserParameters>
    <UserParameter>
      <Name>Residence time</Name>
      <Description>Mean Residence time</Description>
      <Mode>input</Mode>
      <Type>real</Type>
      <DefaultValue>500.0</DefaultValue>
      <LowerBound>0.0</LowerBound>
      <UpperBound>1000.0</UpperBound>
      <MOTVariable>tau</MOTVariable>
    </UserParameter>
    ...
  </UserParameters>

  <MaterialPorts>
    <MaterialPort>
      <Name>FeedStream</Name>
      <Description>Feed stream to the unit</Description>
      <Direction>inlet</Direction>
      <Mandatory>false</Mandatory>
      <Variables>
        <Variable>
          <Name>T</Name>
          <Description>Temperature</Description>
          <MOTVariable>T_0</MOTVariable>
          <Type>real</Type>
          <MaterialObjectAccess>
            <COProperty>temperature</COProperty>
            <COPhase10>Overall</COPhase10>
            <COPhases11/>
            <Type>Mixture</Type>
          </MaterialObjectAccess>
        </Variable>
        ...
      </Variables>
    </MaterialPort>
    ...
  </MaterialPorts>

</UnitOperation>
```


Fig. 8. XML configuration file structure

"variables" sections where the list of variables are specified. These variables are used to transfer data from and to the CAPE-OPEN Material Objects connected to the port and are non constant properties of the mixture or of a pure compound. Note that since the XML configuration file specifies and controls the information shared between the PMC and PME, its construction needs to be done carefully.

A flow-diagram for a Direct Methanol Fuel Cell modeled through multiscale approach is shown in Fig. 9. Multiscale modelling approach basically consists of the division of a complex problem/model into a set of sub-problems/models that are described at different scales of length or/and time, in order to improve the degree of details of the phenomena that the set of mathematical model is describing in product-process design. Furthermore, multiscale approach facilitates the discovery and manufacture of complex products (Cameron et al., 2005); it is possible to observe that two different scales are involved: meso-scale and micro-scale. The meso-scale involves the modelling of the anode and cathode compartments while the micro-scale is employed to model, the behaviour of the anode and cathode catalyst layers and the proton membrane exchange.

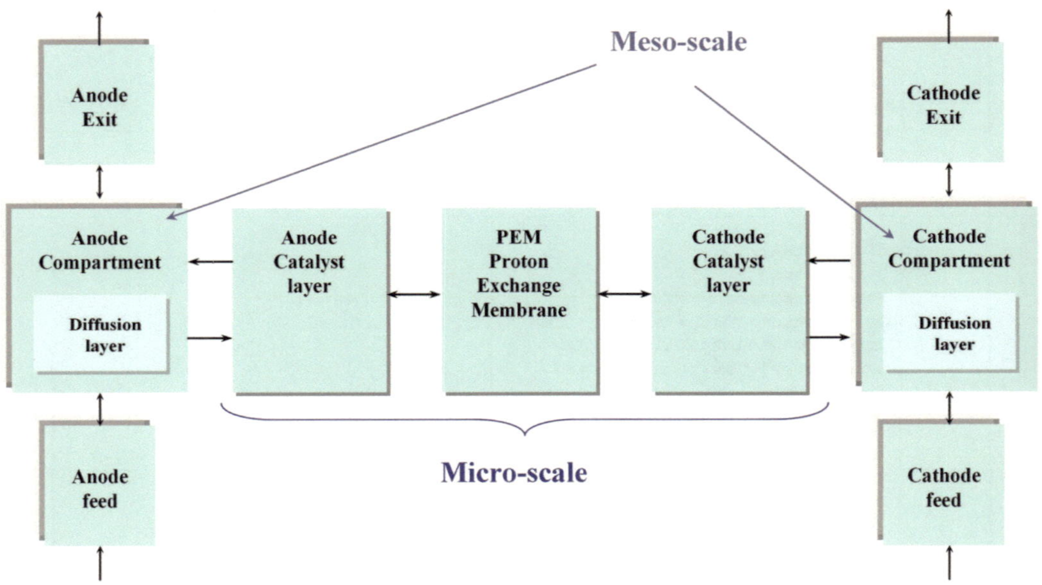

Fig. 9. Direct methanol fuel cell through multiscale approach

The set of equations for the DMFC unit is shown in table 1 where the description of each type of equation and the scale embedded in the unit is illustrated. This model is adapted from Sundmacher et al. (2001) and Xu et al. (2005) and this is more detalied in its solution by Morales-Rodriguez (2009) and Cameron and Gani (2010).

In order to highlight the differences between the results of the multiscale modelling and single-scale modelling, two scenarios have been chosen:

- For the multiscale modelling approach (MS): the entire set of equations shown in table 1 was solved; it means, meso-scale and micro-scale were taken into account. The variables shared between both scales are following (see table 2):

Scale	Description		Equation	
Level 3 (Meso-scale)	Mass Balance	Anode compartment	$\frac{dc_{CH_3OH}}{dt}=\frac{1}{\tau}\left(c_{CH_3OH}^{F}-c_{CH_3OH}\right)-\frac{k^{LS}A^{S}}{V_a}\left(c_{CH_3OH}-c_{CH_3OH}^{CL}\right)$	(1)
			$\frac{dc_{CO_2}}{dt}=\frac{1}{\tau}\left(c_{CO_2}^{FA}-c_{CO_2}\right)-\frac{k^{LS}A^{S}}{V_a}\left(c_{CO_2}-c_{CO_2}^{CL}\right)$	(2)
Level 2 (micro-scale)		Anode Catalyst Layer	$\frac{dc_{CH_3OH}^{CL}}{dt}=\frac{k^{LS}A^{S}}{V_a^{CL}}\left(c_{CH_3OH}-c_{CH_3OH}^{CL}\right)-\frac{A^{S}}{V_a^{CL}}n_{CH_3OH}^{M}-\frac{A^{S}}{V_a^{CL}}r_1$	(3)
			$\frac{dc_{CO_2}^{CL}}{dt}=\frac{k^{LS}A^{S}}{V_a^{CL}}\left(c_{CO_2}-c_{CO_2}^{CL}\right)+\frac{A^{S}}{V_a^{CL}}r_1$	(4)
		Membrane model	$\frac{\partial c_{CH_3OH}^{M}(t,z)}{\partial t}=-D^{M}\frac{\partial\, c_{CH_3OH}^{M}(t,z)}{\partial z}+vc_{CH_3OH}^{M}(t,z)$	(5)
	Charge Balances	Anode Catalyst Layer	$\frac{d\eta_a}{dt}=\frac{1}{C_a}\left(-i_{cell}-6Fr_1\right)$	(6)
		Cathode Catalyst Layer	$\frac{d\eta_c}{dt}=\frac{1}{C_c}\left(-i_{cell}-6F\left(r_5+n_{CH_3OH}^{M}\right)\right)$	(7)
	Constitutive Equations	Electrode kinetics	$r_1=k_1\exp\left(\frac{\alpha_1 F}{RT}\eta_a\right)\left\{\theta_{Pt}^{3}c_{CH_3OH}^{CL}-\frac{1}{K_1}\exp\left(-\frac{F}{RT}\eta_a\right)\theta_{Pt-COH}\right\}$	(8)
			$r_5=k_5\exp\left(\frac{\alpha_5 F}{RT}\eta_c\right)\left\{1-\exp\left(-\frac{F}{RT}\eta_c\right)\left(\frac{p_{O_2}}{p^0}\right)^{3/2}\right\}$	(9)

Table 1. Direct methanol fuel cell model equation divided at the different scales and parts of the unit

Description	Variable
Methanol bulk concentration	c_{CH_3OH}
Concentration in the catalyst layer of methanol	$c_{CH_3OH}^{CL}$
Carbon dioxide bulk concentration	c_{CO_2}
Concentration in the catalyst layer of methanol	$c_{CO_2}^{CL}$

Table 2. Variables shared between meso-scale and micro-scale

- For the single-scale scenario (SS): equations at the meso-scale level were solved (equations 1 and 2) while as far as micro-scale equations are concerned, values for the dependent and explicit variables (equations 3-9) values were given as known (that is, values at a specific point are known or obtained from experiment or simply as data from the literature). In this case study, the values for these variables were taken from the multiscale steady state simulation.

In the next step, the different parts of the model are assembled through the middleware (CAPE-OPEN interface) within the PME (external simulator) as shown in Fig. 10. The construction of the flowsheet is carried out by adding the corresponding identifiers of the unit operation(s) as well as the feed, product and connection streams, available in the

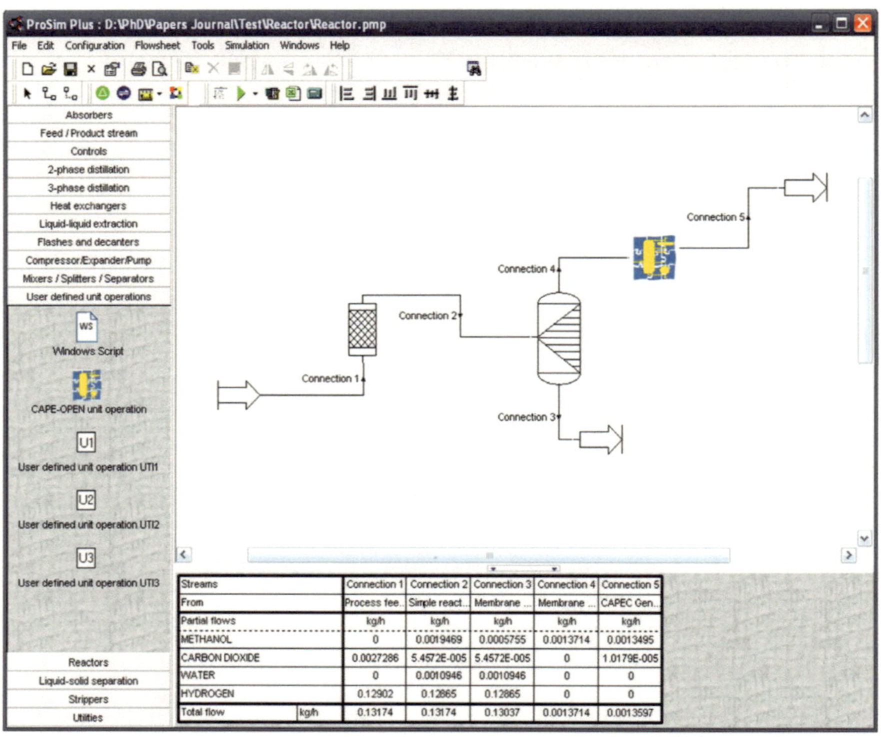

Fig. 10. Flowsheet in the ProSimPlus including the CAPE-OPEN unit operation using an ICAS-MoT model

process flow diagram menu on the left side of the graphic user interface window. Once, the addition of the unit operation has been done, a dialog box window appears to introduce the information, conditions and values for each unit operation that have been chosen previously. It is convenient to highlight the information introduced in the dialog box window for the generic CAPE-OPEN unit operation (in the PME) corresponds to the parameters specified in the XML configuration file in the "UserParameters" block.

The flowsheet shown in Fig. 10 illustrates a small part of the industrial production of methanol where this product is generated in the reactor by chemical reaction between carbon dioxide and hydrogen. The outstream of the reactor is composed of methanol, water (as the reaction products) and the reactants. This mixture is fed to a unit operation representing a membrane unit operation where pure methanol is obtained as the top and the mixture of the four compounds is located on the bottom. The top product is fed in the Generic CAPE-OPEN unit operation that is representing a DMFC unit operation and its mathematical model is generated and represented by ICAS-MoT file. Both scenarios mentioned below and described in separate ICAS-MOT files but work with the same XLM file, thereby providing "plug and play" option.

Dynamic simulation results for the flowsheet presented in Fig. 10 is shown in Fig. 11. Note that the PME simulator is not a dynamic simulator but Direct Methanol Fuel Cell model contained in the ICAS-MoT file is. The simulation strategy involved was changing the final time of integration of the model directly in ICAS-MoT. This was done in order to see the different results that can be obtained for the dynamic simulation, with or without the multiscale feature. Obviously, the two models should give the same steady state values but the path to achieve them would be predicted different.

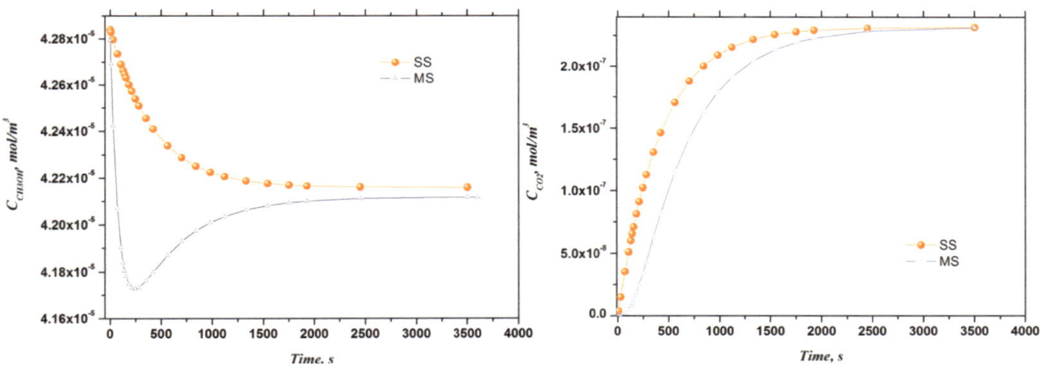

Fig. 11. Results of the CAPE-OPEN unit operation embedding multiscale modeling

Methanol bulk composition and carbon dioxide bulk concentration, which are variables at the meso-scale level were chosen to illustrate the different results obtained for each scenario. Fig. 11 (a) is showing the bulk composition of methanol present in the fuel cell obtained with the mutiscale (MS) and Single-multiscale (SS) models. As it can be seen, the values of these variables at the steady state are quite similar, but they are not for the transient state where some differences in the composition along the time can be noted. Those details are the extra information that can be obtained through the use of the multiscale modelling approach. The bulk compositions for carbon dioxide are also shown in Fig. 11 (b) using the multiscale and single-multiscale models. The simulated results are also different for the two modelling approaches. Certainly, it is possible to add further details of the phenomena that it is occurring in the process depending on whether the model for lower or higher scale are available or necessary for the study. The multiscale modelling framework is useful for the integration, connection and description at different scales, but mathematical model to use will depend on the scenario and objectives of the model-based study.

3.3 ProSimPlus - ICAS-MoT – COFE

A combination between the Generic CAPE-OPEN Unit Operation and COFE (taken from www.cocosimulator.org) have also been tested in order to prove the interoperability issues with another PME different than ProSimPlus. A simple splitter mathematical model is used in this interoperability test. Fig. 12 is showing the different windows when ProSiMPlus-ICAS-MoT - COFE based calculations are carried out: (a) it is showing COFE interface with the generic CAPE-OPEN unit operation that is using an ICAS-MoT model; (b) this window allows the introduction of parameters which are specified in the XML configuration file; (c), (d) and (e) windows show the stream report of the CAPE-OPEN unit operation, and these confirm the possibility of the interoperability between these computational tools.

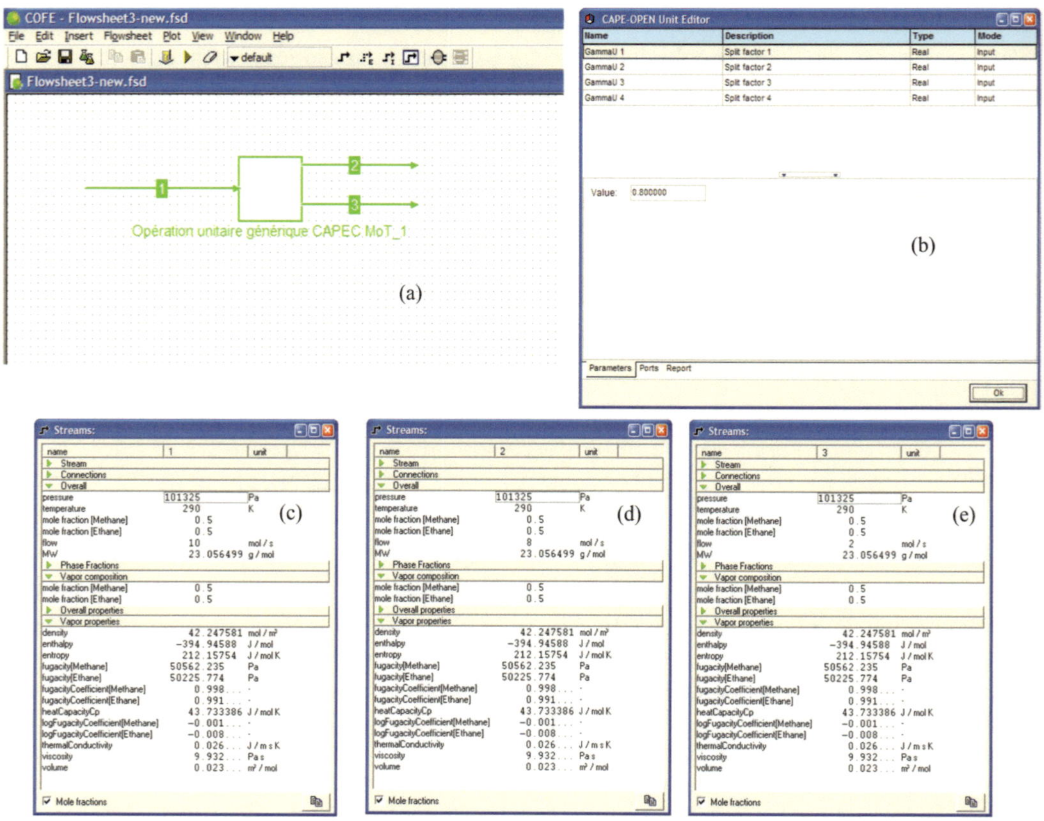

Fig. 12. ProSimPlus-ICAS-MoT – COFE interoperability

Note that the use of the CAPE-OPEN standards facilitates the simulation of new unit operations in PMEs or those that cannot be found in commercial simulators. The use of the multiscale modelling approach embedded in the model of the new unit operation is used to highlight the importance of selecting the appropriate details of the model needed to match the objective.
Other examples for chemical product-process design have been solved in a software so called the *Virtual Product-Process Design Lab*, which also combines different computer-aided tools through the use of COM-objects (Morales-Rodriguez, 2009; Morales-Rodriguez & Gani, 2009).

4. Conclusion

The advantages of the use of CAPE-OPEN standards for the integration of different tools have been tested and highlighted. Furthermore, the use of a multiscale modelling approach in the simulation of unit operations not found in the host PME has been highlighted for different scales and objectives of the model.
Using a standard middleware for thermo-models interoperability aspects of the integrated tools have been illustrated through the calculation of fugacity coefficients and activity coefficients by different property models.
The integration between ICAS-MoT and ProSimPlus for reliable simulation of any unit operation, where models can be supplied from different sources can be easily achieved through plug & play (interoperability) of software tools and models. Here, the CAPE-OPEN interface for unit operations plays an important role.
Current and future work is involved with the development and testing of more middleware and in the creation of a library of models ready to be used through the developed CAPE-OPEN compliant middleware.

5. References

Barrett Jr, W.M. & Yang, J. (2005). Development of a chemical process modelling environment based on CAPE-OPEN interfaces standards and the Microsoft .NET framework. *Computers and Chemical Engineering*, Vol. 30, No. 2, 191-201. ISSN 0098-1354.

Belaud, J.P. & Pons, M. (2002). Open software for process simulation: the current status of CAPE-OPEN standard. *Computer-Aided Chemical Engineering, vol. 10, in Grievink, J. and van Schijndel, J. (eds). European Symposium on Computer Aided Process Engineering – 12, (Elsevier, Amsterdam, Netherlands)*. ISBN 10: 0-444-51109-1.

Braunschweig, B.L., Pantelides, C.C., Britt, H.I. & Sama, S. (2000). Process modelling: The promise of open software architectures. *Chemical Engineering Progress*. Vol. 96, 65-76. ISSN 0360-7275.

Cameron, I.T., & Gani, R. (2010). Product and Process Modelling: A Case Study Approach. *Elsevier Science*.ISBN 13: 978-0444531612.

Cameron, I.T., Wang, F.Y., Immanuel, C.D. & Stepanek, F. (2005). Process system modelling and applications in granulation: A review. *Chemical Engineering Science*, Vol. 60, No. 14, 3723-3750. ISSN 0009-2509.

Morales-Rodriguez, R. (2009). *Computer-Aided multiscale Modelling for Chemical Product-Process Design*. Department of Chemical and Biochemical Engineering, Technical University of Denmark. ISBN: 978-87-92481-01-6.

Morales-Rodriguez, R. & Gani, R. (2007). Multiscale Modelling Framework for Chemical Product-Process Design. *Computer-Aided Chemical Engineering*, Vol. 26. ISSN 1570-7946.

Morales-Rodriguez, R., Sales-Cruz, M., Gani, R., Déchelotte, S., Vacher, A. & Baudouin, O. (2007). Interoperability between modelling tools (MoT) and process simulators (ProSim) through CAPE-OPEN standards, *Paper*489c, Presented at AIChE Annual Meeting, San Francisco, 12-17 November.

Pons, M. (2003). Industrial implementations of the CAPE-OPEN standard. In *AIDIC Conference Series*, Vol 6, 253-262. ISBN 0390-2358.

Pons, M. (2005a). What is the CAPE-OPEN Laboratories Network (CO-LaN)?. *Chemical Processing*. Available at: http://www.chemicalprocessing.com/

Pons, M. (2005b). Introduction to CAPE-OPEN part 1, Presentation in Scandpower, June 14. At: http://www.colan.org/Communication/Y05_SPTGroup_Introduction.pdf

Sundmacher, K., Schultz, T., Zhou, S., Scott, K., Ginkel, M. & Gilles, E.D. (2001). Dynamics of the direct methanol fuel cell (DMFC): experiments and model based analysis. *Chemical Engineering Science*, Vol. 56, No. 2, 333-341. ISSN 0009-2509.

Xu, C., Follmann, P.M. Biegler, L.T. & Jhon, M.S. (2005). Numerical simulation and optimization of a direct methanol fuel cell. *Computers and Chemical Engineering*, Vol. 29, No. 8, 1849-1860. ISSN 0098-1354.